THE YEAR WITHOUT A SUMMER?

WORLD CLIMATE IN 1816

EDITED BY

C. R. HARINGTON

Above and cover:
Medallion struck in southern Germany in memory of the great famine of 1816-1817.

The inscription reads: "Great is the distress, Oh Lord, have pity."

Both faces shown; from **Volcano Weather, The Story of 1816, The Year Without a Summer** by Henry and Elizabeth Stommel

CANADIAN MUSEUM
OF NATURE
OTTAWA, 1992

©1992 Canadian Museum of Nature

Published by the:

Canadian Museum of Nature
Ottawa, Canada K1P 6P4

Catalogue No. NM95-20/1 1991-E

Available by mail order from:

Canadian Museum of Nature
Direct Mail Section
P.O. Box 3443, Station "D"
Ottawa, Canada K1P 6P4

Printed in Canada
ISBN: 0-660-13063-7

Text pages printed on paper
containing recycled fibre.

©1992 Musée canadien de la nature

Publié par le :

Musée canadien de la nature
Ottawa, Canada K1P 6P4

N° de catalogue NM95-20/1 1991-E

L'éditeur remplet les commandes postales
adressées au :

Musée canadien de la nature
Section des commandes postales
C.P. 3443, succursale D
Ottawa, Canada K1P 6P4

Imprimé au Canada
ISBN : 0-660-13063-7

Les pages du texte sont imprimés
sur un papier contenant
des fibres recyclés.

Print of original handwritten copy of Lord Byron's poem "Darkness". Written at Geneva during 1816 (courtesy of Princeton University Library).

DARKNESS.

I had a dream, which was not all a dream.

The bright sun was extinguish'd, and the stars

Did wander darkling in the eternal space,

Rayless, and pathless, and the icy earth

Swung blind and blackening in the moonless air;

Morn came and went—and came, and brought no day,

And men forgot their passions in the dread

Of this their desolation; and all hearts

Were chill'd into a selfish prayer for light:

And they did live by watchfires—and the thrones,

The palaces of crowned kings—the huts,

The habitations of all things which dwell,

Were burnt for beacons; cities were consumed,

And men were gather'd round their blazing homes

To look once more into each other's face;

Printed version of Lord Byron's poem "Darkness" (courtesy of Princeton University Library).

CONTENTS

Acknowledgements — 5

Introduction — 6
   C.R. Harington

General — 9

   Before Tambora: the Sun and Climate, 1790-1830 — 11
      John A. Eddy

   Eyewitness Account of the Distant Effects of the Tambora Eruption of April 1815 — 12
      Michael R. Rampino

   The Eruption of Tambora in 1815: Environmental Effects and Eruption Dynamics — 16
      Haraldur Sigurdsson and Steven Carey

   The Possible Effects of the Tambora Eruption in 1815 on Atmospheric Thermal and Chemical Structure and Surface Climate — 46
      R.K.R. Vupputuri

   Climatic Effects of the 1783 Laki Eruption — 58
      Charles A. Wood

   The Effects of Major Volcanic Eruptions on Canadian Surface Temperatures — 78
      Walter R. Skinner

Northern Hemisphere — 93

North America
   Climate in 1816 and 1811-20 as Reconstructed from Western North American Tree-Ring Chronologies — 97
      J.M. Lough

   Volcanic Effects on Colorado Plateau Douglas-Fir Tree Rings — 115
      Malcolm K. Cleaveland

   1816 in Perspective: the View from the Northeastern United States — 124
      William A. Baron

Expansion of Toronto Temperature Time-Series from 1840 to 1778 Using
Various United States and Other Data   145
  R.B. Crowe

Climate in Canada, 1809-20: Three Approaches to the Hudson's Bay
Company Archives as an Historical Database   162
  Cynthia Wilson

Climatic Change, Droughts and Their Social Impact: Central Canada,
1811-20, a Classic Example   185
  Timothy F. Ball

The Year without a Summer: Its Impact on the Fur Trade and History
of Western Canada   196
  Timothy F. Ball

The Ecology of a Famine: Northwestern Ontario in 1815-17   203
  Roger Suffling and Ron Fritz

The Development and Testing of a Methodology for Extracting Sea-Ice
Data from Ships' Log-Books   218
  Marcia Faurer

River Ice and Sea Ice in the Hudson Bay Region during the Second
Decade of the Nineteenth Century   233
  A.J.W. Catchpole

The Climate of the Labrador Sea in the Spring and Summer of 1816,
and Comparisons with Modern Analogues   245
  John P. Newell

Spatial Patterns of Tree-Growth Anomalies from the North American
Boreal Treeline in the Early 1800s, Including the Year 1816   255
  Gordon C. Jacoby, Jr. and Rosanne D'Arrigo

Early Nineteenth-Century Tree-Ring Series from Treeline Sites
in the Middle Canadian Rockies   266
  B.H. Luckman and M.E. Colenutt

How Did Treeline White Spruce at Churchill, Manitoba Respond
to Conditions around 1816?   281
  David C. Fayle, Catherine V. Bentley and Peter A. Scott

The Climate of Central Canada and Southwestern Europe Reconstructed
by Combining Various Types of Proxy Data: a Detailed Analysis of
the 1810-20 Period*   291
  J. Guiot

CONTENTS

Acknowledgements   5

Introduction   6
C.R. Harington

General   9

    Before Tambora: the Sun and Climate, 1790-1830   11
        John A. Eddy

    Eyewitness Account of the Distant Effects of the Tambora Eruption
      of April 1815   12
        Michael R. Rampino

    The Eruption of Tambora in 1815: Environmental Effects and
      Eruption Dynamics   16
        Haraldur Sigurdsson and Steven Carey

    The Possible Effects of the Tambora Eruption in 1815 on Atmospheric
      Thermal and Chemical Structure and Surface Climate   46
        R.K.R. Vupputuri

    Climatic Effects of the 1783 Laki Eruption   58
        Charles A. Wood

    The Effects of Major Volcanic Eruptions on Canadian Surface Temperatures   78
        Walter R. Skinner

Northern Hemisphere   93

North America
    Climate in 1816 and 1811-20 as Reconstructed from Western North American
      Tree-Ring Chronologies   97
        J.M. Lough

    Volcanic Effects on Colorado Plateau Douglas-Fir Tree Rings   115
        Malcolm K. Cleaveland

    1816 in Perspective: the View from the Northeastern United States   124
        William A. Baron

Expansion of Toronto Temperature Time-Series from 1840 to 1778 Using
Various United States and Other Data   145
R.B. Crowe

Climate in Canada, 1809-20: Three Approaches to the Hudson's Bay
Company Archives as an Historical Database   162
Cynthia Wilson

Climatic Change, Droughts and Their Social Impact: Central Canada,
1811-20, a Classic Example   185
Timothy F. Ball

The Year without a Summer: Its Impact on the Fur Trade and History
of Western Canada   196
Timothy F. Ball

The Ecology of a Famine: Northwestern Ontario in 1815-17   203
Roger Suffling and Ron Fritz

The Development and Testing of a Methodology for Extracting Sea-Ice
Data from Ships' Log-Books   218
Marcia Faurer

River Ice and Sea Ice in the Hudson Bay Region during the Second
Decade of the Nineteenth Century   233
A.J.W. Catchpole

The Climate of the Labrador Sea in the Spring and Summer of 1816,
and Comparisons with Modern Analogues   245
John P. Newell

Spatial Patterns of Tree-Growth Anomalies from the North American
Boreal Treeline in the Early 1800s, Including the Year 1816   255
Gordon C. Jacoby, Jr. and Rosanne D'Arrigo

Early Nineteenth-Century Tree-Ring Series from Treeline Sites
in the Middle Canadian Rockies   266
B.H. Luckman and M.E. Colenutt

How Did Treeline White Spruce at Churchill, Manitoba Respond
to Conditions around 1816?   281
David C. Fayle, Catherine V. Bentley and Peter A. Scott

The Climate of Central Canada and Southwestern Europe Reconstructed
by Combining Various Types of Proxy Data: a Detailed Analysis of
the 1810-20 Period*   291
J. Guiot

Climatic Conditions for the Period Surrounding the Tambora Signal
in Ice Cores from the Canadian High Arctic Islands   309
Bea Taylor Alt, David A. Fisher and Roy M. Koerner

Europe (including Iceland)

1816 - a Year without a Summer in Iceland?   331
A.E.J. Ogilvie

First Essay at Reconstructing the General Atmospheric Circulation
in 1816 and the Early Nineteenth Century   355
H.H. Lamb

Weather Patterns over Europe in 1816   358
John Kington

The Climate of Europe during the 1810s with Special Reference to 1816   372
K.R. Briffa and P.D. Jones

The 1810s in the Baltic Region, 1816 in Particular: Air Temperatures,
Grain Supply and Mortality   392
J. Neumann

The Years without a Summer in Switzerland: 1628 and 1816   416
Christian Pfister

Climatic Conditions of 1815 and 1816 from Tree-Ring Analysis in the
Tatra Mountains   418
Zdzisław Bednarz and Janina Trepińska

Major Volcanic Eruptions in the Nineteenth and Twentieth Centuries
and Temperatures in Central Europe   422
Vladimir Brůžek

Asia

Climate over India during the First Quarter of the Nineteenth Century   429
G.B. Pant, B. Parthasarathy and N.A. Sontakke

Evidence for Anomalous Cold Weather in China 1815-17   436
Pei-Yuan Zhang, Wei-Chyung Wang and Sultan Hameed

Was There a Colder Summer in China in 1816?   448
Huang Jiayou

The Reconstructed Position of the Polar Frontal Zone around Japan
in the Summer of 1816   453
Yasufumi Tsukamura

The Climate of Japan in 1816 as Compared with an Extremely Cool
Summer Climate in 1783   462
T. Mikami and Y. Tsukamura

Southern Hemisphere 477

Evidence for Changes in Climate and Environment in 1816 as Recorded in Ice Cores from the Quelccaya Ice Cap, Peru, the Dunde Ice Cap, China and Siple Station, Antarctica* 479
Lonnie G. Thompson and Ellen Mosley-Thompson

Changes in Southern South American Tree-Ring Chronologies following Major Volcanic Eruptions between 1750 and 1970 493
Ricardo Villalba and Jose A. Boninsegna

Tree-Ring Chronologies from Endemic Australian and New Zealand Conifers 1800-30 510
Jonathan Palmer and John Ogden

New Zealand Temperatures, 1800-30 516
David A. Norton

Summary 521

Workshop on World Climate in 1816: a Summary and Discussion of Results 523
Cynthia Wilson

Index 557

* The geographic sections above are not exact. For example, J. Guiot's paper, although listed under North America, also provides substantial information on southwestern Europe and northwestern Africa. Similarly, the paper by L. Thompson and E. Mosley-Thompson, although listed under Southern Hemisphere, also concerns China.

Acknowledgements

The editor is grateful to his colleagues on the Organizing Committee of the international meeting ("The Year Without a Summer? Climate in 1816", Ottawa, 25-28 June 1988) from which this volume arose: Drs. C. Wilson, A.J.W. Catchpole, T.F. Ball, R.M. Koerner, G.C. Jacoby (Members); Mrs. Gail Rice (Secretary-Treasurer); and Mr. Kieran Shepherd (Coordinator, Poster Presentations). I am also grateful to the following institutions for so firmly supporting the meeting: Canadian Climate Centre; Climatic Research Unit, University of East Anglia; National Center for Atmospheric Research (operated by the University Corporation for Atmospheric Research under the sponsorship of the United States National Science Foundation); and the World Meteorological Organization. I thank the directorate of the museum for its interest in and encouragement of the project.

Joanne Dinn (Paleobiology Division) and Marie-Anne Resiga helped greatly in preparing this book, as did Mireille Boisonneau, Arch Stewart (Canadian Museum of Nature Library) and Daphne Sanderson (Canadian Climate Centre Library). Sharon Helman kindly redrafted several of the figures, and with Bonnie Livingstone (Publications Division) provided strong support during the last phases of preparing this volume.

Finally, I express my sincere thanks to Cynthia Wilson and Tim Ball for their help in organizing the Workshop, as well as to Richard Martin for audiotaping the discussions. Cynthia Wilson performed a particularly useful service by analyzing and summarizing the Workshop results.

Introduction

This book is the last gasp of the National Museum of Natural Sciences (now Canadian Museum of Nature) *Climatic Change in Canada Project*! Because of Canada's vulnerability to climatic change, and the lack of an integrated multidisciplinary program for studying our past climate this project was organized. Since its beginning in 1977, a basic aim of the project has been to publish in our *Syllogeus* series significant data on climatic change in Canada since the peak of the last glaciation (about 20,000 years ago).

We began the project with a general assessment of Quaternary paleoclimatic information available in Canada and techniques that could be used for interpreting it (*Syllogeus* 26, 1980); we later broadened the number of disciplines involved and actually began gathering and interpreting the paleoclimatic data (*Syllogeus* 33, 1981; 49, 1983); and then produced an annotated bibliography on the subject (*Syllogeus* 51, 1984). In May 1983, the project sponsored an international meeting "Critical Periods in the Quaternary Climatic History of Northern North America" (*Syllogeus* 55, 1985). It was clear from papers in *Syllogeus* 55 that several authors had gone well beyond the data-gathering stage: Alan Catchpole not only tested the value of one type of proxy data (climatic records from Hudson's Bay Company documents, including Ships' logs) against another (Marion Parker's tree-ring records) for the Hudson Bay region, but showed that sea-ice conditions were indicative of prevailing northerly or northwesterly winds, pumping cold arctic air over the central and eastern parts of North America in the summer of 1816; and Cynthia Wilson took a magnificent step forward by providing a series of six daily weather maps for early June 1816 - actually showing the tracks of high- and low-pressure areas across central and eastern North America.

These papers prompted me to consider convening an international meeting focusing on global climate during 1816, "the year without a summer". What were conditions like beyond the regions so well documented by John D. Post in *The Last Subsistence Crisis in the Western World* (Johns Hopkins University Press, Baltimore, 1977) and Henry and Elizabeth Stommel in *Volcano Weather, the Story of 1816, the Year without a Summer* (Seven Seas Press, Newport, 1983), among others? Could anything useful on a global basis be added to Lamb's and Johnson's (1966, Figure 5) excellently constructed pressure map for July 1816 extending from western Europe across the Atlantic Ocean to central North America (Lamb, this volume)?

Accordingly, I wrote to Professor Hubert Lamb in Norwich in February 1985 and received an encouraging, constructive reply: "Your idea of holding some sort of a conference or workshop meeting specifically to put together the best possible reconstruction of summer 1816, or of the whole 'year without a summer', or perhaps usefully rather more of that decade particularly aimed at covering the period from just before the atmospheric/radiation budget disturbance caused by the huge volcanic eruption of Tambora in 1815 till the return to the status quo ante, has intriguing possibilities." The proposal for this meeting was approved by the director of the museum in 1986.

The objective of the meeting was, by bringing together workers in various fields (e.g., volcanologists, glaciologists, climatologists, tree-ring experts, geographers, historians and biologists) from various countries, to gain the clearest picture possible of weather and climatic sequences in different parts of the world during 1816, or about that time (e.g., 1810-20), in an effort to discover key factors influencing the unusual weather then. For example, how important was the eruption of Tambora, and what other cooling influences may have been involved? How widespread were the cold summer conditions from a global viewpoint? Did blocking play an important part?

From the beginning, the Workshop was considered to be the heart of the meeting. The attempt to actually plot weather and climatic data from various sources for the Tambora period on base maps proved challenging, frustrating and exciting. Could we really shed more light on the nature of the climatic events, their intensity and timing? Although evidence is circumstantial, it seems that widespread cooling was underway before the eruption of Tambora. Evidently, the massive injection of Tambora aerosols into the atmosphere in 1815 resulted in crossing a threshold to highly anomalous weather (probably involving blocking highs and break monsoons) in many parts of the globe. Certainly "the year without a summer" in 1816 was a regional phenomenon. In the northern hemisphere parts of western North America, eastern Europe and Japan seem to have had average or above-average temperatures, as opposed to the remarkable cold that characterized much of eastern North America, western Europe, and China. Incursion of freezing arctic air southward in one region was offset by poleward flow of tropical air in another. In the southern hemisphere, El Niño may have diminished the cooling reflected in tree-ring records from Argentina in 1816-17, whereas in 1817-18 the tremendous moderating influence of the Pacific Ocean may have effectively damped any cooling recorded there (see Wilson, Workshop section, for more details on the group's findings).

This book is intended for those who are deeply interested in: historical climate (particularly that of the Little Ice Age) and its human impact; relationships between volcanism and climate; and the ways paleoclimatic proxy data are gathered, treated and interpreted. The volume begins with a general section concerning: solar influences on the trend of climate before the eruption of Tambora; a vivid eyewitness account of the eruption; the nature of the eruption, the aerosol produced and its course through the atmosphere - as well as a discussion of the effects of the 1783 eruption of Laki in Iceland on climate for comparative purposes and a consideration of the effects of major volcanic eruptions following Krakatau (1883) on Canadian temperatures. Coverage is then (loosely) geographic, first dealing with the northern hemisphere (North America, Europe, Asia), then the southern hemisphere (South America, Antarctica, Australia and New Zealand). Perhaps readers will gather from these contributions an inkling of the tremendous investment in time that is presently required to distil a useful drop of paleoclimatic data from archival and other sources.

Finally, I hope that this exercise will lead others to look more carefully at the "Tambora period" and similar paleoclimatic problems - adding data in vast expanses of the globe where our evidence is deficient, as well as testing and refining data given here until a more coherent picture emerges. Information presented in this volume may also be food for ravenous paleoclimatic modellers!

C.R. Harington

General

Before Tambora: The Sun and Climate, 1790-1830

John A. Eddy[1]

Abstract

The unusual summer of 1816 is commonly attributed to the increase in atmospheric turbidity that followed the eruption of Mount Tambora (Stommel and Stommel 1979). The awesome eruption occurred, in fact, during a span of several decades of colder climate that had interrupted the gradual global warming that followed seventeenth century extrema of the Little Ice Age (Lamb 1985). These background trends may well explain a particularly severe seasonal response in 1816 to a short-term injection of volcanic dust. The colder climate that characterized the opening decades of the nineteenth century was quite possibly related to a coincident depression in solar activity between about 1790 and 1830, called the "Dalton Minimum" or sometimes the "Little Maunder Minimum" (Siscoe 1980). The probability of a solar connection is strengthened by recent analyses of long-term changes in the level of solar activity and decadal averages of global-surface temperature in the last 100 years (Reid and Gage 1987), as well as in the correspondence of the Maunder Minimum in solar activity (1645-1715). A probable mechanism for solar forcing can be found in recent spaceborne measurements of year-to-year variations in the so-called "solar constant" (Willson et al. 1986). I plan to examine the evidence for solar and climatic anomalies in the period from about 1790-1830 and the recent findings that provide a probable connection between the sun and long-term climatic change.

References

Eddy, J.A. 1977. The case of the missing sunspots. *Scientific American* 236:80-92.

Lamb, H.H. 1985. *Climate History and the Future.* Princeton University Press, Princeton, New Jersey. 884 pp.

Reid, G. and K.S. Gage. 1987. Influence of solar variability on global sea surface temperatures. *Nature* 329(6135):142-143.

Siscoe, G.L. 1980. Evidence in the auroral record for secular solar variability. *Review of Geophysics and Space Physics* 18:647-658.

Stommel, H. and E. Stommel. 1979. The year without a summer. *Scientific American* 240:176-186.

Willson, R.C., H.S. Hudson, C. Frohlich and R.W. Brusa. 1986. Long-term downward trend in total solar irradiance. *Science* 234:1114-1117.

---

[1] University Corporation for Atmospheric Research, Boulder, Colorado 80307, U.S.A.

Eyewitness Account of the Distant Effects of the Tambora Eruption of April 1815

Michael R. Rampino[1]

Abstract

The following is a brief description of the effects of the eruption of Tambora volcano in 1815 on conditions about 800 km away in eastern Java. Evidently Tambora was quite active for at least six days prior to the cataclysmic eruption of 11 April 1815, and direct cooling was associated with the ash cloud.

Introduction

Large explosive volcanic eruptions can have far-reaching effects on the atmosphere. The eruption of Tambora volcano on Sumbawa Island in Indonesia in April 1815 was the largest ash eruption in recent historic times, producing a bulk volume of about 150 km^3 of pumice and ash (Stothers 1984). The loss of life and the destruction of agricultural land on Sumbawa and neighbouring Lombok were catastrophic. In the aftermath of the Tambora eruption, in order to obtain more information about the effects on Java and the surrounding islands, the Lieutenant Governor of Java, Thomas Stamford Raffles, circulated a letter with three brief questions. The following questionnaire was completed by the Resident of Surakarta in eastern Java describing local eyewitness accounts of the effects of the Tambora eruption (catalogued in Blagden 1916). It gives a vivid picture of the effects of the massive eruption some 800 km from the volcano. (The style, punctuation, and spelling of the original handwritten report in the MacKenzie Collection of the British Library has been retained throughout).

Questionnaire and Response by the Resident of Surakarta

Points of Enquiry
Circular of the Honble [T.S. Raffles] the Lieut Governor [of Java]
First, the effects of the eruption of Sumbawa in April 1815 would appear to have been first noticed at Banjuwangie on the 1st and at Batavia on the 6th of April but the atmosphere would appear to have been successively affected by the ashes between the 10th and 14th. On what day and at what hour were they first noticed in different parts of your Residency - how long - when did they continue and what was the nature of them?

At Souracarta the first explosions were heard on Wednesday the 5th of April between the hours of 4 and 6 PM, distinct and separate sounds exceeding the number of twenty were perceived with irregular intervals greatly resembling a military operation, but more that is denominated *mortar* practice than a regular cannonade. On the successive evenings of the 6th, 7th, 8th and 9th, occasional noises were heard which were mistaken for distant thunder. During these days the opacity of the atmosphere, resembling former volcanic eruptions on this Island, first indicated

---

[1] Earth Systems Group, Department of Applied Sciences, New York University, New York, New York 10003, U.S.A. Also at NASA Goddard Space Flight Center, Institute for Space Studies, New York, New York 10025, U.S.A.

the probable cause of the explosions which, by a person unaccustomed to their effects could not be distinguished from the reports of guns or thunder etc.

On Monday the 10th, a very slight fall of dust was perceived, but alone by the most attentive observation, and the explosions continued at intervals in the east.

On Tuesday the 11th the reports were more frequent and violent through the whole day: one of the most powerful occurred in the afternoon about 2 O'Clock, this was succeeded, for nearly an hour by a tremulous motion of the earth, distinctly indicated by the tremor of large window frames; another comparatively violent explosion occurred late in the afternoon, but the fall of dust was scarcely perceptible. The atmosphere appeared to be loaded with a thick vapour: the Sun was rarely visible, and only at short intervals appearing very obscurely behind a semitransparent substance.

The day on which the opacity of the atmosphere first commenced had not been noted accurately - but its continuance was above twelve days, and even at the commencement of the present month it was not entirely dissipated.

From the 5th to the 18th of the last month the Sun was not distinctly perceived, and if his rays occasionally penetrated they appeared as observed through a thick mist. The general darkening of the atmosphere was strikingly exhibited by such objects of which the prospect is familiar; thus for instance at Souracarta the Mountain above continued invisible through all this period, and even near objects were clouded or its appearance obscured by smoke -

On Wednesday the 12th the appearance of day light showed a very copious discharge of dust, this gradually increased till 1 PM and then appeared to diminish but was still very discernible at sunset: the following day (the 13th) it was still rarely perceptible and gradually and successively ceased.

On the 12th a considerable darkness was occasioned by the abundance of the fall of dust: every operation which required strong light was almost impossible within doors. The gloomy appearance caused by the rain of dust *"Udshan abu"* need not be described as it was uniform in every part of this Island to which the discharge extended. It may be remarkable that an unusual sensation of chillings was felt during the whole of the 12th this was in great measure (tho' probably not exclusively) occasioned by the temperature: the thermometer at 10 O'Clock AM stood at 75 and 1/2 degrees of Fahrenheit Scale. It would appear that the subterranean commotion, like the discharge of dust, was propogate or travelled from east to west as the explosions were later perceived in the Western parts of the island: it is likewise highly probable (which must however be determined by a comparison of various and accurate remarks made in different parts of the island) that the most violent explosions were not simultaneous, but that the combustion caused locally more powerful shocks in particular parts from Banju-wangie perhaps to the western extremity. Something like this was remarked during the combustion of the Kloet in 1811, when the explosions were much more violent at Batavia than at Souracarta although the latter was much nearer to the burning Mountain. It would appear from creditable information that effects were more sensibly felt along the Southern Shore of the Island and that the tremulous motion of the earth was there more violent - a very uncommon rising of the water was also perceived about the period of the most violent explosions at Harang bollong, Kadilangu etc. but the day and hour had not been noted with sufficient accuracy for any decided inference. The colour of the dust of the present eruption is ash grey inclining to brown it is a most impalpably fine, divided earthy substance, if water is added it diffuses the peculiar odour of clay; it does not acquire ductility enough to be moulded, but has been observed to improve the quality of the common clay of the Island in the manufacture of pottery. Its chief component parts are Silecious and Aluminous earth. It is evidently a finely divided Lava, the iron of which having by means of gravity subsided in the vicinity of the Volcano. Scarcely any of the particles are attracted by the magnet in this it differs from volcanic dust which was thrown from the Gunung Gunter in 1803 and, being precipitated about Batavia, possessed a blackish colour and was strongly attracted by the magnet. The dust which was exploded by the Gunung Klut in June 1811 differs from the present as far as can be determined without chemical analysis only by having a blueish grey colour, and in being less finely divided; it was supposed to possess superior qualities for the manufacturing of pottery but had not ductility enough to be moulded alone.

Have any injurious consequences resulted from within your Residency as affecting the salubrity of the Country, or in the destruction of the Crops or Cattle respecting the latter, state the particulars, if any and in what manner the injury may have been effected.

If the generality of the discharge of the volcanic dust is considered and the abundance of the substance which covered the earth and of vegetation for many days, its effects on the health of the animals were inconsiderable: instances of mortality among cattle particularly Buffaloes and Cows in this neighbourhood during the continuance of and since the rain of dust are Solitary, and leave it doubtful whether they must be ascribed to this or other accidental cause. In a few cases (within my observation) death was induced suddenly: these may probably be ascribed to this cause, but the inquiries I have made have confirmed the opinion that the health of the Cattle has not been (in a general manner) injuriously affected. It should be kept in view in determining the question, that previously to the rain of dust the Buffaloes in particular districts were affected by an epidemic disease denominated *Puttie* by the Natives of which several died and the mortality has in some degree continued to the present time. Neither Horses, Sheep, or Goats have been sensibly affected.

An injury of a more serious nature threatened the crops of rice - but the forward state of cultivation has preserved this grain in most of the neighbouring districts and such a season of abundance as the present has not been known for many years: it has been observed by various persons who are conversant with the cultivation of this grain, that plantations in which the rice had nearly acquired maturity were not affected, but the dust falling upon the grain newly transplanted in many cases destroyed the young plants. This is in some degree rendered probable by the nature of the volcanic substance, and its effects would be more powerful towards the period of the terminations of the rains or where a deficiency of moisture prevailed. Falling upon the young plants and fields sparingly supplied with water it would from its clayey nature absorb their juices and destroy them.

What was the general opinion at the time regarding the locality of the volcano?

The general opinion at this place ascribed the eruptions to the Mountain *Klut* of which three previously similar "*rains of ashes*" were recollected by all aged inhabitants.

Conclusion

The above report documents that Tambora was quite active for at least six days prior to the cataclysmic eruption of 11 April 1815. Note that the eruption was misidentified with Klut (Kelut) volcano in Java during the ash rain. The reply gives evidence of a direct cooling associated with the ash cloud, and such a cooling effect was observed as far away as Madras, India, where midday temperatures fell below freezing as the cloud passed overhead (Stothers 1984). The anomalous weather of the infamous summer of 1816 was quite likely related to the radiative perturbation by stratospheric H_2SO_4 aerosols generated by the eruption. Without doubt, a similar eruption in Indonesia today would be a regional disaster, and would create a global atmospheric perturbation of a magnitude not seen in almost two centuries.

Acknowledgements

A grant from the American Philosophical Society supported a literature search at the British Library for information pertaining to the aftermath of large volcanic eruptions in the nineteenth century. The author thanks I.A. Baxter of the India Office Library, Blackfriars Road, London, for his help, and S. Self, H. Sigurdsson, and R.B. Stothers for valuable discussions. This is a slightly altered version of a paper published by the author in *EOS* 1989, p. 1559, (copyright by the American Geophysical Union).

References

Mackenzie Collection: Private, Document 2:33, pp. 193-198, 1916. *In:* C.A. Blagden, *Catalogue of Manuscripts in European Languages belonging to the Library of the India Office, Volume 1: The Mackenzie Collection, Part I: the 1822 Collection and Private Collection,* p. 43. Oxford University Press, London.

Stothers, R.B. 1984. The great Tambora eruption in 1815 and its aftermath. *Science* 224:1191-1198.

The Eruption of Tambora in 1815: Environmental Effects and Eruption Dynamics

Haraldur Sigurdsson[1] and Steven Carey[1]

Abstract

New studies of deposits from the 1815 eruption of Tambora volcano provide data on eruption dynamics, mass eruption rate and volcanic volatile emission to the atmosphere. These data form a basis for assessment of the environmental impact of the eruption. Initial phases of activity were two plinian explosive eruptions on 5 and 10 April with column heights of 33 and 43 km, and mass eruption rate of 1.1×10^8 and 2.8×10^8 kg/s respectively. The calculated column heights therefore indicate a major injection of volcanic ash and volatile gases to the stratosphere during the eruption. Rapid transition to pyroclastic flow generation occurred late on 10 April, when the bulk of the material was erupted at a rate of 5.4×10^8 kg/s, producing widespread co-ignimbrite ash fall. A large component of the co-ignimbrite ash fall was produced by explosive interaction of hot pyroclastic flows and sea water, when flows advanced into the ocean around Tambora. Total erupted mass is estimated as 50 km^3 dense-rock equivalent, or 1.4×10^{14} kg. Petrologic estimates of volatile yield to the atmosphere during the eruption indicate that sulphur degassing formed a stratospheric aerosol mass equivalent to 1.75×10^{11} kg sulphuric acid, in agreement with volcanic aerosol estimates based on ice-core evidence. Furthermore, volcanic degassing of 10^{11} kg HCℓ and 7.4×10^{10} kg HF occurred, but the fate of these species in the atmosphere is unknown. Climatological data indicate a short-term northern hemisphere surface temperature decrease of 0.7 °C following the eruption, and this climatic response agrees with the empirical relationship observed between sulphuric acid volcanic aerosol mass and temperature decline observed after several major explosive volcanic events. It is likely, however, that the observed surface temperature decline is not solely due to the Tambora event, as a cooling trend was already in progress prior to the eruption.

Introduction

The 1815 Tambora eruption on the island of Sumbawa in Indonesia exceeded in magnitude any other volcanic eruption in historical times, producing over 50 km^3 of magma. As a measure of the uniqueness of this great natural disaster, it is remarkable that we have to search some 20,000 years back in the geological record to find an explosive eruption of greater magnitude: the Shikotsu eruption in Japan (Katsui 1959). The volume of material erupted from Tambora is an order of magnitude greater than that discharged in the celebrated Krakatau eruption of 1883, and two orders of magnitude greater than in the 1980 Mount St. Helens eruption. Locally, 92,000 people died on Sumbawa and adjacent islands, either directly from effects of the eruption or from the ensuing famine and epidemic. In addition to its significance as a geological process, the eruption had unprecedented impact on the Earth's stratosphere. The eruption injected enormous quantities of sulphur, chlorine and fluorine gases into the stratosphere, leading to a variety of global atmospheric phenomena, and was probably responsible for the marked climatic deterioration of 1816. Thus, although the eruption is of great importance to the study of volcanology, its greatest scientific significance probably relates to the environmental effects, i.e.,

---

[1] Graduate School of Oceanography, University of Rhode Island, Kingston, Rhode Island 02881, U.S.A.

its impact on the chemistry of the atmosphere and on climate (see also Vupputuri, this volume). With the growing realization of connections between the biosphere, atmosphere and geosphere and the recognition of global environmental and climatic change brought about by human activity, the detailed study of the effects of Nature's own large-scale experiments such as the Tambora eruption can greatly aid in our understanding of short- and long-term changes in the global environment.

In this paper we summarize our findings based on a new study of the Tambora deposits, involving two expeditions to the volcano in 1986 and 1988, which provide fresh data on the eruption dynamics and erupted mass. In addition, our recent petrologic study of the sulphur, chlorine and fluorine yield of the eruption to the atmosphere gives quantitative estimates of degassing and provides a framework for modelling the environmental impact of this great volcanic pollution event.

Chronology of the 1815 Eruption

Before 1815, Tambora volcano was conical in form, possibly with two peaks, and the highest mountain in the Sunda Islands. When sailing east from Java, Tambora appeared equally prominent on the horizon as the 3,726-m high Rinjani volcano on Lombok, and Zollinger (1855) estimates that the volcano may have been over 4,000 m before the eruption. His estimates are based on discussions with people in Sumbawa, who maintained that the volcano had lost at least one third of its height. The maximum height of the caldera rim after the eruption is 2,850 m.

Contemporary local sources about the 1815 Tambora eruption are mainly newspapers and government accounts - particularly the *Asiatic Journal*. These accounts are especially useful for establishing the timing of various eruptive phases, and the extent and nature of their effects. This section summarizes important eyewitness observations that are relevant to interpreting the pyroclastic deposits studied in the field (Rampino, this volume).

More than three years before the great eruption, a thick cloud had formed over the peak, which not even the strongest winds could dissipate (Zollinger 1855). It gradually grew darker and larger, and extended farther down the volcano's flanks. Explosions were heard from the volcano during this time; first only a few and weak, but gradually they became more frequent and louder. People living around the volcano sent delegates to the government authorities in Bima on Sumbawa requesting an investigation of these phenomena. The authorities sent a man by the name of Israel, whose brother was still alive at the time of Zollinger's visit. Israel reached the Tambora region on the evening of 9 April, the day before the climax of the eruption, and was killed during the activity the following day.

On the evening of 5 April, the first major eruption began and was heard widely in the Indonesian region. The explosions heard in Java resembled cannonfire and soldiers in Yogyakarta (central Java) combed the land and seas for invaders (Figure 1). Ash fell "like fine snow" in Banjuwangi in eastern Java, accumulating up to one-half inch (1.3 cm) thickness. Minor ash fall also occurred at Besuki in east Java. At Solo (central-eastern Java, 800 km from Tambora) sounds of explosions commenced on the evening of 5 April. The naval vessel *Benares* was in Macassar on 5 April, about 350 km NNE of Tambora (Figure 1). Loud explosions were heard from the south, which continued the entire afternoon. At sunset the explosions grew louder and seemed closer. Listeners suspected that a naval battle was taking place nearby and sent troops to search the region.

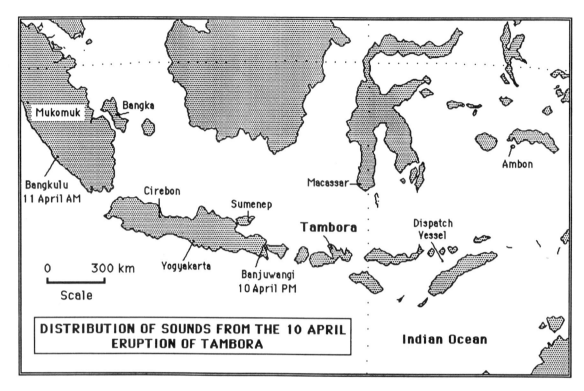

Figure 1: The Indonesian region, showing areas where sounds were heard from the 1815 Tambora eruption.

Ash fell in east Java during the morning of 6 April, but only a trace descended on western Java. The sky gradually cleared during the day, but the air was hot and the atmosphere unusually still. Between 6 and 8 a.m. on 6 April, loud noises were heard at Ternate, where the ship *Teignmouth* lay at anchor some 1,400 km northeast of Tambora.

After four days of minor activity, the volcano became very active again on 10 April. Witnessing the activity from Sanggar on the eastern slopes of Tambora, the Rajah described the second and larger major eruption. At about 7 p.m. three columns of fire rose high from Tambora's crater, uniting in a single firestorm over the volcano. Moments later the entire mountain was a sea of glowing flows, which spread in all directions. Large quantities of ash and stones fell on Sanggar (Figure 2), "up to two fists in size", but most were no larger than a nut. Between 9 and 10 p.m. the ash fall increased, and a strong "whirlwind" descended carrying off houses in Sanggar and nearby villages. In the part of Sanggar nearest to Tambora, the largest trees were uprooted by the windstorm, and carried off with houses, people and livestock. These descriptions are consistent with the passage of a pyroclastic surge through the village. Sea level rose suddenly 12 feet (3.7 m).

At Bima, 80 km east of the volcano (Figure 2), the explosions sounded like heavy mortar fire during the night of 10 to 11 April. The town was in complete darkness from the ash cloud overhead from 7 a.m. on 11 April to 14 April. Ash fall was so heavy, that roofs of most houses collapsed. The air was completely still and there was no wind at sea, but nevertheless the waves were very high and flooded the coast and into the town. All boats were torn from their moorings and tossed ashore.

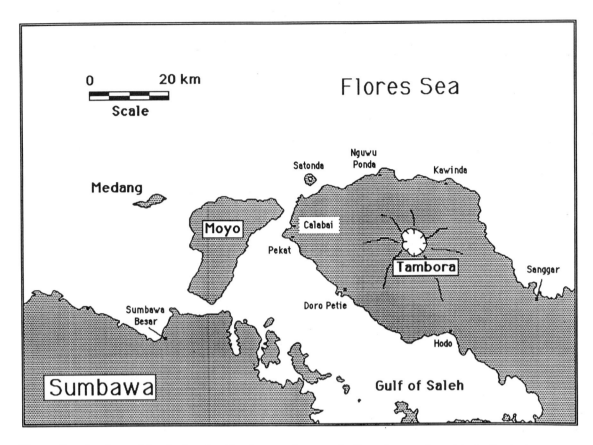

Figure 2: Sumbawa Island, showing Tambora volcano and other places referred to in the text.

Beginning on 10 April, thunderous noises were heard in many parts of Java, which were much louder than on 5 April, especially east of Cirebon (western Java, 1,050 km from Tambora). At Banjuwangi in east Java (400 km distant) the evening noises were very loud and shook the earth. The sounds became somewhat weaker toward morning the next day, but continued until 14 April. At Sumenep (Madura Island, 470 km distant) the noises were like rapid cannonfire. The sky was completely obscured by ash and, in some districts such as Solo and Rembang (central Java, Figure 1), earth vibrations were felt.

During the night of 10 to 11 April the *Benares* reported from Macassar that explosions began again and grew in frequency the next morning, shaking both houses and ships. Lightning flashes were common and the sky was very dark, especially to the south and southwest. The sea rose from five to seven feet (1.6 to 2.2 m) above normal in Besuki, eastern Java, on the night of 10 April.

On 11 April the continuing activity was so severe that houses shook in eastern districts of Java. The coast of Bali was totally invisible from Banjuwangi in eastern Java, where candles were lit at 1 p.m. By 4 p.m. it was pitch-dark and remained dark until 2 p.m. the next day. In Sumenep (Madura) the light was so faint that candles had to be lit before 4 p.m. The following night was indescribably dark. At about 7 p.m. a tidal wave struck Sumenep Bay raising sea level about four feet (1.2 m) for several minutes. Major ash fall also began in Besuki, eastern Java, on 11 April, with darkness extending from 4 p.m. on 11 April until 2 p.m. on 12 April. Explosions were also heard on 11 April at Ambon (*Asiatic Journal*, February 1816, p. 116).

A boat sailing from Timor in the east noted that the sky became very dark as they approached Tambora on 11 April. When they were off Tambora, the base of the volcano was engulfed in flames and the peak was shrouded in a dark cloud, with fires and flames shooting out. They went ashore for water in Sumbawa and found that all boats had been cast ashore by tidal waves. They came across a large number of corpses. As they sailed from Sumbawa, they encountered large rafts of pumice, which formed thick layers on the ocean hindering their passage. Some pumice rafts were so thick that they resembled sandbanks or low cliffs. They were caught in a pumice raft over two feet (0.6 m) thick the entire night of 12 April. The vessel *Dispatch* heard explosions on the night of 11 April, when about 7° east of Bima (about 750 km). Rafts of pumice and timber were so thick along the coast of Flores, that the ship had great difficulty in making way.

Effects of the eruption were noted as far west as Sumatra. On the morning of 11 April, loud noises were heard at Bengkulu on the south coast of Sumatra about 1,800 km west of Tambora, and as far as Terumon in western Sumatra some 2,600 km WNW of Tambora. Explosions were also heard on Bangka Island (1,500 km) off the northeastern coast of Sumatra. People from the interior of Sumatra reported that the leaves of trees and crops were covered with a layer of very fine ash (*Asiatic Journal*, June 1816, p. 600 and August 1816, p. 164).

On 12 April only very faint daylight was visible in eastern Java, and objects were barely visible at a distance of 100 paces in Solo. Some light returned in Banjuwangi about 2 p.m., but the sun was not visible until 14 April. It was unusually cold during this period. Ashfall in Banjuwangi was nine inches (of which eight inches (20.3 cm) had accumulated by 12 April), two inches (5 cm) in Sumenep and somewhat less in Gresik. West of Samarang (central Java) the daylight was little affected.

At 8 a.m. on 12 April it was dark on the *Benares* in Macassar, and by 10 a.m. it was so dark that nearby ships could not be seen. By 11 a.m. the sky was completely dark, except for a small clearing in the east. The ash fell as heavily as snow, and the sea and air were still. By 12 noon the faint light in the east had vanished and it was so dark that a hand held in front of the face could not be seen. Ash fell all night and was so fine that it penetrated all parts of the ship below decks. By 13 April the intensity of the eruption had decreased but its effects were still widespread. At 6 a.m. it was still totally dark in Macassar but faint light returned at 7:30 a.m., and by 8 a.m. one could discern objects. Sounds of the explosion ceased the following day at Banjuwangi but ash fall continued in Macassar, accompanied by a calm and great heat until 15 April.

Not until 17 April did the ash fall cease, and heavy rains spread over the region. In Banjuwangi, many houses had collapsed under the weight of the ash, and fever and epidemics had broken out in several regions affected by the ash fall. On Java the damage to livestock and agriculture was most severe in the eastern district around Banjuwangi, where the destruction of crops and grazing areas was so extensive that many horses and cattle died of hunger.

The devastating effects of the eruption on the local population were first realized when ships reached ports on Sumbawa. *Benares* reached the coast of Sumbawa on 18 April and was trapped in large pumice rafts on the sea. The rafts were so large, that they at first took them for sandbanks or new islands: they were often over a nautical mile in length, and had varied surface features. Large numbers of carbonized and splintered trees were trapped in the rafts. *Benares* dropped anchor at Bima on 19 April, where ash fall was 3¾ inches (9.5 cm) thick. The harbour had changed, and they found eight fathoms (14.9 m) where the depth had been six fathoms (11.2 m) before the eruption.

Some pumice rafts were up to three miles (4.8 km) long, and were still troublesome to navigation between Moyo and Sanggar three years after the eruption. Pumice rafts from the volcano drifted widely over the southern seas in the following months. Between 1 and 3 October 1815, the ship *Fairlie*, in the Indian Ocean on passage to Calcutta, sailed for two days through extensive pumice rafts, about 3,600 km west of Tambora (*Asiatic Journal*, August 1816, p. 161). These rafts travelled at a rate of 0.2 m/s from Tambora and were most likely transported in the South Equatorial Current, driven by the southeast trade winds. Ash fall from the eruption also reached Brunei in Borneo, where the phenomenon so impressed the local people, that they subsequently counted the years from "the great fall of ashes" (Reclus 1871).

On 22 April, the *Dispatch* arrived in Bima. It had first dropped anchor near Sanggar, where the Rajah had told them that all the land was now a desert and all crops and fruits were destroyed. Sanggar Bay was covered with pumice rafts including large trees and remains of houses carried out to sea by the eruption. The volcano was still covered in dense clouds of ash and steam. Smoke emanated in many places from hot flows of ash on the lower flanks, which had also entered the sea.

The British Governor of Java sent Lieutenant Owen Philipps to Sumbawa to study the event and its effects on the people. On the way from Bima to Dompu (Figure 2), Philipps observed a large number of corpses along the road. Villages were abandoned and houses had generally collapsed under the weight of the ashfall. The few survivors wandered about in search of food. The population had been affected by severe diarrhoea, which had caused many deaths. The people blamed this on their drinking water, which was contaminated with the volcanic ash. Horses and other livestock were also killed in large numbers by this disease. The Rajah of Sanggar met with Philipps in Dompu. The misery of his people was much worse than in Dompu and even one of the Rajah's daughters had died of hunger. Coconuts were the only food supply of the ruined village, where starvation was severe. Philipps gave him some rice, for which the Rajah gave thanks with tears in his eyes.

Zollinger (1855) describes the misery of the remaining population. Many continued to wander in search of food and willingly sold themselves as slaves, sometimes for a few pounds of rice. His studies indicate that about 10,100 people died in Sumbawa directly by the effects of the eruption, most likely in pyroclastic flows and surges (Table 1). Contemporary estimates of number of fatalities in several villages vary. Thus, for example, Tobias claims there were 10,000 deaths in Tambora village alone, whereas Philipps claims 12,000 victims in this village. In addition, 37,825 died by starvation and 36,275 migrated from Sumbawa. Zollinger estimates that at least 10,000 died in Lombok from starvation and disease, but the loss there was much more severe according to Van der Broeck (1834), who states that the population of Lombok was reduced from 200,000 to 20,000 by the effects of the eruption. Zollinger claims his numbers are all minimum estimates. Junghuhn (1850) estimates that the fatalities on Sumbawa were 12,000 and that 44,000 died on Lombok, but his estimate does not include the starvation victims on Sumbawa. The most-quoted fatality figures of the eruption are those of Petroeschevsky (1949), who estimates that the total number of victims was 92,000 - 48,000 on Sumbawa and 44,000 on Lombok, or 35 and 22.5% of the estimated total population of these islands, respectively.

Table 1: Fate of the Human Population in Sumbawa.[1]

Village	Eruption Victims	Death by Starvation	Refugees
Pekat	2,000	--	--
Tambora	6,000	--	--
Sanggar	1,100	825	275
Dompu	1,000	4,000	3,000
Sumbawa	--	18,000	18,000
Bima	--	15,000	15,000
Totals:	10,100	37,825	36,275

[1] After Zollinger (1855).

The only village near Tambora that remained undamaged was Tempo, with 40 inhabitants. Of the total population of 12,000 of Tambora and Pekat, only five or six survived. All trees and vegetation north and west of the volcano were completely destroyed, with the exception of a high point near the village of Tambora. Zollinger remarked on the long-term effects of the eruption on Sumbawa's climate and vegetation. Soil became very dry, rainfall decreased, and all vegetation suffered a severe setback, and would take an estimated several hundred years to recover fully.

Pyroclastic Deposits from the 1815 Eruption

As a consequence of the eruption, the upper part of the volcano collapsed to form a 6-km diameter, 1,200-m deep caldera with a total volume of about 28 km^3, and Tambora lost about 1,200 to 1,400 m of its height, corresponding to about 6 km^3 or a total of 34 km^3. The void formed by the caldera collapse represents in part rock formations ejected from the volcano, and in part the subsidence of the volcano's edifice into the underlying magma chamber. The former can be evaluated from proportion of lithics in the fall deposits, which is about 5.5 wt.% (Sigurdsson and Carey in press, Table 2) or less than 4 km^3 of rock. Ejection of solid rock can consequently account for one-tenth of the caldera volume. The total ejected mass of magma is 1.3×10^{14} kg, less the lithics, corresponding to about 50 km^3 of magma withdrawn from the reservoir - substantially larger than the observed caldera volume. Subsidence into the emptying magma reservoir is regarded as the dominant mechanism of caldera formation.

The deposits laid down outside the caldera during the eruption reflect two major processes: (1) early explosive activity (plinian and phreatomagmatic) producing high eruption columns and four tephra or ash-fall deposits; and (2) subsequent ignimbrite phase activity during collapse of the eruption column, producing at least seven pyroclastic flows and surge deposits, with associated large-volume co-ignimbrite ash falls.

Table 2: Tambora 1815; Composition of Glass Inclusions in Plagioclase Phenocrysts.[1]

	1	2	3	4	5
SiO_2	57.37 (1.28)	57.01 (.51)	56.37 (.29)	56.88 (1.24)	56.58 (1.13)
TiO_2	0.6 (.07)	0.56 (.06)	0.73 (.01)	0.60 (.10)	0.72 (.31)
Al_2O^3	19.66 (.41)	19.58 (.22)	19.43 (.06)	19.88 (.45)	20.17 (.21)
FeO	4.56 (.39)	4.47 (.20)	4.66 (.07)	4.73 (.48)	5.23 (.94)
MnO	0.27 (.05)	0.28 (.08)	0.25 (.07)	0.19 (.08)	0.24 (.05)
MgO	1.33 (.17)	1.09 (.07)	1.18 (.03)	1.37 (.15)	1.75 (.44)
CeO	2.71 (.35)	3.09 (.08)	2.87 (.01)	2.85 (.25)	2.50 (.25)
Na_2O	6.19 (.25)	5.5 (.50)	5.89 ___	6.44 ___	3.15 (.54)[2]
K_2O	5.09 (.47)	5.59 (.27)	5.69 (.09)	5.35 (.86)	6.00 (.48)
P_2O_5	0.06 (0)	0.04 (.02)	0.31 (.13)	0.36 (.15)	___
Total	97.68	97.21	97.38	98.65	96.34
Number of Inclusions	6	2	1	7	___
Number of Analyses	10	6	2	11	7
Volatiles by Difference	2.32	2.79	2.62	1.35	___
Water by Difference[3]	1.95	2.42	2.25	0.98	___
Sulphur (ppm)	512±62	589±94	613±157	___	381± 44
Chlorine (ppm)	1,747±337	2,057±732	2,375±532	2,817±1253	2,106±163
Fluorine (ppm)	___	___	___	___	1,185± 87

[1] 1 - glass inclusions in plagioclase from plinian fall layer F-2, sample TB-42; 2 - glass inclusions in plagioclase from lower part (0 to 5 cm) of plinian fall F-4, sample TB-86; 3 - glass inclusions in plagioclase from upper part (10 to 15 cm) of plinian fall F-4, sample TB-88; 4 - glass inclusions in plagioclase from co-ignimbrite fall deposit F-5, sample TB-136; 5 - glass inclusions in plagioclase from tephra fall, sample T58-A (Devine et al. 1984).

[2] Not corrected for sodium loss during microprobe analysis. All other values are corrected.

[3] Water by difference is calculated as volatiles by difference minus S, Cl and F (0.37 wt.%).

The four initial explosive events produced widespread tephra fall deposits, which can be traced at least to Lombok, 150 km west of the caldera (Sigurdsson and Carey in press). The basal F-1 ash fall is the product of phreatomagmatic explosions, resulting from interaction of magma with the hydrothermal system of the volcano (Figure 3). Historical evidence (Petroeschevsky 1949) indicates that the volcano was mildly active in the period 1812-15, when "rumblings and dense clouds" were noted. The F-1 ash fall probably originated during this early activity, as magma was making its way from a deep reservoir toward the surface and periodically erupting in small outbursts. The total volume of tephra erupted during the phase of activity was about 0.1 km^3. Evidence from our excavations in the ruins of the ancient Tambora village, 2 km east of Tambora Coffee Estate, indicates that the F-1 event took place long before the subsequent activity. The F-1 layer is absent from the village, indicating its complete erosion before the 5 April eruption.

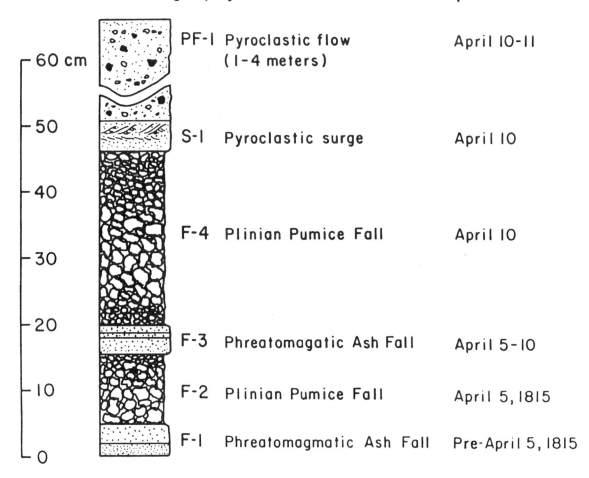

Figure 3: Stratigraphy of the 1815 pyroclastic deposits in a typical section at Gambah on the northwestern slopes of Tambora volcano, 25 km from the caldera. Dates on the right of the stratigraphic column indicate the timing of successive eruptive phases, based on historical reports.

The F-2 pumice fall layer marks the first major explosive eruption during 1815. The distribution, lithology and grain-size of this deposit indicate typical plinian activity. About 1.2 km^3 of material was ejected. We correlate this plinian eruption with the explosion of 5 April 1815 that was heard in Jakarta (1,250 km away) and Ternate (1,400 km away) and caused ash fall as far as Besoeki in East Java (Raffles 1835).

After the F-2 plinian event, Tambora lapsed into a state of low-level activity from 5 to 10 April. During this period, several smaller explosions produced tephra fall, which forms layer F-3 (Figure 3). The deposit is highly fragmented, like the first layer of the eruption (F-1), but the evidence of the phreatomagmatic activity is not as compelling.

A second major plinian eruption produced the F-4 pumice fall layer. Grading of this deposit shows a rapid rise of the eruption column during the first third of the eruption, followed by a slow decline. The F-4 layer is much thicker and coarser than the earlier F-2 plinian fall, although similar in lithology. Despite its high intensity, this phase of the eruption ejected only a moderate amount of material (3 km^3 of tephra). The F-4 fall deposit is clearly from the beginning of the 10 April paroxysmal event.

The Rajah of Sanggar reported an intensification of activity at about 7 p.m. on 10 April, followed by a rain of pumice on Sanggar, east of the volcano, at approximately 8 p.m. Tephra fall continued until about 10 p.m. when the village experienced winds that uprooted trees and buildings. The whole volcano appeared like a flowing mass of "liquid fire". This event is marked clearly in the volcanic stratigraphy everywhere on Sanggar Peninsula by the abrupt transition from F-4 plinian pumice fall to the overlying charcoal-bearing surge and pyroclastic flows (Figure 3). The change in the eruption mechanism may have been primarily due to continued vent erosion during F-4 plinian activity, leading to eruption column collapse, with resulting pyroclastic flows and surges. No significant time break may have occurred during the transition.

In distal localities the F-4 plinian fall is overlain by a 12- to 25-cm thick greyish-brown, poorly-sorted, silty-sandy ash (F-5). Unlike the other fall deposits from the 1815 eruption that show systematic thinning with distance from source, the F-5 ash fall retains a constant thickness to a remarkable degree - in fact thickening appreciably to the west, away from the volcano. Consequently the F-5 layer represents an increasing proportion of the total fall with distance from source, increasing from about 25% of the total fall deposit thickness at 40 km, to about 80% beyond 90 km (Figure 4).

The F-5 layer does not correspond to any fall deposits in the proximal area, but is stratigraphically equivalent to the surges and pyroclastic flows. We therefore consider the F-5 deposit formed primarily from ash and pumice fallout from an eruption column generated during the surge and pyroclastic flow phase, i.e., a co-ignimbrite and co-surge ash fall deposit. The co-ignimbrite ash fall was not only generated by glass elutriation from the convecting eruption columns and flows, but also by wholesale depletion of the fine fraction of crystals and glass alike from the column and flows. We attribute this depletion to explosive interaction between pyroclastic flows and the sea along the coast of the Sanggar Peninsula (Sigurdsson and Carey in press), based on comparative grain-size studies of inland and coastal pyroclastic-flow deposits. Our model proposes the creation of large secondary eruption columns around the peninsula of the volcano, where high-temperature pyroclastic flows were discharged into, and reacted explosively with seawater. The secondary plumes consisted mostly of fine-grained (<200 micron) ash and steam.

Total Erupted Mass

It is generally recognized that the Tambora eruption involved an exceptionally large volume of magma, although quantitative estimates have varied greatly. Thus, Zollinger (1855) estimated the ash fall volume at >1,000 km^3, Junghun (1850) 318 km^3, Verbeek (1885) 150 km^3, Sapper (1917) 140 km^3, Pannekoek van Rheden (1918) 30 km^3, and Petroeschevsky (1949) estimated total ash fall of 100 km³ on the basis of observed thicknesses. A reassessment of the ash fall volume by Stothers (1984) led to an estimate of 150 km³, and Self *et al.* (1984) estimate 175 km³. New estimates can now be made on basis of our recent field work, Sigurdsson and Carey (in press).

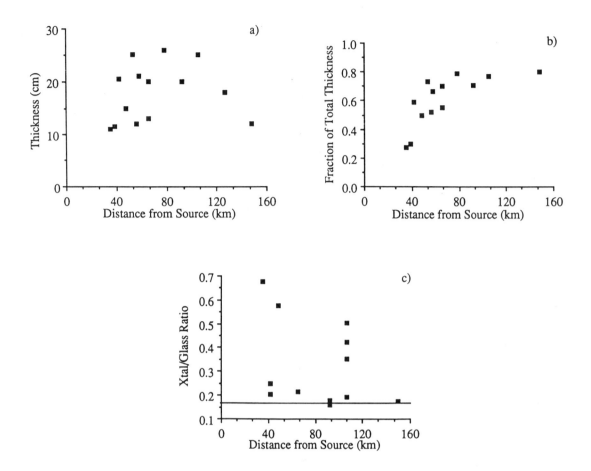

Figure 4: Characteristics of the F-5 co-ignimbrite tephra fall deposit as a function of distance from source, showing: (a) variation in thickness in cm; (b) thickness of F-5 co-ignimbrite ash fall as a fraction of total ash fall thickness; (c) crystal/glass ratio of the co-ignimbrite ash fall. Horizontal line in (c) is the crystal/glass ratio in the erupted magma, as determined in artificially-crushed pumices from the pyroclastic flows.

As shown above, the products of the eruption form a multi-layer deposit, reflecting several processes in action. The four early fall deposits produced during activity from 5 to 10 April (F-1 to F-4), have a total volume of 4.6 km³, corresponding to 1.8 km³ of dense rock, or about 4.3×10^{12} kg of magma. While this represents an eruption larger than the 1980 Mount St. Helens event, and comparable in volume to the 1982 El Chichón event, these early April falls from Tambora represent only 5% of the total erupted mass in 1815. The co-ignimbrite ash-fall layer

F-5 is the dominant part of the deposit. Volume of the total ash fall can be estimated from contemporary accounts of ash fall and thickness, and deep-sea core evidence (Neeb 1943). Using the isopach map compiled by Self *et al.* (1984) and shown in Figure 7, we estimate that 90 km^3 of tephra was deposited within the 1 micron isopach, corresponding to 22 km^3 dense-rock equivalent of distal fall (5.5×10^{13} kg). The F-5 fall must represent about 92% of this volume, as the combined volume of the earlier F-1 to F-4 fall layers is only 1.8 km^3 DRE. An estimate of density of the deposit is required in order to assess the erupted mass. During the eruption, ash fell on decks of the ship *Benares* near Macassar in Sulawesi. A pint of the ash was reported to weigh 12¼ oz, corresponding to a deposit density of 611 kg/m^3 of the fresh-fallen ash (*Asiatic Journal* 2, 1816, p. 166). A minimum mass of 5.8×10^{13} kg is therefore represented by the fall deposit.

Studies of the volcano and its deposits indicate that a large mass of pyroclastic flows entered the ocean during the eruption (Sigurdsson and Carey in press). We estimate a pyroclastic flow deposit volume of about 30 km^3, equivalent to 8.2×10^{13} kg. Thus, the total erupted mass is of the order 1.4×10^{14} kg of magma. No historical eruptions have produced as large a mass of magma as Tambora, which emitted more than twice the mass of the nearest large-magnitude event, i.e., the 1783 Laki eruption in Iceland.

Mass Eruption Rate and Column Height

Pumice and lithic isopleth maps of the F-2 and F-4 layers are presented by Sigurdsson and Carey (in press). The area encompassed by a specific isopleth is considerably larger for the F-4 layer, demonstrating the greater intensity and dispersal of that plinian event. The distribution of the two Tambora plinian layers is compared with several other well-documented plinian fall deposits (Figure 5). Our new isopleth data indicate that the two Tambora plinian fall deposits had greater dispersal than any plinian eruption in historic times. Despite the fact that their dispersal compares with some of the largest known plinian fall deposits in the geological record, the thicknesses and thus volumes of the two Tambora fall deposits are relatively small (Figure 5). The great dispersal of clasts during the two plinian eruptions of Tambora is noteworthy and has important implications for existing models of the 1815 eruption.

The dispersal characteristics of the pumice fall preserve information about the dynamics of the eruption column and the atmospherically-dispersed plume. Thus, the geometry of lithic isopleths can be used to determine the maximum eruption-column height and average wind speed for a specific fall layer (Carey and Sparks 1986). The half-width of an isopleth measured perpendicular to the main dispersal axis is primarily a function of the eruption column height, whereas the maximum downwind range along the axis is controlled by both column height and average wind speed. Data from the 3.2-cm diameter lithic isopleths of the F-2 and F-4 layers indicate eruption-column heights of 33 and 43 km, respectively (Figure 6). This places the F-2 column higher than the maximum height achieved by the 79 A.D. plinian eruption of Vesuvius (Carey and Sigurdsson 1987), and the F-4 column is slightly higher than the great 1956 Bezyminanny eruption (Gorshkov 1959). The F-4 plinian phase is thus the most energetic plinian activity ever recorded in historic times, and is exceeded in intensity by only one eruption in the geological record - the "ultraplinian" Taupo pumice fall in New Zealand (Walker 1980).

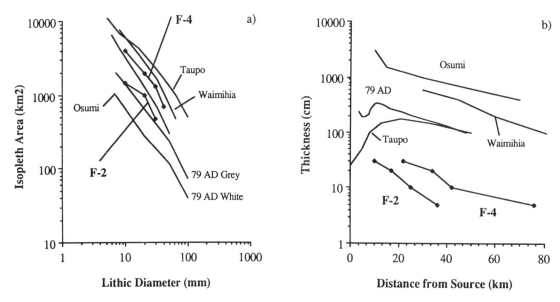

Figure 5: Comparison of 1815 Tambora fall deposits with characteristics of deposits from other major volcanic eruptions. (a) Plot of lithic isopleth area versus lithic diameter for the F-2 (5 April) and F-4 (10 April) plinian fall deposits compared with the plinian falls from the eruptions of Vesuvius, Italy (79 AD), Osumi, Japan, and Taupo and Waimihia, New Zealand. (b) Plot of thickness versus distance from source for the F-2 and F-4 Tambora plinian fall layers compared to other well-known plinian deposits, as in Figure 5 (a). Note that despite the fact that Tambora layers are very widely dispersed, they are substantially thinner than other major plinian fall deposits.

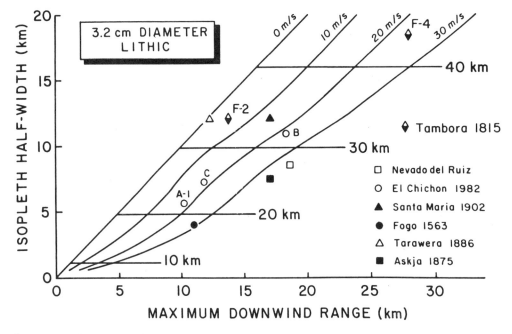

Figure 6: Plot of isopleth half-width versus maximum downwind range of the 3.2-cm diameter lithics isopleths for the F-2 and F-4 plinian Tambora fall deposits, compared with other well-known plinian falls. Diagonal lines are wind-velocity contours in m/second, and horizontal lines are maximum eruption-column heights in kilometres. Note the 43 km high F-4 plinian column from 10 April 1815 above Tambora volcano, and the 33 km high column from the F-2 eruption on 5 April.

The estimates of eruption-column height can be used to calculate the eruption rate by using relations for a tropical atmosphere (Sparks 1986). Our calculations indicate a rate of 1.1×10^8 kg/s for the F-2 phase and 2.8×10^8 kg/s for the more energetic F-4 event. With these values it is possible to estimate the duration of the events by simply dividing the total mass of tephra in each layer by the perspective rate of magma discharge. Assuming that the maximum magma-discharge rate was active throughout the plinian eruptions, each layer would have been ejected in 2.8 hours.

Duration of the co-ignimbrite fall was about three days, judging from the historical reports, e.g., from Madura Island, 500 km WNW of the volcano (Figure 7). This is the period of the sedimentation of tephra from the atmosphere and thus represents the maximum duration of the eruption which began on 10 April. In order to accommodate the total ignimbrite and co-ignimbrite mass in this period (1.4×10^{14} kg), we infer a minimum ignimbrite mass-eruption rate of the order of 5.4×10^8 kg/s, or about three times the peak rate during the eruption of the preceding F-4 plinian fall. This is about half the rate of the highest intensity event known: the Taupo eruption in 130 AD, with an eruption rate of 1.1×10^9 kg/s (Walker 1980).

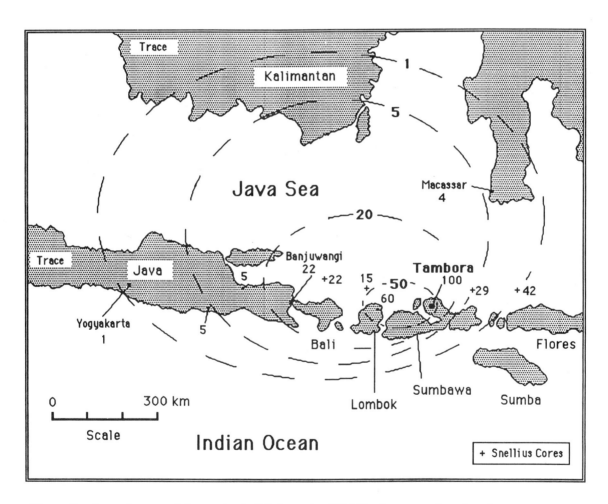

Figure 7: Isopach map of the total ash fall from the 1815 Tambora eruption, based on contemporary reports of ash fall and evidence from bottom samples collected during the Snellius Expedition.

Volatile Emission from the Tambora Eruption

Recent studies have shown that quantitative estimates can be made by petrological methods of the mass and type of volatiles (e.g., sulphur, chlorine and fluorine) released during volcanic eruptions. The potential of trapped glass inclusions as recorders of pre-eruption volatile content of magmas was first recognized by Anderson (1974), who applied this method in estimating the volcanic volatile contribution to the sulphur and chlorine budget of the oceans. The method was also applied in the 1976 St. Augustine eruption by Johnston (1980), who demonstrated the potentially great contribution of volcanic eruptions to the chlorine budget of the stratosphere. These studies paved the way for the petrologic estimates of volcanic degassing during earlier and prehistoric eruptions (Sigurdsson 1982; Devine *et al.* 1984; Palais and Sigurdsson 1989). When compared with other determinations of volatile emission based on ice-core acidity and atmospheric observations, the petrologic estimates yield similar results for the same eruptions (Sigurdsson *et al.* 1985).

In the first petrologic study of volcanic volatiles from the 1815 Tambora eruption, Devine *et al.* (1984) found that seven glass inclusions in feldspar phenocrysts from a single pumice sample contained on the average 380 ppm sulphur, 2,100 ppm chlorine, 1,190 ppm fluorine. We have analyzed glass inclusions in plagioclase phenocrysts and matrix glasses in five tephra samples from the 1815 eruption, representing all major deposits produced during the event. Our results (Tables 2, 3) show that the eruption tapped a homogenous body of trachyandesite magma, with no systematic chemical gradients. We find that the average pre-eruption concentration of volatiles is 570 ppm sulphur, 2,220 ppm chlorine and 1,190 ppm fluorine, whereas the degassed matrix glass has on the average 266 ppm sulphur, 1,486 ppm chlorine and 680 ppm fluorine. These results indicate, that about 53% of the pre-eruption sulphur content of the magma was lost to the atmosphere during the eruption, accompanied by loss of 33% of the chlorine and 43% of the fluorine. In addition, the results indicate a pre-eruption water content of about 2 to 2.4 wt.% in the magma.

Table 3: Matrix Glass Composition of Tambora 1815 Tephra (parts per million).[1]

	Sample Numbers					
	TB-42	TB-86	TB-88	TB-136	TB-87	T58-A
Sulphur	363±57	126±17	241±67	196±48	362±33	309±7
Chlorine	1,523±174	1,460±142	1,621±22	1,476±69	1,627±139	1,211±50
Fluorine	-	-	-	-	-	679±69

[1] Errors are one standard deviation of the average.

With a known total erupted mass of magma of 1.4×10^{14} kg, the minimum mass of volatiles emitted to the atmosphere can be estimated from the difference in volatile concentration between glass inclusions and matrix glasses. These calculations show that about 4.3×10^{10} kg of sulphur were released to the atmosphere, 1×10^{11} kg chlorine, and 7×10^{10} kg fluorine. These improved estimates are somewhat lower than the preliminary values of Devine *et al.* (1984) for Tambora volatile degassing, but still place Tambora as the pre-eminent volcanic pollution event in historic time, with a total mass of 2.1×10^{11} kg of sulphur, chlorine and fluorine released to the atmosphere. Further studies of the poorly-constrained volume of the distal ash fall will probably lead to an increase in these estimates. In addition, we infer that about 2.8×10^{12} kg of magmatic H_2O was introduced into the atmosphere during the eruption, or equivalent to more than doubling the stratospheric water-vapour content. Further addition of large quantities of meteoric water vapour to the stratosphere resulted from the large-scale convective flow of humid tropospheric air, entrained in the ascending eruption column.

No measurements have been made of carbon dioxide levels in the Tambora products, but some inferences can be made of CO_2 output from the eruption. Magma of the type erupted from Tambora in 1815 is likely to have CO_2 levels of the order 500 ppm, judging from the solubility data of Stolper and Holloway (in press). Degassing of magma of this type would then yield about 10^{14} g CO_2 to the atmosphere during the 1815 eruption. Thus, the carbon dioxide output from Tambora would be roughly equivalent to the annual output from the Earth's mantle, and only about 1% of the current annual anthropogenic output of CO_2.

Sulphur Aerosol

Sulphur output from Tambora during the three-day period in 1815 was more than double the current annual total sulphur output of volcanoes, which has been estimated as 0.9 to 1.2×10^{10} kg/yr (Berresheim and Jaeschke 1983; Stoiber *et al.* 1987). In comparison, the annual global anthropogenic emission rate of sulphur dioxide is estimated as 1.3×10^{11} kg (Bach 1976). The fates and atmospheric effects of anthropogenic and volcano-derived sulphur aerosols are, however, quite different. The anthropogenic emission, caused by burning of fossil fuels, is mostly confined to the troposphere, where its residence time is short. In contrast, highly energetic explosive volcanic eruptions transport sulphur and other volatile species rapidly to the upper troposphere and lower stratosphere. In the case of Tambora, the early plinian events in April 1815 had sustained eruption columns of 33 to 43 km height above the volcano, but the convective columns during the main ignimbrite phase were probably in the 15 to 20 km range. With estimated magma source rate of 5.4×10^8 kg/s during the 10 April eruption, the source rate of volatiles to the atmosphere during the is period is calculated as 1.7×10^5 kg/s for sulphur, 4×10^5 kg/s for chlorine, 2.7×10^5 kg/s for fluorine and about 10^7 kg/s for magmatic water vapour.

Sulphur emitted by Tambora was initially in the gaseous state, probably dominantly as SO_2 and lesser amounts of H_2S and OCS, which are the precursor gases to sulphate aerosols and consume OH radicals. The large mass of magmatic and atmospheric water vapour injected into the stratosphere during the eruption (2.8×10^{12} kg) is a major potential source of the OH. Upon mixing with air, the sulphur dioxide would undergo oxidation to SO_3 and react with water vapour in the atmosphere to form an aerosol of sulphuric acid droplets. Reactions of the following type may account for the conversion of sulphur gases to sulphuric acid aerosol particles in the atmosphere:

$$SO_2 + OH \rightarrow HOSO_2 + O_2 \rightarrow SO_3 + HO_2$$
$$SO_2 + 1/2\ O_2 \rightarrow SO_3$$
$$SO_3 + H_2O \rightarrow H_2SO_4\ (liq)$$
$$H_2S + 3/2\ O_2 \rightarrow H_2O + SO_2$$
$$H_2S + 2O_2 \rightarrow H_2SO_4\ (liq)$$

The above mass estimates of volatile output from the eruption refer to elemental concentration of sulphur, chlorine and fluorine. Direct analysis of modern volcanic aerosols shows that they are typically composed of a 75% H_2SO_4 aqueous solution (Hofmann and Rosen 1983). Converting the above petrologic estimate of 4.3×10^{10} kg elemental sulphur to sulphuric acid aerosol, we therefore estimate the Tambora sulphur-rich aerosol mass as 1.75×10^{11} kg, or an order of magnitude larger than the 1982 El Chichón aerosol (McCormick and Swissler 1983). By comparison, Hammer *et al.* (1980) estimate a Tambora volcanic aerosol of 1.5×10^{11} kg on the basis of the 1816 acidity layer in Greenland ice cores, and Stothers (1984) estimates 2×10^{11} kg based on observed atmospheric effects. The difference in these estimates is within the uncertainties of the methods, but several factors make the petrologic estimate a minimum value. Firstly, further studies of the thickness and distribution of the distal tephra fall deposit preserved on the ocean floor may conceivably double the total erupted mass estimate and thus double the estimate of sulphur yield to the atmosphere. Secondly, the Tambora gas emission also involved about 1×10^{11} kg HCℓ and 7.4×10^{10} kg HF, and the possible involvement of these gases in aerosol formation cannot be ruled out. Thirdly, the petrologic estimate is only of volatiles exsolved from the magma at the time of eruption, and does not include a possible separate volatile phase. Finally, the Tambora stratospheric aerosol or "dust cloud" also contained some particles of volcanic glass, as demonstrated by the recent identification in a South Pole ice core of Tambora glass fragments by microprobe analysis (J. Palais, personal communication).

The Halogens

The large-scale introduction of odd-chlorine species into the stratosphere during the 1815 Tambora eruption is important because of the potential of chlorine in catalyzing the removal of O_3 and thus damaging the Earth's ozone layer. That layer shields the biosphere from the effects of damaging solar ultraviolet radiation, such as effects on DNA and the immune-system response, skin cancer and sunburn. It is generally believed that diffusion of anthropogenic chlorofluoromethanes (CFC) from the troposphere is currently the principal source of stratospheric chlorine, but the importance of volcanic emissions as a potential source of stratospheric chlorine was first pointed out by Stolarski and Cicerone (1974).

Stolarski and Butler (1978) estimated a stratospheric injection rate for volcanic chlorine of 1.3×10^7 kg/yr, or more than three orders of magnitude less than the 10^{11} kg HCℓ emission during the 1815 eruption alone. By comparison, the annual release of chlorofluorocarbons is about 7×10^8 kg/yr, and the budget of stratospheric chlorine is about 10^9 kg/yr. HCℓ is generally the principal chlorine molecule in volcanic gases, but studies of the 1980 Mount St. Helens stratospheric cloud show that concentrations of methyl chloride ($CH_3Cℓ$) were as high or higher than concentrations of HCℓ (Inn *et al.* 1981). HCℓ is highly soluble in water, so possibly large quantities of the emitted HCℓ are dissolved in eruption-cloud water and returned to the surface of the Earth as precipitation during or shortly after eruption.

Although large quantities of chlorine and fluorine are shown to be emitted by Tambora, it should not be assumed that these gases form aerosols in the stratosphere, as physical and chemical data indicate that HCl and HF gases are unlikely to form liquid aerosols under normal stratospheric conditions (Miller 1983; Solomon and Garcia 1984). As shown by Oskarsson (1980) (Figure 8), however, halogen aerosols may conceivably form at higher temperatures in the eruption column, and the presence of elevated concentrations of HCl and HF in volcanic-acidity layers in Greenland ice cores suggests that halogens have indeed become incorporated into some volcanic aerosols. Thus, the acidity layer from the 934 A.D. Eldgja eruption in a Greenland ice core contains at least 65% HCl (Hammer 1980). Herron (1982) has also shown high levels of both Cl and F in another Greenland ice-core layer from this eruption. Similarly, Herron (1982) and Hammer (1977) have both noted elevated Cl levels in the Greenland ice-core acidity layer from the 1783 Laki eruption. Finally, very high Cl concentration in a northwestern Greenland ice core, which was attributed by Herron (1982) to early nineteenth century volcanic activity, may conceivably represent material from the Tambora eruption. The ice-core data thus suggest that Cl and possibly F may enter the volcanic aerosol. This may not imply the formation of a discrete halogen aerosol, but rather that HCl and HF may be absorbed and dissolved in the sulphuric acid aerosol.

HCl is inert toward ozone, but reaction of HCl with OH leads to formation of atomic chlorine, followed by the catalytic decomposition of the ozone by the Cl. Thus in the stratosphere, Cl can be released from HCl by reactions of the type:

$$HCl + OH \rightarrow H_2O + Cl$$

Similarly, methyl chloride can produce atomic chlorine by photolytic decomposition and attack by OH. Several reactions involving gaseous chlorine have the effect of converting odd-oxygen molecules (including ozone) to diatomic oxygen by ClO catalysis. They are reactions of the type:

$$O_3 + Cl \rightarrow O_2 + ClO$$
$$O + ClO \rightarrow O_2 + Cl$$

The only attempt to model the effects of large volcanic chlorine emission on the ozone layer was made by Stolarski and Butler (1978), who concluded that a Krakatau-size emission, involving 3×10^8 kg Cl_x would result in about 7% depletion of the ozone layer. Chlorine output was two orders of magnitude higher than this value during the Tambora eruption, and major ozone depletion cannot be ruled out. Given the great importance of the ozone layer to the biosphere and climate, the modelling of the potential impact on atmospheric chemistry by a Tambora-size eruption is timely.

Nothing is known about the possible atmospheric or environmental effects of the large (7.4×10^{10} kg) HF gas emission during the eruption indicated by our petrologic study. In general, HF is assumed to be very inert in the stratosphere. The photolysis of HF is shielded by oxygen, and the reaction of HF with OH is endothermic, so that it is believed that F atoms do not play the same role in stratospheric chemistry as chlorine atoms (Sze 1978). Furthermore, fluorine and to some extent chlorine, are known to adsorb onto tephra particles and thus may be rather rapidly removed from the atmosphere in the tephra fallout (Rose 1977; Oskarsson 1980).

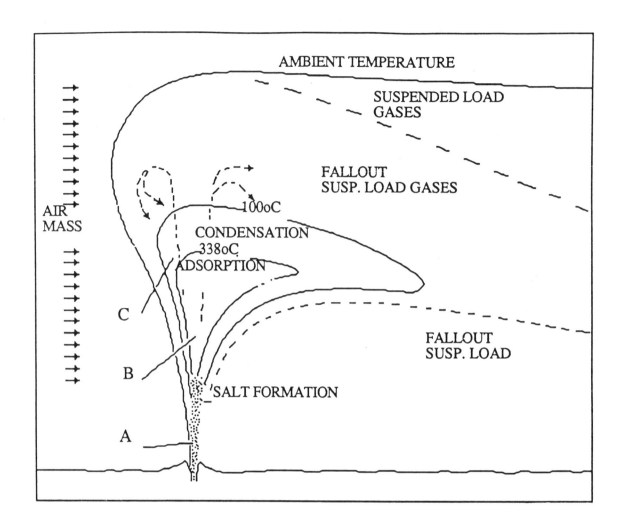

Figure 8: Evolution of volcanic volatiles within an explosive eruption column, showing volcanic volatile reaction zones (from Oskarsson 1980). In the salt formation zone A, aerosol salt particles are formed at magmatic temperatures during high-temperature degassing of magma in the vent region. At temperatures in the range 338° to 700°C, surface adsorption of halogen gases occurs as they react with silicate material (adsorption zone B). At temperatures below 338°C sulphuric acid condenses as an aerosol in the condensation zone C.

Fate of Volatiles in the Eruption Column and Atmosphere
An explosive volcanic eruption represents a rapid transfer of heat and mass into the Earth's atmosphere, resulting in a major thermal and chemical perturbation. In the case of the Tambora eruption, the thermal energy release alone was equivalent to about 1.3×10^{27} ergs, most of which was introduced into the atmosphere over a period of about three or four days. Most of this energy was expended in convective mixing of the eruption column with ambient air and heating of the entrained air, resulting in the buoyant rise of the eruption column to heights of 43 km, as a mixture of pyroclastic fragments, volcanic gases and humid tropospheric air.

Observations and theory (Sparks and Wilson 1982) shows that the solid particle weight fraction in high-eruption columns ($1-n_c$) is only of the order 0.018; the remainder being almost entirely entrained atmospheric air and expanding volcanic gases. Assuming that most of the tephra that generated the fallout deposit (5.8×10^{13} kg) had entered the lower stratosphere, the mass of associated air lofted to the stratosphere would then be about equal, and equivalent to approximately 7×10^{13} m^3 at the surface. The water content of saturated air at 1 atm and 14°C is about 0.01 kg H_2O/kg air. Thus the total mass of atmospherically-derived water entrained into the stratospheric eruption column could have been as high as 5×10^{11} kg. A portion would condense with rise in the eruption column and cause precipitation, but some would enter the stratosphere. Although large, this figure is only one-third of the mass of magmatic water introduced into the atmosphere (1.7×10^{12} kg), as discussed previously. Normally the content of water vapour decreases with height due to lowering of both temperature and saturation vapour pressure and condensation. However, water vapour is likely to be introduced to high levels under the conditions of elevated temperatures and turbulence within a buoyantly rising eruption column.

Water vapour introduced to the stratosphere by an eruption column could be a major source of OH radicals by reaction of water vapour with photodissociated oxygen atoms. Evidence from ground-based spectroscopic measurements of OH during the 1982 El Chichón eruption indicates that water vapour was injected at the level of 20 ppm (two to four times normal), and may have been responsible for the large ozone depletion observed in 1982-1983 (Burnett and Burnett 1984). Elevated levels of volcanically-derived OH from Tambora may have played a major role in generation of H_2SO_4 by reaction with SO_2, in the regeneration of free Cl atoms from HCl and in direct reactions with stratospheric ozone.

The field evidence indicates that during the main ignimbrite phase of the Tambora eruption, transport to the atmosphere was effected by two processes: the eruption column rising above the centre of the volcano, and secondary eruption columns rising from the coastline around the volcano as hot pyroclastic flows entered the ocean and flashed seawater to steam. About 35% of the erupted products entered the ocean in this manner and contributed to the secondary columns, probably resulting in a coastal ring of composite eruption columns around the entire Sanggar Peninsula, some 50 km in diameter.

Evidently, extremely variable conditions exist in the eruption column, with great range in temperature, and mixing proportions of ambient air, condensed water vapour, volcanic gases and pyroclasts. Temperatures will range from magmatic (about 950°C) to stratospheric air (-60°C). The fate of the volcanic gases in the eruption column depends on temperature-dependent reactions in the atmosphere, and although conditions can clearly be highly variable, Oskarsson (1980) has recognized three zones (Figure 8).

Salt Formation Zone
A spontaneous non-equilibrium degassing occurs during a rapid pressure drop such as an explosive eruption. In the hottest core of the eruption column, within the jet-like mixture of pyroclasts and volcanic gases, aerosol salt particles are formed at near-magmatic temperatures. These are solids condensing from a magmatic gas phase and the solid reaction products of magmatic gas and its surroundings (Oskarsson 1980). Major sources for the salts are alkali metals, calcium, aluminum and silica from the silicate melt, and the reactive gases SO_2, HCl, HF and NH_3. The dominant products in the salt formation zone are chlorides, fluorides and sulphates of calcium and the alkali metals. Owing to the high vertical mass flow rates, particles formed in this zone are likely to represent a small fraction of total aerosol production, and they

will be transported as suspended load high into the eruption column and downwind from the volcano.

Surface Adsorption Zone
As shown experimentally (Oskarsson 1980), the halogen gases react with silicate ash by surface adsorption and condensation of the gas phase at temperatures below 700°C. The reactions of the halogen gases with the glassy tephra will produce components such as calcium fluorosilicates, sodium and calcium chlorides and sodium fluoride. Halogens adsorbed on tephra particles in this zone will be removed relatively quickly from the eruption column during fallout. Thus, Rose (1977) has demonstrated that 17% of the Cl released in the 1974 Fuego eruption was stripped from the eruption plume by adsorption onto tephra particles.

Experiments and observations of the 1970 Hekla eruption show that a large fraction of the fluorine is stripped from the high-temperature region of the eruption column (338 to 700°C) by adsorption onto tephra and thus incorporated in the fallout deposit near source (Oskarsson 1980). Most of the Tambora fluorine emission may have been removed by this process, leading to fluorosis and thus accounting for the observed death of livestock. During the 1970 Hekla eruption in Iceland, fluorine-rich fallout led to poisoning of large numbers of livestock up to 200 km from the volcano (Thorarinsson and Sigvaldason 1972). The tephra fall from the eruption was unusually rich in adsorbed fluorine (up to 2,000 ppm). The concentration of the adsorbed fluorine in the fallout deposit was directly dependent on surface area of the tephra grains, and thus the concentration increased with decreasing grain size. The total mass of fluorine deposited is estimated as 3×10^7 kg, corresponding to 700 ppm of the total erupted mass from Hekla (Oskarsson 1980).

Condensation Zone
As temperature in the eruption column falls below 338°C, sulphuric acid can condense as an aerosol by a process controlled by the rate of oxidation of SO_2 by atmospheric oxygen and reaction with water vapour. Below 120°C the halogen acids condense and may form an aerosol prior to condensation of water. The sulphuric acid aerosol droplets can act as a medium in which other acid components, such as HF, HCl and water vapour can be dissolved. In the presence of tephra particles, a portion of the condensed aerosol can be stripped with fallout from the eruption plume by adsorption onto the silicate ash. Rose (1977) estimates that up to 33% of the sulphur released by the Fuego 1974 eruption was removed from the atmosphere in this manner.

Effect of the Sulphuric Acid Aerosol on Climate
Pollack *et al.* (1976) have shown that the optical properties of tephra are distinct from those of volcanic aerosols such as sulphuric acid, derived from conversion of volcanic gases. The importance of this was demonstrated during the 1980 Mount St. Helens eruption, when it was observed that the atmospheric cloud was composed dominantly of a sulphuric acid aerosol a few days after the eruption. During this eruption, the causes for the relatively short atmospheric residence time of even the finest-grained tephra were discovered to be due to silicate particle aggregation (Carey and Sigurdsson 1982). Because of these effects, apparently the potential climatic impact of a volcanic eruption is not primarily governed by the degree of explosivity or the volume of erupted magma, but more importantly by the chemical composition of the magma. Thus, recent studies indicate that the climatological effects of volcanic aerosol emission from large basaltic fissure eruptions may in fact be more important than the effects of explosive eruptions of silicic magmas (Sigurdsson 1982).

It is generally accepted that the remarkable global meteorological and optical phenomena, observed months and years after the Tambora eruption, had a strong connection with activity of the volcano (Figure 9). Most of these phenomena can be attributed to the effect of the stratospheric volcanic aerosol. Owing to the sparse meteorological data available, the annual deviation of the global mean temperature due to the eruption is not well known, but spotty data indicate a minimum deviation in 1816 of -0.7°C in the northern hemisphere (Stothers 1984). In a reconstruction of long time series of temperature data from the eastern United States (Landsberg *et al.* 1968), the great climatic anomaly of the year 1816 is a unique event that also persists in 1817 (Figure 10). Summer temperature was about 1.5°C below the 200-year average, and the June 1816 temperature about 3°C below average.

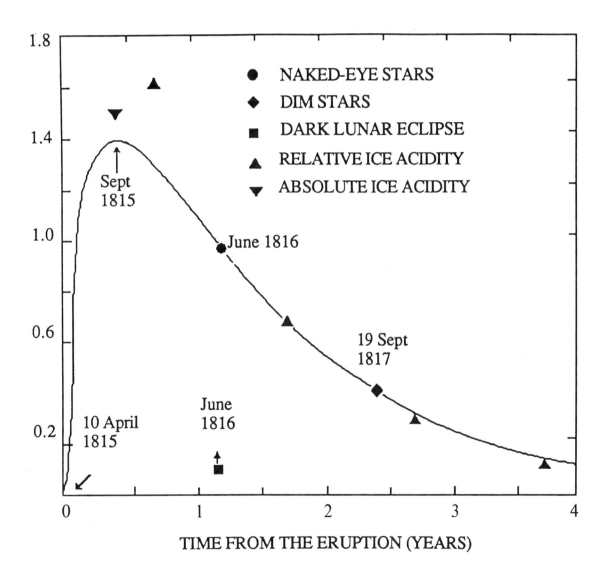

Figure 9: Change in excess visual extinction (in astronomical magnitude units) following the 1815 Tambora eruption at northern latitudes (after Stothers 1984).

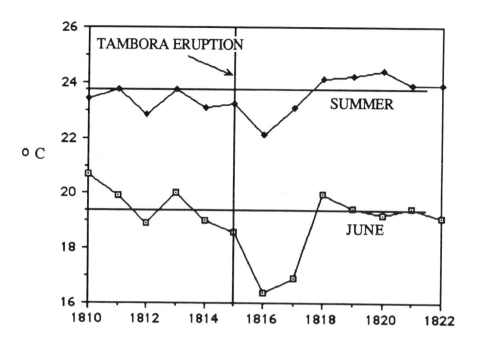

Figure 10: Observed climatic response following the Tambora 1815 eruption. Upper curve is annual summer temperature data for the eastern United States, at the latitude of Philadelphia, Pennsylvania, based on several long temperature series. The solid horizontal line shows the 224-year average summer temperature (after Landsberg *et al.* 1968). Lower curve is annual June temperature data for New Haven, Connecticut. The lower horizontal solid line shows the 145-year New Haven June mean temperature (World Weather Records 1927).

Devine *et al.* (1984) and Palais and Sigurdsson (1989) evaluated the possible effect of volcanic eruptions on climate, and proposed a relationship between the mass yield of sulphur to the atmosphere from an eruption and the observed decrease in mean northern hemisphere surface temperature in the one to three years following the eruption, on basis of published temperature data (Figure 11). Palais and Sigurdsson (1989) found that the mean surface temperature decrease was related to the estimate of sulphur yield by a power function ($r=0.92$), with the power to which the sulphur mass is raised being equal to 0.308. Although these results appear to confirm a relationship between volcanic sulphur aerosol formation and climatic change, we emphasize that the temperature deviations are associated with large errors.

As pointed out by Eddy (1988; this volume), the Tambora eruption was coincident with a depression in solar activity between about 1790-1830, i.e., the Dalton Minimum or the Little Maunder Minimum in sunspot numbers and aurorae (Figure 13). During these decades the characteristic 11-year cycle in solar activity persists, but the amplitude is reduced by an order of magnitude or more (Siscoe 1980). Variations in sunspot frequency have been linked to changes in the solar "constant", and in turn related to climatic changes. Thus the great reduction in surface temperature on Earth between 1650 and 1730 ("the Little Ice Age") corresponds to the Maunder Minimum, when there was a sudden reduction in sunspot numbers, almost to zero. It therefore appears likely that climate was already deteriorating by the beginning of the nineteenth century, due to reduction in solar activity. This climatic trend was then greatly amplified by the impact of the Tambora volcanic aerosol, culminating in the "year without summer" in 1816. Both

ΔO^{18} data on ice cores and northern hemisphere decadal temperature trends support the contention that a climatic change had set in by the first decade of the nineteenth century (Figure 12). Thus, for example, evidence from Peruvian ice cores shows that the decade 1810-20 is characterized by the most negative ΔO^{18} values (coldest temperatures) of the entire record (Figure 12), culminating in the southern hemisphere wet season of 1819-20 (Thompson *et al.* 1986; Thompson and Mosley-Thompson, this volume). The relative contribution of solar variability versus volcanic aerosol to the deterioration occurring after 1815 is unknown, but John A. Eddy (personal communication) estimates that solar variability may account for at most 10 to 50%.

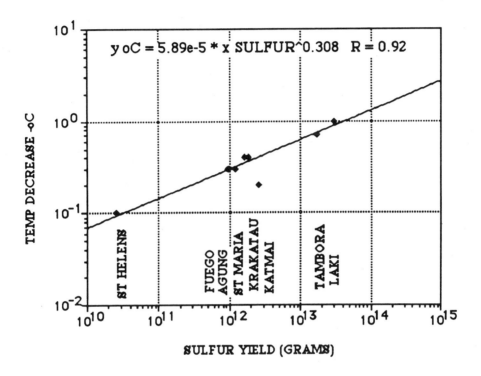

Figure 11: The observed relationship between sulphur yield to the atmosphere during large volcanic eruptions and the northern hemisphere temperature decline following the event. Sulphur data are from Devine *et al.* (1984) and Palais and Sigurdsson (1989). Climatological data are from Rampino and Self (1982) and other sources cited in the text. The equation describes the best fit to the data, with a correlation coefficient of 0.92.

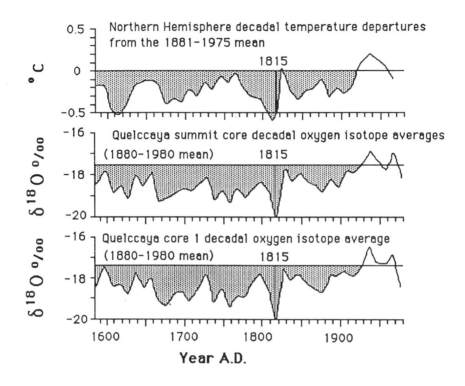

Figure 12: Variations in ΔO^{18} in Peruvian ice cores and northern hemisphere surface temperature trends, showing surface temperature decline in progress before the onset of the Tambora 1815 eruption (after Thompson *et al.* 1986).

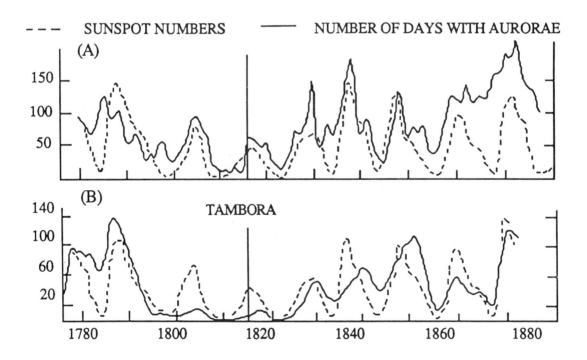

Figure 13: Auroral and sunspot trends from 1780 to 1880, showing the "Dalton Minimum" or "Little Maunder Minimum" in solar activity between 1790 and 1830 (after Siscoe 1980). (A) Sunspot numbers and number of days per year on which aurorae were recorded in Norway. (B) Sunspot and auroral data in the United States and Europe, south of 54°N Latitude.

Conclusions

The dynamics of the Tambora 1815 eruption columns and source rates of magma and volatiles can be determined by studying the deposits and petrology of the products. The initial plinian eruption of Tambora on 5 April was a brief but highly energetic event with eruption rate of 1.1×10^8 kg/s producing a column height of 33 km. In the early phase of the paroxysmal eruption on 10 April, a plinian column rose to 43 km, with eruption rate of 2.8×10^8 kg/s. The buoyant column was only sustained for about three hours, before column collapse occurred due to increasing eruption rate. Subsequent ignimbrite-phase activity during a three-day period was at rates of about 5.4×10^8 kg/s, producing a co-ignimbrite fall deposit of 5.8×10^{13} kg and a pyroclastic flow deposit of 8.2×10^{13} kg, or a total deposit of 1.4×10^{14} kg. The convective column above the volcano during the main ignimbrite phase was at least 20 km high, judging from grain-size data of the deposit, and thus injected material into the lower stratosphere. Although the ash fallout affected a broad area, the dispersal was dominantly to the west of Tambora, over Java and as far as Sumatra. This spread of the eruption plume is consistent with 10-year average rawinsonde data for Surabaja in eastern Java, which shows dominant easterly upper troposphere and lower stratosphere winds for the spring months, with mean velocities ranging from 5 to 10 m/s. The column height evidence indicates that only about 2% of the erupted mass was emplaced into the middle stratosphere, up to 43 km, and that the vast majority of the erupted products were injected in the lower stratosphere and upper troposphere.

The Tambora magma was enriched in volatile components, with 2 to 2.4 wt.% H_2O, 570 ppm sulphur, 2,220 ppm chlorine and 570 ppm fluorine. Judging from the difference in volatile concentration in glass inclusions and in matrix glasses of the tephra, the yield of sulphur to the atmosphere was 4.3×10^{10} kg, 10^{11} kg of chlorine, and 7×10^{10} kg fluorine. Magmatic water evolved from the volcano was about 2.8×10^{12} kg, whereas the mass of atmospheric water entrained in the eruption columns is estimated at 5×10^{11} kg. Source rates of the volatile species were about 1.7×10^5 kg/s for sulphur, 4×10^5 kg/s for chlorine, 2.7×10^5 kg/s for fluorine and 10^7 kg/s for magmatic water.

Generation of the sulphuric acid aerosol by gas to particle conversion was probably greatly facilitated by OH radicals in the eruption cloud, derived dominantly from reactions between excited atomic oxygen and magmatic water vapour. Assuming a typical volcanic aerosol composed of 75% H_2SO_4 and 25% water, the petrologic data indicate a minimum Tambora aerosol mass of 1.75×10^{11} kg. This compares closely to aerosol estimates based on the ice-core acidity layer (Hammer 1980) and atmospheric phenomena (Stothers 1984).

The very high proportion of halogens released by the Tambora eruption is typical of volcanic activity of such trachytic magmas in subduction-zone environments. The fate of volcanic halogens in the atmosphere is unclear at this stage. Fluorine and chlorine most likely form HCℓ and HF gas molecules upon degassing from the magma. The latter is relatively inert in the stratosphere, as HF photolysis is shielded by oxygen and HF is also relatively indifferent to OH abundance.

Chlorine was probably also removed in significant amounts from the high-temperature region of the Tambora eruption column by adsorption onto tephra. Studies of the 1974 Fuego eruption indicate that up to 17% of the chlorine was removed by this process (Rose 1977). Although HCℓ is not known to form stratospheric aerosols, chlorine may conceivably enter other aerosol droplets. Studies of ice cores cited above indicate that acidity layers from some eruptions contain significant chlorine, requiring incorporation of this species into the aerosol by some process. While HCℓ is relatively inert in the stratosphere, reaction with OH or by photolytic reactions leads to formation of atomic chlorine. As the Tambora eruption cloud was dominantly in the region below 30 km, which is photolytically inactive, formation of Cl and ClO by the latter process would have been minor. On the other hand, we contend that water vapour was injected in large quantities, involving both magmatic and atmospheric water. Thus, OH radicals were abundant in the eruption column and available for reaction with HCl to produce atomic chlorine. Reactions of atomic chlorine with ozone are catalytic, and a single chlorine atom may destroy thousands of ozone molecules before it becomes inert and enters the HCl reservoir. Independent of their role in generation of single chlorine atoms, OH radicals from the eruption cloud would also lead directly to ozone destruction.

The dominant environmental effects of the Tambora eruption were therefore probably of three types: (1) formation of a sulphuric acid aerosol, leading to a northern hemisphere temperature reduction of at least 0.7°C at the surface and stratospheric heating; (2) adsorption of fluorine onto tephra, leading to very high fluorine levels in the fallout on the ground in Indonesia and resulting in widespread fluorosis; and (3) extensive ozone depletion as a consequence of generation of odd chlorine atoms and high levels of volcanically-derived stratospheric OH radicals.

Acknowledgements

This research was carried out with funding from the National Science Foundation (grants EAR-8607336 and EAR-8804117), and field studies in Indonesia were made possible by funding from the National Geographic Society (grant NGS 3390-86). We thank the Volcanological Survey of Indonesia for collaboration in the field, and the Indonesian Research Council (LIPI), for permission to undertake research in Sumbawa. The assistance of David Browning in electron microprobe and grain-size analysis is gratefully acknowledged.

References

Anderson, A.T. 1974. Chlorine, sulfur, and water in magmas and oceans. *Geological Society of America Bulletin* 85:1483-1492.

Bach, W. 1976. Global air pollution and climatic change. *Review of Geophysics and Space Physics* 14:429-474.

Berresheim, H. and W. Jaeschke. 1983. The contribution of volcanoes to the global atmospheric sulfur budget. *Journal of Geophysical Research* 88:3732-3740.

Burnett, C.R. and E.B. Burnett. 1984. Observational results on the vertical column abundance of atmospheric hydroxyl: description of its seasonal behaviour 1977-1982 and of the 1982 El Chichón perturbation. *Journal of Geophysical Research* 89:9603-9611.

Carey, S.N. and H. Sigurdsson. 1982. Influence of particle aggregation on deposition of distal tephra from the May 18, 1980 eruption of Mount St. Helens volcano. *Journal of Geophysical Research* 87:7061-7072.

_____. 1987. The eruption of Vesuvius in A.D. 79, temporal variations in column height and magma discharge rate. *Geological Society of America Bulletin* 99:303-314.

Carey, S.N. and R.S.J. Sparks. 1986. Quantitative models of fallout of pyroclasts from explosive eruptions. *Bulletin of Volcanology* 48:109-125.

Devine, J., H. Sigurdsson, A. Davis and S. Self. 1984. Estimates of sulfur and chlorine yield to the atmosphere from volcanic eruptions and potential climatic effects. *Journal of Geophysical Research* 89:6309-6325.

Eddy, J.A. 1988. Before Tambora: The Sun and Climate in 1790 to 1830. *In: The Year Without A Summer? Climate in 1816.* C.R. Harington (ed.). An International Meeting Sponsored by the National Museum of Natural Sciences, Ottawa, 25-28 June 1988. Program and Abstracts. p. 21.

Gorshkov, G.S. 1959. Gigantic eruption of the volcano Bezymianmy. *Bulletin Volcanologique* 20:77-109.

Hammer, C.U. 1977. Past volcanism revealed by Greenland ice sheet impurities. *Nature* 270:482-486.

_____. 1980. Acidity of polar ice cores in relation to absolute dating, past volcanism, and radioechoes. *Journal of Glaciology* 25:359-372.

Hammer, C.U., H.B. Clausen and W. Dansgaard. 1980. Greenland Ice Sheet evidence of post-glacial volcanism and its climatic impact. *Nature* 288:230-235.

Herron, M.M. 1982a. Impurity sources of F, $C\ell$, NO_3 and SO_4 in Greenland and Antarctic precipitation. *Journal of Geophysical Research* 87:3052-3060.

_____. 1982b. Glaciochemical dating techniques. *In: Nuclear and Chemical Dating Techniques. American Chemical Society Series* 176:303-318.

Hofmann, D.J. and J.M. Rosen. 1983. Stratospheric sulfuric acid fraction and mass estimate for the 1982 volcanic eruption of El Chichón. *Geophysical Research Letters* 10:313-316.

Inn, E.C.Y. J.F. Vedder, E.P. Condon and D. O'Hara. 1981. Gaseous constituents in the plume from eruptions of Mount St. Helens. *Science* 211:821-823.

Johnston, D.A. 1980. Volcanic contribution of chlorine to the stratosphere: more significant to ozone than previously estimated? *Science* 209:491-493.

Junghuhn, F. 1850. *Java.* Volume 2. pp. 1249-1264.

Katsui, Y. 1959. On the Shikotsu pumice fall deposit; special reference to the activity just before the depression of the Shikotsu caldera. *Bulletin of the Volcanological Society of Japan* 2(4):33-48.

Landsberg, H.E., C.S. Yu and L. Huang. 1968. Preliminary reconstruction of a long time series of climatic data for the eastern United States. Report II: On a Project Studying Climatic Changes. *University of Maryland, Institute for Fluid Dynamics and Applied Mathematics, Technical Note BN-571.* 31 pp.

McCormick, M.P. and T.J. Swissler. 1983. Stratospheric aerosol mass and latitudinal distribution of the El Chichón eruption cloud for October 1982. *Geophysical Research Letters* 10:877-880.

Miller, E. 1983. Vapor-liquid equilibria of water-hydrogen chloride solutions below 0°C. *Journal of Chemical and Engineering Data* 28:363-367.

Neeb, G.A. 1943. *The Snellius Expedition in the Eastern Part of the Netherlands East Indies 1928-1929*. Volume 5. Brill, Leiden. pp. 55-268.

Oskarsson, N. 1980. The interaction between volcanic gases and tephra: fluorine adhering to tephra of the 1970 Hekla eruption. *Journal of Volcanology and Geothermal Research* 8:251-266.

Palais, J. and H. Sigurdsson. 1989. Petrologic evidence of volatile emissions from major historic and pre-historic volcanic eruptions. *In: Understanding Climate Change*. A. Berger, R.E. Dickinson and J.W. Kidson (eds.). *American Geophysical Union, Washington, D.C., Geophysical Monograph* 52:31-53.

Pannekoek van Rheden, J.J. 1918. Geologische Notizen uber die Halbinsel Sanggar, Insel Sumbawa. *Zeitschrift für Vulkanologie* 4:185-192.

Petroeschevsky, W.A. 1949. A contribution to the knowledge of the Gunung Tambora (Sumbawa). Koninklijk Nederlands Aardrijkskundig Genoostschap, *Tijdschrift* 66:688-703.

Pollack, J.B., O.B. Toon, C. Sagan, A. Summers, B. Baldwin and W. Van Camp. 1976. Volcanic explosions and climatic change: a theoretical assessment. *Journal of Geophysical Research* 81:1071-1083.

Raffles, S.H. 1835. *Memoir of the Life and Public Services of Sir Thomas Stamford Raffles*. Volume 1. Duncan, London. 267 pp.

Reclus, E. 1871. *The Earth*. G.P. Putnam and Sons, New York. 666 pp.

Rose, W.I. 1977. Scavenging of volcanic aerosol by ash: atmospheric and volcanologic implications. *Geology* 5:621-624.

Sapper, K. 1917. *Katalog der Geschichtlichen Vulkanausbruche*. Truber, Strasbourg. 358 pp.

Self, S., M. Rampino, M. Newton and J. Wolff. 1984. Volcanological study of the great Tambora eruption of 1815. *Geology* 12:659-663.

Self, S. and J.A. Wolff. 1987. Comments on "The petrology of Tambora volcano, Indonesia: a model for the 1815 eruption" by J. Foden. *Journal of Volcanology and Geothermal Research* 31:163-170.

Sigurdsson, H. 1982. Volcanic pollution and climate: the 1783 Laki eruption. *EOS* 63:601-602.

Sigurdsson, H., J.D. Devine and A.N. Davis. 1985. The petrologic estimation of volcanic degassing. *Jökull* 35:1-8.

Sigurdsson, H. and S. Carey. (in press). The 1815 eruption of Tambora, Indonesia: generation of co-ignimbrite ash fall during entrance of pyroclastic flows into the ocean. *Bulletin of Volcanology*.

Siscoe, G.L. 1980. Evidence in the auroral record for secular solar variability. *Review of Geophysics and Space Physics* 18:647-658.

Solomon, S. and R.R. Garcia. 1984. On the distribution of long-lived tracers and chlorine species in the middle atmosphere. *Journal of Geophysical Research* 89:11,633-11,644.

Sparks, R.S.J. 1986. The dimensions and dynamics of volcanic eruption columns. *Bulletin of Volcanology* 48:3-15.

Sparks, R.S.J. and L. Wilson. 1982. Explosive volcanic eruptions - V. Observations of plume dynamics during the 1979 Soufriere eruption, St. Vincent. *Geophysical Journal of the Royal Astronomical Society* 69:551-570.

Stoiber, R.E., S.N. Williams and B. Huebert. 1987. Annual contribution of sulfur dioxide to the atmosphere by volcanoes. *Journal of Volcanology and Geothermal Research* 33:1-8.

Stolarski, R.S. and D.M. Butler. 1978. Possible effects of volcanic eruptions on stratospheric minor constituent chemistry. *Pure and Applied Geophysics* 117:486-497.

Stolarski, R.S. and R.J. Cicerone. 1974. Stratospheric chlorine: a possible sink for ozone. *Canadian Journal of Chemistry* 52:1610-1615.

Stolper, E.M. and J.R. Holloway. (in press). Experimental determination of the solubility of carbon dioxide in molten basalt at low pressure. *Earth and Planetary Science Letters*.

Stothers, R.B. 1984. The great Tambora eruption in 1815 and its aftermath. *Science* 224:1191-1198.

Sze, N.D. 1978. Stratospheric fluorine: a comparison between theory and measurements. *Geophysical Research Letters* 5:781-783.

Thompson, L.G., E.M. Thompson, W. Dansgaard and P.M. Grootes. 1986. The Little Ice Age as recorded in the stratigraphy of the tropical Quelccaya Ice Cap. *Science* 234:361-364.

Thorarinsson, S. and G.E. Sigvaldason. 1972. The Hekla eruption of 1970. *Bulletin Volcanologique* 36(2):269-288.

Verbeek, R.D.M. 1885. *Krakatau*. Batavia, Indonesia. 495 pp.

Van der Broeck. 1834. *Oosterling* 1:183.

Walker, G.P.L. 1980. The Taupo pumice: products of the most powerful known (ultraplinian) eruption? *Journal of Volcanology and Geothermal Research* 8:69-94.

Zollinger, H. 1855. *Besteigung des Vulkanes Tambora auf der Insel Sumbawa und Schilderung der Erupzion Desselben in Jahr 1815*. J. Wurster (ed.). Winterthur. 20 pp.

The Possible Effects of the Tambora Eruption in 1815 on Atmospheric Thermal and Chemical Structure and Surface Climate

R.K.R. Vupputuri[1]

Abstract

A coupled 1-D radiative-convective-photochemical diffusion model that takes into account the influence of ocean inertia on global radiative perturbation is used to investigate the possible climatic and other atmospheric effects of the Tambora eruption of 1815. The volcanic cloud was introduced in the model stratosphere between 20-25 km, and the global average peak aerosol optical thickness was assumed to be 0.25. Both the aerosol optical thickness and aerosol composition determining the optical properties were allowed to vary in the model atmosphere during the life cycle of the volcanic cloud. The results indicate that the global average surface-temperature decreases steadily from the date of eruption in 1815 with maximum cooling of 1°K occurring in spring 1816. The calculations also show significant warming of the stratosphere, with temperature increasing up to 15°K at 25 km in less than six months after the date of eruption. The important effects of the Tambora eruption on stratospheric ozone and UV-B radiation at the surface are also discussed.

Introduction

Modelling studies of global radiative perturbations caused by volcanic eruptions are extremely important to understand the nature of past and present changes in the Earth's climate. As pointed out by Kondratyev (1983), the primary mechanism by which volcanic activity influences the net radiation (and consequently the climatic system) is through alterations in the aerosol content of the atmosphere. It is well known that volcanic aerosols scatter and absorb solar radiation and absorb and emit infrared radiation. The net effect causes general cooling of the troposphere and the surface, and warming in the stratosphere. The extent of the cooling and warming, however, depends upon the composition, size distribution and the morphological structure of aerosols.

Several prominent volcanic eruptions took place during the past 200 years (Laki, 1783; Krakatau, 1883; Mount Agung, 1963; and El Chichón, 1982). The largest and deadliest was that of Mount Tambora in April 1815, on the island of Sumbawa, Indonesia (8°S, 118°E). It was also the world's greatest ash eruption since the end of the last ice age. The dust veil index (a measure of increase in the atmospheric turbidity arising from small particles injected into the stratosphere) has been estimated to be more than twice that of Agung (Lamb 1970; Robock 1981, Mitchell 1982). The Tambora eruption is blamed by some studies in the literature for the cold summer of 1816 on the east coast of North America, where average temperature was the lowest on record with 1.5 to 2.5°C below the seasonal norm (Landsberg and Albert 1974). Indeed 1816 was called "the year without a summer" in New England and eastern Canada, where daily minimum temperatures were abnormally low from late spring through early autumn. It was also very cold and wet in western Europe in the summer of 1816, although it was milder at some stations in eastern Europe. Despite these earlier claims of strong cooling on a regional basis, the more recent analysis of climatic data for 1781-1983 by Angell and Korshover (1985) suggests that there is no

---

[1] Canadian Climate Centre, 4905 Dufferin Street, Downsview, Ontario M3H 5T4, Canada.

clear evidence of strong cooling on a hemispheric basis following the Tambora eruption. As pointed out by Angell (1988), the reason for the lack of evidence of strong volcanically-induced cooling on a hemispheric basis is that such cooling may or may not have been observed depending upon the extent of sea-surface warming in the eastern equatorial Pacific due to an El Niño event following the volcanic eruption.

It is clear from the above arguments that the injection of ash, sulphur and dust into the stratosphere by a large volcano could alter the existing radiative-photochemical balance of the Earth's atmosphere, in turn leading to changes in the vertical temperature and chemical structure and surface climate. In this respect a volcanic event as large as the Tambora eruption provides a unique opportunity for a case study of the response of the climatic system to a global radiative-photochemical perturbation. It also allows testing of our ability to model and understand the nature of the climatic system and climatic change. Several model calculations have been made in the past to study the climatic impact of Agung and El Chichón eruptions (Hansen et al. 1977; Robock 1984; McCracken and Luther 1984; Vupputuri and Blanchet 1984). All these calculations showed warming of the stratosphere and cooling in the troposphere and at the surface, although the amplitudes of warming and cooling differ depending upon the assumed peak aerosol optical depth, altitude of peak aerosol concentration and optical properties. In this paper a 1-D time dependent radiative-convective-photochemical diffusion model (RCPD model) taking into account the thermal inertia of oceans is used to investigate the thermal and chemical response of the atmosphere and surface climate to radiative-photochemical perturbations caused by the Tambora eruption in 1815.

The Climatic Model

The coupled one-dimensional climatic model extending from the surface to 60.5 km is described in detail in Vupputuri (1985). It involves combining the radiative-convective model of the type developed by Manabe and Wetherald (1967) with a photochemical transport model. Starting from the assumed vertical temperature distribution and chemical composition, the basic procedure is to compute the local net radiative heating and cooling and photochemical sources and sinks at each altitude to determine the vertical temperature and trace-constituent structure with the time marching method. The upward heat transfer by atmospheric motions is taken into account implicitly by a simple numerical procedure called convective adjustment - first introduced by Manabe and Strickler (1964). Using this numerical procedure, the vertical lapse rate is restored to a pre-assigned stable value ($6.5°$ km^{-1}) whenever it becomes greater due to radiative heat transfer. The relative humidity is fixed in the model, and the mixing ratio of water vapour is computed as a function temperature. Cloud-top altitude is assumed to be at 6.5 km, with a fractional cloud cover kept at 50%. The climatic model is coupled to the underlying surface through the energy balance equation at the surface. For climatic change calculations in this study, thermal heat capacity of the underlying surface is assumed to zero for the land and the value appropriate for the upper mixed layer in the case of the ocean. For several other computational details of the model, see Vupputuri (1985).

The Radiative Transfer Model

The solar radiation code used to compute the short wave solar heating within the atmosphere is based on the delta-Eddington method, which is computationally efficient and fairly accurate (Joseph et al. 1976). It considers the absorption and scattering by atmospheric gases (H_2O, CO_2, O_3 and NO_2), aerosols and cloud droplets. To compute the infrared cooling due to H_2O, O_3 and

CO_2, the analytical formulae for the mean transmissivities of finite frequency intervals derived by Kuo (1977) have been adopted. The mean transmission functions take into account the temperature effect and overlapping absorption between gases, and the computed transmissivities have been shown to be in good agreement with line-by-line calculations. For other trace gases, such as N_2O, CH_4 and CFCs, empirical expressions for the mean band absorptivities based on laboratory and spectroscopic data (Burch *et al.* 1962; Ramanathan *et al.* 1985) were adopted. Also considered in long-wave calculations are the heating or cooling-rate contributions due to aerosols in the atmospheric window band. Both IR and solar radiation codes have been validated by comparing them with other standard radiation codes through participation in the workshop on the intercomparison of radiative codes in the climatic models (World Meteorological Organization 1984).

The Photochemical Model

The photochemical system considered here includes the important reactions affecting the concentration of ozone and other relevant trace constituents in the atmosphere above 10 km: they are listed in Table 1. The chemical species considered are: O_x (O, O(^{1}D), O_3), HO_x (H, HO, HO_2), NO_x(N, NO, NO_2, HNO_3), CH_4, N_2O, $CF_2C\ell_2$, $CFC\ell_3$ and $C\ell_x$($C\ell$, $C\ell O$, $C\ell NO_3$) chemical species. The concentrations of H_2O, CH_4 and H_2 are specified based on observations. The chemical kinetics and photochemical data used are based on NASA (1985) recommendations. Computed photodissociation rates take into account the effects of Rayleigh scattering and absorption and scattering by aerosols and cloud droplets. The chemistry of the global troposphere below 10 km is considered to be much more complex, and therefore to have more uncertainties than in the stratosphere due to the presence of higher hydrocarbons, heterogeneous processes and long photochemical relaxation times for the chemical species. In view of some of these uncertainties, the chemistry is frozen in the troposphere by prescribing the ozone concentration below 10 km for the purpose of this study.

The Perturbed Aerosol Model for the Tambora Eruption

To calculate possible climatic and other atmospheric effects of the Tambora eruption, observational information on perturbed aerosol concentration and optical properties as a function of time starting from the date of eruption are needed. It is not, however, possible to obtain such detailed information for Mount Tambora. For this study it is assumed that the dust veil index for Mount Tambora is roughly twice that of Agung. Using the peak optical depth of 0.125 as a representative global average value for the added aerosols in the case of Agung (Hansen *et al.* 1978), the maximum optical thickness for Mount Tambora is estimated to be 0.25. The assumed shape of the vertical aerosol profile producing this maximum optical thickness is shown in Figure 1. The vertical distribution of perturbed aerosols and the altitude of peak aerosol concentration are similar to those of Agung and El Chichón eruptions. Since the added aerosol concentration and the optical properties are expected to vary with time during the lifetime of the volcanic cloud, it is not realistic to assume fixed optical thickness and properties for the model calculations. In the present calculations, both the aerosol optical thickness and properties are varied as a function time starting from the date of volcanic eruption. The assumed variation of perturbed aerosol optical thickness is illustrated in Figure 2. For the first four months after the eruption, the perturbation optical thickness is allowed to increase linearly to a maximum 0.25, and during this period ash optical properties are assigned for the added aerosols. From four to

Table 1: The Principle Chemical and Photochemical Reactions Used in the Model.

$O_2 + h\nu \rightarrow O + O$	$OH + O_3 \rightarrow HO_2 + O_2$
$O + O_2 + M \rightarrow O_3 + M$	$OH + O \rightarrow H + O_2$
$O + O_3 \rightarrow O_2 + O_2$	$HO_2 + O_3 \rightarrow OH + 2O_2$
$O_3 + h\nu \rightarrow O_2 + O$	$H + O_2 + M \rightarrow HO_2 + M$
$O_3 + h\nu \rightarrow O_2 + O(^1D)$	$H + O_3 \rightarrow OH + O_2$
$O(^1D) + O_3 \rightarrow O_2 + O_2$	$HO_2 + HO_2 \rightarrow H_2O_2 + O_2$
$O(^1D) + M \rightarrow O + M$	$H_2O_2 + h\nu \rightarrow OH + OH$
$NO_2 + h\nu \rightarrow NO + O$	$H_2O_2 + OH \rightarrow H_2O + HO_2$
$NO + O + M \rightarrow NO_2 + M$	$OH + HO_2 \rightarrow H_2O + O_2$
$NO_2 + O \rightarrow NO + O_2$	$HO + NO_2 + M \rightarrow HNO_3 + M$
$NO + O_3 \rightarrow NO_2 + O_2$	$OH + HNO_3 \rightarrow H_2O + NO_3$
$N_2O + h\nu \rightarrow N_2 + O$	$HO_2 + NO \rightarrow OH + NO_2$
$N_2O + O(^1D) \rightarrow N_2 + O_2$	$CF_2Cl_2 + h\nu \rightarrow CF_2Cl + Cl$
$N_2O + O(^1D) \rightarrow NO + NO$	$CFCl_3 + h\nu \rightarrow CFCl_2 + Cl$
$NO + h\nu \rightarrow N + O$	$Cl + H_2 \rightarrow HCl + H$
$N + O_2 \rightarrow NO + O$	$Cl + CH_4 \rightarrow HCl + CH_3$
$N + NO \rightarrow N_2 + O$	$Cl + HO_2 \rightarrow HCl + O_2$
$H_2O + O(^1D) \rightarrow OH + OH$	$OH + HCl \rightarrow H_2O + Cl$
$H_2O + h\nu \rightarrow H + OH$	$HCl + O \rightarrow OH + Cl$
$CH_4 + O(^1D) \rightarrow OH + CH_3$	$HCl + h\nu \rightarrow H + Cl$
$CH_3 + O_2 + M \rightarrow CH_3O_2 + M$	$Cl + O_3 \rightarrow ClO + O_2$
$CH_3O_2 + NO \rightarrow CH_3O + NO_2$	$ClO + O \rightarrow Cl + O_2$
$CH_3O + O_2 \rightarrow H_2CO + HO_2$	$ClO + NO \rightarrow Cl + NO_2$
$H_2CO + h\nu \rightarrow H + HCO$	$ClO + NO_2 + M \rightarrow ClNO_3 + M$
$HCO + O_2 \rightarrow CO + HO_2$	$ClNO_3 + h\nu \rightarrow ClO + NO_2$
$OH + CO \rightarrow CO_2 + H$	$ClNO_3 + HCl \rightarrow Cl_2 + HNO_3$
$H_2 + O(^1D) \rightarrow H + OH$	$ClNO_3 + O \rightarrow ClO + NO_3$
$HNO_3 + h\nu \rightarrow OH + NO_2$	

10 months the peak optical thickness remains the same but the optical properties are changed from ash to sulphuric acid. After 10 months the optical thickness decreases exponentially until it reaches the background stratospheric value while the optical properties change from sulphuric acid to background stratospheric aerosols. The optical parameters (extinction coefficients, single scatter albedo, asymmetry factors) for ash, sulphuric acid and background stratospheric aerosols vary as a function of wavelength both in solar and infrared spectra. Sulphuric acid properties chosen for this study are those reported in Bundeen and Fraser (1982), and they are derived assuming the aerosol particles are composed of 75% H_2SO_4 and 25% H_2O. The ash optical properties are determined by assuming an imaginary refractive index of 0.002 (Patterson and Pollard 1983). Both the sulphuric acid and ash properties are similar to those used for the El Chichón volcanic eruption.

Results and Discussion

Before discussing the results of atmospheric response to radiative-photochemical perturbations due to the Tambora eruption, it should be pointed out that for the prescribed annual average insolation and background stratospheric aerosols, the coupled 1-D model produced reference atmosphere simulations of minor trace constituents and temperature that are representative of natural background atmosphere in tropical latitudes. A detailed comparison of reference atmosphere model simulations of ozone and temperature, and discussion on some of the deficiencies in 1-D model calculations, were given in Vupputuri (1985).

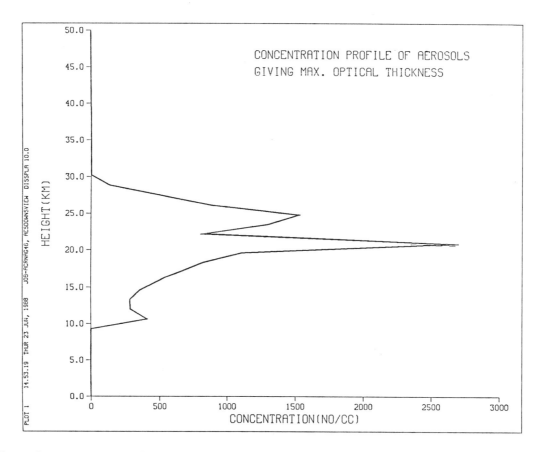

Figure 1: Assumed aerosol concentration profile (NO/CC) for Mount Tambora volcanic eruption which produces the maximum optical depth of 0.25.

Effects on Solar and Infrared Radiation

Figure 3 shows the calculated change in direct, diffuse and net solar radiation at the surface as a function of time beginning from the date of eruption of Tambora in 1815, while the corresponding effects on infrared flux and planetary albedo are illustrated in Figure 4. Direct solar flux decreases by about 15% following the eruption (Figure 3). However this decrease is compensated in large measure by an increase in diffuse radiation, leaving a net decrease of solar flux at the surface of about 4%. There are no observational data on direct solar radiation following the Tambora eruption. The visual extinction curve constructed by Stothers (1984) suggests that the excess zenithal visual extinction increases rapidly during the first four to five months from the eruption date, and then returns to normal gradually within four to five years. The time variation of visual extinction is quite consistent with the variation of calculated direct solar radiation following the eruption of Tambora. As indicated in Figure 4, the infrared flux at the surface also decreases by up to 4%, while planetary albedo increases by about 7% following the Tambora eruption.

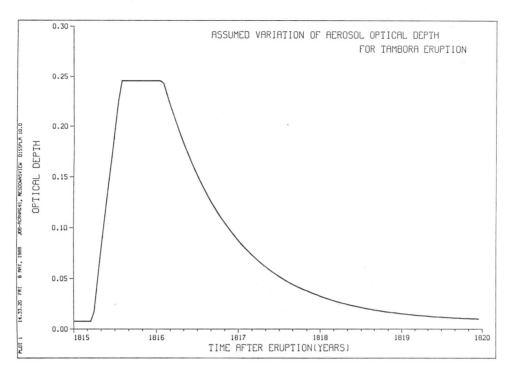

Figure 2: The assumed variation of perturbed stratospheric aerosol optical thickness with time following the Mount Tambora eruption in 1815.

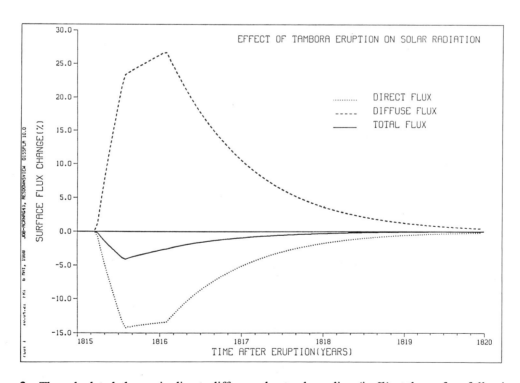

Figure 3: The calculated change in direct, diffuse and net solar radion (in %) at the surface following the Mount Tambora eruption.

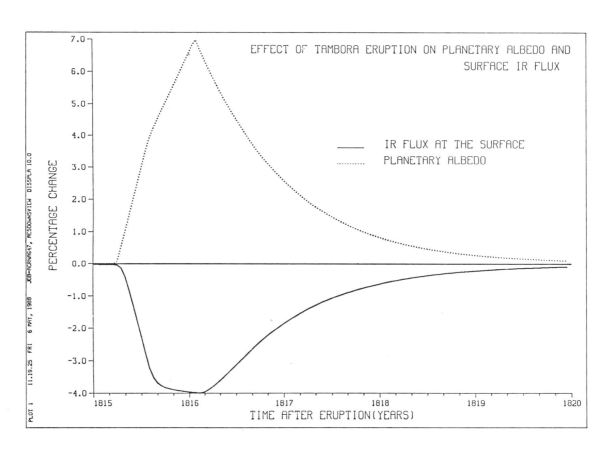

Figure 4: The calculated percentage change in the infrared flux at the surface and planetary albedo following the Mount Tambora eruption.

Effects on Global Climate

Figure 5 shows the calculated response of stratospheric and surface temperature as a function of time following the Tambora eruption. The response is shown for two different assumed heat capacities for the underlying lower-boundary surface. The solid curve represents the calculated surface-temperature response assuming that the lower boundary has no heat capacity (valid assumption for the land surface). The dashed curve, on the other hand, corresponds to the heat capacity of the underlying surface equivalent to that of 70 m of ocean water. The combined response of land and ocean is given by the dotted curve. In the stratosphere, the ocean heat capacity has no effect on temperature response due to volcanic forcing caused by the Tambora eruption (dash-dot curve in Figure 3). Note however that the stratosphere responds much more quickly to the global radiative perturbation due to volcanic forcing. As indicated in Figure 5, the eruption of Tambora could have resulted in a peak warming of about 15°K within less than six months after the eruption. But in the case of the troposphere and the surface it takes almost twice as long to reach the maximum cooling. The calculated maximum cooling for the land surface following the Tambora eruption is about 2°K, and for the combined land and ocean it is roughly 1°K. It may be seen from Figure 5 that not only the land surface cools faster than the ocean surface but it also warms faster than ocean as the temperature recovers to its pre-volcanic state.

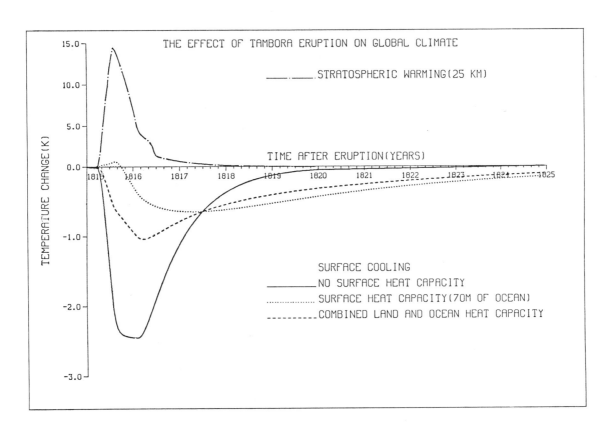

Figure 5: The calculated stratospheric and surface-temperature change (O K) following the Tambora eruption. The surface-temperature response is shown for two different assumed heat capacities for the underlying lower boundary surface.

The physical explanation for the calculated temperature effects in the present model is straightforward. For the aerosol concentrations and optical properties assumed in this study, the net effect of the added aerosols in the stratosphere from the Tambora eruption would be to increase the planetary albedo and decrease both the solar and thermal radiation at the surface (Figures 3, 4). This leads to cooling in the troposphere and at the surface. The local heating in the stratosphere, on the other hand, is caused by both the absorption of thermal radiation emanating from the warmer lower atmosphere and *in situ* absorption of solar radiation (in the near infrared and UV part of the spectrum) by the added aerosols. Due to low air density in the stratosphere, only a small change in radiational energy is needed to cause a large change in local air temperature.

As pointed out earlier, although surface temperatures were abnormally low in the summer of 1816 (following the Tambora eruption) in New England, eastern Canada and parts of western Europe, the evidence for a strong land-surface temperature cooling on a global average basis (as indicated by the results of this study) following the Tambora eruption is rather weak. As Angell and Korshover (1985) indicate, the extent of surface cooling on an hemispheric or global basis depends critically upon the timing and strength of El Niño in relation to the time of a large volcanic eruption. Indeed Quinn *et al.* (1978) have found some evidence for a moderate El Niño one year after the Tambora eruption. This might partially explain the reason for the lack of evidence of strong cooling on a global or hemispheric basis in 1816 following the Tambora eruption.

Effects on Stratospheric Ozone and UV-B Radiation at the Surface

As mentioned earlier, the absorption of solar and thermal radiation by the added aerosols in the stratosphere from a large volcanic eruption such as Tambora can lead to a large increase in the stratospheric temperature. The altered temperature in turn effects the concentration of ozone and other minor constituents through temperature-dependent reaction-rate coefficients. Due to the inverse relationship between ozone and temperature in the middle stratosphere, the temperature increase in that region causes ozone concentration to decrease. Ozone is also destroyed in the stratosphere by enhanced photodissociation resulting from the backscattered UV radiation from the added aerosols. Figure 6 shows the computed ozone column reduction and UV-B radiation increase at the surface following the Tambora eruption. It is seen from Figure 6 that the added aerosols in the stratosphere from the Tambora eruption could have resulted in up to about 7% decrease in total ozone, which translates into up to 15% increase in the UV-B radiation at the surface following the eruption event. Although there were no observational data following the Tambora eruption to verify the computed ozone depletion, the analysis of Umkehr observations following El Chichón in 1982 by DeLuisi *et al.* (1984) clearly indicates the evidence of volcanically-induced ozone depletion. This lends support for the theoretical calculations shown in Figure 6.

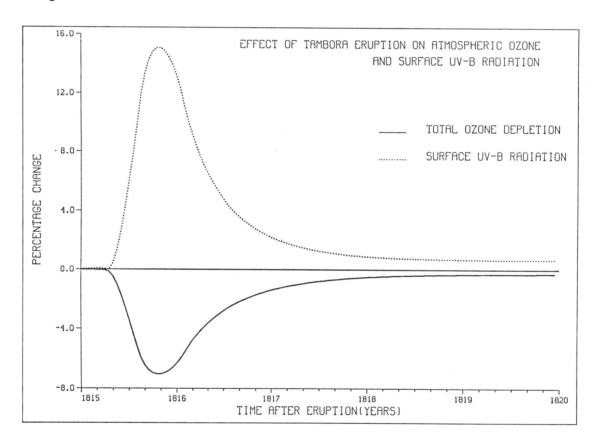

Figure 6: The calculated changes in the total ozone column and UV-B radiation at the surface (in %) following the Mount Tambora eruption.

Concluding Remarks

The Tambora eruption in 1815 - the largest and deadliest eruption in recorded history - also injected the greatest amount of ash, sulphur and dust into the stratosphere. The event that produced the largest dust veil index provided a unique opportunity to investigate climatic and other atmospheric response to global radiative perturbation, and to understand the effects of volcanic eruptions on past and present climate.

Despite the simplicity of a 1-D radiative-convective-photochemical diffusion model that does not include the interaction of radiative heating perturbation with atmospheric dynamics and other uncertainties regarding the input data, the magnitude of land-surface temperature decrease generally agrees with the observed cooling in the east coast of North America, where 1816 was called "the year without a summer". However there is lack of strong evidence from the observational data to support the computed combined land-ocean surface cooling on hemispheric or global bases. The computed cooling may or may not have been observed depending upon the timing and extent of the El Niño event following the Tambora eruption. There were no observational data in 1816 to verify the computed changes in ozone and temperature in the stratosphere following this eruption. However, the well-documented observational evidence of temperature warming (Quiroz 1983) and ozone depletion (DeLuisi *et al.* 1984) in the stratosphere following the El Chichón eruption lends support for the calculated changes in the case of the Tambora eruption.

One should exercise caution in interpreting either the observations or the model results presented here. Temperature and ozone observations are not as detailed as desired for accurate determination of observed climatic and total ozone changes on a global average basis. Further, the observed climatic change is not a part of other natural events such El Niño, QBO and the sun-spot cycle, or simply noise in the climatic system (Hansen *et al.* 1978). Climatic calculations in the model are too simplified and, in particular, the model is not capable of handling the complex interactions between radiative heating and large-scale dynamics and other cloud feedback mechanisms. Nevertheless, the calculated amplitudes of climatic and ozone perturbations resulting from Tambora's eruption are large enough to believe that volcanic eruptions do indeed strongly affect the Earth's climate and the ozone layer. As pointed out by Angell (1988), the reason that the evidence for volcanically-induced cooling of the Earth's surface in the past was so uncertain is that the cooling may or may not have been observed depending upon the extent of warming due to El Niño following the volcanic eruption. Detailed observations and careful data analysis taking into account the impact of other natural events following a major eruption such as Tambora would enable us to understand better the role of volcanic aerosols in altering the radiative-photochemical balance of the global atmosphere and climate.

Acknowledgements

The author thanks: Dr. G.J. Boer and the Director General of the Canadian Climate Centre for encouragement and support; Mr. Frank Szeckeli and Lynda Smith for programming and manuscript preparation support, respectively.

References

Angell, J.K. 1988. Impact of El Niño on the delineation of tropospheric cooling due to volcanic eruptions. *Journal of Geophysical Research* 93:3696-3704.

Angell, J.K. and J. Korshover. 1985. Surface-temperature changes following the six major volcanic episodes between 1780 and 1980. *Journal of Climate and Applied Meteorology* 24:937-951.

Bundeen, W.R. and R.S. Fraser. (eds.) 1983. Radiative effects of the El Chichón volcanic eruption: preliminary results concerning remote sensing. *NASA Technical Memorandum* 84959, Goddard Space Flight Center, Greenbelt, Maryland.

Burch D.E., D. Grynak, E.B. Singleton, W.L. France and D. Williams. 1962. Infrared absorption by CO_2, H_2O and minor atmospheric constituents. *AFCRL-62-688*, Ohio State University Contract AF19(604)-2366.

DeLuisi, J.J., C.L. Mateer and W.D. Komhyr. 1985. *Effects of the El Chichón stratospheric aerosol cloud on Umkehr measurements at Mauna Loa, Hawaii.* Atmospheric ozone. C.S. Zerefos and A. Ghazi. (eds.). D. Reidel Publishing Co., Dordrecht.

Hansen, J.E., W.C. Wang and A.A. Lacis. 1978. Agung eruption provides test of a global climate perturbation. *Science* 199:1065-1068.

Joseph, J.H., W.J. Wiscombe and J.A. Weinman. 1976. The delta-Eddington approximation for radiative flux transfer. *Journal of Atmospheric Sciences* 33:2452-2459.

Kondratyev, K. Ya. 1983. *Volcanoes and Climate.* R.D. Bojkov and B.W. Boville (eds.). World Meteorological Organization, WCP-54.

Kuo, H.L. 1977. Analytic infrared transmissivities of the atmosphere. *Beitrage zur Physik der Atmosphaere* 50:331-349.

Lamb, H.H. 1970. Volcanic dust in the atmosphere, with a chronology and assessment of its meteorological significance. *Philosophical Transactions of the Royal Society of London* A226:425-533.

Landsberg, H.E. and J.M. Albert. 1974. The summer of 1816 and volcanism. *Weatherwise* 27:63-66.

Manabe, S. and R.F. Strickler. 1964. Thermal equilibrium of the atmosphere with a convective adjustment. *Journal of Atmospheric Sciences* 21:361-385.

Manabe, S. and R.T. Wetherald. 1967. Thermal equilibrium of the atmosphere with a given distribution of relative humidity. *Journal of Atmospheric Sciences* 24:241-259.

McCracken, M.C. and F.M. Luther. 1984. Preliminary estimates of the radiative and climatic effects of the El Chichón eruption. *Geofisica Internacional* 23(3):385-401.

Mitchell, J.M., Jr. 1982. El Chichón, weather-maker of the century. *Weatherwise* 35:252-259.

NASA Panel for Data Evaluation. 1985. Chemical kinetics and photochemical data for use in stratospheric modelling. *California Institute of Technology, Jet Propulsion Laboratory, Publication 85-37*, Pasadena, California. 217 pp.

Patterson, E.M. and C.O. Pollard. 1983. Optical properties of the ash from El Chichón volcano. *Geophysical Research Letters* 10:317-320.

Quinn, W.H., D.O. Zopf, K.S. Short and R.T.W. Kuo Yank. 1978. Historical trends and statistics of Southern Oscillation, El Niño and Indonesian droughts. *Fishery Bulletin* 76:663-678.

Quiroz, R.S. 1983. The isolation of stratospheric temperature change due to the El Chichón volcanic eruption from non-volcanic signals. *Journal of Geophysical Research* 88:6773-6780.

Ramanathan, V., H.B. Singh, R.J. Cicerone and J.T. Kiehl. 1985. Trace gas trends and their potential role in climate change. *Journal of Geophysical Research* 90:5547-5566.

Robock, A. 1981. A latitudinally dependent volcanic dust veil index and its effect on climate simulations. *Journal of Volcanology and Geothermal Research* 11:67-80.

Stothers, R.B. 1984. The great Tambora eruption in 1815 and its aftermath. *Science* 224:1191-1198.

Vupputuri, R.K.R. 1985. The effect of ozone photochemistry on atmospheric and surface temperature changes due to increased CO_2, N_2O, CH_4 and volcanic aerosols in the atmosphere. *Atmosphere-Ocean* 23:359-374.

Vupputuri, R.K.R. and J.-P. Blanchet. 1984. The possible effects of El Chichón eruption on atmospheric thermal and chemical structure and surface climate. *Geofisica Internacional* 23(3):433-447.

World Meteorological Organization. 1984. The intercomparison of radiation codes in climate models (ICRCCM). Longwave clear-sky calculations. F.M. Luther (ed.). *World Meteorological Organization, WCP-93*.

Climatic Effects of the 1783 Laki Eruption

Charles A. Wood[1]

Abstract

From 8 June 1783 to 7 February 1784, 12 km^3 of lava poured from a series of volcanic vents in southern Iceland, devastating farmland and ultimately causing a severe famine that decimated the island's human and animal populations. This Laki eruption apparently had much more widespread consequences, however, for its ash and sulphurous gases were transported in the lower atmosphere across Europe causing a remarkably warm summer, which was followed in Europe, eastern North America and at least some parts of Asia by one of the most severe winters on record. Poor weather continued through the summer and winter of 1784. Scarce meteorological measurements and abundant written records and proxy data graphically document these climatic anomalies.

Conventional volcano-climate theories cannot readily explain these apparent climatic effects. Great eruptions such as Tambora, 1815 or Krakatau, 1883 explosively emplace volcanic aerosols into the stratosphere where, during a two- to three-year period before they are finally flushed out, they absorb incoming radiation, thus depriving the lower atmosphere of a portion of its heat. For the Laki eruption there is no direct evidence that significant quantities of sulphuric aerosols reached the stratosphere. If the continuing climatic deterioration of 1784 was related to the clearly volcanic weather of 1783, then a new mechanism needs to be identified.

Introduction

The first recognition that volcanism may effect climate was Benjamin Franklin's (1784) speculation that the non-explosive Laki eruption in Iceland could be responsible for the poor climate in Europe and North America during the summer and winter of 1783. Although the Laki eruption was the largest effusive volcanic activity in historic times, it, and its possible climatic effects, have been studied only at the reconnaissance level. Nonetheless, general information on the nature and chronology of the eruption, and compilation of reports of anomalous weather thereafter, provide strong support for Franklin's prescient theory, and lead to puzzling aspects of conventional eruption-climate relationships.

Laki Eruption of 1783

The only accessible accounts of the Laki eruption are Thorarinsson's (1969) 20-year old preliminary report and a recent abstract by Thordarsson et al. (1987), which incorporate eyewitness descriptions and reconnaissance geological mapping. Most of the 1783 activity occurred along a 25-km-long series of fissures, creating large fields of lava flows with small cones marking their vents. The eruption began on 8 June 1783, following three weeks of earthquakes, and continued for eight months (until 7 February 1784). Thorarinsson estimates that ~ 12 km^3 of tholeiitic lava flows were emplaced, with the majority (10 km^3) produced during the first 50 days (8 June to 28 July 1783). From examination of Thorarinsson's map (Figure 1)

---

[1] NASA Johnson Space Center, SN2, Houston, Texas 77058, U.S.A.

showing the extent of the lava at different dates, it appears that a significant portion of the flows had formed by 21 June and the majority of the 10 km³ by 22 July.

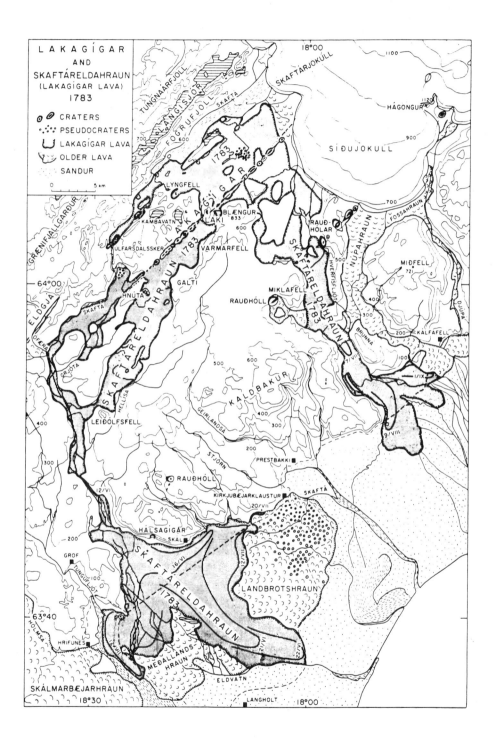

Figure 1: Map of Laki from Thorarinsson (1969). The Laki lavas of 1783 are lightly stippled.

Thus, the extrusion rate during the first two weeks may have been considerably greater than the 50-day average of 2,200 m^3/sec estimated by Thorarinsson. Following this initial, high-rate extrusion, activity continued at a much slower rate for the next six months. The total area covered by the flows is 565 km^2 (Thoroddsen 1925), which implies an average thickness of ~20 m.

A recent development has been the recognition that the Laki fissure activity was part of volcanic-tectonic eruption centred on the Grimsvotn caldera, northeast of Laki (Thordarsson *et al.* 1987; Thordarsson and Self 1988). During and after the Laki fissure eruptions, Grimsvotn had a series of explosive eruptions between 18 July 1783 and 26 May 1785.

From the perspective of climatic effects it is important to know how much explosive activity occurred at the beginning of the eruption. According to eyewitnesses the eruption was very violent during the first few days, with enormous lava fountains. Thorarinsson pointed out that groundwater may have been involved in some of these eruptions because phreatomagmatic tephra cones were formed. Based upon his field measurements of buried ash from Laki, Thorarinsson believed the volume of explosively erupted material was 0.3 km^3; he discounted earlier estimates of 3 km^3. Thordarsson *et al.* give a similar value (0.21 km^3) for the tephra.

Effects of the Laki Eruption in Iceland

Thorarinsson (1969) states that the Laki eruption was the greatest catastrophe in Icelandic history, and the mortality statistics bear him out. Gases from the eruption stunted the growth of grass so that it was insufficient to feed livestock. As a result, 50% of the cattle, 79% of the sheep, and 76% of the horses starved. During the next three years 24% (9,000 people) of the human population died of starvation, and the population did not return to earlier levels until 1780 (Jackson 1982).

Ogilvie (1986) cites contemporary diaries that provide graphic information on the effects of the eruption. From 8 June to at least 26 August, "the air was full of ash and smoke. On the rare occasions that we have had a glimpse of the sun it has looked like the reddest blood". Grass turned yellow and white due to "sulphurous rain" and it withered to the roots. Fishermen were not able to go to sea because of "murmurings" (earthquakes?) and continuous smoke that reduced visibility to less than a mile. In northeastern Iceland, one writer recorded, "From early June, and to this time (13 August) we have lived in continual smoke and fog, sometimes accompanied by sulphur-steam and ashfalls."

Weather in Iceland during the eruption seemed variable from location to location according to written records analyzed by Ogilvie (1986), but she concludes that 1783-84 winter began very early, and was very severe and long-lasting. Various accounts state that the ground was frozen with hard ice from 2 October until the end of April 1784. All fiords were reported to be frozen over in late February 1784 (for the first time in 39 years), and sea ice was very widespread and long-lasting. The summer of 1784 was also cold and wet, with occasional periods of frost and sleet. Ogilvie's summary of seasonal weather shows that there was uniformly cold weather across Iceland for five seasons (summer 1783 through summer 1784) after the onset of the Laki eruption. Also, 1782 was unusually cold, although apparently not as uniformly so as following the eruption.

Haze and Dust

The tremendous quantity of volcanic gases and dust released during the Laki eruption was reported from many locations in the northern hemisphere. The English rector Gilbert White (1977) wrote:

> The summer of the year 1783 was an amazing and portentous one, and full of horrible phenomena...the peculiar haze, or smoky fog, that prevailed for many weeks in this island, and in every part of Europe, and even beyond its limits, was a most extraordinary appearance, unlike anything known within the memory of man. ...The sun at noon looked as blank as a clouded moon, and shed a rust-coloured ferruginous light...and was particularly lurid and blood-coloured at rising and setting.

Icelandic accounts describe the volcanic haze from early June to the end of August (Ogilvie 1986), and Lamb (1970) reported the following first sightings of dry fog or haze in Europe and eastward:

Copenhagen	29 May
France	6 June
North Italy	18 June
Syria	1 July
Altai, central Russia	1 July.

Evidently, the dry fog spread eastward and southeastward at an average rate of approximately 250 km/day during the first month of the eruption. This is only about 10% of the rate of propagation for the Krakatau haze (2,700 km/day; Russell and Archibald 1888). The difference in velocity may be due to the differences in direction (east for Laki, west for Krakatau) or the altitude (tropospheric for Laki, stratospheric for Krakatau). Benjamin Franklin (1784) also noted that the haze was seen in North America, although the original sources and details of this observation were not reported. Nonetheless, the haze persisted long enough, or rose to different atmospheric heights with differing wind directions, so that it was carried both eastward to Europe and westward to North America.

Volcanic dust also fell out of the sky in Europe. Lamb (1971) reported that tulips in Holland were damaged by the dust and sulphurous smells during 18-24 June 1783. In Scotland the dust was thick enough to destroy crops in June. The detection of sulphurous odours in Europe proves that the haze was volcanic and not from some unknown forest fire, for example. The odour and eye irritation imply that the haze was at low altitudes. Volcanic dust that fell in Holland 11 days after the eruption started, apparently was transported much more slowly than dust from other Icelandic eruptions. Ash from the 1875 eruption of Askja reached Europe within a day (Thorarinsson 1963).

The Summer of 1783

White's (1977) account quoted above continues:

> All the time the heat was so intense that butchers' meat could hardly be eaten on the day it was killed; and the flies swarmed so in the lanes and amid hedges that they rendered the horses half frantic, and riding irksome.

Instrumental temperature records reveal that 1783 was the warmest English July on record (Kington 1978). Other early thermometer data for six other major European cities allow quantification of how extreme the summer heat was in 1783. *World Weather Records* data (Figure 2) for Stockholm, Copenhagen, Edinburgh, Berlin, Geneva, and Vienna for a 31-year period centred on 1783 demonstrate that July 1783 was 1.6 to 3.3°C warmer than the 31-year average. In general the amount of the July temperature anomaly is closely correlated with the distance of each city from Laki (Figure 3). Thus, however the haze raised the summer temperatures in Europe, the effect was most pronounced where the haze was thickest, and the excess heating declined where the haze was probably less intense. Temperature data from eastern North America (Landsberg *et al.* 1968) reveal that the summer of 1783 was significantly hotter than the 225-year average (Sigurdsson 1982). Figure 4 shows the same data graphically.

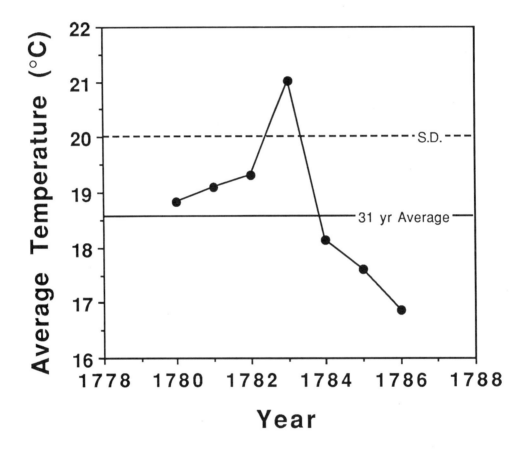

Figure 2: Average July temperatures (data from *World Weather Records*) for six European cities (Stockholm, Copenhagen, Edinburgh, Berlin, Geneva, and Vienna) for the seven-year period centred on 1783, the year of Laki's eruption. The 31-year average is based on recorded temperatures for the 31 years centred on 1783.

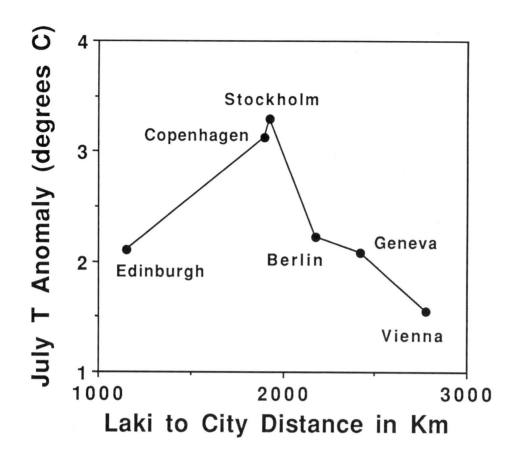

Figure 3: Deviation of July 1783 temperatures from 31-year averages as a function of the distance from Laki to six European cities. Edinburgh's anomaly is less than expected based on the other cities, suggesting that the temperature increases were not latitudinally uniform. Perhaps these anomalies - from only one month after the start of the eruption - were due to tropospheric dust, which would not be as uniformly distributed as stratospheric aerosols.

There are various sources of proxy weather information for this period; e.g., in Switzerland the summer was drier as well as warmer than normal (Pfister 1981), and there was a drought and poor harvest in Finland (Schove 1954).

Additional circumstantial evidence that the summer of 1783 was warm includes a severe drought in the Yangtze region of China (Wang and Zhao 1981). The Yangtze drought continued into 1784, but in both years there were floods in southeastern China and the Hwang Ho (Yellow) River Basin. An extraordinarily severe famine throughout Japan in the summer of 1783, however, was not caused by drought: Mikami (1988; Mikami and Tsukamura, this volume) has shown that an excess of rain destroyed many crops, and that, in fact, the summer of 1783 was wettest and coolest in Japanese history. Based on the high price of wheat in Delhi, India, the rains probably failed in 1783 with a consequent famine (Pant et al. 1988). These extremes in Asian weather during the summer of 1783 exhibit regional variations in the response to volcanism that are similar to previously-documented patterns in North America (Lough and Fritts 1987).

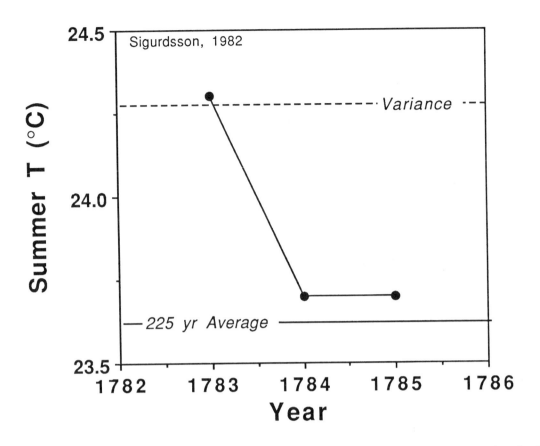

Figure 4: Average summer temperatures in the eastern United States in the 1780s compared to the 225-year average. Data from Landsberg *et al.* (1968) as reported by Sigurdsson (1982).

Winter of 1783-84

Scant temperature measurements, abundant proxy data and anecdotal accounts demonstrate that the winter of 1783-84 was one of the most severe on record in Europe and North America. The average January temperature for six European cities was 3° below the 31-year average centred on 1784 (Figure 5). Proxy temperature data (from viticulture/agriculture) indicate Switzerland had two extremely severe winters in 1783-84 and 1784-85, with the first year being the worst (Pfister 1981). The longest period of sea ice around Iceland also occurred during the winter of 1783-84 when temperatures were nearly 5° colder than the 225-year average (Sigurdsson 1982). The next two winters were also significantly colder than normal (Figure 6). Information compiled by Ludlum (1966) includes the following records for the winter of 1783-84 in the eastern United States:

>   Longest in early American history (last snow in late April),
>   Near record depth of snowcover,
>   Near record low temperatures,
>   Greatest seasonal snowfall ever in New Jersey,
>   Longest period of below zero temperatures ever in New England,

Longest freezing ever of Chesapeake Bay,
Longest and coldest winter in Maine,
One of the greatest southern snowstorms (18-19 December),
Freezing of Charleston Harbour (ice skating occurred),
Freezing of Mississippi River at New Orleans (13-19 February 1784),
Ice floes in Gulf of Mexico 100 km south of New Orleans.

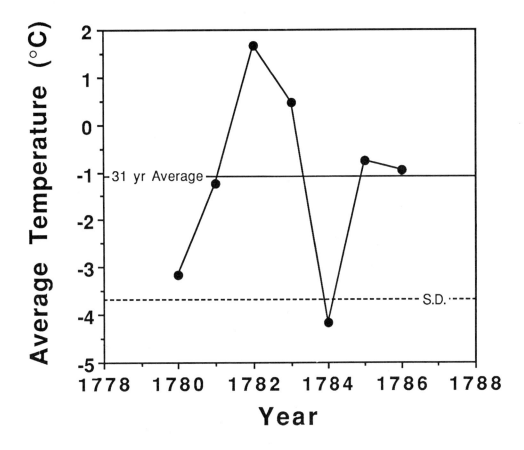

Figure 5: Average January temperatures (data from *World Weather Records*) for six European cities (Stockholm, Copenhagen, Edinburgh, Berlin, Geneva, and Vienna) for the seven-year period centred on 1783, the year of Laki's eruption. The 31-year average is based on recorded temperatures for the 31 years centred on 1783.

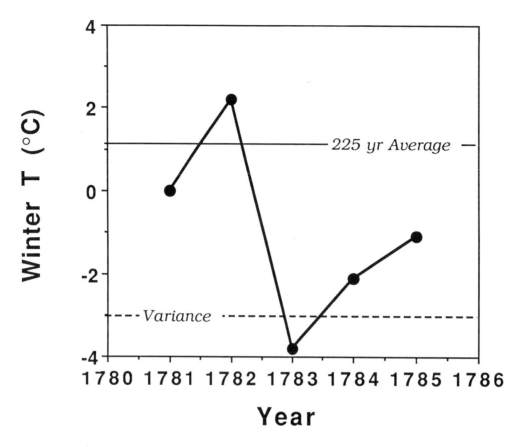

Figure 6: Average winter temperatures in the eastern United States in the 1780s compared to the 225-year average. Data from Landsberg *et al.* (1968) as reported by Sigurdsson (1982).

Ludlum (1966) provides graphic evidence of the severity of the winter by quoting contemporary letters and newspapers. Following a series of early and frequent storms, the worst weather of the winter occurred in mid-February, when minimum recorded temperatures for eight nights at Hartford, Connecticut were about 12°C or colder. From Virginia, James Madison wrote on 11 February 1784, "We had a severer season and particularly a greater quantity of snow than is remembered to have distinguished any preceding winter." In another letter of 5 March 1784, George Washington complained that he, "arrived at this Cottage [Mount Vernon, Virginia] on Christmas eve, where I have been locked up ever since in frost and snow."

The February cold spell froze the western end of Long Island Sound, and at New York City the Narrows between Staten Island and Long Island were blocked by ice for 10 days, preventing ships in Manhattan harbours from reaching the sea. Baltimore harbour was frozen by 2 January 1784 and remained closed until 25 March. Chesapeake Bay was nearly completely frozen, and the Delaware River at Philadelphia froze on 26 December 1783 and was icebound until 12 March 1784. Ludlum reports that even the southern harbour of Charleston, South Carolina was frozen in February, "having produced ice strong enough for skating on, which is very uncommon there." The most amazing phenomenon of the winter was the freezing of the Mississippi River at New Orleans, which Ludlum (1966, p. 154) reports has happened only once before (1899):

> On the 13th of February, 1784, the whole bed of the river, in front of New Orleans, was filled up with fragments of ice, the size of most of which was from twelve to thirty feet, with a thickness of two to three. This mass of ice was so compact, that it formed a field of four hundred yards in width, so that all communications was interrupted for five days between the two banks of the Mississippi. On the 19th, these lumps of ice were no longer to be seen. "The rapidity of the current being then at the rate of two thousand and four hundred yards an hour," says Villars, "and the drifting of the ice by New Orleans having taken five days, it follows that it must have occupied in length a space of about one hundred and twenty miles. These floating masses of ice were met by ships in the 28th degree of latitude [in the Gulf of Mexico].

That the unusually cold winter was not just confined to the eastern United States is clear from the Hudson's Bay Company records indicating 1783-84 had the fifth worse ice blockage of Hudson Strait on record (Catchpole 1988).

Summer of 1784

Summer temperature in England averaged 0.5°C, and as much as 1.6°C, below the long-term norm during 1784. The driest 12 months in English history began in August 1784 (Kington 1978). Temperatures in the eastern United States tended to be below average: <53.9°F (12.1°C) in Philadelphia and <48.0°F (8.9°C) in New Haven (Bray 1978). Tree-rings indicate a marked growth minimum in the growing season of Douglas fir in Nevada, Utah, and Wyoming in 1784 and 1785 (Woodhouse 1988). Tree-ring densities from the Mackenzie Delta region of Canada indicate that 1784 had a very cold summer (Parker 1988). Light coloured rings in black spruce at the treeline near Quebec indicate low temperatures shortened the growing season in 1784 (Filion *et al.* 1986). Similarly, tree rings from Alaska show that the cool weather of 1784 extended far to the north (Oswalt 1957).

Winter of 1784-85

In Switzerland, the long duration of snowcover during the 1784-85 winter resulted in widespread growth of the snow mold *Fusarium nivale*, which led to harvest failure of the spring grain crops (Pfister 1981). In Bern the winter was also very severe, with snow on the ground for more than 150 days (Pfister 1978). Winter, spring, and early summer of 1784 in Brittany were disastrous, with a cold winter, hail at the end of April, spring floods, and a drought until the end of June (Sutherland 1981). In England, 1785 tied with 1674 as the coldest March on record (Kington 1978). The date of freezing of Lake Suwa, Japan (Figure 7), occurred 22 days earlier than the long-term average (Arakawa 1954).

Subsequent Seasons

1785 was the worst year of the decade in Brittany, and one of the worst of the century. In some areas no rain fell between January and August (Sutherland 1981). The summer was also cool in England (Bray 1978), and the autumn of 1786 was one of the three coldest in English history (Kington 1978).

In the eastern United States, summer temperatures were lower than normal in New Haven in 1785 but returned to normal in 1786 (Bray 1978).

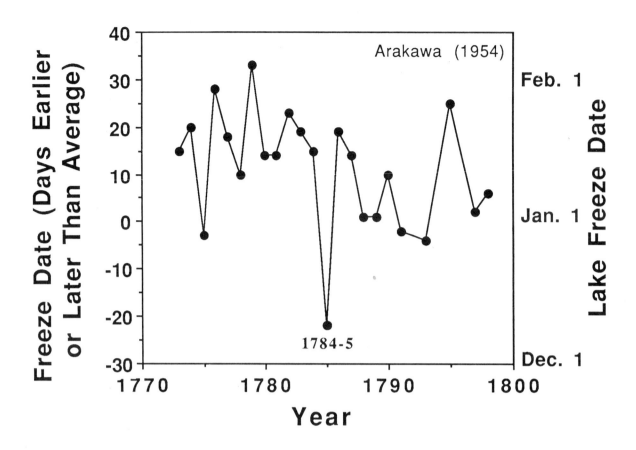

Figure 7: 23-year record of the date of freezing of Lake Suwa, Japan centred on 1783 (data from Arakawa 1954). In 1783 freezing occurred 22 days earlier than normal.

Mechanisms to Explain Observed Climatic Anomalies

Each of the unusual weather records reported above can be dismissed as a freak occurrence within the normal range of variation, and thus not requiring a special origin. Consideration of the entire list of anomalies - and these are only the items that were found during a brief examination of secondary and tertiary historic records - suggests, however, that a period of unusual weather affected various places in the northern hemisphere from the summer of 1783 through 1785. The principal observations to be accounted for are:

> Early summer 1783: Dry fog over Europe, western Asia, and the United States
> Summer 1783: Hot in Europe, United States and China; cold in Iceland
> Winter 1783-84: Exceptionally cold in Europe, United States and Japan
> Winter 1784-85: Very cold in Europe and Japan
> Summer 1785: Cool and dry in Europe and United States

Benjamin Franklin's famous 1784 communication to the Literary and Philosophical Association of Manchester was the first suggestion that volcanic eruptions might effect climate (see Sigurdsson 1982):

> During several of the summer months of the year 1783, when the effect of the sun's rays to heat the earth in these northern regions should have been greatest, there existed a constant fog over all of Europe, and a great part of North America. This fog was of a permanent nature; it was dry, and the rays of the sun seemed to have little effect toward dissipating it, as they easily do a moist fog, arising from water. They were indeed rendered so faint in passing through it, that when collected in the focus of a burning glass, they would scarce kindle brown paper. Of course, their summer effect in heating the earth was exceedingly diminished.
>
> Hence the surface was early frozen.
>
> Hence the first snows remained on it unmelted, and received continual additions.
>
> Hence the air was more chilled, and the winds more severely cold.
>
> Hence perhaps the winter of 1783-84 was more severe, than any that had happened for many years.
>
> The cause of this universal fog is not yet ascertained. Whether it was adventitious to this earth, and merely a smoke, proceeding from the consumption by fire of some of those great burning balls or globes which we happen to meet with in our rapid course around the sun, and which are sometimes seen to kindle and be destroyed in passing our atmosphere, and whose smoke might be attracted and retained by our earth: *or whether it was the vast quantity of smoke, long continuing to issue during the summer from Hecla in Iceland, and that other volcano which arose out of the sea near that island, which smoke might be spread by various winds, over the northern part of the world, is yet uncertain.*[1]

Franklin's first speculation, that the summer fog and winter coldness could be due to smoke from a meteor is rather bizarre, and has been forgotten (except perhaps as an unremembered contribution to the idea that a comet or asteroid collision with the Earth 65 million years ago resulted in extinction of the dinosaurs). His second option, that smoke from "Hecla in Iceland, and that other volcano which rose out of the sea near that island" caused the observed weather anomalies is more enduring.

The general concept that volcanic activity can affect the climate has been developed since the obvious weather anomalies following the eruption of Krakatau in 1883 (Russell and Archibald 1888). Mitchell (1961); Lamb (1971); Self *et al.* (1981); Devine *et al.* (1984) and others have demonstrated statistically that years in which volcanic aerosols are ejected into the stratosphere by volcanic eruptions are typically followed by one to three years of temperatures 0.2 - 0.5°C below average. Thus, although there are doubters, there is a widely-promoted model that explosive eruptions of sulphur-rich magmas can implant sulphuric-acid aerosols into the stratosphere where they remain suspended for a few years. The aerosols spread around the globe,

---

[1] Italics added by editor.

absorbing incoming solar radiation, thus heating the stratosphere and, by a reduction in the amount of radiation reaching the ground, cooling the Earth's surface (Sigurdsson 1982; Rampino and Self 1984).

The Laki eruption does not appear to fit this general model because it was not a classical eruption, such as Krakatau or Tambora, which explosively ejected aerosols into the stratosphere. Effusive eruptions, like Laki and typical of activity at Hawaiian volcanoes, are thought to have only minimal explosive activity, with nearly all magma flowing quietly across the Earth's surface as lava flows. Wood (1984a); Devine *et al.* (1984); and Stothers *et al.* (1986) proposed, however, that aerosols from Laki may have entered the stratosphere even though the eruption was largely quiescent. The last two groups proposed that heat from fire fountains and lava fields could generate convective plumes that would rise into the stratosphere. This effect would be enhanced by normal atmospheric mixing across the tropopause which replaces the entire air mass in the lower and middle stratosphere with tropospheric air every one to two years (Flohn 1968). Following my previous suggestion (Wood 1984a), this mixing could result, over the prolonged period of the Laki eruption and with the possibly high altitude of convectively-transported materials, in substantial deposition of sulphur aerosols in the stratosphere. Thus, the Laki eruption may have emplaced sufficient material in the stratosphere to produce the multi-year climatic effects.

Most of the observed climatic effects in the early to mid-1780s can be explained by the volcanic hypothesis (Lamb 1970). The dry fog in Europe, the Near East, and North America, and the sulphurous smells, burning of eyes, and singeing of tulips in western Europe resulted from the tropospheric movement of Laki dust and sulphur aerosols mainly to the east but apparently also to the west. Gilbert White reported that the dry fog lasted one month, which coincided with the period of maximum volcanic activity (Wood 1984a). The hot summer weather in 1783 in Europe, United States and China is an unusual occurrence; no other volcanic eruption is associated with such hot weather. Perhaps the heated gases in the mid-troposphere hindered normal convection so that heat was trapped near the surface. The cold summer in Iceland, however, was presumably caused by the blockage of sunlight by persistent dense haze and smoke from Laki. Similar immediate cooling near dense volcanic plumes occurred at Tambora (Rampino, this volume), as well as Krakatau and Mount St. Helens (10° and 8°C below normal, respectively; reported in Simkin and Fiske 1983).

The exceptionally cold winter of 1783-84 in Europe, North America and Japan is proposed to have resulted from the standard volcanic mechanism of stratospheric warming and hence tropospheric cooling due to the abundance of volcanic sulphuric-acid aerosols in the stratosphere. The very cold winter in Europe and Japan during 1784-85 and the following (1785) cool and dry summer in Europe and United States can only be explained if large numbers of aerosols remained in the stratosphere through 1785. This would be consistent with other large explosive eruptions, such as Tambora and Krakatau which were followed by lower than normal temperatures for one to three years (Rampino and Self 1984).

Uncertainties

These explanations would be impossible if aerosols from Laki did not enter the stratosphere. One significant piece of evidence suggests they did not. Sulphuric acid droplets from volcanic-eruption clouds fall to the Earth everywhere under the passing cloud, but the existence of the droplets is recorded only when they fall on permanent ice fields, as in Greenland and Antarctica, where they leave an acidic trace in that year's ice layer (Hammer 1977). The largest acid spike in the Camp Century ice core in Greenland occurs in 1783 (Hammer 1977); but as pointed out by Sigurdsson (1982) there is no acid anomaly for 1784. The 1783 anomaly could be due to either tropospheric or stratospheric transport of aerosols from Laki (only about 1200 km to the east). An anomaly for 1784, which could only occur if significant aerosols were stored in the stratosphere for a year, would be strong evidence that Laki materials reached the stratosphere; the lack of a 1784 anomaly is most consistent with no stratospheric contribution. If this is true then the present understanding of volcanic influences on climate require that the cold winter of 1784-85 is unrelated to the Laki eruption and the cold winter of 1783-84.

A second uncertainty is whether Laki was actually the volcano that produced the anomaly recorded in the Camp Century ice core and that caused the observed climatic effects. One confusing piece of evidence is the report (in Lamb 1970) that the dry fog was first observed on 29 May 1783 in Copenhagen and on 6 June in France. Yet the Laki eruption is recorded to have started only on 8 June 1783. Was another, earlier eruption responsible for the dry fog and other effects?

Eleven eruptions are recorded to have begun in 1783 (Simkin *et al.* 1981), and two other little known ones may have occurred (Table 1). In terms of volume, Laki was the largest eruption and Asama, in Japan was the second largest. All other eruptions of the year are thought to have been considerably smaller, but one of them may have been important.

Nyey

As Franklin (1784) noted there was another "...volcano which rose out of the sea near..." Iceland. This volcano was the temporary island of Nyey or Noyöe (New Island) which formed over the Mid-Atlantic Ridge some 50 km southwest of the Reykanes Peninsula in southwestern Iceland. Nyey began erupting by 1 May 1783 and produced a large deposit of pumice that floated on the sea for about 250 km around the volcano, causing great hardship for sailors (Lyell 1969). By autumn, when the Danish government sent an expedition to lay claim to the island, Nyey had been destroyed by wave action. One of the few descriptions (and a drawing, Figure 8) of the eruption is reproduced in Thorarinsson (1967). The Danish Captain Mindelberg of the brig *Boesand* first saw a smoke column on 1 May and wrote in his ship log, "At three o'clock in the morning we saw smoke rising from the sea and thought it to be land; but on closer consideration we concluded that this was a special wonder wrought by God and that a natural sea could burn...When I caught sight of this terrifying smoke I felt convinced that Doomsday had come." (quoted by Thorarinsson 1967). On 3 May *Boesand* approached the area of the smoke plume, but ..."when we had come within half a mile of the island we had to turn away for fear that the crew might faint owing to the enormous sulphur stench."

Two, perhaps similar, eruptions near Iceland during the last 25 years provide comparisons. In 1963 the island of Surtsey formed off the southern coast, and in 1973, a new cone and lava flow was constructed near Surtsey at Heimaey on the Vestmann Islands. Both eruptions were similar to the account of Nyey in that explosive eruptions produced scoria cones, but both Surtsey and Heimaey were armoured by lava flows and have been able to withstand wave erosion. Neither of the recent island eruptions produced as large a pumice field as reported for Nyey, and only minor amounts of ash fell in Europe. Based on the modern examples, it seems unlikely that Nyey could have caused the widespread effects commonly attributed to Laki, but the reported widespread pumice and lack of detailed information makes it impossible to reject completely the notion that Nyey contributed to the 1783 climatic phenomena.

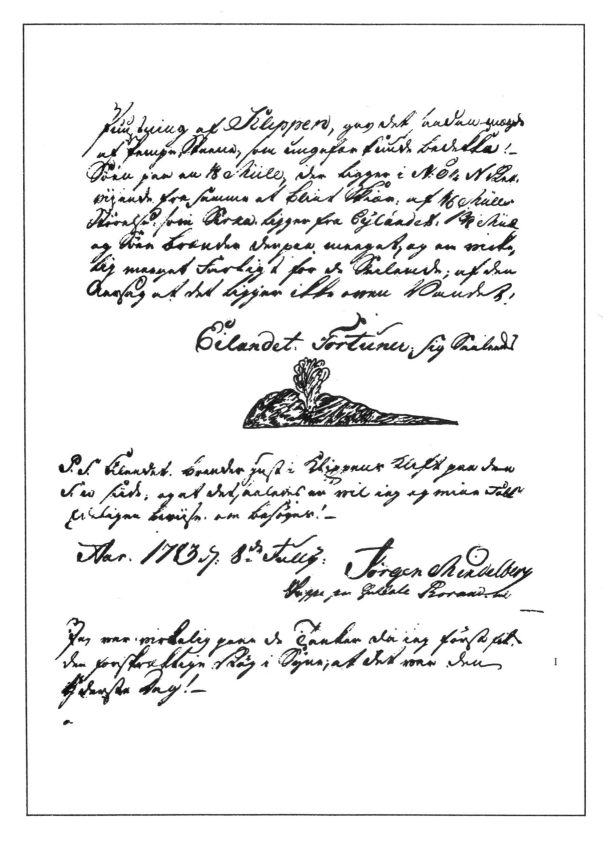

Figure 8: Drawing and last page of text from Captain Mindelberg's report on the Nyey eruption southwest of Iceland in May 1783. Reproduced from Thorarinsson (1967).

Table 1: Volcanic Eruptions, 1783.

Start Date	Volcano	Location	VEI	Comment
05 May	Nyey	Off Iceland	2	
09 May	Asama	Japan	4	Biggest eruption in August
12 May	Barren Island	Andaman Islands	2	
08 June	Laki	Iceland	4	till 8 Feb. 1784
? July	Izalco	El Salvador	0	
18 August	Vesuvius	Italy	2	>3 years
03 September	Sakurajima	Japan	2?	
03 December	Iwaki?	Japan	?	
?	Kurikoma	Japan	?	
?	Kanaga	Alaska	?	>3 years
??	Unnamed	Greenland Sea	2?	
??	Unnamed	North Atlantic	2?	

[1] From Simkin *et al.* (1981). ? = Unknown date within 1783; ?? = uncertainty if eruption occurred in 1783. VEI 4 = Volcanic eruption index (0-8); VEI 4 = 10^8 to 10^9 m^3 of ejecta.

Asama

The largest historic eruption of Asama volcano in Japan began on 9 May 1783. Bullard (1976) and others have suggested that this eruption caused the climatic anomalies of 1783. I have previously summarized (Wood 1984b) recent Japanese literature on the 1783 Asama eruption which tends to discount it as the source of the dry fogs and other early summer climatic effects. The main argument is that although eruptive activity began in early May, nearly half of the total of 0.5 km^3 of ejecta was deposited during two days of intense eruptions on 3 and 4 August 1783, and most of the remainder formed during the next five days (Imai and Mikada 1982), two months after the dry fogs were reported. Asama may have contributed to the generally cool winter of 1783, but it did not contribute to the strong atmospheric effects of early summer.

Laki

The observation that dry fog was reported in Europe 10 days before the onset of activity of Laki is well dated by eyewitness accounts. Thordarsson and Self (1988) discovered by studying old Icelandic maps that the Grimsvotn basaltic caldera, about 50 km northeast of Laki along the fissure trend, erupted repeatedly throughout the Laki eruption. As proposed by Sigurdsson and Sparks (1978), activity along the Laki fissure system was probably intimately tied to activity at the Grimsvotn caldera. Thordarsson and Self (1988) suggest that there may have been an eruption at Grimsvotn in May, before the first Laki activity. Thus, the Laki/Grimsvotn system may have produced all the dry fogs of the summer of 1783.

Summary

There are many loose ends in the story of Laki and its possible climatic effects. In this report a variety of readily available observations of unusual climatic phenomena occurring during the two years following the eruption is presented. The simplest assumption is that these anomalies are related to Laki, just as similar types and durations of climatic phenomena are clearly accepted as being associated with Tambora's eruption in 1815. It is most likely that eruptions of the Laki/Grimsvotn system caused the dry fogs and hot summer of 1783 and the cold winter of 1783-84. In order to cause the cold winter of 1783-84 volcanic aerosols must have reached the stratosphere. And probably the cold winter of 1784-85 was due to the same stratospheric aerosols, which however, left no trace in the Greenland ice core for 1784. If Laki produced all of these effects, present volcano-climate models are inadequate to explain how. If Laki was not responsible, then a major eruption 200 years ago is completely missing from our records.

In compiling the historical data for the 1780s it became obvious that most reports are from the eastern United States and western Europe. A much greater effort is required to search the historical (and proxy) archives of Africa, Asia, South and Central America, and central and western United States to further define possible climatic effects of Laki and other eruptions.

Acknowledgement

I thank Michael Helfert for sharing information concerning the unusual weather following the eruption Laki.

References

Arakawa, H. 1954. Fujiwhara on five centuries of freezing dates of Lake Suwa in central Japan. *Archív für Meteorologie, Geophysik und Bioklimatologie* 6:152-166.

Bray, J.R. 1978. Volcanic eruptions and climate during the past 500 years. In: *Climatic Change and Variability: A Southern Perspective*. A.B. Pittock *et al.* (eds.). Cambridge University Press, Cambridge. pp. 256-262.

Bullard, F.M. 1976. *Volcanoes of the Earth*. University of Texas, Austin. 579 pp.

Catchpole, A.J.W. 1988. Presentation at *The Year Without A Summer? Climate in 1816*. An International Meeting Sponsored by the National Museum of Natural Sciences, 25-28 June 1988, Ottawa.

Devine, J.D., H. Sigurdsson, A.N. Davis and S. Self. 1984. Estimates of sulfur and chlorine yield to the atmosphere from volcanic eruptions and potential climatic effects. *Journal of Geophysical Research* 89:6309-6325.

Filion, L., S. Payette, L. Gauthier and Y. Boutin. 1986. Light rings in subarctic conifers as a dendrochronological tool. *Quaternary Research* 26:272-279.

Flohn, H. 1968. *Climate and Weather*. McGraw Hill, New York. 256 pp.

Franklin, B. 1784. Meteorological imaginations and conjectures. (Reprinted in Sigurdsson, 1982).

Hammer, C.U. 1977. Past volcanism revealed by Greenland ice sheet impurities. *Nature* 270:482-486.

Imai, H. and H. Mikada. 1982. The 1783 activity of Asama Volcano inferred from the measurements of bulk density of tephra (pumice) and the old documents. *Bulletin of the Volcanological Society of Japan* 27(1):27-43.

Jackson, E.L. 1982. The Laki eruption of 1783: impacts on population and settlement in Iceland. *Geography* 67:42-50.

Kington, J.A. 1978. Historical daily synoptic weather maps from the 1780s. *Journal of Meteorology* 3(27):65-70.

Lamb, H.H. 1970. Volcanic dust in the atmosphere: with a chronology and assessment of its meteorological significance. *Proceedings of the Royal Society of London* 266:425-533.

Landsberg, H.E., C.S. Yu and L. Huang. 1968. Preliminary reconstruction of a long time series of climatic data for the eastern United States. Report II: On a project studying climatic changes. *University of Maryland, Institute of Fluid Dynamics, Technical Note* BN-571.

Lough, J.M. and H.C. Fritts. 1987. An assessment of the possible effects of volcanic eruptions on North American climate using tree-ring data, 1602 to 1900 A.D. *Climatic Change* 10:219-239.

Ludlum, D.M. 1966. *Early American Winters, 1604-1820*. American Meteorological Society, Boston. 283 pp.

Lyell, C. 1969. *Principles of Geology*, Vol. 1. Johnson Reprint Collection, New York. pp. 371-374. (A reprint of 1830-33 edition).

Mikami, T., 1988. Climatic reconstruction in historical times based on weather records. *Geographical Review of Japan* 61(1):14-22.

Mitchell, J.M. 1961. Recent secular changes of global temperature. *Annals of the New York Academy of Sciences* 95:235.

Ogilvie, A.E.J. 1986. The climate of Iceland, 1701-1784. *Jökull* 36:57-73.

Oswalt, W.H. 1957. Volcanic activity and Alaskan spruce growth in A.D. 1783. *Science* 126:928-929.

Pant, G.B., B. Parthasarathy and N.A. Sontakke. 1988. Climate over India during the first quarter of the nineteenth century. In: *The Year Without A Summer? Climate in 1816*. An International Meeting Sponsored by the National Museum of Natural Sciences, 25-28 June 1988, Ottawa. Abstracts p. 52

Parker, M.L. 1988. Presentation at *The Year Without A Summer? Climate in 1816.* An International Meeting Sponsored by the National Museum of Natural Sciences, 25-28 June 1988, Ottawa.

Pfister, C. 1978. Fluctuations in the duration of snow cover in Switzerland since the late 17th century. *Danish Meteorological Institute, Climatological Papers* 4:1-6.

_____. 1981. An analysis of the Little Ice Age climate in Switzerland and its consequences for agricultural production. *In: Climate and History.* T.M.L. Wigley, M.J. Ingram and G. Farmer (eds.). Cambridge University Press, Cambridge. pp. 214-248.

Rampino, M.R. and S. Self. 1984. Sulphur-rich volcanic eruptions and stratospheric aerosols. *Nature* 310:677-679.

Russell, F.A.R. and E.D. Archibald. 1888. On the unusual optical phenomena of the atmosphere, 1883-1886, including twilight effects, coronal appearances, sky haze, coloured suns, moons. (Reprinted in T. Simkin and R.S. Fiske [1983 pp. 397-404]).

Schove, D.J. 1954. Summer temperatures and tree-rings in North Scandinavia, A.D. 1461-1950. *Geografiska Annaler* 37:40-80.

Self, S., M.R. Rampino and J.J. Barbera. 1981. The possible effects of large 19th and 20th century volcanic eruptions on zonal and hemispheric surface temperatures. *Journal of Volcanology and Geothermal Research* 11:41-60.

Sigurdsson, H. 1982. Volcanic pollution and climate: the 1783 Laki eruption. *EOS* 63:601-602.

Simkin, T. and R.S. Fiske. 1983. *Krakatau 1883, the Volcanic Eruption and Its Effects.* Smithsonian Institution Press, Washington, D.C. 464 pp.

Simkin, T., L. Siebert, L. McClelland, D. Bridge, C. Newhall and J.H. Latter. 1981. *Volcanoes of the World.* Hutchinson Ross, Stroudsburg. 240 pp.

Stothers, R.B., J.A. Wolff, S. Self and M.R. Rampino. 1986. Basaltic fissure eruptions, plume heights, and atmospheric aerosols. *Geophysical Research Letters* 13:725-728.

Sutherland, D. 1981. Weather and the peasantry of Upper Brittany. *In: Climate and History.* T.M.L. Wigley, M.J. Ingram and G. Farmer (eds.). Cambridge University Press, Cambridge. pp. 434-449.

Thorarinsson, S. 1963. *Askja on Fire.* Almenna Bokafelagid, Reykjavik. 48 pp.

_____. 1967. *Surtsey, The New Island in the North Atlantic.* Viking Press, New York. 47 pp.

_____. 1969. The Lakagígar eruption of 1783. *Bulletin volcanologique* 38:910-929.

Thordarsson, Th., S. Self, G. Larson, and S. Steinthorsson. 1987. Eruption sequence of the Skaftar Fires 1783-1785, Iceland. *EOS* 68:1550.

Thordarsson, Th. and S. Self. 1988. Old maps help interpret Icelandic eruption. *EOS* 69:86.

Thoroddsen, Th. 1925. *Die Geschichte der islandischen Vulkane. Kongelige Danske Videnskabernes Selskabs Skrifter, Naturvidenskabelig og Mathematisk.* Afdeling 8, IX.

Wang, S.W. and Z.C. Zhao. 1981. Droughts and floods in China, 1470-1979. *In: Climate and History.* T.M.L. Wigley, M.J. Ingram and G. Farmer (eds.). Cambridge University Press, Cambridge. pp. 271-288.

White, G. 1977. *The Natural History of Selbourne.* Penguin, New York (reprint of original 1788-89 edition).

Wood, C.A. 1984a. Amazing and portentous summer of 1783. *EOS* 65:409.

_____. 1984b. Asama 1783: lost in the rush to remember Krakatau. *Volcano News* 16:5.

Woodhouse, C. 1988. Tree-ring chronologies in the Great Salt Lake Basin and paleoclimatic implications. *Proceedings of the Annual Meeting of the Association of American Geographers.* (in press).

The Effects of Major Volcanic Eruptions on Canadian Surface Temperatures

Walter R. Skinner[1]

Abstract

The superposed epoch method of analysis was used to detect changes in Canadian surface temperatures due to large volcanic dust veils in the atmosphere. This method accentuates weak signals that are present in a data series, as a temperature signal caused by a volcanic dust veil is expected to be of the same magnitude as the background noise level. Lamb's Dust Veil Index (DVI), a measure of the amount of volcanic material injected into the atmosphere, was used to select the volcanic-eruption dates beginning with the eruption of Krakatau in 1883. The DVI is directly related to the total loss of solar radiation reaching the Earth's surface, and has the advantage of not being calculated from temperature information. Surface-temperature records for up to 20 Canadian stations were analyzed on national, regional (Arctic) and seasonal (summer and winter) bases for both equatorial and mid-latitude eruptions. A small sample test of significance was applied, and all suspected temperature signals proved to be significant at the 0.01 level or better. The annual temperature depression following a mid-latitude eruption was about 0.4°C, and occurred during the eruption year and lasted no longer. A decline in annual surface temperature of about 1.0°C occurred in the first year after an equatorial eruption and persisted to a lesser degree, for another year or so. The difference appears to be directly related to the substantially greater mean DVI for the equatorial eruptions. The annual temperature drop in the Arctic was slightly greater than that for the country as a whole. Summer-temperature signals were stronger than those in winter, and in almost all cases were of a greater magnitude than the annual signals. There was a marked drop in winter temperatures of about 1.0°C following an equatorial eruption.

Introduction

The eruption of El Chichón in southern Mexico between 28 March and 4 April 1982 ejected huge concentrations of gases and particles into the upper atmosphere. By mid-November 1982, detailed solar radiation measurements in Fairbanks, Alaska began to display distinct differences from the previous five-year normal (Wendler 1984). Clear days during the 15 November 1982 to 31 May 1983 period, when compared to clear-day data for the previous five years, showed a decrease in the direct beam of almost 25% and a decrease in global radiation of about 5%. Mass and Schneider (1977) previously determined that large volcanic dust veils in the atmosphere can reduce direct solar radiation by as much as 10%. This is simultaneously accompanied by an increased scattering effect that could substantially change the total amount of solar radiation reaching the Earth's surface.

Many theoretical investigations (Schneider and Mass 1975; Pollack *et al.* 1976) and empirical studies (Lamb 1970; Oliver 1976; Mass and Schneider 1977; Taylor *et al.* 1980) have been made in an attempt to determine the possible influence of large volcanic dust veils on surface weather

---

[1] Canadian Climate Centre, Atmospheric Environment Service, 4905 Dufferin Street, Downsview, Ontario M3H 5T4, Canada.

and climate. Most of these investigations were conducted on either a global or hemispheric scale. Taylor *et al.* (1980) also searched for volcanic signals on latitudinal, continental/marine and seasonal bases. A drop in annual average surface temperature of between 0.5 and 1.0°C in the first or second year following a large volcanic eruption was found in most of the empirical studies.

Canada, having an extensive area in mid- and high latitudes, should experience volcanic dust veil influences at varying times after a major eruption depending upon both the location and the time of year of the eruption. Oliver (1976) estimated a mid-latitude eruption to have a same year impact on northern hemisphere mean temperatures, while a similar impact by an equatorial eruption would be delayed for about a year. Lamb (1970) states that the transfer of upper-level dust veils from equatorial to mid-latitudes is accomplished mainly in autumn, and to a lesser extent in spring, with the great seasonal circulation changes.

In this investigation, surface-temperature records for up to 20 Canadian stations were analyzed on national, regional (Arctic) and seasonal (summer and winter) bases for both equatorial and mid-latitude eruptions.

Methodology and Data

The Superposed Epoch Method
The superposed epoch method of analysis, as outlined by Panofsky and Brier (1965) and employed by Mass and Schneider (1977) and Taylor *et al.* (1980), was used to detect changes in Canadian surface temperatures due to large volcanic dust veils in the atmosphere. This method accentuates weak signals that are present in a data series. A temperature signal caused by a volcanic dust veil is expected to be of the same magnitude as the background noise level, or the variability of the atmosphere (Taylor *et al.* 1980).

Volcanic Eruption Dates
Volcanic eruptions were selected on the basis of amounts of material ejected into the atmosphere, latitude of the eruption and the isolation in time of the eruption from any other major volcanic event. Lamb's (1970) Dust Veil Index (DVI) was used as a basis for selecting most of the eruption dates. This is a measure of the amount of volcanic material injected into the atmosphere, and is directly related to the loss of solar radiation reaching the Earth's surface. The eruption of Krakatau in 1883 was given a value of 1000, and all other eruptions were adjusted to it. The DVI has the advantage of not being calculated from temperature information (Mass and Schneider 1977).

Five of the six volcanic eruptions selected have a total DVI greater than 150, and are classed as major volcanic events (Table 1). The 1956 eruption was chosen because of its mid-latitude location and isolation in time from any other major volcanic event. Three equatorial and three mid-latitude eruptions were selected in an attempt to isolate both the temporal dimensions and the magnitudes of the temperature signals in Canada following a major volcanic event.

Selected volcanic events had to be separated by at least five years from any other major volcanic event. This was done to avoid the problem of cumulative dust veils that might obscure resulting signals. The separation of the 1907 event from the preceding event (1902) and the following event (1912) is exactly five years. Incorporating the 1907 event was called for as it was a major mid-latitude event for which ample records were available.

Table 1: Date, Location and Dust Veil Index (DVI) of Selected Major Volcanic Eruptions.

#	Eruption	Location	Dates	Key Dates	DVI	Total DVI
1.	Krakatau, Indonesia	6.0° S 105.0° E	Aug 1883	Aug 1883	1000	1000
2.	Mont Pelée, Martinique	15.0° N 61.0° W	May 1902	- -	100	-
	Soufrière, St. Vincent	13.5° N 61.0° W	May 1902	- -	300	-
	Santa Maria, Guatemala	14.5° N 92.0° W	Oct 1902	- -	600	-
	Cumulative Data:			May 1902	-	1000
3.	Shytubelya Sopka, Kamchatka	52.0° N 157.5° W	Mar 1907	Mar 1907	500	500
4.	Katmai, Alaska	58.0° N 155.0° W	Jun 1912	Jun 1912	150	150
5.	Bezymjannaja, Kamchatka	56.0° N 160.5° E	Mar 1956	Mar 1956	10	10
6.	Gunung Agung, Bali	8.5° S 115.5° E	Mar 1963	Mar 1963	800	800
	Equatorial Eruptions (# 1, 2 and 6)			Average Total DVI =		933
	Mid-Latitude Eruptions (# 3, 4 and 5)			Average Total DVI =		220

Eruption year key months were also determined from Lamb (1970). The key month was the month during which the volcano entered its most explosive phase. In the case of two or more eruptions in the same year, such as 1902, the month of the first eruption was used. Table 1 includes the key eruption date for each selected event.

Composite Key Dates and Composited Temperatures

The key volcanic eruption date was defined as the 12-month period beginning with the month during which the eruption occurred. The use of this period results in a cleaner volcanic signal than using the actual calendar year of the eruption (Taylor *et al.* 1980). This 12-month period was termed the "eruption year", or year "0". Sequences of four preceding years, or the four 12-month periods prior to the eruption year and the four following years, or the four 12-month periods after the eruption year, were then determined. These sequences provided the bases for both individual and multiple composites. The five annual periods, the eruption year and the four following years, were analyzed because a volcanic dust veil produced by a single eruption exists for only a few years (Lamb 1970).

Average temperature values, for selected Canadian stations, were calculated for each month of each of the 12-month periods associated with an eruption year. The resulting 12 monthly values were then summed and averaged to yield an annual value for that particular year. Graphs based on individual volcanic events were then plotted and studied in an attempt to define climatic signals.

Annual temperature values for each individual eruption were then associated with the corresponding values for all other individual eruptions. In addition, values for equatorial eruptions were isolated and inter-associated. The same was done for mid-latitude eruption values. These corresponding values were then summed and averaged to yield a "superposed epoch". Graphs, based upon these multiple volcanic events, were plotted and analyzed in a comparable manner to the analyses of the individual events.

Data

The database used consisted of mean monthly temperature values for up to 20 Canadian weather stations over common time periods. Stations were selected on the basis of length and completeness of record and upon location. Thirteen stations were available for the 1883 eruption date. There were no long-term records available for this date west of Winnipeg. Four stations were added for the 1902 eruption date to provide east to west coast spatial coverage. Another station was added for the 1907 and 1912 eruptions. The lack of long-term records for stations in northern Canada restricts the study of the first four eruptions to more southerly Canadian latitudes. Northern stations were added for the last two eruptions. This brought the total to 20 stations for the 1963 event. Table 2 shows the stations used for the 1883 eruption. Table 3 shows the stations added for the 1902, 1907 and 1912 eruptions. Table 4 lists the stations used for the 1956 and 1963 eruptions. In some cases, such as Quebec City and Winnipeg, weather-observation sites were moved during the 1940s from city to airport locations. However, none of the eruptions used in this study occurred during this period. In addition, some long-term temperature records, such as those from Toronto and Montreal, have been subjected to an artificial warming due to the influence of urban expansion. It was hoped that this would have only minor influence on the results and that the method of analysis would subdue the apparent noise in this small portion of the data.

Missing monthly values were estimated for each station by calculating the 30-year mean for that particular month. In most cases only one of the 13 to 20 values was absent. The resulting estimate had little effect on the overall monthly composite. There were never more than two missing values in any monthly composite.

Canadian Analysis

Taylor *et al.* (1980) found it necessary to use data from a group of stations rather than just individual stations when searching for a temperature signal related to a volcanic eruption. This is due to the year-to-year and station-to-station variability when dealing with single station superpositions. Thus, the superposed epoch method outlined previously was applied to some or all of the 20-station temperature database selected for this study.

Individual Eruption Composites

Figures 1 to 6 show the individual eruption dust veil temperature composites for the selected Canadian stations. The 1883, 1902, 1956 and 1963 composites each display a marked dip in average annual temperature either in the eruption year or in the following two years. The 1907 and 1912 composites show no such dip during these years. The low values for the "-4" and "-3" years for the 1907 composite might be the result of a large 1902 dust veil. However, there is no such dip in the early years of the 1912 composite that might similarly be attributed to a 1907 dust veil.

Graphs, based on the multiple volcanic events, were then plotted and examined in an attempt to define volcanic signals.

Table 2: Weather Stations Used in Studying the Influence of Volcanic Dust Veils on Canadian Surface Temperatures for the 1883 Krakatau Eruption.

Weather Station	Location	Period	Years	AES No.
1. Winnipeg, Manitoba	49° 53' N, 97° 07' W	1872-1938	67	5023243
2. Port Arthur, Ontario	48° 26' N, 89° 13' W	1877-1941	65	6046588
3. Ottawa, Ontario	45° 24' N, 75° 43' W	1872-1935	64	6105887
4. Beatrice, Ontario	45° 08' N, 79° 23' W	1876-1979	104	6110605
5. Woodstock, Ontario	43° 07' N, 80° 45' W	1870-1981	112	6149625
6. Toronto, Ontario	43° 40' N, 79° 24' W	1840-1981	142	6158350
7. Québec City, Quebec	46° 48' N, 71° 13' W	1872-1959	88	7016280
8. Montréal, Québec	45° 30' N, 73° 35' W	1871-1981	111	7025280
9. Chatham, New Brunswick	47° 03' N, 65° 29' W	1873-1947	75	8100990
10. Fredericton, New Brunswick	45° 57' N, 66° 36' W	1871-1952	82	8101700
11. Halifax, Nova Scotia	44° 39' N, 63° 36' W	1871-1933	63	8202198
12. Sydney, Nova Scotia	46° 09' N, 60° 12' W	1870-1941	72	8205698
13. St. John's, Newfoundland	47° 34' N, 52° 42' W	1874-1956	83	8403500

Table 3: Weather Stations Added to Those Used for the 1883 Krakatau Eruption for the 1902, 1907 and 1912 Eruptions.

Weather Station[1]	Location	Period	Years	AES No.
1. Victoria, British Columbia	48° 25' N, 123° 22' W	1898-1981	84	1018610
2. Medicine Hat, Alberta	50° 01' N, 110° 43' W	1883-1981	99	3034480
3. Banff, Alberta	51° 11' N, 115° 34' W	1887-1981	95	3050520
4. Regina, Saskatchewan	50° 26' N, 104° 40, W	1883-1981	99	4016560
5. Ottawa (CDA), Ontario	45° 23' N, 75° 43' W	1889-1981	93	6105976

[1] Ottawa (6105887) not used for 1902 eruption.

Table 4: Weather Stations Used In Studying the Influence of Volcanic Dust Veils on Canadian Surface Temperatures for the 1956 Bezymjannaja and 1983 Gunung Agung Eruptions.

Weather Station[1]	Location	Data Period	Years of Record	AES No.
1. Victoria, British Columbia	48° 25' N, 123° 22' W	1898-1981	84	1018610
2. Medicine Hat, Alberta	50° 01' N, 110° 43' W	1883-1981	99	3034480
3. Banff, Alberta	51° 11' N, 115° 34' W	1887-1981	95	3050520
4. Regina, Saskatchewan	50° 26' N, 104° 40' W	1883-1981	99	4016560
5. Winnipeg, Manitoba	49° 53' N, 97° 07' W	1938-1981	44	5023243
6. Churchill, Manitoba	58° 45' N, 94° 04' W	1943-1981	39	5060600
7. Ottawa CDA, Ontario	45° 23' N, 75° 43' W	1889-1981	93	6105976
8. Beatrice, Ontario	45° 08' N, 79° 23' W	1876-1979	104	6110605
9. Woodstock, Ontario	43° 07' N, 80° 45' W	1870-1981	112	6149625
10. Toronto, Ontario	43° 40' N, 79° 24' W	1840-1981	142	6158350
11. Quebec City A, Quebec	46° 48' N, 71° 23' W	1843-1981	39	7016294
12. Montreal, Quebec	45° 30' N, 73° 35' W	1871-1981	111	7025280
13. Chatham A, New Brunswick	47° 01' N, 65° 27' W	1943-1981	39	8101000
14. Fredericton CDA, N.B.	45° 55' N, 66° 37' W	1913-1981	69	8101600
15. Halifax, Nova Scotia	44° 39' N, 63° 34' W	1939-1974	36	8202200
16. Sydney A, Nova Scotia	46° 10' N, 60° 03' W	1941-1981	41	8205700
17. St. John's, Newfoundland	47° 35' N, 52° 44' W	1957-1975	19	8403501
18. Cambridge Bay, N.W.T.	69° 07' N, 105° 01' W	1929-1981	53	2500600
19. Mould Bay, N.W.T.	76° 14' N, 119° 20' W	1948-1981	34	2502700
20. Kuujjuaq, Quebec	58° 06' N, 68° 25' W	1947-1981	35	7112400

[1] St. John's (8403501) not used for the 1956 eruption.

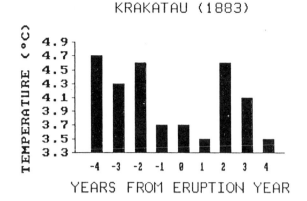

Figure 1: Dust veil temperature composite. Thirteen Canadian station events.

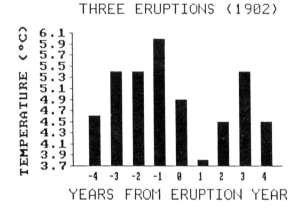

Figure 2: Dust veil temperature composite. Seventeen Canadian station events.

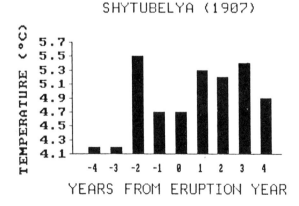

Figure 3: Dust veil temperature composite. Eighteen Canadian station events.

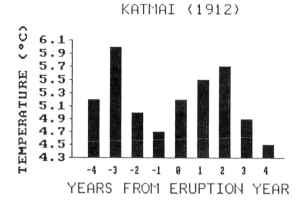

Figure 4: Dust veil temperature composite. Eighteen Canadian station events.

Figure 5: Dust veil temperature composite. Nineteen Canadian station events.

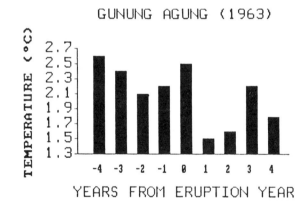

Figure 6: Dust veil temperature composite. Twenty Canadian station events.

The apparent significance of these graphs must be viewed with caution. The first four eruptions were embedded in a hemispheric-warming trend, whereas the last two eruptions occurred during a hemispheric-cooling trend (Mass and Schneider 1977). The year-to-year variability, or noise, found by Taylor *et al.* (1980) is evident in a Canadian context. The compositing of several volcanic events should reduce this noise level and accentuate a volcanic dust veil signal.

Multiple Eruption Composites

Figure 7 shows the temperature composite for all stations and all eruptions. There is an obvious temperature dip during the eruption year and the "+1" year. The temperature dip during these two years is about 0.4°C below the level of years "-4" to "-1". Figure 8 shows the composite for the three equatorial eruptions. There is a well-marked dip in the "+1" year, about 1.1°C below the level of years "-4 to "-1". Figure 9 shows the composite for the three mid-latitude eruptions. Here the temperature dip is in the actual eruption year, about 0.5° C below the levels of years "-4" to "-1".

Arctic Analysis

The data noise level, or year-to-year variability, when based upon different groupings of stations, should vary randomly while the volcanic signal should remain fairly constant (Taylor *et al.* 1980). A regional analysis was one step in determining the significance of the possible volcanic signals outlined previously. It also provided the basis for volcanic signal investigation into a part of Canada which can be extremely sensitive to small alterations in surface temperature.

The solar-radiation deficit produced by volcanic dust veils must be greatest in Arctic areas where dust veils persist longer and the sun's rays travel obliquely through the layers of dust (Lamb 1970). Reduced surface temperatures result in an accumulation of both sea ice and land snow. The increased albedo would produce a radiation deficit long after the dust veil has disappeared (Lamb 1970). It would also affect the general atmospheric circulation, possibly having far-reaching spatial effects.

The superposed epoch analysis method was applied to four Canadian Arctic stations for the 1956 and 1963 eruptions. There were no Canadian Arctic station records for the earlier eruptions. The stations used were Churchill, Manitoba, Cambridge Bay, Northwest Territories, Mould Bay, Northwest Territories and Kuujjuaq, Quebec.

Figure 10 shows the average annual temperature composite for the 1956 and 1963 eruptions. The temperature dip in the "0" and "+1" years is similar to that in Figure 7 for all Canadian stations. It is about 1°C below the levels of years "-4" to "-1". The surrounding noise level, however, is quite different than that in Figure 7. Years "+2" to "+4" hint at Arctic temperature stability following a volcanic eruption.

Seasonal Analyses

Summer and winter investigations were made in an attempt to determine the relative magnitudes of the dust-veil signals. The key summer season was defined as the first three-month period (June to August) to follow an eruption. The key winter season was defined as the first three-month period (December to February) to follow an eruption. Sequences of four preceding and four following seasons were determined in the same manner outlined previously. Seasonal averages were calculated for all Canadian stations and for each year associated with a volcanic eruption.

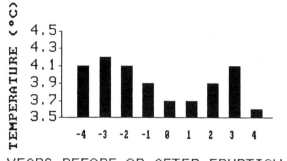

Figure 7: Dust veil temperature composite. All Canadian station events.

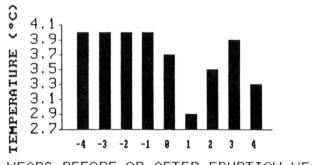

Figure 8: Dust veil temperature composite. Fifty Canadian station events for three equatorial eruptions.

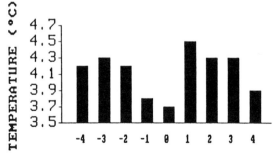

Figure 9: Dust veil temperature composite. Fifty-four Canadian station events for three mid-latitude eruptions.

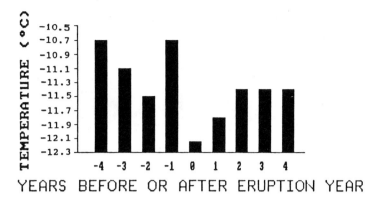

Figure 10: Dust veil temperature composite. Canadian Arctic station events.

Summer
Figures 11 to 13 show the summer season temperature composites. All but one graph show a distinct drop of up to several tenths of a degree in either the eruption year or the following year. These composites display a close resemblance to those in Figures 7 to 9. The magnitude of each temperature drop, however, is at least equal to or greater than that of the corresponding annual composite.

Winter
Figures 14 to 16 show the winter season temperature composites. There is a higher degree of year-to-year variability than there was in the summer composites. This makes it more difficult to detect a possible volcanic signal. There is a distinct drop in temperature during the first winter following an equatorial eruption (Figure 15). However, there is no such drop in temperature following a mid-latitude eruption (Figure 16).

Significance Tests
The fact that the regional Arctic analysis identified much the same volcanic signals as those of the national study is a supportive indication of significance. A more rigorous small-sample test, however, is desirable.

The Student t-test was applied to the multiple eruption composites to determine whether the sample mean, or the mean of the one or two years during which the volcanic signal is evident, is significantly different than the population mean, or the mean of the nine years from which it was taken. Mass and Schneider (1977) applied this test to volcanic dust veil composites for northern hemisphere stations. The basic formula used was:

$$t = \frac{(\bar{x} - \mu)\sqrt{N}}{\sigma}$$

where, $\bar{x}$ = sample mean
μ = population mean
σ = population standard deviation
and, N = average number of stations, and x = number of eruptions in the composite.

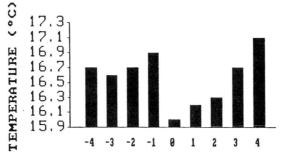

Figure 11: Dust veil summer season temperature composite. All Canadian station events.

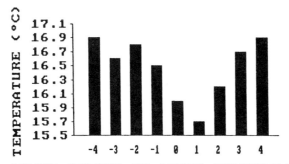

Figure 12: Dust veil summer season temperature composite. All Canadian station events.

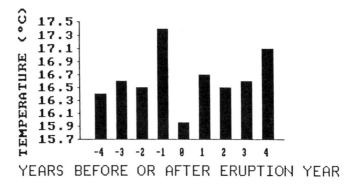

Figure 13: Dust veil summer season temperature composite. All Canadian station events.

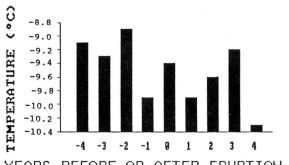

Figure 14: Dust veil winter season temperature composite. All Canadian station events.

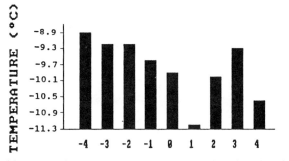

Figure 15: Dust veil winter season temperature composite. All Canadian station events.

Figure 16: Dust veil winter season temperature composite. All Canadian station events.

Thus, for a mid-latitude composite for all stations

$$N = 18 \times 3 = 54$$

The degrees of freedom (Gregory 1963) are

$$d.f. = (n-1) + (n-1)$$
$$= n_a + n_b - 2$$

where, n_a = number of population years
n_b = number of sample years
d.f. (1 sample year) = 8
d.f. (2 sample years) = 9

Table 5: Student t-Test Calculations for Composites Having an Apparent Volcanic Signal.

Figure	Composite		$\bar{x}$	Years	μ	σ	N	d.f.	t	α
7	All Stations All Events		3.70	(0,1)	3.91	0.21	108	9	10.4	0.001
8	All stations Equatorial Events		2.91	(1)	3.71	0.37	51	8	15.2	0.001
9	All Stations Mid-Latitude Events		3.68	(0)	4.13	0.25	54	8	13.2	0.001
10	Arctic	(a)	-11.97	(0,1)	-11.34	0.44	8	9	4.1	0.01
		(b)	-12.15	(0)	-	-	-	8	5.2	0.001
11	Summer All Stations All Events		16.09	(0,1)	16.57	0.33	108	9	15.1	0.001
12	Summer All Stations Equatorial Events		15.93	(1,2)	16.48	0.41	51	9	9.6	0.001
13	Summer All Stations Mid-Latitude Events		15.95	(0)	16.64	0.39	54	8	13.0	0.001
15	Winter All Stations Equatorial Events		-10.63	(1,2)	-9.76	0.72	51	9	8.6	0.001

The problem encountered earlier concerning the low number of stations and eruptions when dealing with the Arctic composites needs to be discussed. The fewer the stations and eruptions used, the greater the difference between the means must be, in order to attain a given level of significance. Table 5 shows the calculated Student-t values and the associated levels of significance for all composites where a volcanic signal was apparent. The test results are similar to those of Mass and Schneider (1977). In all cases, there is a difference between the sample population and the entire set at a significance level ($\propto$) of at least 0.01.

Conclusions

Climatic variation is complex and influenced by many factors. It is therefore difficult to clearly identify possible volcanic influences. As a result, caution must be exercised when interpreting apparent historical evidence and using it to predict future events. However, the results of this study do provide some evidence of the effects of volcanic dust veils on surface temperatures. This allows some tentative conclusions to be made.

The magnitude of the annual temperature drop, for all Canadian stations, was at least 0.5°C greater after the equatorial eruptions analyzed than after the mid-latitude eruptions analyzed. The average total DVI for the selected equatorial eruptions was 933, while it was 220 for the mid-latitude events. The mid-latitude temperature depression was about 0.4°C, occurring during the eruption year and lasting no longer. The equatorial signal of about 1.0°C occurred in the first year after the eruption year and persisted to a lesser degree, for another year or so.

The annual temperature drop in the Arctic was slightly greater than that for the country as a whole. It was approximately 1.0°C, and occurred in both the eruption year and the year following. The lower significance levels for the Arctic signals reflects the small number of stations and events used. Further investigation of this region, using more stations and events, might be appropriate.

Temperature signals were stronger in the summer than in the winter. In addition, the summer drops in temperature were, in almost all cases, of a greater magnitude than the annual drops. There was a marked drop in winter temperature of about 1.0°C in the year following an equatorial eruption.

This investigation did not take trends of temperature into account. No technique, other than the compositing of several volcanic events, was used to eliminate trends. The first four eruptions selected occurred during hemispheric-warming trends, whereas the latter two occurred during hemispheric-cooling trends. An accurate assessment of volcanic dust veil signals would eliminate these trends before applying the compositing technique. The temperature results found in this investigation are quantitatively similar to the empirical results found by Mass and Schneider (1977) and Taylor *et al.* (1980) and to the theoretical results of Pollack *et al.* (1976).

Acknowledgements

This project was undertaken in the Applications and Impact Division of the Canadian Climate Centre, Environment Canada. Mr. M.O. Berry provided project supervision.

References

Gregory, S. 1963. *Statistical Methods and the Geographer.* Longmans, Green and Co. Ltd., London. 240 pp.

Lamb, H.H. 1970. Volcanic dust in the atmosphere; with a chronology and assessment of its meteorological significance. *Philosophical Transactions of the Royal Society, London* 266:425-533.

Mass, C. and S.H. Schneider. 1977. Statistical evidence on the influence of sunspots and volcanic dust on long-term temperature records. *Journal of Atmospheric Science* 34:1995-2004.

Oliver, R.C. 1976. On the response of hemispheric mean temperature to stratospheric dust: an empirical approach. *Journal of Applied Meteorology* 15:933-950.

Panofsky, H.A. and G.W. Brier. 1965. *Some Applications of Statistics to Meteorology.* First Edition. Pennsylvania State University. pp. 159-161.

Pollack, J.B., O.B. Toon, C. Sagan, A Summers, B. Baldwin and W. Van Camp. 1976. Volcanic eruptions and climatic change: a theoretical assessment. *Journal of Geophysical Research* 81:1071-1083.

Schneider, S.H. and C. Mass. 1975. Volcanic dust, sunspots and temperature trends. *Science* 190:741-746.

Taylor, B.L., T. Gal-Chen and S.H. Schneider. 1980. Volcanic eruptions and long-term temperature records: an empirical search for cause and effect. *Quarterly Journal of the Royal Meteorological Society* 106:175-199.

Wendler, G. 1984. Effects of the El Chichón volcanic cloud on solar radiation received at Fairbanks, Alaska. *Bulletin of the American Meteorological Society* 65:216-218.

Northern Hemisphere

North America

Climate of 1816 and 1811-20 as Reconstructed from Western North American Tree-Ring Chronologies

J.M. Lough[1]

Abstract

Reconstructed temperature and sea-level pressure anomalies are presented for the year 1816 and the decade 1811-20. The reconstructions were developed from western North American semi-arid site tree-ring chronologies. The reconstructed climatic conditions for North America and the North Pacific were not very anomalous for either 1816 or 1811-20. More unusual conditions were reconstructed in years other than 1816 between 1811-20, and for decades other than 1811-20 in the first half of the nineteenth century. The factors responsible for the unusual climatic conditions of the "year without a summer" do not appear to have affected surface climate of western North America to the extent that these conditions are translated into the climatic reconstructions.

Introduction

The exceptionally large eruption of Tambora in April 1815 has frequently been speculated to have been the cause of the unusual climatic conditions experienced in 1816 - "the year without a summer". Anomalous weather was recorded in that year in eastern North America and Europe (Milham 1924; Rampino and Self 1982; Stommel and Stommel 1983; Stothers 1984; and elsewhere in this volume). The extent of climatic anomalies outside of the regions bordering the North Atlantic has not, as yet, been appraised satisfactorily.

Although empirical studies have provided evidence of large-scale area-averaged surface temperature decreases following major volcanic eruptions (e.g., Oliver 1976; Taylor *et al.* 1980; Self *et al.* 1981; Kelly and Sear 1984; Sear *et al.* 1987) and the results of a variety of models have supported the role of volcanic eruptions as a source of thermal forcing (e.g., Hunt 1977; Robock 1981; Gilliland 1982; Gilliland and Schneider 1984), the importance of volcanic eruptions (such as Tambora in 1815) as a major source of climatic variability is still disputed (e.g., Deirmendjian 1973; Landsberg and Albert 1974; Parker 1985; Ellsaesser 1986). Difficulties in assessing the role of volcanic eruptions in climatic variability arise for a number of reasons. Theoretical (e.g., Baldwin *et al.* 1976; Pollack *et al.* 1976) and empirical studies (e.g., Rampino and Self 1982, 1984) indicate that the amount of sulphate aerosols produced by an eruption is of more importance than the amount of silicate ash in determining the subsequent climatic impact. Unfortunately, most historical chronologies of volcanic eruptions (e.g., Lamb 1970; Hirschboeck 1979-80; Newhall and Self 1982) do not provide measures of sulphate aerosols, only of the explosive magnitude of the eruptions, which is often assessed by the amount of ash produced. Acidity profiles from ice cores (e.g., Hammer *et al.* 1980; Legrand and Delmas 1987) can provide records of eruptions that produced considerable amounts of sulphuric acid aerosols. The ice-core records tend, however, to be biased towards eruptions occurring at higher latitudes at the expense of those occurring at lower latitudes, and so such records tend to be incomplete.

---

[1] Australian Institute of Marine Science, PMB 3, Townsville M.C., Queensland 4810, Australia.

Other problems result from the small number of possibly climatically important volcanic eruptions that have occurred during the period for which extensive instrumental climatic records are available. The small sample size limits the statistical inferences that can be made regarding the impact of volcanic eruptions on climate. Consequently, most empirical studies have examined temperature series averaged over zonal or hemispheric space scales and little attention has been given to the possible regional variations of a climatic response. For periods before the mid-nineteenth century, instrumental records can provide information for geographically limited regions, usually those bordering the North Atlantic (e.g., Angell and Korshover 1985). For periods prior to the introduction of widespread instrumental climatic records we must rely on proxy climatic information from documentary, geological and biological sources. Lough and Fritts (1987), for example, identified a possible spatial response of North American temperatures to low-latitude volcanic eruptions. The response comprised warming in the western states and cooling in the central and eastern states. This study was based on temporally and spatially detailed reconstructions of North American temperatures derived from western North American tree-ring chronologies, and covered the period from 1602 to 1900 A.D. Some verification of the reconstructed climatic response was provided by independent sources of proxy climatic information both within and outside of the study area. This is important as each proxy climatic record is an imperfect record of past climate. Each series contains bias and error terms which may be unrelated to climate. In addition, different series may respond to different climatic variables, in different seasons and with different frequency responses. The most comprehensive description and understanding of past climatic variations (and their possible causes) will, therefore, only be obtained by the careful comparison and integration of independent sources of information (e.g., National Academy of Science 1975; National Science Foundation 1987).

As a contribution to the improved description and understanding of the climate of 1816 and the decade 1811-20, I present reconstructions of seasonal climate for North America and the North Pacific developed from western North American tree-ring width chronologies.

Data

The reconstructions used in this study were developed by H.C. Fritts and co-workers at the Laboratory of Tree-Ring Research, Tucson, Arizona, following the methods outlined by Fritts *et al.* (1979) and described in detail by Fritts (in press). Only a general description of some of the characteristics of these reconstructions is given here. Fritts (1976), Hughes *et al.* (1982) and Stockton *et al.* (1985) describe the general principles and procedures applied in dendroclimatology.

An array of 65 low-altitude, semi-arid site tree-ring chronologies (Figure 1; Fritts and Shatz 1975) was used to estimate, by canonical regression, seasonal values of temperature at 77 stations, precipitation at 96 stations in the United States and southwestern Canada, sea-level pressure at 96 stations in the United States and southwestern Canada and sea-level pressure at 96 gridpoints between 100°E and 80°W, 20°N and 70°N. Because of the general west to east movement of weather systems across North America it was possible to attempt reconstruction of climate outside the area covered by the tree-ring predictor grid (see also Kutzbach and Guetter 1980). The temperature and precipitation models were calibrated over the period 1901-63, and those for sea-level pressure from 1899-1963. The temperature and precipitation estimates were verified with data independent of that used for model calibration. The general form of the final sea-level pressure models was verified using a subsample replication technique (Gordon 1982).

The final reconstructions, representing the average of the two or three best-calibrated and verified models, were for each variable, station or gridpoint and season for the years from 1602 to 1961. The seasons were December to February (DJF), March to June (MAMJ), July to August (JA) and September to November (SON). The annual series were the average of the four seasonal reconstructions and, therefore, were from December to November. In retrospect, the use of the four- and two-month seasons has proved a drawback in comparing these reconstructions to other sources of information.

The annual calibration and verification statistics can provide some insight into the reliability of these reconstructions. More than 30% of the temperature variance was explained over North America with values exceeding 50% over much of the central United States (Figure 2). Most of the region also showed reliability through positive reduction of error (RE) statistics and the majority of verification tests passed. Positive values of RE indicate that, over the verification period, the estimates are an improvement over simply assuming mean climatic conditions (Gordon 1982). Areas of poor temperature reliability occurred in the northeastern United States, Florida and parts of Nevada and Colorado.

Generally, less variance was calibrated for precipitation than temperature (Figure 2), with more than 30% variance explained only in an area extending along the eastern edge of the tree-ring predictor grid. Verification of the precipitation estimates was also poor over much of the region. The precipitation reconstructions appeared to be of much lower reliability throughout much of the southern and eastern United States. The canonical regression transfer function (which is based on matching of the large-scale patterns of climate and tree-growth represented by the major principal components of the respective grids) does not appear to be well-suited to the reconstruction of precipitation. This variable is dominated by small-scale processes and variability that are not well captured by this regression technique. This is despite the fact that the tree-ring chronologies used are most directly sensitive to precipitation (Fritts 1974).

The calibration and verification statistics for sea-level pressure (Figure 3) show that more than 30% of the variance was calibrated over a large part of the grid. The reconstructions tended to be least reliable over northeastern Asia - the area farthest removed from the tree-ring predictor sites.

Other general features of these reconstructions were (Fritts, in press): (a) sea-level pressure tended to be biased towards lower frequency climatic variations at the expense of high frequency, year-to-year variations; (b) autumn climate was poorly reconstructed for all three variables; (c) precipitation was least well reconstructed, and temperature was probably the most reliably reconstructed; (d) all reconstructions deteriorated in reliability downstream from the tree-ring predictor grid over eastern North America (where Atlantic influences outweigh those of the Pacific) and, for sea-level pressure, over eastern Asia; (e) the large-scale regional patterns of climatic variation were calibrated at the expense of precision at individual stations or gridpoints; and (f) the reliability of the reconstructions was enhanced by averaging over space and filtering through time.

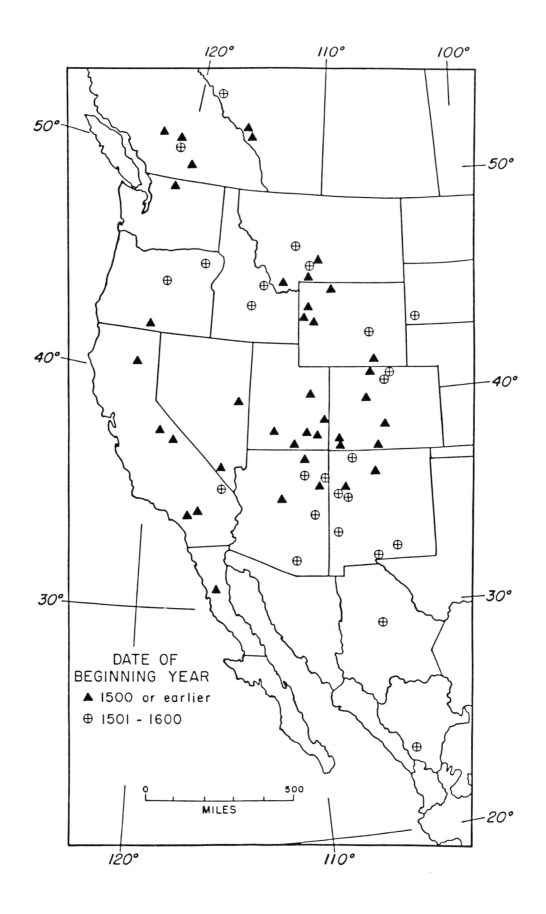

Figure 1: Locations of 65 semi-arid site tree-ring chronologies in western North America.

A full description of these reconstructions and their development is provided by Fritts (in press). The reconstructions have been applied in a number of studies into the nature of climatic variations in North America and the North Pacific and also compared with independent sources of climatic information (e.g., Fritts and Lough 1985; Gordon *et al.* 1985; Lough and Fritts 1985, 1987; Lough *et al.* 1987). These studies, together with analyses of the reconstructions themselves (Fritts in press), have provided insights into the strengths and weaknesses of this particular set of climatic estimates. In the words of H.C. Fritts (in press): "The specific conclusions regarding the climate from 1602 to 1960 are presented as tentative hypotheses derived from one dendroclimatic analysis and test. They must be compared to data from other independent paleoclimatic sources that can reveal changes on seasonal and decadal time scales with accurate yearly dates".

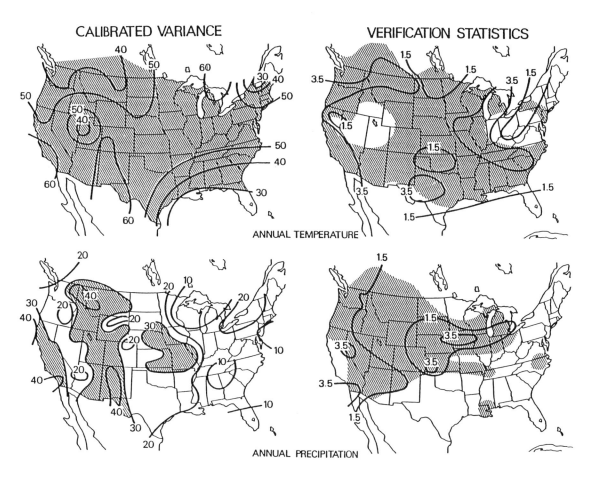

Figure 2: Calibration and verification statistics for annual temperature (top) and annual precipitation (bottom). Percent variance explained with areas of greater than 30% shaded (left-hand figures); number of verification tests passed out of a total of five, and areas with RE statistic greater than zero shaded (right-hand figures).

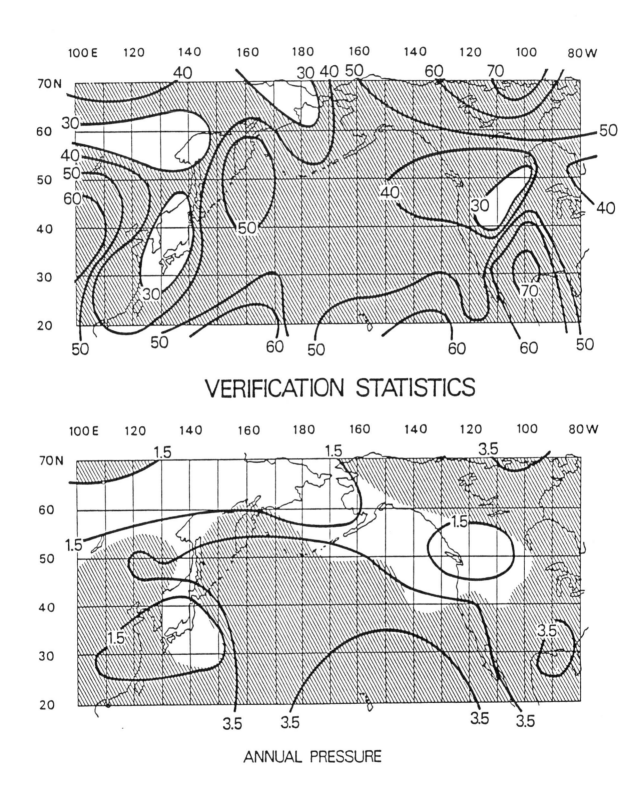

Figure 3: Calibration and verification statistics for annual sea-level pressure. Notation as for Figure 2.

Results

The seasonal and annual reconstructed values of temperature and sea-level pressure for 1816 and 1811-20 were compared with the reconstructed mean climate of 1901-60. The temperature reconstructions were standardized by the 1901-60 standard deviation (s.d.). The reconstructions and original 65 tree-ring chronology series were also compared with the mean of the whole period, 1602 to 1960, to assess how unusual 1816 and 1811-20 were in the longer-term context.

1816

The seasonal and annual reconstructions of temperature and sea-level pressure are presented in Figure 4. In the **winter** of 1815-16, the Aleutian Low was reconstructed to be displaced southeastwards, with slightly higher pressure reconstructed over the Canadian Arctic. Temperatures were reconstructed to be warmer in the western states (associated with enhanced southerly air flow) and cooler over the central and eastern states. Temperature departures up to 2 s.d. below recent-period means were reconstructed over the Great Lakes.

The large sea-level pressure anomalies reconstructed in **spring** over eastern Asia were in an area of low reconstruction reliability and were not, therefore, considered to be significant. The main reconstructed feature was a slight deepening of the Aleutian Low. The reconstructed temperature field did not exhibit very large anomalies, though temperatures were still warmer in the west and cooler in the central states compared to the 1901-60 normals.

Discounting the sea-level pressure anomalies over eastern Asia, the reconstructed sea-level pressure field for **summer** did not show marked departures from the twentieth century mean values. Slightly lower pressure was reconstructed over western Hudson Bay. Temperatures were reconstructed to be slightly above the average over a large part of the United States, with below average conditions in the far western states. Temperatures were reconstructed to be close to the 1901-60 mean over the northeastern United States, the area of extensively documented climatic anomalies for the summer of 1816.

In **autumn**, a positive pressure anomaly was reconstructed over the eastern North Pacific that was linked with the colder temperatures reconstructed in the Pacific Northwest. Elsewhere in the United States, temperatures were reconstructed to be warmer than the 1901-60 mean values by up to 2 s.d. in the northeastern and southern states. However, the reconstructions are least reliable in autumn.

In the **annual** average, discounting sea-level pressure anomalies over Asia, the major reconstructed feature was an area of higher pressure to the west of Hudson Bay. Higher sea-level pressure extended out over the Pacific, and lower pressure was reconstructed to the south. Thus, 1816 seems to have been characterized by a weakened zonal circulation over the North Pacific. Temperatures were reconstructed to be warmer in the western states and cooler in the most southerly states. Although temperatures were reconstructed to be up to 1 s.d. below the 1901-60 mean near the Great Lakes and northeastern United States, the main contribution to this appears to come from the temperatures reconstructed for the winter 1815-16.

1811-20

Figure 5 shows the reconstructed seasonal and annual sea-level pressure and temperature values averaged for the decade of 1811-20. The reconstructions were expressed as departures from the reconstructed mean of 1901-60, and those for temperature were standardized.

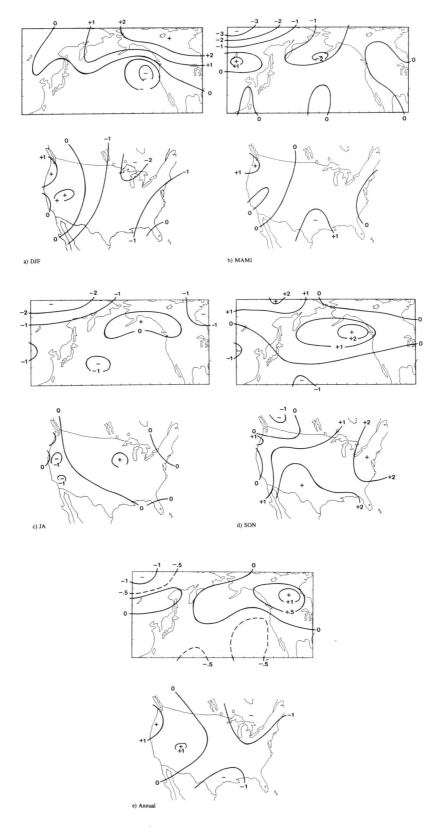

Figure 4: Reconstructed sea-level pressure (mb) and temperatures (s.d. units) expressed as departures from the 1901-60 means the year 1816 for: (a) winter; (b) spring; (c) summer; (d) autumn; and (e) annual data.

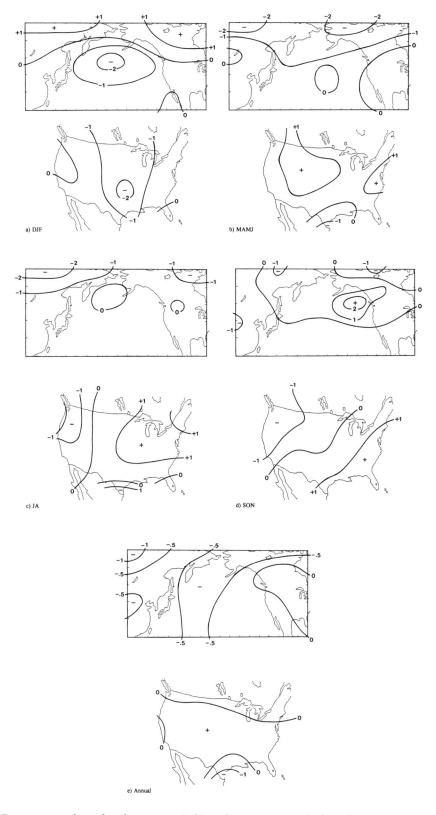

Figure 5: Reconstructed sea-level pressure (mb) and temperatures (s.d. units) expressed as departures from the 1901-60 means for the decade 1811-20 for: (a) winter; (b) spring; (c) summer; (d) autumn; and (e) annual data.

In **winter**, the Aleutian Low was reconstructed to be deeper than the average, with positive sea-level pressure departures over the Canadian Arctic. Temperatures were reconstructed to be cooler than the average through the central United States.

In **spring** a negative sea-level pressure anomaly was reconstructed over Alaska with near-average conditions reconstructed elsewhere. Temperatures were reconstructed to be warmer than the average throughout most of the United States. These departures were significantly different from the 1901-60 mean, at the 5% level, for 73% of the 77 temperature stations.

The **summer** sea-level pressure anomalies were reconstructed to be of small magnitude, with the exception of northeastern Asia. Temperatures were reconstructed to be cooler than average in the northwestern states and generally warmer than average in the central and eastern regions. There was no evidence in these reconstructions of negative temperature anomalies in the eastern United States.

The **autumn** sea-level pressure field was characterized by a positive anomaly in the northeastern North Pacific. Temperatures were reconstructed to be cooler than the average in the northwestern and western regions and warmer in the southeastern and eastern regions.

In the **annual** average, the sea-level pressure anomalies (outside of Asia) were estimated to be of small magnitude. Slightly below average pressure was found over the North Pacific. Temperatures were reconstructed to be slightly warmer than the average over most of the United States, though at only 5% of the 77 stations were these values significantly different from the 1901-60 mean values.

Thus, the climate of 1811-20, as reconstructed from western North American tree-ring chronologies, did not appear to be particularly anomalous when compared to the mean climate of 1901-60. In the annual average, sea-level pressure was slightly lower and temperature slightly higher than the 1901-60 mean, but none of these departures was very large.

The reconstructed climate of 1811-20 was compared with that reconstructed for the other four decades of the first half of the nineteenth century (Figure 6). These data were expressed as departures from the instrumental record mean of 1901-70, and precipitation was included, expressed as a percentage of the mean. In this context, 1811-20, appeared to have been the least unusual of the five decades. Extensive cooling was, for example, reconstructed in 1821-30, 1831-40 and 1841-50. Similarly, sea-level pressure anomalies of greater than 1 mb were evident in all decades except 1811-20. The climate as reconstructed from the western North American tree-ring chronologies for the decade 1811-20 was not very different from the recent mean conditions. More extreme climatic conditions were reconstructed for other decades in the first half of the nineteenth century.

Comparisons with 1602-1960 Mean Conditions
In the preceding sections the reconstructed climate of 1816 and 1811-20 was compared to recent, twentieth century mean conditions. The reconstructed climate did not appear to be very different from this mean. I examined the data with respect to the long-term 1602-1960 reconstruction mean. I also considered the nature of the anomalies of the original tree-ring chronologies which were used to develop these climatic reconstructions.

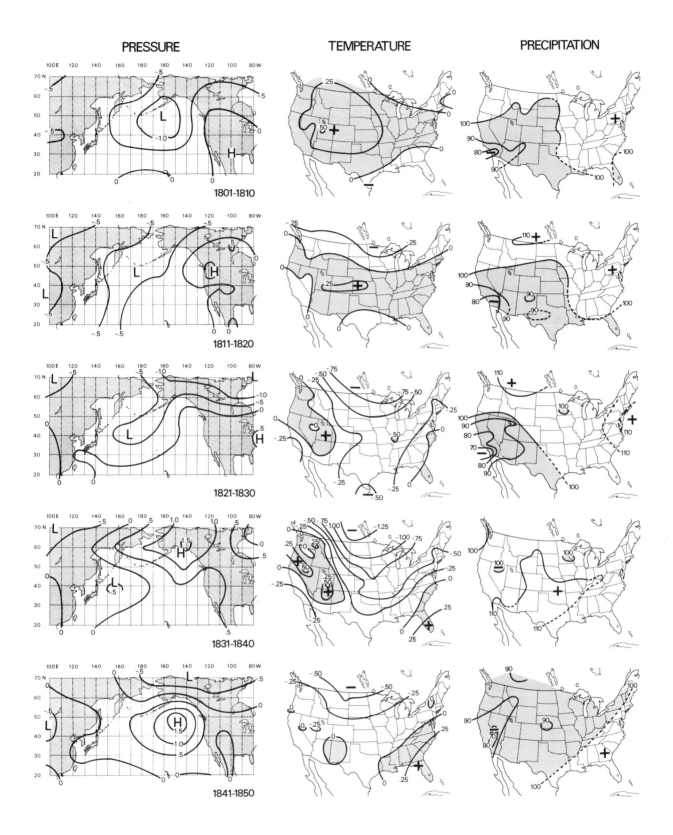

Figure 6: Reconstructed annual sea-level pressure (mb), temperatures (°C) and precipitation (percent of mean) expressed as departures from the 1901-70 instrumental record means for the first five decades of the nineteenth century. Dashed contour lines for the precipitation maps are through areas where the verification statistics indicate that the reconstructions are unreliable.

The percentage of the 77 annual temperature stations and 65 tree-ring chronologies with departures of +1 s.d. and -1 s.d. of the 1602-1960 mean were calculated for each year of the decade 1811-20 (Table 1). The reconstructed temperature field was close to average conditions with only 3% of the stations with reconstructed values ±1 s.d. of the mean in 1816. The years 1811, 1818 and 1819 all had more than 45% of stations with departures ±1 s.d. from the mean. In 1811, the departures were about equally above and below the mean, but in 1818 and 1819, they were mainly positive, indicating warmer conditions.

Forty percent of the original tree-ring chronologies had departures of at least ±1 s.d. of the 1602-1960 mean in 1816, though this was not the most extreme year of the decade. The most extreme years were 1819 with 48% and 1818 with 42% of the 65 sites with departures ±1 s.d. of the mean. For the last two years, the departures were mainly negative, indicating that conditions were generally unfavourable for tree growth. In contrast, in 1816, 35% of the 65 stations had departures of at least +1 s.d. of the mean, indicating conditions were generally favourable for wide growth-ring formation in western North America. This was the most favourable year for the tree growth of the decade 1811-20. The term favourable for tree growth cannot be simply interpreted, as the 65 chronologies cover a range of tree species from different sites in western North America. Factors influencing the width of the annual growth ring vary considerably, and can also operate over a number of growing seasons (Fritts 1976). For semi-arid sites, wider annual rings are often, however, associated with moister and cooler conditions near the trees.

The decade mean for each reconstructed variable and the tree-ring chronologies were compared to the long-term mean for 1602-1960 for each decade between 1602-10 (1602 was the first year of the reconstructions) and 1951-60 (Table 2). Evidently 1811-20 was not particularly unusual in these data. For temperature, 12% of the stations had departures significantly different from the long-term mean in 1811-20 compared to 62% of stations in the most extreme decade of 1681-90. For sea-level pressure, 1811-20 had 33% of the 96 gridpoints with significant departures compared to the most unusual decade of 1881-90 with 66%. None of the 96 precipitation stations was reconstructed to have values significantly different from the long-term mean in 1811-20, compared to 70% in 1611-20. For the original tree-ring chronologies, 19% were significantly different in 1811-20, compared to 60% for the most extreme decade of 1911-20.

Table 1: Percentage of 77 Temperature Stations and 65 Tree-Ring Chronologies with Values ± 1 Standard Deviation of 1602 to 1960 Mean for Each Year of the Decade 1811-20.

	Annual Temperature		
Year	s.d. +1	s.d. -1	s.d. ±1
1811	29	25	53
1812	27	10	37
1813	9	1	10
1814	0	8	8
1815	14	12	26
1816	3	0	3
1817	10	0	10
1818	47	0	47
1819	38	8	46
1820	26	7	33

	Tree-Ring Chronologies		
	s.d. +1	s.d. -1	s.d. ±1
1811	15	6	22
1812	9	6	15
1813	6	23	29
1814	12	11	23
1815	14	9	23
1816	35	5	40
1817	22	5	27
1818	11	31	42
1819	14	34	48
1820	8	31	39

Table 2: Percentage of Stations, Gridpoints or Chronologies for Which Decade Mean is Significantly Different from Long-Term (1602-1960) Mean at the 5% Significance Level for Reconstructed Temperature (T), Sea-Level Pressure (SLP), Precipitation (PPT) and Tree-Ring Chronologies (TREES).

DECADE	T	SLP	PPT	TREES
1602-1610	16	23	19	25
1611-1620	40	21	70	40
1621-1630	52	53	40	28
1631-1640	22	50	52	20
1641-1650	17	31	48	23
1651-1660	12	52	5	19
1661-1670	49	2	19	28
1671-1680	39	46	5	22
1681-1690	62	51	17	9
1691-1700	1	11	4	14
1701-1710	3	9	2	14
1711-1720	0	15	0	9
1721-1730	0	0	2	11
1731-1740	0	16	3	28
1741-1750	5	16	0	23
1751-1760	9	1	30	20
1761-1770	23	8	1	22
1771-1780	30	71	1	26
1781-1790	6	21	0	20
1791-1800	31	17	7	23
1801-1810	31	24	0	22
1811-1820	**12**	**33**	**0**	**19**
1821-1830	12	11	6	23
1831-1840	39	18	41	42
1841-1850	12	14	36	29
1851-1860	0	8	1	11
1861-1870	38	17	40	20
1871-1880	3	15	7	25
1881-1890	32	66	4	19
1891-1900	1	30	0	15
1901-1910	6	52	4	17
1911-1920	47	47	24	60
1921-1930	19	5	5	35
1931-1940	55	38	27	32
1941-1950	14	36	21	31
1951-1960	19	48	-	39
#	77	96	96	65

Summary and Conclusions

As reconstructed by western North American semi-arid site tree-ring chronologies, the climate of North America and the North Pacific does not appear to have been very unusual in 1816 or the decade 1811-20. This is when compared to both the 1901-60 and the 1602-1960 reconstructed data means.

For winter, spring and the annual average of 1816, temperatures were reconstructed to be cooler in the eastern and central United States and warmer in the western United States. In summer and autumn of 1816, temperatures were reconstructed to be warmer in the central and eastern regions and cooler in the west. The pattern of temperature departures for winter, spring and the annual average are similar to the average pattern identified by Lough and Fritts (1987) to characterize the years 0 to 2 after eight low-latitude volcanic eruptions between 1602 and 1900. The Tambora eruption of 1815 was one of eight eruptions used in that analysis. The summer temperature field for 1816 does not resemble the average pattern identified by Lough and Fritts (1987). In the annual average there was reconstructed to be a weakening of the westerly zonal flow pattern over the North Pacific. Sea-level pressure anomalies were not, however large. Thirty-five percent of the original tree-ring chronologies had growth departures of +1 s.d. or more above the 1602-1960 mean in 1816, indicating that conditions, at least in parts of western North America were generally favourable for wide tree-ring formation.

The decade 1811-20, in the annual average, was reconstructed to be slightly warmer than the 1901-60 mean over North America, with lower sea-level pressure reconstructed over the North Pacific. It was, perhaps, the least unusual of the first five decades of the nineteenth century. Relatively large negative temperature departures were reconstructed over the central northern United States in 1821-30, 1831-40 and 1841-50. The decade 1811-20 did not appear to be very unusual when compared to long-term mean conditions for any of the reconstructed variables nor the original tree-ring chronologies.

The evidence from this particular set of climatic reconstructions from western North American semi-arid site tree-ring chronologies is for near-normal climatic conditions in 1816 and 1811-20. Reconstructed climatic anomalies were small in magnitude when compared to the recent, 1901-60, and long-term 1602-1960, mean conditions. Most references to the "year without a summer" in North America tend to come from eastern regions. Because this particular set of reconstructions is known to be less reliable in the east, where Atlantic and Arctic influences outweigh those of the Pacific, the lack of large reconstructed anomalies in this region was not surprising. What was surprising was a lack of evidence for large-magnitude climatic anomalies in areas where the reconstructions are known to be reliable, over the western United States and the North Pacific. Analysis of the original tree-ring chronology series suggested that 1816 was a year favourable for tree-growth in parts of the western states, possibly associated with moister and cooler conditions. Large-scale climatic anomalies are not, however, apparent in the climatic reconstructions from these tree-ring data. This suggests that whatever the nature of the anomalies of climate in 1816 and the decade 1811-20, they were not large enough to significantly influence climatic conditions in the western United States either for good or bad.

Acknowledgements

This study is based on the results of many years of work by Hal Fritts and co-workers at the Laboratory of Tree-Ring Research, University of Arizona.

References

Angell, J.K. and J. Korshover. 1985. Surface temperature changes following the six major volcanic events between 1780 and 1980. *Journal of Climate and Applied Meteorology* 24:937-951.

Baldwin, B.B., J.B. Pollack, A. Summers, O.B. Toon, C. Sagan and W. van Camp. 1976. Stratospheric aerosols and climatic change. *Nature* 263:551-555.

Deirmendjian, D. 1973. On volcanic and other particulate turbidity anomalies. *Advances in Geophysics* 16:267-296.

Ellsaesser, H.W. 1986. Comments on "Surface temperature changes following the six major volcanic episodes between 1780 and 1980". *Journal of Climate and Applied Meteorology* 25:1184-1185.

Fritts, H.C. 1974. Relationships of ring widths in arid-site conifers to variations in monthly temperature and precipitation. *Ecological Monographs* 44:411-440.

Fritts, H.C. 1976. *Tree Rings and Climate*. Academic Press, London. 567 pp.

Fritts, H.C. (in press). *Reconstructing Large-Scale Climatic Patterns from Tree-Ring Data: A Diagnostic Study*. University of Arizona Press.

Fritts, H.C. and D.J. Shatz. 1974. Selecting and characterizing tree-ring chronologies for dendroclimatic analysis. *Tree-Ring Bulletin* 35:31-40.

Fritts, H.C. and J.M. Lough. 1985. An estimate of average annual temperature variations for North America, 1602-1961. *Climatic Change* 7:203-224.

Fritts, H.C., G.R. Lofgren and G.A. Gordon. 1979. Variations in climate since 1602 as reconstructed from tree rings. *Quaternary Research* 12:18-46.

Gilliland, R.L. 1982. Solar, volcanic and CO_2 forcing of recent climatic changes. *Climatic Change* 4:111-131.

Gilliland, R.L. and S.H. Schneider. 1984. Volcanic, CO_2 and solar forcing of northern and southern hemisphere surface temperatures. *Nature* 310:38-41.

Gordon, G.A. 1982. Verification of dendroclimatic reconstructions. *In: Climate from Tree Rings*, M.K. Hughes, P.M. Kelly, J.R. Pilcher and V.C. LaMarche (eds.). Cambridge University Press, Cambridge. pp. 58-61.

Gordon, G.A., J.L. Lough, H.C. Fritts and P.M. Kelly. 1985. Comparison of sea-level pressure reconstructions from western North American tree rings with a proxy record of winter severity in Japan. *Journal of Climate and Applied Meteorology* 24:1219-1224.

Hammer, C.U., H.B. Clausen and W. Dansgaard. 1980. Greenland ice sheet evidence of post-glacial volcanism and its climatic impact. *Nature* 228:230-235.

Hirschboeck, K.K. 1979-80. A new worldwide chronology of volcanic eruptions. *Palaeogeography, Palaeoclimatology and Palaeoecology* 29:223-241.

Hughes, M.K., P.M. Kelly, J.R. Pilcher and V.C. LaMarche. (Editors). 1982. *Climate from Tree Rings*. Cambridge University Press, Cambridge. 223 pp.

Hunt, B.G. 1977. A simulation of the possible consequences of a volcanic eruption on the general circulation of the atmosphere. *Monthly Weather Review* 105:247-260.

Kelly, P.M. and C.B. Sear. 1984. Climatic impact of explosive volcanic eruptions. *Nature* 311:740-743.

Kutzbach, J.E. and P.J. Guetter. 1980. On the design of paleoenvironmental networks for estimating large-scale patterns of climate. *Quaternary Research* 14:169-187.

Lamb, H.H. 1970. Volcanic dust in the atmosphere: with a chronology and assessment of its meteorological significance. *Philosophical Transactions of the Royal Society of London* 266:425-433.

Landsberg, H.E. and J.M. Albert. 1974. The summer of 1816 and volcanism. *Weatherwise* 27:63-66.

Legrand, M. and R.J. Delmas. 1987. A 220-year continuous record of volcanic H_2SO_4 in the Antarctic ice sheet. *Nature* 327:671-676.

Lough, J.M. and H.C. Fritts. 1985. The Southern Oscillation and tree rings: 1600-1961. *Journal of Climate and Applied Meteorology* 24:952-966.

_____. 1987. An assessment of the possible effects of volcanic eruptions on North American climate using tree-ring data, 1602 to 1900 A.D. *Climatic Change* 10:219-239.

Lough, J.M., H.C. Fritts and Wu Xiangding. 1987. Relationships between the climate of China and North America over the past four centuries: a comparison of proxy records. *In: The Climate of China and Global Climate: Proceedings of the Beijing International Symposium on Climate, 30 October-3 November 1984, Beijing, China*. Ye Duzheng, Fu Congbin, Chao Jiping and M. Yoshino (eds.). China Ocean Press, Springer-Verlag (foreign distributor). New York and Berlin. pp. 89-105.

Milham, W.I. 1924. The year 1816 - the causes of abnormalities. *Monthly Weather Review* 52:563-570.

National Academy of Sciences. 1975. *Understanding Climatic Change*. National Academy Press, Washington, D.C. 239 pp.

National Science Foundation. 1987. *Climate Dynamics Workshop on Paleoclimate Data-Model Interaction*. Climate Dynamics Program. National Science Foundation, Washington, D.C.

Newhall, C.G. and S. Self. 1982. The volcanic explosivity index (VEI): an estimate of explosive magnitude for historical volcanism. *Journal of Geophysical Research* 87C2:1231-1238.

Oliver, R.C. 1976. On the response of hemispheric mean temperature to stratospheric dust: an empirical approach. *Journal of Applied Meteorology* 15:933-950.

Parker, D.E. 1985. Climatic impact of explosive volcanic eruptions. *Meteorological Magazine* 114:149-161.

Pollack, J.B., O.B. Toon, C. Sagan, A. Summers, B.B. Baldwin and W. van Camp. 1976. Volcanic explosions and climatic change: a theoretical assessment. *Journal of Geophysical Research* 81:1071-1083.

Rampino, M.R. and S. Self. 1982. Historic eruptions of Tambora (1815), Krakatau (1883), and Agung (1963), their stratospheric aerosols, and climatic impact. *Quaternary Research* 18:127-143.

_____. 1984. Sulphur-rich volcanic eruptions and stratospheric aerosols. *Nature* 310:677-679.

Robock, A. 1981. A latitudinally dependent volcanic dust veil index and its effect on climate simulations. *Journal of Volcanology and Geothermal Research* 11:67-80.

Sear, C.B., P.M. Kelly, P.D. Jones and C.M. Goodess. 1987. Global surface temperature responses to major volcanic eruptions. *Nature* 330:365-367.

Self, S., M.R. Rampino and J.J. Barbera. 1981. The possible effects of large nineteenth and twentieth century volcanic eruptions on zonal and hemispheric surface temperatures. *Journal of Volcanology and Geothermal Research* 11:41-60.

Stockton, C.W., W.R. Boggess and D.M. Meko. 1985. Climate and tree rings. *In: Paleoclimate Analysis and Modeling.* A.D. Hecht (ed.). John Wiley and Sons, New York. pp. 71-151.

Stommel, H. and E. Stommel. 1983. *Volcano Weather. The Story of 1816, the Year Without a Summer.* Seven Seas Press, Newport, Rhodes Island. 177 pp.

Stothers, R.B. 1984. The great Tambora eruption in 1815 and its aftermath. *Science* 224:1191-1198.

Taylor, B.L., T. Gal-Chen and S.H. Schneider. 1980. Volcanic eruptions and long-term temperature records: an empirical search for cause and effect. *Quarterly Journal of the Royal Meteorological Society* 106:175-199.

Volcanic Effects on Colorado Plateau Douglas-Fir Tree Rings

Malcolm K. Cleaveland[1]

Abstract

The explosion of Tambora in April 1815, is the largest volcanic eruption in recorded history. Based on measured temperatures, the significant North American climatic effects of Tambora appear to have been limited to the northeastern United States and eastern Canada. However, diameter growth of conifers on the Colorado Plateau in the southwestern United States was extremely large from 1815 to 1817, and the largest regionally-averaged late season (latewood) growth in 491 years occurred in 1816. This abnormal growth is probably not coincidental. Above-normal growth occurs when moisture stress is reduced, and the eruption probably resulted in abnormally low growing-season temperatures and/or abundant precipitation over the Colorado Plateau. Latewood density was above average from 1815 to 1817, also indicating reduced moisture stress. Trees located on marginal sites that are usually subject to the greatest moisture stress showed the most favourable growth response from 1815 to 1817. These growth changes are postulated to be effects of abnormally cool growing seasons that reduced evapotranspiration and delayed onset of drought-induced late summer dormancy. Delayed dormancy would favour development of an anomalously large latewood zone and increased latewood density. No comparable growth responses are apparent for other known large eruptions, indicating that regional climatic response to volcanic forcing is highly variable. Long-lived, climatically-responsive trees are widely distributed in the northern hemisphere. Analyses of these tree-ring data during recent centuries when instrumental climatic data are sparse may help reveal the impact of known volcanic eruptions on northern hemisphere climate, and may also help identify and date extremely large prehistoric eruptions.

Introduction

The April 1815 explosion of Mount Tambora at Latitude 8°S was the "largest and deadliest volcanic eruption in recorded history..." (Stothers 1984). Lamb's (1970, Tables 7a,b) Dust Veil Index is larger for 1815 than any year since 1500. The 1815 eruption also has the highest Volcanic Explosivity Index since 1500 (Newhall and Self 1982). If such volcanic eruptions do affect global climate, then this huge eruption should have left evidence in instrumental climatic records and perhaps in other proxy climatic records such as ice cores and tree rings (Shutts and Green 1978; Bryson and Goodman 1980; Gilliland and Schneider 1984; Kelly and Sear 1984; Angell and Korshover 1985; Bradley 1988).

Volcanic aerosols reduce solar radiation at the surface on a regional or global basis by increasing the albedo of the upper atmosphere. Temperature effects of aerosols are more closely related to the quantity of sulphates injected into the upper atmosphere than to the quantity of fine particulate ejecta. Sulphuric acid created as a result of sulphur-rich volcanic eruptions plays a major role in reducing transmission of direct solar radiation (Harshvardhan and Cess 1976; Pollack et al. 1976). If the Tambora eruption had a global climatic impact, it must have created a sulphuric acid aerosol. In fact, the 1815 Tambora episode coincides with a very large acidity peak from 1815

---

[1] Department of Geography, University of Arkansas, Fayetteville, Arkansas 72701, U.S.A.

to 1818 in Greenland ice cores, even before correction for losses in transport from equatorial to polar latitudes (Hammer *et al.* 1980; Rampino and Self 1984; Stothers 1984).

Despite abundant evidence pointing to the 1815 Tambora eruption as having all the requisite characteristics for a major influence on global climate, Stothers (1984) and Angell and Korshover (1985) found little evidence of significantly lower surface temperatures following this eruption. Most of the available long instrumental-temperature records that form the basis of this conclusion, however, are confined to Europe and the northeastern United States. In addition, available records show that a period of well-below-average temperature started at least five years before 1815, which may mask some of the volcanic effects (Stothers 1984; Angell and Korshover 1985; Baron, this volume). While 1816 is cooler than 1815 in most of these records, the drop is small compared to the general cooling shown in the period. The long temperature record from New Haven, Connecticut does show signs of considerable cooling, in keeping with the New England reputation of 1816 as "the year without a summer" (Stommel and Stommel 1983; Angell and Korshover 1985), but the northeastern United States and eastern Canada (Wilson 1985) are considered exceptions. Horstmeyer (1989) compiled Cincinnati, Ohio daily weather records from 1814 to the present and found that, "No year since [1816] has even come close to having such a cold summer". This demonstrates for the first time that an abnormally cold summer of 1816 occurred in the American Midwest, as well as in eastern North America.

There is a large and growing network of old, climate-sensitive tree-ring chronologies available that might provide more spatially complete evidence for volcanic effects on climate (Stockton *et al.* 1985). One such network from western North America has been used to estimate seasonal and annual temperature variation for the United States (Fritts and Lough 1985; Lough, this volume). The reconstructed temperature estimates were then used to study average climatic response to selected volcanic events (1602 to 1900) by comparing temperature before and after the events with superimposed epoch analysis (Lough and Fritts 1987). After low-latitude eruptions like Tambora, cooling was especially pronounced in the spring and summer during the trees' growing season. The spatial effects on United States climate were often different, and directly out of phase, between the west coast and the rest of the country. During the spring following eruptions, the Pacific Northwest experienced warming while the rest of the country cooled down. In summer the area of warming expanded down the west coast, while the central and eastern United States remained relatively cool. Lough and Fritts (1987), however, did not specifically report on the Tambora eruption. In this paper a set of climate-sensitive tree-ring width and density chronologies from the Colorado Plateau are used to investigate the possible impact of the 1815 eruption on the climate of the southwestern United States.

Methods

Three sets of Douglas-fir (*Pseudotsuga menziesii*) tree-ring radial samples were collected and crossdated with standard methods (Stokes and Smiley 1968). The total number of radii in the collections ranged from 14 to 20, taken from 10 to 14 trees per site. The samples were collected at Ditch Canyon (DIT) on the Colorado/New Mexico border, at Mesa Verde National Park in Spruce Canyon (SPC) during 1978 (Cleaveland 1983, 1986, 1988) and in Bobcat Canyon (BOB) in 1972 (Drew 1976) (Figure 1).

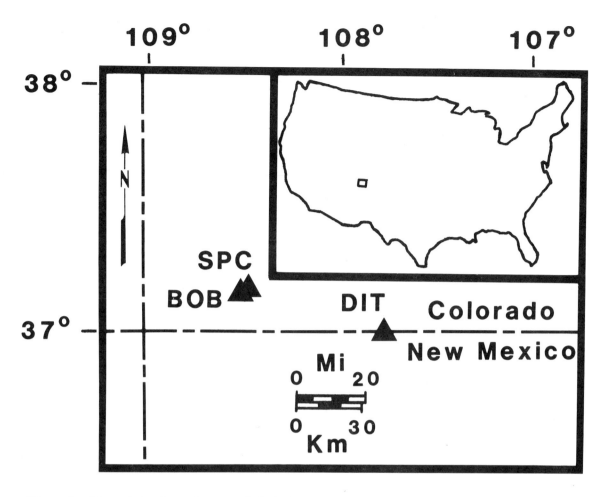

Figure 1: Map of the three sites sampled: Bobcat Canyon (BOB), Spruce Canyon (SPC), and Ditch Canyon (DIT).

Conifer tree rings are divided into earlywood and latewood zones, formed first and last in the growing season, respectively. Earlywood formation is most strongly influenced by climate in the spring and latewood formation by late spring to mid-summer climatic conditions. Typical earlywood cells become large, with relatively thin cell walls surrounding large cavities or lumens. Typical latewood cells are smaller than earlywood cells, with thick cell walls and small lumens. For this reason the earlywood part of a ring is less dense than the latewood portion. The transition to latewood is abrupt in Douglas-fir (Panshin and de Zeeuw 1970). The width of the two zones was measured optically from the BOB and DIT specimens. Characteristics of the SPC samples, including latewood width and average latewood density, were measured by X-ray densitometry (Parker *et al.* 1980; Cleaveland 1983, 1986).

Time series of ring widths are often not statistically stationary because the mean and variance may both change with increasing age and diameter of the tree. The most common form of the growth function approximates an exponential curve declining to a constant value, but linear regression lines or more flexible polynomial or spline curves are also often used to remove growth trend (Stokes and Smiley 1968; Fritts 1976; Cook and Peters 1981). To transform the measurements into stationary time series, a curve is fitted to the measurements from each sample, and each annual value is divided by the corresponding annual curve value. This transforms measurement series into indices with a mean of 1.0, removing the effects of differences in mean growth from tree to tree, and rendering the variance quasi-stationary. The indices for each radial series from a site are averaged on an annual basis into a site chronology. The site chronology has a mean equal to 1.0 and a minimum value greater than 0.0, and represents a selected statistical sample of the macro-environmental factors that control the radial growth of a given species on a certain site through time.

Results and Discussion

BOB and DIT are lower forest-border sites that often experience high levels of moisture stress, whereas the SPC site is more mesic (Drew 1976; Cleaveland 1983, 1986). One measure of response to climate is the mean sensitivity statistic, that is, the average first difference of chronology indices (Fritts 1976). The mean sensitivities of ring-width chronologies at BOB, DIT, and SPC are 0.45, 0.44, and 0.28, respectively. This statistic indicates that the BOB and DIT chronologies should show greater response to departures from normal growing-season conditions than the SPC chronology.

A width or density index greater than 1.0 indicates above average growth that is usually attributable to a cool and/or moist growing season in the Southwest (Fritts *et al*. 1965; Fritts 1976). When latewood width at the BOB, DIT, and SPC sites are averaged for each year, the average index is larger for 1816 than for any year since 1487, a period of almost five centuries (Figure 2). In addition, the ring-width, latewood-width, and latewood-density indices for the three collection sites all equal, or greatly exceed, average growth (1.00) for 1815, 1816, and 1817 (Table 1). These anomalies indicate that the growing seasons were substantially cooler and/or wetter than normal (Cleaveland 1983, 1986).

The very large values of latewood growth in the decade of the 1490s (Figure 2) are probably artifacts of a small number of samples, and end effects of curve fitting. The best replicated chronology, SPC, has an index of only 1.33 in 1491 (Figure 2). If the poorly-replicated 1490s are ignored, 1815 and 1816 summed are the largest average latewood total of two consecutive years, and 1816 and 1817 are the second largest. In addition, if the 1490s are not considered, the 1815-17 period has the largest total latewood growth of three consecutive years.

All chronologies are well replicated after about 1700, giving greater confidence in the estimated index means after 1700. It would certainly be possible to increase the sample depth of long series at many sites in the western United States to improve estimates in the early part of the chronologies. This should be an important consideration before using these chronologies to investigate earlier volcanic eruptions.

Lower forest-border Douglas-fir trees in southwestern Colorado generally become dormant in June or July - forced into dormancy by moisture stress long before photoperiod or low temperatures could become responsible (Fritts *et al*. 1965). Also, a conditional probability analysis of 89 southwestern conifer chronologies (e.g., Stockton and Fritts 1971) indicates that the influence of temperature on tree growth late in the growing season is stronger than the influence of precipitation (Cleaveland, unpublished data). Chronologies at those sites showing the highest degree of inferred moisture stress (BOB and DIT) show a stronger response to the

1815-17 climatic anomaly than the more mesic SPC site. It seems probable, therefore, that the growth anomaly is linked in some way to below-normal temperature and/or above normal precipitation that drastically reduced moisture stress on Colorado Plateau trees during those growing seasons. The greatly enlarged latewood zone in the 1815-17 rings of these Colorado Plateau conifers could be interpreted as evidence for a longer-than-normal growing season extended by below-normal air temperatures during the summer. Normal or above-average precipitation probably also occurred during the extended growing seasons from 1815 to 1817.

Table 1: Southwestern Colorado Tree-Ring Chronology Indices (1810-20).[1]

Year	Bobcat Canyon Ring Width	Bobcat Canyon Latewood Width	Ditch Canyon Latewood Width	Spruce Canyon Latewood Width	Spruce Canyon Latewood[2] Density	Average Latewood Width
1810	0.68	0.70	0.77	1.03	1.01	0.83
1811	1.01	1.07	2.14	1.32	1.05	1.51
1812	1.12	0.87	1.09	0.87	1.01	0.94
1813	0.41	0.36	0.75	0.31	0.88	0.47
1814	0.92	0.92	1.03	0.69	0.97	0.88
1815[3]	1.24	1.73	2.29	1.50	1.06	1.84
1816	2.12	3.47	4.58	1.57	1.10	3.20
1817	2.28	2.23	1.78	1.00	1.02	1.67
1818	0.47	0.42	0.28	0.83	0.97	0.51
1819	0.31	0.77	0.65	0.68	0.90	0.70
1820	0.24	0.32	0.64	0.34	0.86	0.43

[1] Indices greater than 1.0 indicate above-average growth, and indices less than 1.0 represent below-average growth.

[2] Latewood density variability was multiplied by 3.0 to increase the range of variation relative to the other variables.

[3] The year Tambora erupted. Table 1

Other historic eruptions are believed to have affected climate, and might have influenced the growth of trees on the Colorado Plateau. Rampino and Self (1984) list selected eruptions with estimates of the sulphuric acid aerosol generated and the estimated northern hemispheric temperature change. The eruption of Laki in 1783 is estimated to have caused greater cooling than Tambora, but no effect can be detected in Figure 2. Laki is a high-latitude (64°N) volcano, however, and Lough and Fritts (1987) found that volcanic eruptions in low latitudes resulted in the greatest climatic response across the United States.

There appears to have been no growth response of Colorado Douglas-fir to Krakatau (6°S), unless there was a weak effect delayed until 1885. The eruption of Santa Maria (15°N) occurred in 1902, a year of intense drought on the Colorado Plateau (Cleaveland 1983, Appendix 1). The growing season of 1903 had adequate precipitation, and growth was slightly above average, but 1904 was very dry resulting in low growth (Figure 2). The lack of adequate precipitation would certainly curtail the possible response of tree growth to volcanic cooling. Climatic effects from the eruptions of Katmai (58°N) in 1912 and Agung (8°S) in 1963 are also not discernible in these chronologies (Figure 2). The growth anomaly at 1816 is clearly the largest apparent in these data, and is probably the only one that can definitely be attributed to a known major volcanic eruption. However, there are other pronounced increases in growth that may be associated with undocumented volcanic activity. The possible detection of other volcanically created climatic effects in Colorado Plateau tree growth deserves further study.

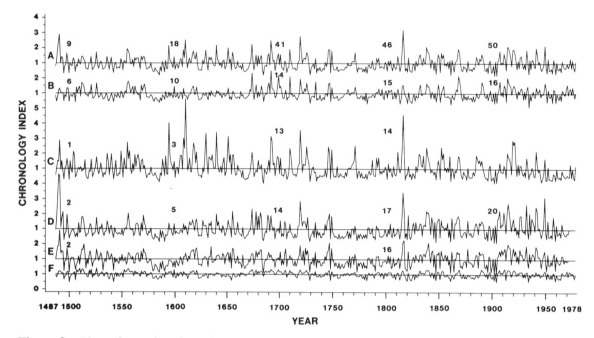

Figure 2: Plots of tree-ring chronology index series from southwestern Colorado. A = average latewood width from Bobcat, Ditch, and Spruce Canyon sites; B = Ditch Canyon latewood width; C = Spruce Canyon latewood width; D = Bobcat Canyon latewood width; E = Bobcat Canyon ring width; F = Spruce Canyon latewood density (with variability multiplied by 3.0 to increase variability relative to the other variables). The numbers above each chronology are the sample depth at that point.

The atmospheric mechanisms that may create regional climatic anomalies in response to volcanic influences are not well understood. Atmospheric and oceanic conditions at the time an eruption occurs may determine climatic response. It is believed, for example, that sea-surface temperatures and the El Niño-Southern Oscillation phenomenon have mediated climatic effects of several recent volcanic eruptions (Angell and Korshover 1985; Angell 1988).

Conclusions

The use of moisture-stressed conifers to investigate spatial patterns of historic volcanic eruption effects on climate may partially compensate for the limited distribution of instrumental climatic records prior to the twentieth century. Latewood width is particularly sensitive to the growing season moisture budget. The 1816 annual rings investigated in this report have the largest amount of latewood growth on the Colorado Plateau in 491 years. The pattern of greatly increased ring width, latewood width, and latewood density in these Colorado Plateau conifers from 1815 to 1817 indicates a reduction of growing season moisture stress unique in the last five centuries. The climatic effect that apparently reached a maximum over the Colorado Plateau in 1816 probably began shortly after the April 1815 eruption of Tambora, and persisted into the growing season of 1817. The cause of this extraordinary growth anomaly was probably a reduction of mid-summer evapotranspiration demand, which appears to have extended the growing season, resulting in extremely large latewood growth. No effects of other known large volcanic eruptions were detected in these tree-ring chronologies. The receptivity of the general circulation to volcanic forcing may partly explain the apparently strong climate-tree growth response on the Colorado Plateau to the Tambora eruption, and the absence of a large growth response to other major eruptions during the historic period. Effects of eruptions on regional temperature and tree growth might be masked by existing regional climatic conditions such as drought, or by other climatic-forcing mechanisms such as sea-surface temperatures and/or the phase of the El Niño-Southern Oscillation. This study has focused on annual ring data from a small set of chronologies in a small part of the Colorado Plateau, but it demonstrates a potential application of tree-ring data to the analysis of volcanic effects on climate.

Acknowledgements

Thanks are due David W. Stahle, University of Arkansas Tree-Ring Laboratory, for suggested improvements to the manuscript, and to Thomas Harlan, University of Arizona Laboratory of Tree-Ring Research, who assisted in the collection and dating of the samples. Part of the data comes from my doctoral dissertation, which was supported by the Laboratory of Tree-Ring Research and the Department of Geosciences, University of Arizona. Additional support from the National Science Foundation, Climate Dynamics Program (grant ATM-8612343) is also acknowledged.

References

Angell, J.K. 1988. Impact of El Niñõ on the delineation of tropospheric cooling due to volcanic eruptions. *Journal of Geophysical Research* 93D:3697-3704.

Angell, J.K. and J. Korshover. 1985. Surface temperature changes following the six major volcanic episodes between 1780 and 1980. *Journal of Climate and Applied Meteorology* 24:937-951.

Bradley, R.S. 1988. The explosive volcanic eruption signal in northern hemisphere continental temperature records. *Climatic Change* 12:221-243.

Bryson, R.A. and B.H. Goodman. 1980. Volcanic activity and climatic changes. *Science* 207:1041-1044.

Cleaveland, M.K. 1983. X-ray densitometric measurement of climatic influence on the intra-annual characteristics of southwestern semiarid site conifer tree rings. Ph.D. dissertation. University of Arizona, Tucson. 177 pp.

_____. 1986. Climatic response of densitometric properties in semiarid site tree rings. *Tree-Ring Bulletin* 46:13-29.

_____. 1988. Corrigendum to: Climatic response of densitometric properties in semiarid site conifer tree rings. *Tree-Ring Bulletin* 48:41-47.

Cook, E.R. and K. Peters. 1981. The smoothing spline: a new approach to standardizing forest interior tree-ring width series for dendroclimatic studies. *Tree-Ring Bulletin* 41:45-53.

Drew, L.G. (ed.). 1976. *Tree-ring Chronologies for Dendroclimatic Analysis*. University of Arizona Laboratory of Tree-Ring Research, Tucson. 64 pp.

Fritts, H.C. 1976. *Tree Rings and Climate*. Academic Press, London. 567 pp.

Fritts, H.C., D.G. Smith and M.A. Stokes. 1965. The biological model for paleoclimatic interpretation of Mesa Verde tree-ring series. *American Antiquity* 31:101-121.

Fritts, H.C. and J.M. Lough. 1985. An estimate of average annual temperature variations for North America, 1602 to 1961. *Climatic Change* 7:203-224.

Gilliland, R.L. and S.H. Schneider. 1984. Volcanic, CO_2 and solar forcing of northern and southern hemisphere surface air temperatures. *Nature* 310:38-41.

Hammer, C.U., H.B. Clausen and W. Dansgaard. 1980. Greenland Ice Sheet evidence of post-glacial volcanism and its climatic impact. *Nature* 288:230-235.

Harshvardhan and R.D. Cess. 1976. Stratospheric aerosols: effect upon atmospheric temperature and global climate. *Tellus* 28:1-9.

Horstmeyer, S.L. 1989. In search of Cincinatti's weather. *Weatherwise* 42:320-327.

Kelly, P.M. and C.B. Sear. 1984. Climatic impact of explosive volcanic eruptions. *Nature* 311:740-743.

Lamb, H.H. 1970. Volcanic dust in the atmosphere; with a chronology and assessment of its meteorological significance. *Philosophical Transactions of the Royal Society of London* A266:425-533.

Lough, J.M. and H.C. Fritts. 1987. An assessment of the possible effects of volcanic eruptions on North American climate using tree-ring data, 1602 to 1900 A.D. *Climatic Change* 10:219-239.

Newhall, C.G. and S. Self. 1982. The volcanic explosivity index (VEI): an estimate of explosive magnitude for historical volcanism. *Journal of Geophysical Research* 87C:1231-1238.

Panshin, A.J. and C. de Zeeuw. 1970. *Textbook of Wood Technology*, Vol. I, Third Edition. McGraw-Hill Inc., New York. 705 pp.

Parker, M.L., R.D. Bruce and L.A. Jozsa. 1980. X-ray densitometry of wood at the W.F.P.L. Forintek Canada Corporation Western Laboratory, *Technical Report No. 10*:1-18.

Pollack, J.B., O.B. Toon, C. Sagan, A. Summers, B. Baldwin and W. Van Camp. 1976. Volcanic explosions and climatic change: a theoretical assessment. *Journal of Geophysical Research* 81:1071-1083.

Rampino, M.R. and S. Self. 1984. Sulphur-rich volcanic eruptions and stratospheric aerosols. *Nature* 310:677-679.

Shutts, G.J. and J.S.A. Green. 1978. Mechanisms and models of climatic change. *Nature* 276:339-342.

Stockton, C.W. and H.C. Fritts. 1971. Conditional probability of occurrence for variations in climate based on width of annual tree rings in Arizona. *Tree-Ring Bulletin* 31:3-24.

Stockton, C.W., W.R. Boggess and D.M. Meko. 1985. Chapter 3, Climate and tree rings. *In*: *Paleoclimate Analysis and Modelling*. A.D. Hecht (ed.). John Wiley and Sons, New York. pp. 71-161.

Stokes, M.A. and T.L. Smiley. 1968. *An Introduction to Tree-Ring Dating*. University of Chicago Press, Chicago. 73 pp.

Stommel, H. and E. Stommel. 1983. *Volcano Weather*. Seven Seas Press, Newport, Rhode Island. 177 pp.

Stothers, R.B. 1984. The great Tambora eruption in 1815 and its aftermath. *Science* 224:1191-1198.

Wilson, C. 1985. The Little Ice Age on eastern Hudson/James Bay: the summer weather and climate at Great Whale, Fort George and Eastmain, 1814-1821, as derived from Hudson's Bay Company records. *In*: *Climatic Change in Canada* 5. C.R. Harington (ed.). *Syllogeus* 55:191-218.

1816 in Perspective: the View from the Northeastern United States

William R. Baron[1]

Abstract

The year 1816 is remembered in the northeastern United States as one of the harshest, coldest years on record, and continues even to the present to be one of the most widely known folk-climate episodes of the region. A study of the period 1790-1839 helps to place 1816 in its climatological context. New evidence supports the case that 1816 had a particularly cold and dry growing season but was by no means the coldest or driest year of the period. Several other years during the second and fourth decades of the century climatically rivaled the abnormal conditions of the "year with no summer". 1816's claim to fame rests on the severe impact that the cold and drought had on the area's then extensive agricultural operations. For this reason 1816 came to represent the abnormal climatic conditions of not only a single year but also most of the second decade of the nineteenth century.

Introduction

The year with no summer, 1816, is one of the best known folk-weather occurrences of the northeastern United States. During the nineteenth century, "eighteen-hundred-and-froze-to-death" was a subject for many newspaper articles, autobiographical reminiscences, and local histories (Mussey and Vigilante 1948). Even now, some 170 years after that frosty summer, it continues to be a topic of great popular interest, still commanding feature articles in the region's newspapers and periodicals (Fichter 1971; Leach 1974; Reichmann 1978; Parsons 1980).

1816 also has not escaped the attention of scientific investigators, and has been a subject of considerable debate since the early nineteenth century (Skeen 1981). During the twentieth century, research appearing on this topic included that by W.I. Milham (1924), J.B. Hoyt (1958), H.E. Landsberg and J.M. Albert (1974), H. and E. Stommel (1979, 1983), and R.B. Stothers (1984), and centred on the issue of what factors contributed to 1816's abnormal summer. Most researchers concluded that the great Tambora eruption of 1815 and low sunspot activity were the major factors involved; although a minority, including Landsberg, have questioned the influence of volcanic dust in the atmosphere as a major contributor. Of late, historians have begun to assess the climatic impact of the 1810s on society in both Europe and North America (Post 1977; Skeen 1981). Our continuing interest in and fascination with 1816 finally led to the international conference that produced the papers included in this volume.

The purpose of this study is to present new and additional evidence for the northeastern United States covering the period 1790 through 1839, in order to help place 1816 in its proper climatological context. Data presented include instrument readings for temperature, precipitation, and wind direction. Additional reconstructions for snowfall, seasonal precipitation, cloud cover, growing-season length, thunderstorm frequency, river freeze-up and ice-out, and phenology records (based on the analysis of qualitative materials such as diaries), weather journals and newspaper reports, also are discussed. However, before proceeding, I will present a

---

[1] Historical Climate Records Office, Center for Colorado Plateau Studies, Northern Arizona University, Box 5613, Flagstaff, Arizona 86011, U.S.A.

reconstruction of the weather history of 1816 based on the observations of 74 diarists from northeastern United States.

1816, "The Year With No Summer"

Appropriately enough, 1816 began, at least in Phillipstown, Massachusetts, with enough snow on the ground for sleighing. All over New England, January was a snowy, stormy month until the very end when a sudden thaw caused localized flooding such as the one reported by Isaiah Thomas at Worcester, Massachusetts on 23 January where some mill dams were carried off and some items stored in a warehouse were destroyed.

According to among others, Leonard Hill of East Bridgewater, Massachusetts, February was a mild and pleasant month with only three snows reported. By the beginning of March there was little deep snow anywhere with the exception of most of northern New England. Early March was clear and cold, and was followed by a series of three snow storms around mid-month that produced a few days of sleighing but soon melted. On 28 and 30 March, warm air returned producing thunder and lightning as reported by Elijah Kellogg at Portland, Maine and Thomas at Worcester.

April quickly turned cold again with frequent frosts and some snow. However, by 14 April, there was little snow left at Hallowell, Maine. By 19 April, Alexander Miller of Wallingford, Vermont had begun to plough his fields; Stephen Longfellow of Gorham, Maine was already planting wheat; and Theodore Lincoln of Dennysville (in far downeast Maine) was reporting ice-out on the local streams - a sure sign of coming spring. At the end of the month, Joshua Lane of Sanbornton, New Hampshire already was reporting the start of a drought that would later plague all of northern New England.

In early May, farmers throughout the region completed planting their major crop, corn (maize). By mid-month the weather had become "backward" again with a "heavy black frost" that froze the ground to at least one-half inch (1.3 cm) reported on 15 May as far south as Trenton, New Jersey. Miller, at Wallingford, Vermont, reported snow on 14, 17 and 29 May while Lane, over at Sanbornton, saw a large frost on 24 May, and ended the month with further complaints about the continuing drought. B.F. Robbins, visiting Concord, New Hampshire noted that May ended with two days of "remarkable cold" that froze the ground "to near an inch".

June is the month most remembered for its outbreak of cold weather. On 4 June, there were frosts at Wallingford, Vermont and Norfolk, Connecticut. By 5 June the cold front was reported over most of northern New England. On 6 June, snow was reported at Albany, New York and Dennysville, Maine, and there were killing frosts at Fairfield, Connecticut.
7 June brought reports of severe killing frosts from across the region, and as far south as Trenton, New Jersey.

Typical of comments by diarists concerning this day are those by George W. Featherstonhaugh of Albany, New York, who wrote that the frost killed most of the fruit, as many apple trees were then just finishing blossoming. Leaves on most of the trees were "blasted" by the cold. Corn and vegetable crops were injured. He also feared many of the sheep that had just been sheared might die of cold.

Cold weather continued through the night of 10 June. By the end of the month most observers were reporting the return of warm weather, but by then most crops were either killed or "backward" and stunted in their growth. In northern New England, those crops that survived the frosts were hit by what was now a very serious drought, greatly reducing production of one of the area's primary crops, hay.

In early July there was another outbreak of cold weather in northern New England. On 5 July, at Gorham, Maine, there was a very hard frost. Benjamin Kimball of Concord, New Hampshire and Thomas Robbins of Norfolk, Connecticut reported hard frost on 7 July. There was frost on 8 July at Portland, Maine and on the following day at Sanbornton, New Hampshire. Thereafter the cold held off for the remainder of the month. Throughout the entire month dry conditions, generally reported earlier in northern areas, persisted.

Frosts returned on the morning of 21 August, being reported at York and Portland, Maine and Wallingford, Vermont. By 22 August hard frosts were noted all over the region and as far south as Trenton where buckwheat crops were killed. Thomas, at Worcester, Massachusetts, reported that these frosts "cut off Indian corn in many places", while others such as Hill at East Bridgewater, Massachusetts observed that frosts did little or no damage.

The frosts continued into September. In northern New England there were frosts on 10 and 11 September and throughout New England during 25-27 September. On 28 September, there was a killing frost throughout the region extending as far south as Trenton. It killed any vegetation that had somehow survived to that date. The drought in northern New England was finally broken by rains in the last week of the month.

The remainder of autumn was very mild with very few snowfalls or storms. December was also mild, until the last 10 days or so, when it turned cold enough to freeze the harbour at Beverly, Massachusetts. The year ended with enough snow on the ground at Phillipstown, Massachusetts to use a sleigh.

The place of 1816 in the memory of the regional population has been summed up well by the historian H.F. Wilson when he wrote that, in 1816, farmers experienced an "almost total failure" of major crops. There was a fair yield of winter grain, but other crops such as corn and hay failed leading to the loss of many sheep and cattle for lack of feed during the following winter. As a result 1816 has come down to us as the "cold year", "the famine year" and "eighteen hundred and froze to death".

Of great interest to climatologists and historians alike is the fact that 1816 was not the only difficult, abnormal year of the second decade of the nineteenth century. Based on statistical analysis of climatological data, other years might justifiably claim a portion of 1816's notoriety. An analysis of the 50 years surrounding 1816 serves to locate some of these years and to place the "year with no summer" in its climatological context.

Databases and Methodologies

The analysis that follows is based upon records assembled from several databases. The first of these is a set of eight yearly mean temperature records comprised of instrument readings for periods of 26 years or more, that overlap 1816 by at least three years and that, in composite, cover the 50 years from 1790 through 1839. From north to south these records include: Castine, Maine, 1809-39 (Baron *et al.* 1980); Brunswick, Maine, 1807-36 (Cleaveland 1867); Salem, Massachusetts, 1790-1829 (Holyoke 1833); New Bedford, Massachusetts, 1813-39 (Rodman

1905); Williamstown, Massachusetts, 1811-38 (Milham 1950); New Haven, Connecticut, 1790-1839 (Landsberg 1949); New Brunswick, New Jersey, 1790-1839 (Reiss et al. 1980); and Philadelphia, Pennsylvania, 1790-1839 (Landsberg et al. 1968). All but one of these stations, Williamstown, are located in the present coastal climatic zone as computed by the United States National Oceanic and Atmospheric Administration. Williamstown's inland situation makes it more vulnerable to outbreaks of Arctic cold coming from the northern interior of the continent.

Instrumental records of precipitation for the period are more limited than those for temperature. The three best include those for New Bedford, 1814-39 (Rodman 1905); New Haven, 1804-21 (Kirk 1939); and Philadelphia, 1790-1839 (Landsberg et al. 1968). No long precipitation records for inland locations are available.

To assure the representativeness of these long-run temperature and precipitation records, and to enhance the record density and distribution throughout the region, a database of short-duration instrumental records was assembled. The location of these records is shown in Figure 1. The number of records available increases, by decade, in a steady progression from the 19 available in the 1790s to 57 for the 1830s.

Yet another database (qualitative materials from diaries, journals, newspapers, and local histories - many available only in manuscript form) was compiled to supplement the instrumental records. At that time, only qualitative materials for New England were sufficiently organized for inclusion. The database is comprised of 174 sources representing 55 of New England's 67 counties. Coastal and intermediate interior locations are well represented whereas data for some upland and western interior counties are missing. From this database, frequency counts for days with precipitation, fair skies, thunder and lightning storms, westerly winds, and snowfalls, as well as the yearly dates for spring and autumn killing frosts, length of snowfall seasons, dates of apple-tree blossoming and lengths of droughts were compiled. The methodology used to reconstruct these various records is explained in Baron (1988, 1989), Baron and Gordon (1985) and Baron et al. (1984).

Compilations of killing-frost reports were further refined by computing the year and day of each frost and calculating the length of the growing season. To assess the possible impact of these frosts on a major crop, information from an agricultural database made up of over 60 farm journals was used to provide the mean dates for the planting and harvesting of Indian corn.

Analysis of Records

As can be seen in part from Figure 2, moving from north to south, 1816's yearly mean temperature is not the coldest for the period. In Maine it was only the fourth lowest; while farther south at Salem, New Haven, and New Bedford it was third or second lowest for the record. Even farther south at New Brunswick and Philadelphia, 1816 was close to the mean for the period 1790 through 1839. Farther inland at Williamstown, 1816 was the seventh coldest year (Figure 3). In New England the years noted to be as cold or colder than 1816 include: 1790, 1812, 1817, 1818, 1823, 1835, 1836 and 1837. 1836 and 1812 appeared most often as the two coldest years. South of New England, 1836 and 1817 were the coldest years.

Figure 1: Northeastern United States. Instrumental record locations, 1790-1840.

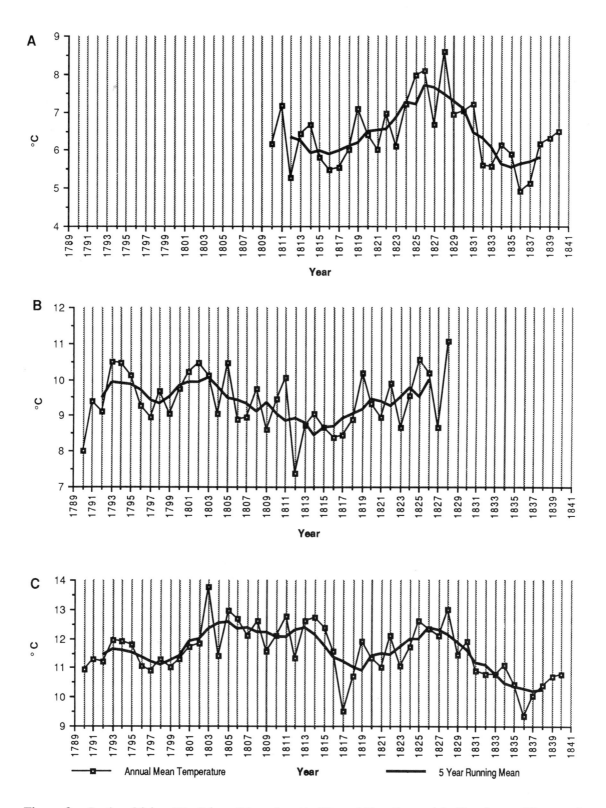

Figure 2: Castine, Maine (A); Salem, Massachusetts (B); and New Brunswick, New Jersey (C); annual mean temperatures, 1790-1840.

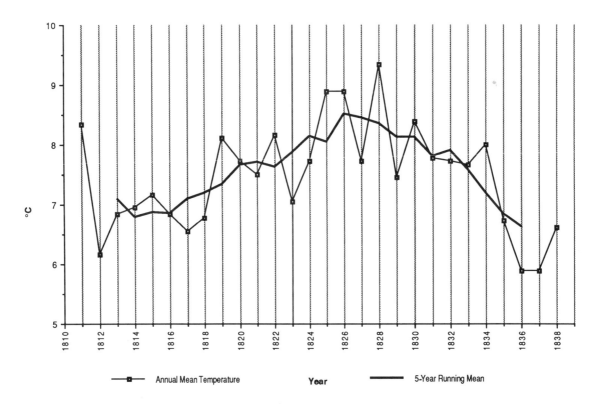

Figure 3: Williamstown, Massachusetts. Yearly mean temperatures, 1811-38.

Taking all temperature records together for the 50-year span, apparently the years 1790 through 1798 were either normal or slightly cooler than normal. 1799 through 1805 was warmer than normal. A period of variability between 1806 and 1811 followed. Starting in 1812 (with some variation until 1818) it was much cooler than normal. This cool period was followed by a series of variable years from 1819 through 1823. From 1824 through 1830 it was somewhat warmer than normal, and from 1831 through 1837 it was again very cool. The last two years of the decade show a warming trend that extended into the 1840s.

Figure 4 shows an analysis of seasonal mean temperatures for Brunswick and New Haven. Winter mean temperatures were calculated from December, January and February monthly means; spring means from March, April and May; summer means from June, July and August; and autumn means from September, October and November. The Brunswick record shows the winter of 1816 was very cold, whereas the summer was cool, the spring about average and the autumn a little warmer than average. Summer mean temperatures in 1812 were as low as those for 1816. New Haven presents a somewhat different picture with a mild winter, average autumn and cool spring and summer. As far as New Haven is concerned, the summer mean temperatures for 1812 and 1817 were far lower than that for 1816.

Figure 5, a daily mean temperature record for 1816 kept at Brunswick, Maine shows why this year is so well remembered. The outbreaks of cold in much of May, early June, early July, late August and late September tell much of the story. These cold periods, all well below the monthly mean temperature for the entire record (1807-36), doomed many a farmer's crops to failure. The key to 1816's infamy lies in the extreme shortness of its growing season - the primary reason why it, and not 1812 or 1836, has gone down in the regional weather lore as "the year with no summer".

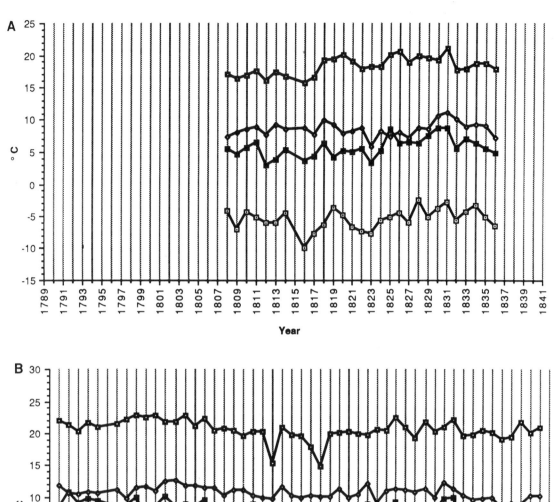

Figure 4: Brunswick, Maine (A) and New Haven, Connecticut (B); winter, spring, summer, and autumn season mean temperatures, 1790-1840.

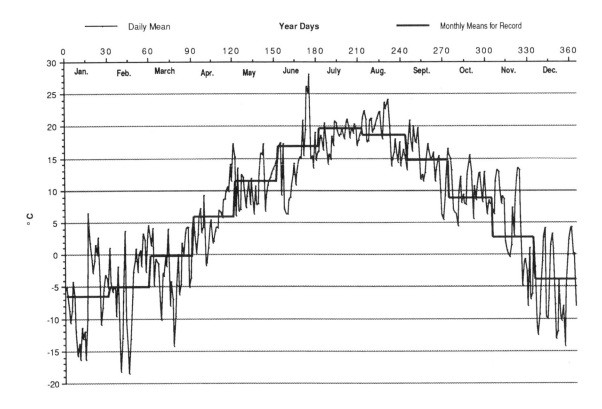

Figure 5: Daily mean temperatures at Brunswick, Maine for 1816.

The plots of growing-season lengths for eastern Massachusetts, southern New Hampshire and southern Maine (Figure 6) leave one unmistakable impression - 1816, by far, has the shortest growing season. Other particularly short growing seasons occurred in 1808, 1824, 1829, 1834 and 1836. With the exception of 1816 and 1836, a number of these short seasons can probably be attributed to one-night radiational cooling under clear skies during either spring or autumn, and not to prolonged outbreaks of cold weather; otherwise these years would have appeared in our lists of cool yearly and seasonal temperatures. Of course clear skies in combination with cold fronts also contributed to some frosts during 1816, as that year has one of the highest percentages of days with fair skies (Figure 7).

Figure 8, showing spring and autumn killing-frost dates in combination with corn-planting and maturation dates, further illustrates the importance of growing-season data in understanding the notoriety of 1816. For eastern Massachusetts, 1816 is the only year in which young corn was killed in the spring after it had sprouted and in which corn that survived replanting was killed in the autumn, before it could reach maturity. Under these circumstances, it is safe to assume that in most places in New England corn crops were an almost total failure. The story for 1816 is the same for New Hampshire and Maine. These reconstructions also show there were a number of years when corn crops were hit by late spring or early autumn frosts. Particularly difficult periods include: 1793-96, 1812-17, and 1835-36.

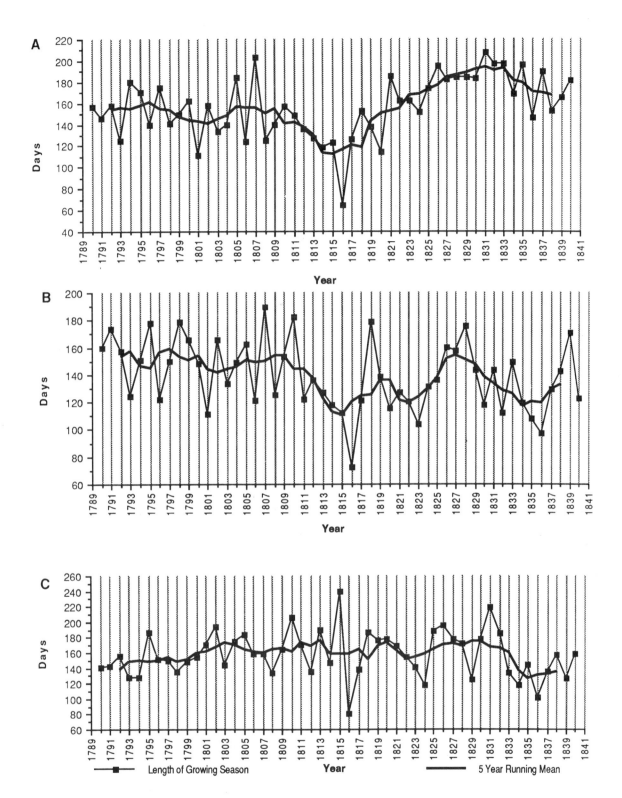

Figure 6: Growing-season lengths for southern Maine (A), southern New Hampshire (B) and eastern Massachusetts (C), 1790-1840.

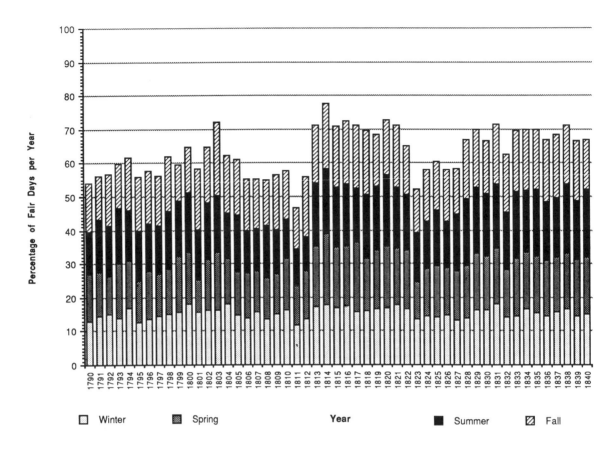

Figure 7: Percentage of days with fair skies for eastern Massachusetts, 1790-1840.

Reconstructions indicating the onset of spring-like weather, such as dates of last snowfalls and blossoming dates of various fruit trees such as apples (Figure 9), show that 1816's spring was cooler and more unsettled than normal. However, it was far more satisfactory for agricultural pursuits than 1812, 1832, or 1836 through 1838, when conditions were extremely "backward".

While the major part of the 1816 story lies in its growing-season record, there are several other types of records in which 1816's position is worthy of mention. The first of these is the precipitation record (Figure 10). 1816 was a year of about average precipitation, with the exception of the summer, which was particularly droughty. Reconstructions concerning the period and intensity of agriculturally-defined droughts show that the magnitude of dry conditions increased significantly the farther north within the region one looks. For the 50 years from 1790 through 1840, the periods from 1791-1806 and 1813 to 1820 saw numerous growing-season droughts. After 1820 the number of reported droughts decreased markedly. This apparent increase in precipitation also can be seen in Figure 11, illustrating the mean number of days per year with precipitation for southern New England.

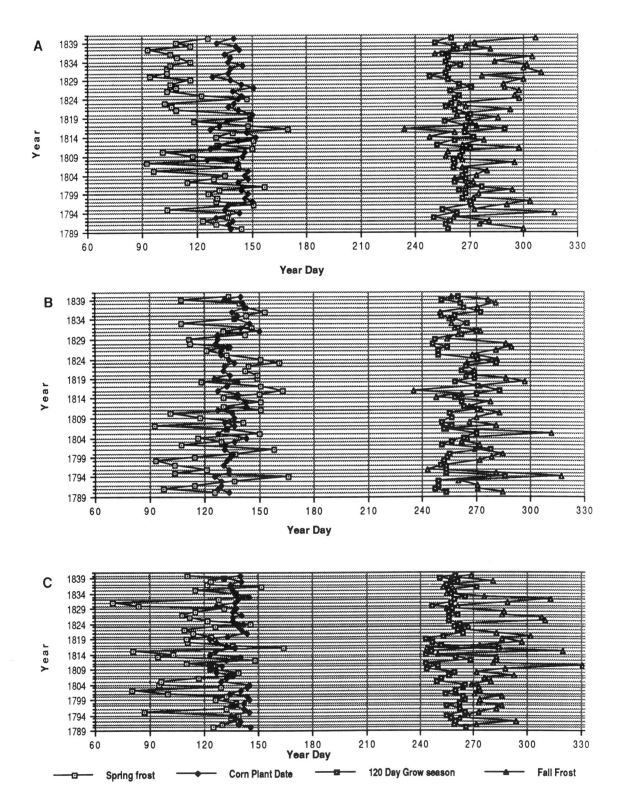

Figure 8: Killing frost and corn plant/harvest dates for southern Maine (A), southern New Hampshire (B), and eastern Massachusetts (C), 1790-1840.

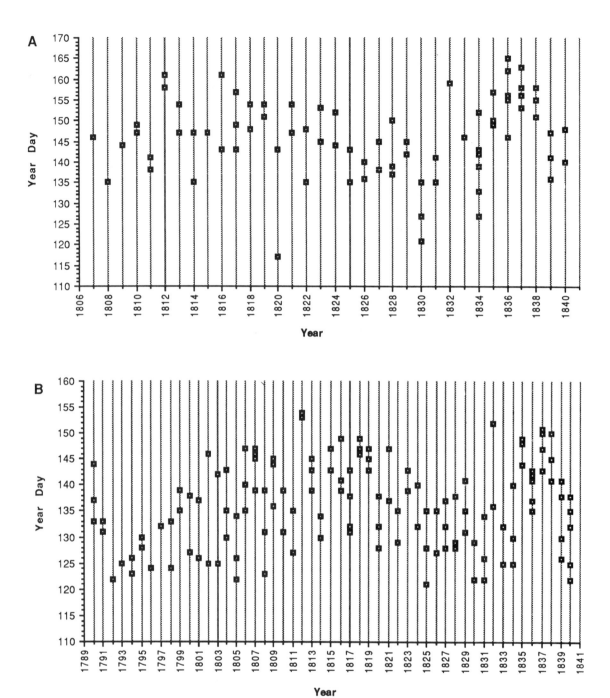

Figure 9: Apple-blossom dates for northern (A) and southern (B) New England, 1790-1840.

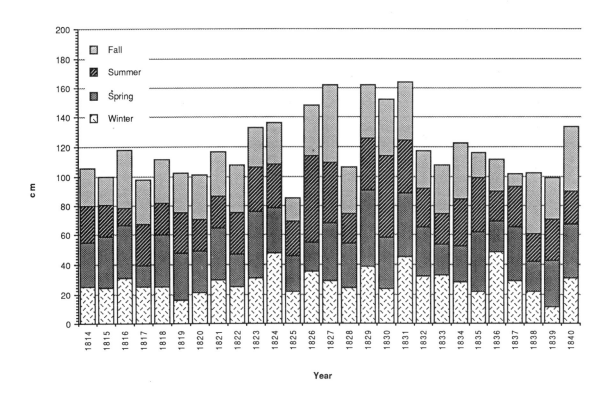

Figure 10: New Bedford winter, spring, summer and autumn precipitation, 1814-40.

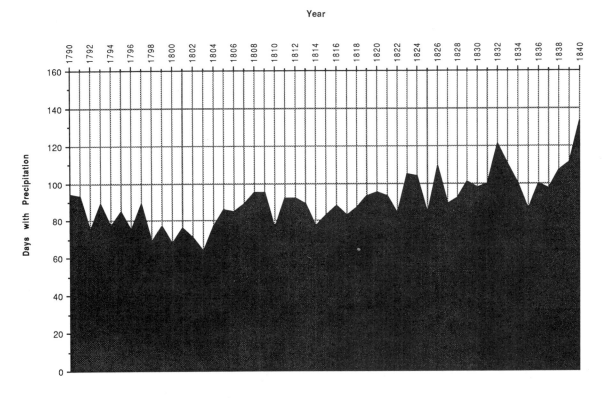

Figure 11: Mean number of days per year with precipitation for southern New England, 1790-1840.

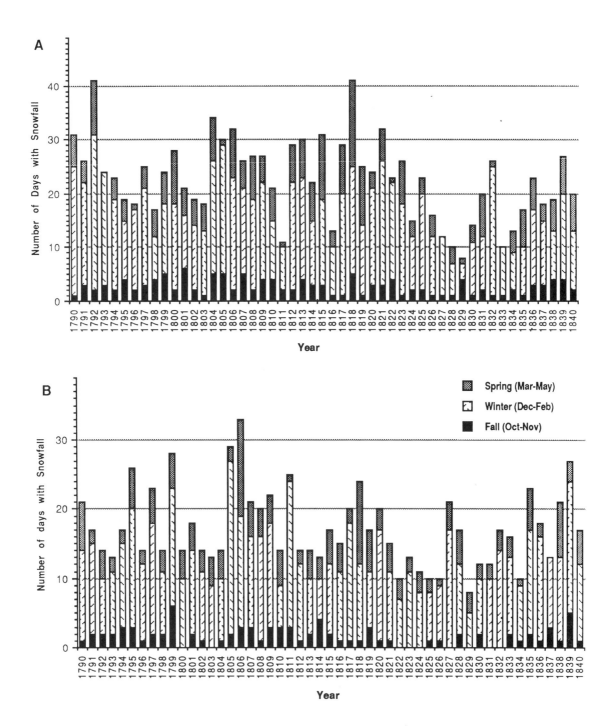

Figure 12: Number of days with snowfall in southern Maine (A) and eastern Massachusetts (B), 1790-1840.

Snowfall records show that during the winter of 1815-16 there were relatively few days when it snowed, especially in northern New England (Figure 12), but the New Bedford seasonal precipitation record shows average or slightly above average precipitation for that winter. However, under no circumstances can 1816 be viewed as a particularly snowy winter. Available records led me to believe that in northern New England, during 1816, winter conditions were drier and colder than normal; while farther south in Massachusetts the season was warmer and wetter. Among the years with the greatest number of snowfalls were: 1792, 1804, 1805, 1806, and 1818. Those years with the least number included: 1828, 1829 and 1834.

Reconstructions bearing evidence concerning the storminess of the period [e.g., the percentage of days with westerly fair-weather winds (Figure 13) and the number of days in which thunder and lightning storms occurred (Figure 14)], show that 1816 was rather stormy. The frequency of westerly "fair weather bearing" winds was somewhat below the record mean; while days with thunder and lightning storms numbered close to the record mean. Evidently there was a decrease in storm activity during the 1790s. In the early 1800s, there was a small increase followed by another decrease late in the decade. From 1809 through 1822, there was considerable year-to-year variability but an overall increase in storminess. During the late 1820s and all of the 1830s, there was a general decrease in storm frequency.

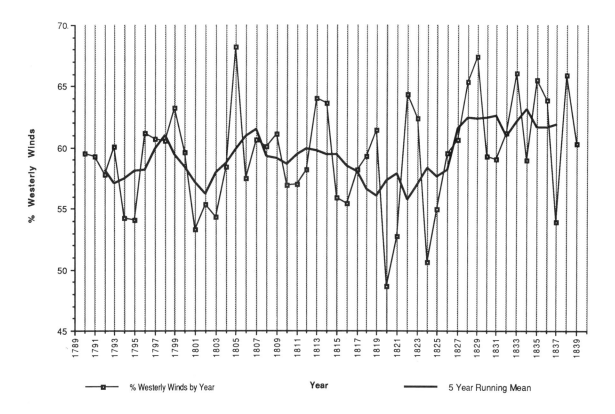

Figure 13: Westerly winds per year over southern New England, 1790-1839.

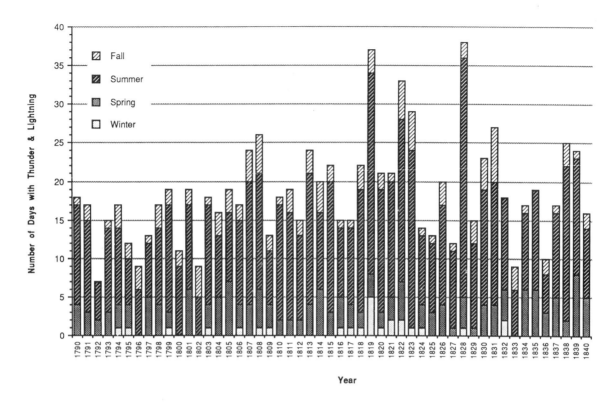

Figure 14: Number of days per year with thunder and lightning storms for eastern Massachusetts, 1790-1840.

Conclusions

1790 through 1839 featured two abnormally cold periods (1812 through 1818 and 1832 through 1838) and two warmer, relatively more stable periods (1799 through 1810 and 1819 through 1830). The warmth and stability of the latter two decades, compared with cold and relative storminess of the 1810s, heightened peoples' awareness of the contrast between the two climatic regimes. Especially in northern New England, where considerable farming took place on climatically-marginal lands, the cold years brought disaster. To make matters worse, the swing from warm to cold in the 1810s coincided with an increase in economic competition from the midwestern United States and central Canada. The additional stress of crop shortfalls due to shortened growing seasons forced many farmers to leave New England for what they believed were more hospitable climates to the west (Smith *et al* 1981).

1816 was only one of several abnormal years that occurred during 1790 through 1839. When viewed from this perspective, 1816's abnormality pales. Why then is 1816 so well remembered while 1812 or 1836 are assigned to the second rank of the region's fabled years of climatic adversity?

The answer lies not in our careful compilation of climatological records (for statistically 1816 does not measure up) but in the nature of 1816's abnormality and the impact of its greatly shortened growing season on New Englanders' capability to raise food. A harsh, snowy winter or severe spring flood have a great impact on certain segments of society, but a series of killing frosts accompanied by a severe drought (especially in northern New England) hit nearly the entire society by forcing up food prices for the rich and by reducing the available larder for the poor. For the average New Englander, particularly the farmer, 1816 was the worst year of a series of bad years. As time passed, 1816, in New England's folk memory, came to stand for the 1810s as a whole. This idea has been passed down from generation to generation as the story of "the year with no summer".

Acknowledgements and a Note on the Availability of Climatic Data

I thank the staff of the Historical Climate Records Office for their assistance. This research was partially supported through Northern Arizona University's Organized Research Fund; I thank particularly Henry O. Hooper, Associate Vice President for Academic Affairs, Research and Graduate Studies for his support and assistance.

All databases discussed here are kept in the Historical Climate Records Office, part of the Center for Colorado Plateau Studies at Northern Arizona University, Flagstaff. The Office has on file and computer disc a large number of United States records collected for the seventeenth through nineteenth centuries. There are particularly strong record groups for the northeastern United States and the Colorado Plateau. The Office was founded by the author with the intention of making these climatic materials available to other researchers. Record collection was done by the members of the now disbanded Northeast Environmental Research Group centred at the University of Maine and, after 1985, by the staff of the Historical Climate Records Office. Record collection undertaken through 1985 was supported by grants from the National Science Foundation and the Northeast Regional Experiment Stations to the University of Maine. A listing of available records may be obtained by writing me.

References

Baron, W.R. 1988. Historical climates of the northeastern United States: seventeenth through nineteenth centuries. *In*: *Holocene Human Ecology in Northeastern North America*. G.P. Nicholas (ed.). Plenum Press, New York. pp. 29-46.

_____. 1989. Retrieving climate history: a bibliographical essay. *Agricultural History* 63. (in press).

Baron, W.R., D.C. Smith, H.W. Borns, Jr., J. Fastook and A.E. Bridges. 1980. *Long-Time Series Temperature and Precipitation Records for Maine, 1808-1978. In: Life Sciences and Agriculture Experiment Station Bulletin* 771. University of Maine Press, Orono, Maine. p. 97.

Baron, W.R., G.A. Gordon, H.W. Borns, Jr. and D.C. Smith. 1984. Frost-free season record reconstruction for eastern Massachusetts 1733-1980. *Journal of Climate and Applied Meteorology* 23:317-319.

Baron, W.R. and G.A. Gordon. 1985. A reconstruction of New England climate using historical materials 1620-1980. *In: Climatic Change in Canada 5.* C.R. Harington (ed.). *Syllogeus* 55:229-245.

Cleaveland, P. 1867. Results of meteorological observations made at Brunswick, Maine, between 1807 and 1859, reduced and discussed by C.A. Schott. *Smithsonian Contributions to Knowledge* 16:2-25.

Fichter, G.S. 1971. Eighteen-hundred-and-froze-to-death: snowfilled summer of 1816. *Science Digest* 69(2):62-66.

Holyoke, E.A. 1883. A meteorological journal from the year 1786 to the year 1829, inclusive. *Memoirs of the American Academy of Arts and Sciences*, New Series 1:107-216.

Hoyt, J.B. 1958. The cold summer of 1816. *Annals of the Association of American Geographers* 48:118-131.

Kirk, J.M. 1939. *The Weather and Climate of Connecticut. State Geological and Natural History Survey Bulletin* 61. Connecticut Geological and Natural History Survey, Hartford. 242 pp.

Landsberg, H.E. 1949. Climatic trends in the series of temperature observations at New Haven, Connecticut. *Geografiska Annaler* 1(2):125-132.

Landsberg, H.E., C.S. Yu and L. Huang. 1968. Preliminary reconstruction of a long time series of climatic data for the eastern United States. *Institute for Fluid Dynamics and Applied Mathematics, University of Maryland, Technical Note* B14-571:1-42.

Landsberg, H.E. and J.M. Albert. 1974. The summer of 1816 and volcanism. *Weatherwise* 27(4):63-66.

Leach, A. 1974. 1816 was a year for complaints. *Barre-Montpelier Vermont, Times Argus*, 22 July.

Milham, W.I. 1924. The year 1816 - the causes of abnormalities. *Monthly Weather Review* 52:563-570.

_____. 1950. *Meteorology in Williams College*. McClelland Press, Williamstown. 40 pp.

Mussey, B. and S.L. Vigilante. 1948. "Eighteen-hundred-and-froze-to-death". The cold summer of 1816 and westward migration from New England. *Bulletin of the New York Public Library* 1948:565.

Parsons, M. 1980. An eyewitness report of 1816, the cold year. *Bittersweet* 4-1:39-41.

Post, J.D. 1977. *The Last Great Subsistence Crisis in the Western World*. Johns Hopkins University Press, Baltimore. 240 pp.

Reichmann, R. 1978. 1816 recalled as year without a summer. *Providence, Rhode Island Sunday Journal*, 17 September.

Reiss, N.M., B.S. Groveman and C.M. Scott. 1980. Seasonal mean temperatures for New Brunswick, N.J. *Bulletin of the New Jersey Academy of Science* 1980:1-10.

Rodman, T.R. 1905. Monthly, annual and average temperatures and precipitation at New Bedford, Mass., 1813-1904. *Climate and Crops: New England Section, Annual Summary* 1905:9.

Skeen, C.E. 1981. The year without a summer: a historical view. *Journal of the Early Republic* 1:51-67.

Smith, D.C., H.W. Borns, Jr., W.R. Baron and A.E. Bridges. 1981. Climatic stress and Maine agriculture, 1785-1885. *In: Climate and History*. T.M.L. Wigley, M.J. Ingram and G. Farmer (eds.). Cambridge University Press, Cambridge. pp. 450-464.

Stommel, H. and E. Stommel. 1979. The year without a summer. *Scientific American* 240:176-186.

_____. 1983. *Volcano Weather: The Story of 1816, The Year Without a Summer*. Seven Seas Press, Newport. 177 pp.

Stothers, R.B. 1984. The great Tambora eruption in 1815 and its aftermath. *Science* 224(4654):1191-1198.

Wilson, C.M. 1970. The year without a summer. *American History Illustrated* 5(6):24-29.

Wilson, H.F. 1967. *The Hill Country of Northern New England: Its Social and Economic History 1790-1939*. AMS Press, New York. 455 pp. (Reprint of 1936 edition).

Sources Directly Mentioned in 1816 Weather History

Bascom, R.H. 1816. Manuscript diary for Phillipstown, Massachusetts. American Antiquarian Society, Worcester, Massachusetts.

Dickerson, M. 1816. Manuscript diary for Morristown and Trenton, New Jersey. New Jersey Historical Society, Newark, New Jersey.

Featherstonhaugh, G.W. 1816. Manuscript diary for Albany, New York. Albany Institute, Albany, New York.

Hill, B.T. (Editor). 1910. The diary of Isaiah Thomas 1805-1828. *American Antiquarian Society Transactions and Collections* 10:251-263.

Hill, L. 1869. Meteorological and Chronological Register. L. Hill, Plymouth, Massachusetts. pp. 61-66.

Kellogg, E. 1816. Manuscript notes in Old Farmer's Almanack for 1816. Maine Historical Society, Portland, Maine.

Kimball, B. 1856. Extracts of a diary. *In: The History of Concord, 1725-1853*. N. Bouton (ed.). B.W. Sanborn, Concord, New Hampshire. p. 771.

Lane, J. 1816. Manuscript diary for Sanbornton, New Hampshire. New Hampshire Historical Society, Concord, New Hampshire.

Larcom, J. 1951. Diary of Jonathan Larcom of Beverly, Massachusetts. *Essex Institute Historical Collections* 87:65-95.

Lincoln, T. 1816. Manuscript record book of weather reports. Maine Historical Society, Portland, Maine, 1: (not paginated).

Longfellow, S. 1816. Manuscript diary for Gorham, Maine. Maine Historical Society, Portland, Maine.

Miller, A. 1816. Manuscript diary for Wallingford, Vermont. Vermont Historical Society, Montpellier, Vermont.

Robbins, B.F. 1816. Manuscript travel diary. Maine Historical Society, Portland, Maine.

Robbins, T. 1986. *Diary of Thomas Robbins, DD, 1796-1854*. Beacon Press, Boston. pp. 236-247.

Sewall, H. 1816. Manuscript diary for Hallowell, Maine. Maine State Archives, Augusta, Maine.

Weare, J. 1912. The diary of Jeremiah Weare, York, Maine. *New England Historical and Genealogical Register* 66:77-79.

Wheeler, W. 1930. *Diary of William Wheeler*. Yale University Press, New Haven. p. 221.

Extension of Toronto Temperature Time-Series from 1840 to 1778 Using Various United States and Other Data

R.B. Crowe[1]

Abstract

Daily maximum and minimum temperatures for the city of Toronto are archived from 1 March 1840 to the present day. This lengthy time-series can be extended considerably by using standard differences in mean monthly temperatures between Toronto and some United States stations, the earliest of which dates from July 1778. In addition, there are considerable temperature data taken three times a day from another station in Toronto in the 1830s. These data were adjusted and monthly mean temperatures calculated. Mean July temperature for Toronto in 1816 is calculated to have been remarkably low (nearly 16°C).

Introduction

In December 1839, the British Government established a meteorological and magnetic observatory at Toronto, Ontario. Some sporadic observations began late in the month at Fort York (on the shore of Lake Ontario just west of the town, then called York). Fixed hourly observations of temperature commenced in the new year, but not until 1 March 1840 were regular daily maximum and minimum values recorded: on this date daily Archive readings begin. On 5 September 1840, the observation site was moved to the University of King's College (now the University of Toronto) about 2 km north of the lakefront. Although a number of small changes in location occurred in later years, a relatively homogenous, high-quality data set extends from September 1840 to the present day. However, the rather large urban heat-island effect, which influences the records of all such large cities, is evident.

The Toronto record from 1840 to the present comprises the longest continuous temperature time-series in the Canadian climatological Archive, and thus it is frequently used in analyses of long-term temperature trends.

The purpose of this paper is to present a method of extrapolating the Toronto temperature time-series backwards from 1840, using various United States and other data. The earliest American data used in the analysis were taken in 1778. The significance of these data for 1816 is mentioned.

Sources of Early Climatic Data Used for Comparative Purposes

Toronto Area
The first fragmentary climatic data for Toronto were taken in the year 1801. These are contained in the *Hodgins Papers* in the Archives of Ontario, Toronto, dating from the late nineteenth century. The data are identical to those published in the *Upper Canada Gazette* at the time the observations were taken, so presumably Dr. Hodgins merely copied long-hand this original

---

[1] Canadian Climate Centre, Environment Canada, 4905 Dufferin Street, Downsview, Ontario M3H 5T4, Canada.

source. The data include the temperature and weather or sky condition at three fixed times a day. Similar data were published for a number of months in the same newspaper around 1820. Neither data set was lengthy enough to be used in this study.

Later a longer, more useful data set was taken by Dade (1831-41) from January 1831 to April 1841. Reverend Dade was the Headmaster of Upper Canada College, then situated close to the centre of town on the lake shore, not far from the later Fort York station. The thermometer was read two or three times a day at fixed times, but slight changes in reporting hours occurred during the decade, and occasionally only one observation was taken in a day. Only a few months were incomplete, except for an extended period from October 1838 to June 1839 when Dade returned to England for the winter.

Periods of record for the various early Toronto stations used for comparative purposes are shown (Figure 1). Data from Fort York are combined in the Archive with those from the University station and identified as "Toronto" (no modifier), but is unofficially called "Toronto City". Data later than 1855 were not considered.

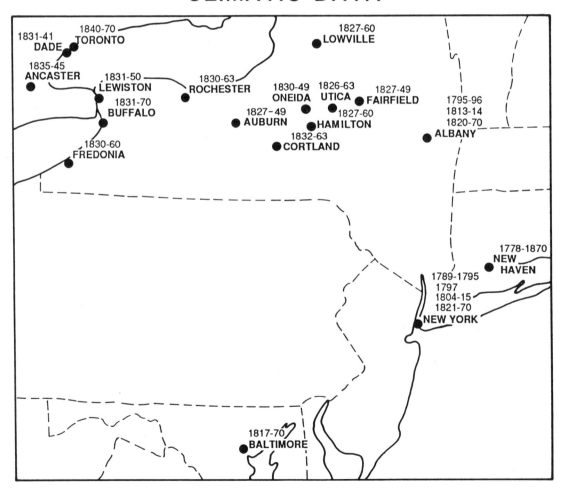

Figure 1: Early eastern North America climatic data.

Remainder of Southern Ontario

Data for Ancaster (about 65 km southwest of Toronto) were taken by Craigie (1835) from January 1835 to December 1845, and proved to be of limited use in the Toronto data extension. William Craigie was a surgeon who apparently tabulated daily maximum and minimum temperatures as well as fixed-hour readings. His thermometers "were in a northern exposure, five feet from the ground, and shaded from the effects of direct insolation and radiation from the sky". However, only newspaper tabulations of monthly means of the 9 a.m. and 9 p.m. observations survive.

American Stations

Mean monthly temperature data were abstracted from publications of the Smithsonian Institution (1927) for Albany, New Haven and New York City and the United States Weather Bureau (1932-37) for Albany, Baltimore and Rochester. Considerable monthly data were also available from grammar schools in New York State (Hough 1855, 1872). Data for Auburn, Buffalo, Cortland, Fairfield, Fredonia, Hamilton, Lewiston, Lowville, Oneida, Rochester (College) and Utica were used, other stations listed in the above publications having insufficient useful data.

All stations actually used in the study are shown in Figure 2. Many months and years for most stations were noted in the New York State grammar school records, and only the first and last years of data are shown. In all cases, data later than 1855 were not used.

In the case of most of these early data, excepting Toronto (city), observations were taken with the thermometer attached to the north wall of a building. Recording maximum and minimum thermometers were not generally used. Monthly means were computed from two, three or more observations a day, and the time and number of daily observations frequently changed and were not consistent, either at a site or from one station to another. In addition, thermometers may not have been calibrated accurately or sufficiently shielded from insolation, and changes in exposure or siting may not have been recorded.

Method of Estimation of Toronto Mean Temperatures

Three distinct methods were employed in the calculation of Toronto mean temperatures due to significant differences in the form of the source data: monthly means at the American stations (calculated by a variety of methods depending on the station); daily data for Dade; and monthly means for 9 a.m. and 9 p.m. in the case of Ancaster. These were labelled Method "S", Method "D", and Method "A" (Figure 3).

All three methods were employed whenever data permitted. In deriving the final Toronto estimates, however, Method "D" was chosen whenever Dade information was available. Thus, Method "S" was used up to December 1830, but Method "D" from January 1831 to February 1840. For missing Dade months, Method "S" was substituted before 1835, but from this year on, a linear regression equation was used based on the 52 months when the calculated American data could be compared with both Dade and Ancaster calculations:

$$T = -0.126 + 0.6045 \, T_A + 0.4108 \, T_S,$$

where T is the estimated Toronto monthly mean (°F),
T_A is the estimated Toronto mean using Method "A",
and T_S is the estimated Toronto mean using Method "S".

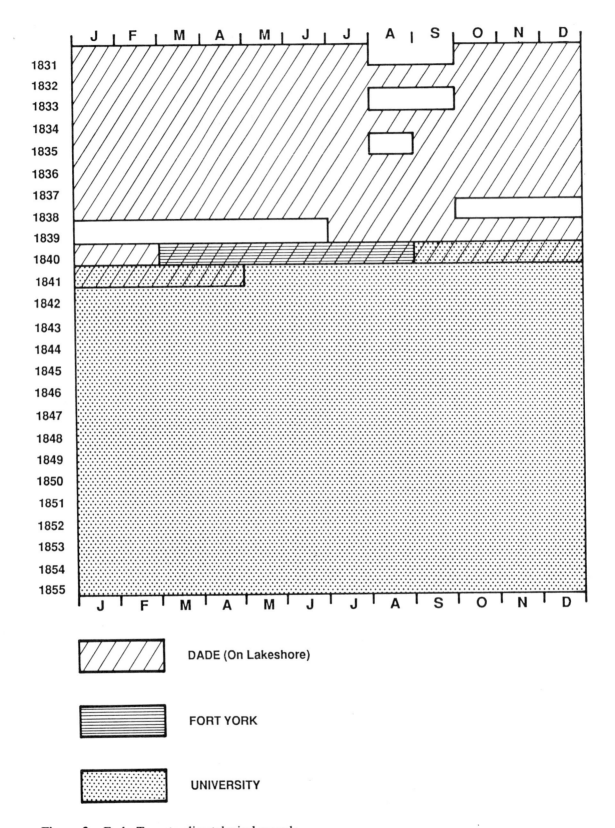

Figure 2: Early Toronto climatological records.

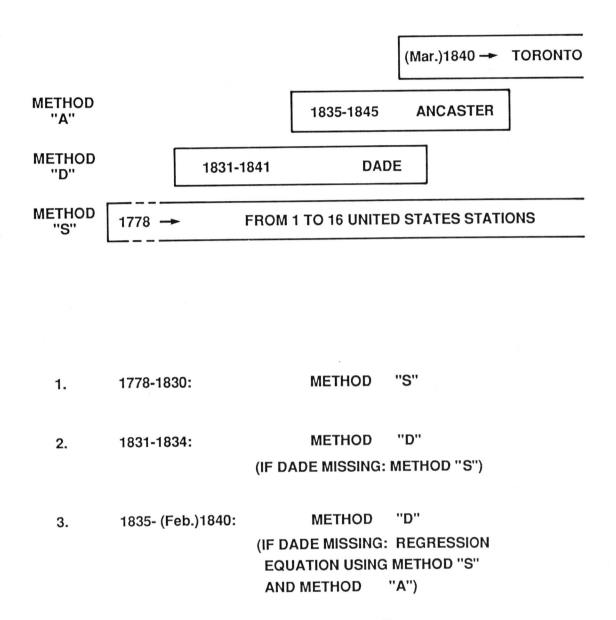

Figure 3: Data used to estimate mean monthly temperatures at Toronto.

Method "S"

This method essentially involved calculating standard differences between various American stations and Toronto. For example, since August is normally 4.5°F (-15.3°C) cooler at Toronto than at Albany, it was assumed that all missing August means at Toronto for which data were available at Albany could be reasonably estimated by subtracting 4.5° from the Albany values, no matter whether the month was near normal or significantly below or above normal.

Method "S" is outlined in Figure 4. There are five distinct steps:

1. **Difference calculations.** Mean monthly temperature data for Toronto (city), 1840 (March-December) - 1870, and for all available American stations within about 400 miles (644 km) of Toronto having significant data before 1870 were tabulated. Sixteen distinct United States stations were available. Most stations did not have data before 1820, but New Haven had data as early as 1778 (Figure 2). In order to facilitate comparison of data from all stations, for each individual month, differences were calculated for all possible stations pairs, for example, Toronto minus Rochester, Toronto minus Albany, Toronto minus New Haven, Rochester minus Albany, etc.

2. **Correction and deletion of bad data.** The differences for each station pair were tabulated by month and year, and the overall monthly-mean differences calculated. Also, standard deviations of the mean monthly temperatures for each station were calculated. Then, for each station pair for each of the 12 months, an average standard deviation (s.d.) was calculated, and all differences greater or less than one standard deviation were identified. For example, the s.d. of the August means at Toronto is 1.8°, Albany 2.2°, for an average of 2.0°. The mean difference for the month, Toronto-Albany, is -4.5°, so that the 1 s.d. range is -2.5° to -6.5°. By identifying differences outside the 1 s.d. range, it was easy to spot unusual months or questionable data. For any month for which no station-pair differences lay outside the 1 s.d range, the data were assumed to be reasonably good. Otherwise, a subjective assessment was made by the areal plotting of the means and departures from normal. In a few cases, it was possible to correct a value when it was obvious that a typographical error of 10°F (12.2°C) had been made in the printed source. Most questionable data, however, were discarded.

3. **Choosing of stations and periods for analysis.** Following the corrections and the discarding of questionable months, a new data set for each of the 16 American stations was prepared. Toronto data were assumed to be "good", and the aim was to choose as many of the 16 United States stations as possible for the 1840-70 period for comparative purposes. Many stations had missing or discarded data in the last half of this period, so it was necessary to use only 1840-55. Within this period, not enough data were available for the computation of reasonable monthly means at Auburn or Buffalo, and data for Rochester College were identical to, or varied by only a small constant from, the Rochester data and hence was suspect. The number of American stations useful for comparative purposes was therefore reduced to 13.

4. **Standard difference calculations.** For each of the 12 months, mean differences in monthly temperatures between Toronto and each of the 13 American stations were calculated. The calculations were based on all months, September 1840 to December 1855, inclusive. I considered that the early data for Toronto (March to August 1840), taken near the lake shore at Fort York, were not homogeneous with the later University site observations. Because of

some missing months for most United States stations, standard differences were calculated from means based on 10 to 15 years in most cases, and which varied from month-to-month and from station-to-station. The standard differences between Baltimore and Toronto ranged between 10.0° and 13.5°, depending upon the month. Because of its great distance from Toronto, and the resulting high and variable differences in the monthly mean temperatures, I decided to eliminate Baltimore. Thus, only 12 stations were left for further analysis.

5. **Calculation of Toronto means.** For each month, July 1778 to February 1840, an estimated mean temperature for Toronto was calculated separately based on each American station for which a mean monthly temperature was available. Thus, in June 1831, the mean monthly temperature at Albany was 72.8°F (22.7°C), and since the standard difference, Toronto-Albany (based on 1840-55) is -6.8°, the Toronto mean was estimated at 66.0°. Similarly, New Haven was 71.1° in the same month, and the standard difference, Toronto-New Haven, is -5.5°, so that Toronto's mean was calculated at 65.6°. The overall Toronto mean was calculated as the unweighted average of all the individual estimates from the various American stations, which for any month would vary from one to 12. This method of estimating Toronto means was checked against the actual means from September 1840 to December 1855. Although some monthly errors were as great as 4°F, the standard deviations of the errors varied from 0.94 to 1.75 for individual months, averaging 1.17°.

Method "D"
In the case of Dade data, mean monthly temperatures had to be calculated from two or three hourly observations a day. Then, standard difference calculations were made by comparing Dade monthly means with those of several American stations. Toronto monthly means were also compared to the same United States stations for the period 1840-55. In this way, a first approximation was made of Toronto-Dade differences. The period of the Dade observations was somewhat cooler on average than that of the later Toronto observations. Consequently, a second approximation of the Toronto-Dade differences was made, allowing for mean temperature differences between the two periods. When Toronto means were calculated month-by-month using these second approximation differences, a comparison with the means obtained by using Method "S" showed a consistent low bias. Hence, a final approximation of Toronto-Dade differences, and therefore, calculation of Toronto means, was made to allow for this low bias.

Method "D" is outlined in Figure 5. There are eleven distinct steps:

1. **Estimation of daily maximum and minimum temperatures.** When Reverend Dade began observations on 1 January 1831, he took readings at 9 a.m., 3 p.m and 6 p.m. However, this pattern did not continue in later months and years. Sometimes the morning reading was at 7 or 8 a.m., but the mid-day observation was usually taken at noon and the evening one at 5 p.m. To complicate matters, while there was always a morning reading (excepting the months with partial days, which were excluded from analysis), sometimes there was in addition only a mid-day reading, sometimes only an evening one, sometimes both and sometimes neither. It was necessary, therefore, to estimate daily maximum and minimum readings. This was done by using mean hourly temperatures in comparison with mean daily maxima and minima for each month at Toronto's Pearson International Airport (Atmospheric Environment Service 1978). Thus, a correction factor was calculated to be subtracted from the morning reading to estimate the daily minimum and to be added to the mid-day and evening observations to estimate the daily maximum. These applied to Pearson Airport, so corrections for Dade were computed by multiplying the Pearson corrections by the ratio of

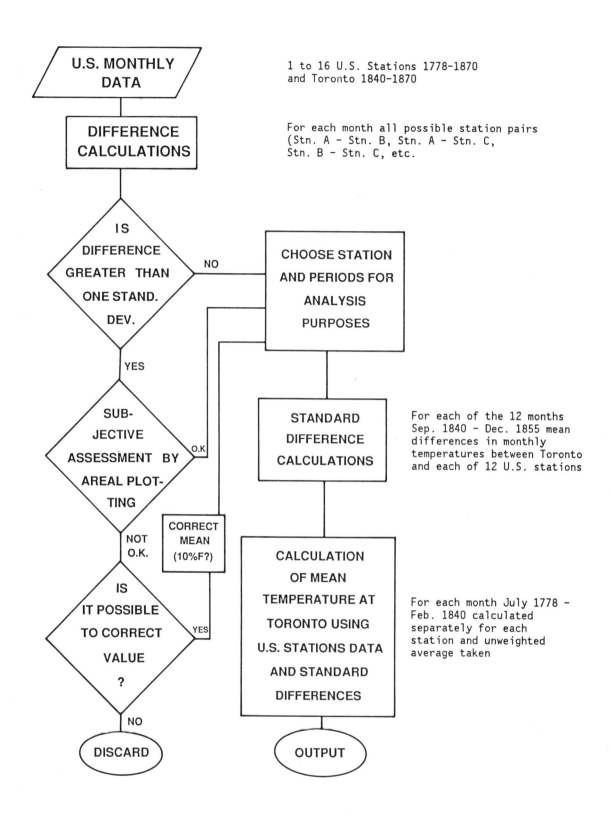

Figure 4: Method "S".

the monthly mean daily range at Toronto (city) to that at Pearson. These figures were rounded to the nearest whole Fahrenheit degree. According to modern practices, the "climatological day" for a climatological station that takes only observations in the morning and evening ends with the morning observation as far as the daily maximum for the previous day is concerned. The daily minimum for Dade observations was calculated as the lowest of: (a) the actual evening reading the day before; (b) the morning observation corrected for the diurnal minimum; (c) the mid-day actual reading; and (d) the evening actual reading. Similarly, the daily maximum for Dade was calculated as the highest of: (a) the morning actual reading; (b) the mid-day observation corrected for the diurnal maximum; (c) the evening observation corrected for the diurnal maximum; and (d) the morning actual reading the following day. In those rare cases where neither mid-day nor evening observations were taken, the minimum temperature for the day was calculated as above. Then a maximum was estimated using the mean daily range for the month at Toronto (city). This value was checked against the morning temperature the following day, and the higher of the two taken. Hence, the daily maximum and minimum calculations took under consideration abnormal diurnal temperature trends.

2. **Calculation of mean monthly temperatures.** For each month, the mean daily maximum and mean daily minimum were calculated from the daily values. The mean monthly temperature was then simply the mean of the mean daily maximum and the mean daily minimum.

3. **Standard difference calculations, Dade minus United States stations.** In order to estimate Toronto mean temperatures by using Dade data, it was necessary to compare Dade monthly means as computed above to those of as many American stations as possible. For each of the 12 months for the period January 1831 to April 1841, mean differences in monthly temperature were calculated between Dade and each of nine American stations: Albany, Cortland, Fredonia, Lewiston, New Haven, New York, Oneida, Rochester, and Utica. Data for Fairfield, Hamilton and Lowville were not used in this analysis due to many missing months of information during the decade.

4. **Standard difference calculations, Toronto minus United States stations.** Similarly, for each of the 12 months for the period March 1840 to December 1855, mean differences in monthly temperatures were calculated between Toronto and each of the nine American stations used in the Dade standard differences above.

5. **First approximation of Toronto-Dade differences.** Since both Toronto and Dade means are compared to the same nine American stations, the first approximation of Toronto-Dade differences was obtained by subtracting the Toronto standard differences above from the Dade standard differences above. These were calculated separately for each of the nine American stations, and the overall mean taken for each month. This analysis indicated that, for every month of the year, Dade values were high, and that correction values ranging from -0.6° (February) to -4.0° (July) had to be applied to Dade means to give a reasonable estimate of Toronto means. For the bitterly cold December of 1831, as an example, the Dade calculated mean was 15.8°. Since the correction value for December is -2.2°, the first approximation of the Toronto mean for the month would be 13.6°F (10.2°C).

6. **Comparison of mean temperatures for 1831-41 with 1840-55.** There was no reason to assume that the whole period 1831-55 was climatologically homogeneous. In order to obtain a measure of the differences in mean temperature between the early period of the Dade

observations (1831-41) and the later period of the Toronto observations (1840-55), calculations were performed for the three United States stations with the best and most continuous observations, Albany, New Haven and New York. Means were calculated for each month separately for each of the three stations and for both periods, allowing for those months when Dade observations were missing. For each month, an overall mean difference (unweighted average of the three stations) between the period means was obtained. The earlier period was colder than the later at each of the three stations for each of the 12 months. The overall monthly differences ranged from 0.4°F (-17.6°C) for April to 2.5°F (-16.4°C) for December.

7. **Second approximation of Toronto-Dade differences.** The second approximation considers the fact that the earlier 1831-41 period was significantly colder than the later 1840-55 period. The first approximation Toronto-Dade difference in mean temperature is -4.0°. The difference between the two periods for the same month is 1.2°, so that the total correction applied to Dade means to obtain Toronto means is -5.2°F. Because of the variability from month to month, Fourier smoothing was applied to the monthly values. As a result, the second approximation of Toronto-Dade differences ranged from -2.7° in September and October to -4.1° in December. Again, in the case of the frigid December of 1831, the Dade mean of 15.8° with a correction of -4.1° results in a Toronto mean of 11.7°F (-11.3°C). This is 1.9° lower than the first approximation calculation.

8. **Preliminary calculation of mean Toronto temperatures using Dade and second approximation differences.** For each month for which Dade means were available in the period January 1831 to April 1841, a Toronto mean was calculated using the second approximation differences (Fourier-smoothed) above.

9. **Comparison of mean temperatures at Toronto by using Dade calculations above and by using Method "S".** For each month for which Dade means were available in the period January 1831 to April 1841, the Toronto mean using the Fourier-smoothed second approximation differences with Dade were compared with means as calculated by Method "S" (using all available American data). The overall mean differences in the two methods were compiled for each month and it was found that Method "S" gave higher values than the Dade method in all months - ranging from 0.4° in March to 2.6° in February. The standard deviation of the monthly differences between the two methods ranged from 0.6° in June to 1.7° in January. In the case of the frigid December of 1831, Method "S" indicated a Toronto mean of 14.1°F (-9.9°C), 2.4° higher than the preliminary calculation using Dade.

10. **Final approximation of Toronto-Dade differences.** Since Method "S" indicated somewhat higher means for Toronto for every month of the year than those by using the preliminary Dade calculations, apparently the second approximation allowing for the mean temperature differences between the two periods was based on differences that were too great. Consequently, the final approximation of Toronto-Dade differences was calculated by reducing the second approximation differences by the differences indicated between Method "S" and the preliminary Dade calculations above. Thus, the December second approximation Toronto-Dade differences, Fourier-smoothed, is -4.1°, the correction due to the Method "S" comparison is +1.7°, so the final Toronto-Dade correction for the month is -2.4°. Again, because of the month-to-month variation in the correction values, a Fourier smoothing was applied. The final approximation of Toronto-Dade differences then ranged from -1.1°F in October to -3.2°F in June. In the case of bitterly cold December 1831, the

smoothed final Toronto-Dade difference is -1.9°, so that when applied to the Dade mean of 15.8°, the Toronto estimate works out to be 13.9°, very close to the 14.1°F (-9.9°C) in the case of Method "S".

11. **Final calculation of Toronto means.** For each month for which Dade means were available in the period January 1831 to February 1840, a Toronto mean was calculated using the final approximation differences (Fourier-smoothed) above.

Method "A"

In the case of Ancaster data, published monthly means based on two observations per day, 9 a.m. and 9 p.m., had to be corrected to standard monthly means based on daily maxima and minima. Then, standard difference calculations were made between the overlapping records of Ancaster and Toronto . From these, estimates were made of Toronto means for those months before records began in March 1840.

Method "A" is outlined in Figure 6. There are three distinct steps:

1. **Correction of mean monthly temperatures.** No daily observations are available for Ancaster, only published monthly means of 9 a.m. and 9 p.m. observations, from which a simple average was computed to produce a monthly mean. Correction values were calculated for each month in order to provide monthly means based on the modern practice of using daily maxima and minima. These were done by comparing means produced by averaging 9 a.m. and 9 p.m. monthly means at Toronto's Pearson International Airport (Atmospheric Environment Service 1978) with the monthly means at the same station, which are calculated by the usual mean of daily maximum and minimum values.

2. **Standard difference calculations.** Ancaster and Toronto data overlap for the period March 1840 to December 1845. For each of the 12 months during this period, mean differences were calculated between the corrected Ancaster monthly means and the official Toronto means.

3. **Calculation of Toronto means.** For each month, January 1835 to February 1840, a mean temperature was calculated for Toronto using the corrected Ancaster mean and the standard differences above.

The Reconstructed Toronto Temperature Time-Series

By using Methods "S", "D" or "A" as appropriate, monthly mean temperatures were estimated for Toronto from July 1778 to February 1840. No American data were available for September 1778 or February, July, August, October, November and December 1779, so that no means could be estimated for these months. Beginning with January 1780, a complete set of monthly values was obtained. *All calculations were done using the Fahrenheit scale, and then the whole set was converted to Celsius* and combined with Atmospheric Environment Service Archive values that begin March 1840 and continue with no breaks until the present day.

Statistical F tests were applied to monthly and seasonal values to test the homogeneity of variance between various periods. In the first instance, three 30-year periods were chosen: (A) 1780-1809; (B) 1810-39; and (C) 1840-69. Period A involves only Method "S": through much of this period

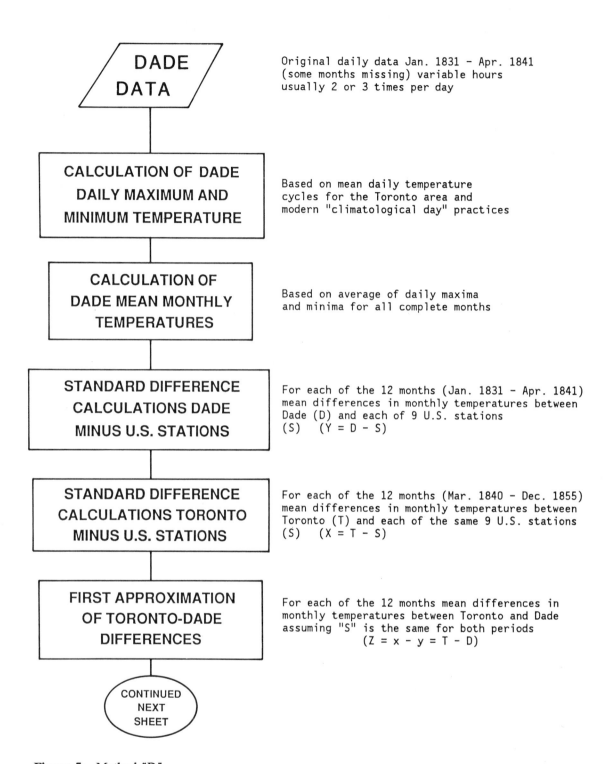

Figure 5: Method "D".

Figure 5: (cont'd)

METHOD "D" - CONTINUED

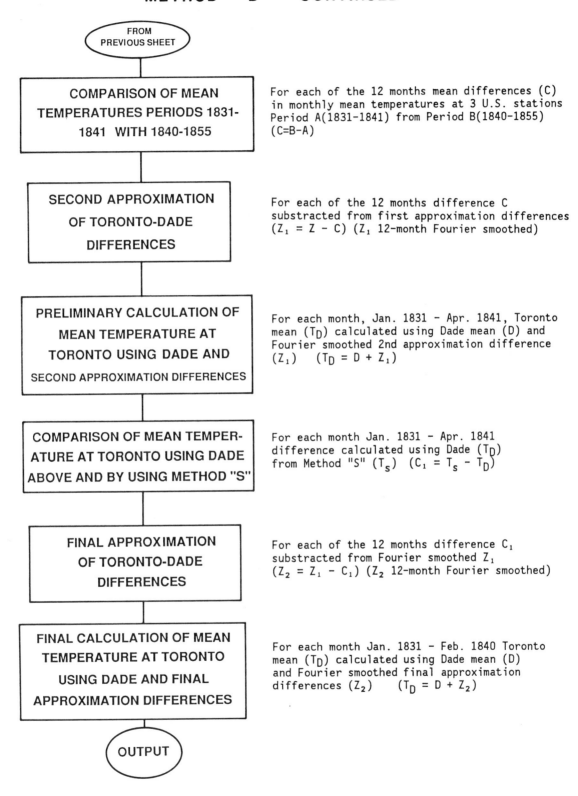

FROM PREVIOUS SHEET

COMPARISON OF MEAN TEMPERATURES PERIODS 1831-1841 WITH 1840-1855

For each of the 12 months mean differences (C) in monthly mean temperatures at 3 U.S. stations Period A(1831-1841) from Period B(1840-1855) (C=B-A)

SECOND APPROXIMATION OF TORONTO-DADE DIFFERENCES

For each of the 12 months difference C substracted from first approximation differences ($Z_1 = Z - C$) (Z_1 12-month Fourier smoothed)

PRELIMINARY CALCULATION OF MEAN TEMPERATURE AT TORONTO USING DADE AND SECOND APPROXIMATION DIFFERENCES

For each month, Jan. 1831 - Apr. 1841, Toronto mean (T_D) calculated using Dade mean (D) and Fourier smoothed 2nd approximation difference (Z_1)   ($T_D = D + Z_1$)

COMPARISON OF MEAN TEMPERATURE AT TORONTO USING DADE ABOVE AND BY USING METHOD "S"

For each month Jan. 1831 - Apr. 1841 difference calculated using Dade (T_D) from Method "S" (T_S) ($C_1 = T_S - T_D$)

FINAL APPROXIMATION OF TORONTO-DADE DIFFERENCES

For each of the 12 months difference C_1 substracted from Fourier smoothed Z_1 ($Z_2 = Z_1 - C_1$) (Z_2 12-month Fourier smoothed)

FINAL CALCULATION OF MEAN TEMPERATURE AT TORONTO USING DADE AND FINAL APPROXIMATION DIFFERENCES

For each month Jan. 1831 - Feb. 1840 Toronto mean (T_D) calculated using Dade mean (D) and Fourier smoothed final approximation differences (Z_2)   ($T_D = D + Z_2$)

OUTPUT

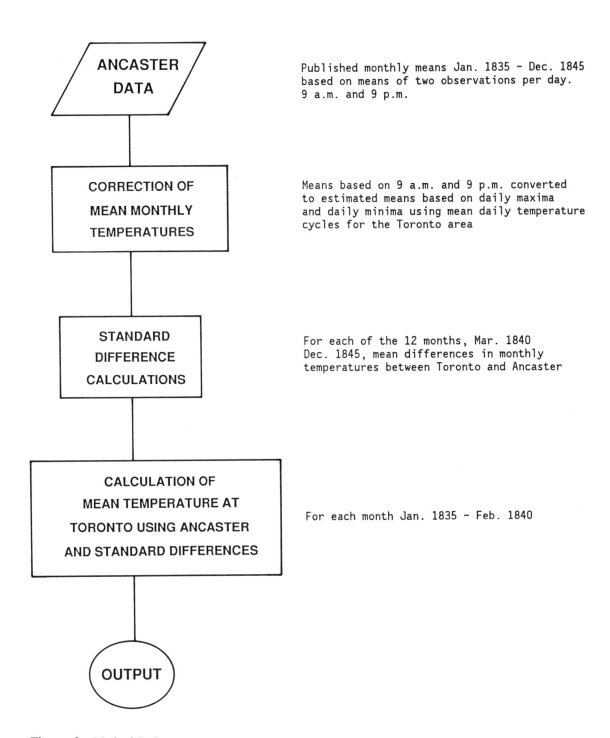

Figure 6: Method "A".

only one, two or three American stations had data available for comparative purposes. Period B involves Dade data as well as an increasing number of United States stations. Period C involves the early instrumental record at Toronto before urban warming was significant. Only June temperature variances are significantly different at the 99% level between periods A and B and between A and C. In the second instance, two 40-year periods were chosen: (A) 1801-40 and (B) 1841-80. Period A represents the last 40 years of reconstruction prior to the official observations beginning in March 1840, whereas Period B contains the first full 40 years of instrumental data. Only January variances are significantly different at the 99% level.

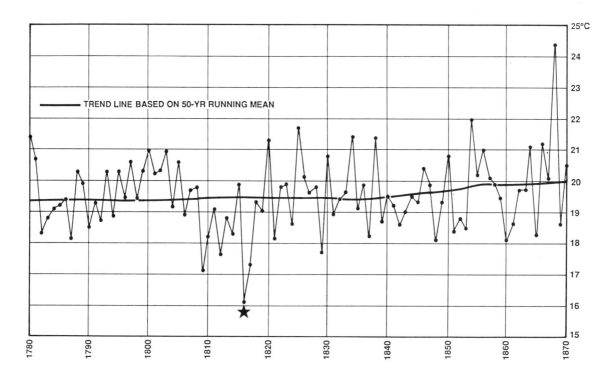

Figure 7: Mean July temperatures at Toronto.

Individual plotted mean monthly July temperatures from 1780-1870 are shown (Figure 7). *The cold July of 1816 (the year without a summer) is immediately evident.* The trend line is based upon a 50-year running mean.

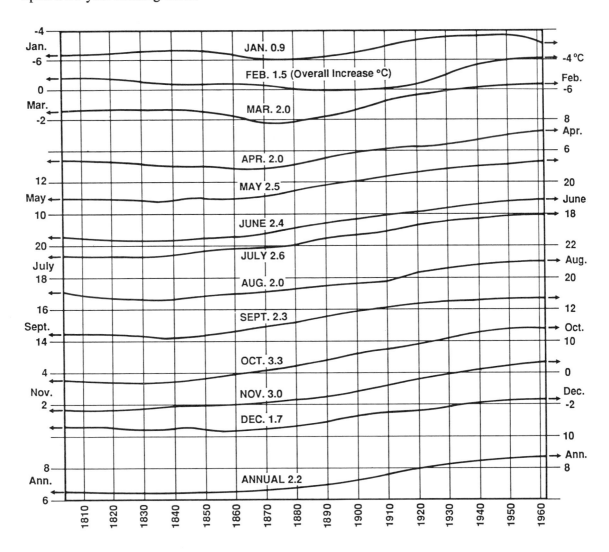

Figure 8: Trend lines based on 50-year running means of mean monthly and annual temperatures at Toronto (values before 1804 and after 1962 are considered constant).

In Figure 8 trend lines based upon 50-year running means plotted for the middle year are shown for all months. Fifty-year means for 1780-1987 can be plotted only from 1805 to 1962. Temperatures were considered constant before and after these dates. The overall rise in temperature during the period ranges from 0.9°C in January to 3.3°C in October, with an annual average of 2.2°C. A significant amount of this increase is no doubt due to the urban heat-island effect, which became increasingly significant from the 1880s on.

References

Atmospheric Environment Service. 1978. *Hourly Data Summaries - No. 3R, Toronto International Airport, Ontario.* Climatological Services Division, Atmospheric Environment Service, Downsview, Ontario. p. 28.

Craigie, W. 1835. Mean results for each month of eleven years (1835 to 1845, inclusive) of a Register of the Thermometer and Barometer, kept at Ancaster, C.W. Clippings from newspapers, names and dates unknown. (Unpublished manuscript, Atmospheric Environment Service, Downsview, Ontario).

Dade, Reverend C. 1831-41. Temperatures at Toronto. 47 pp. (Unpublished manuscript, Atmospheric Environment Service, Downsview, Ontario).

Hough, F.B. 1855. *Results of a series of Meteorological Observations, Made in Obedience to Instructions from the Regents of the University, at Sundry Academies in the State of New York, from 1826 to 1850, Inclusive.* Weed, Parsons and Company, Albany. 502 pp.

_____. 1872. *Results of a Series of Meteorological Observations, Made under Instructions from the Regents of the University, at Sundry Stations in the State of New York, Second Series, From 1850 to 1863, Inclusive.* Weed, Parsons and Company, Albany. 406 pp.

Smithsonian Institution. 1927. *World Weather Records.* Smithsonian Miscellaneous Collections 79. The Lord Baltimore Press, Baltimore. 1199 pp.

United States Weather Bureau. 1932-37. *Climatic Summary of the United States; Climatic Data Herein from the Establishment of Stations to 1930, Inclusive.* Third Edition. United States Government Printing Office, Washington, D.C.

Climate in Canada, 1809-20: Three Approaches to the Hudson's Bay Company Archives as an Historical Database

Cynthia Wilson[1]

Abstract

The Hudson's Bay Company archives are rich in weather information. Material includes: Meteorological Registers; descriptive entries encapsulating the day's weather or seasonal comment in the Post Journals, Correspondence and Annual Reports; and proxy weather data. As a database for studying past climate, the strength of the archive is its diversity, permitting cross-checking of results and the convergence of evidence. But the problem of fragmentary evidence has to be overcome.

This paper briefly describes three approaches I have taken to integrate the different material, in studying May-October climate during the nineteenth century along the east coast of Hudson/James Bay, and over east/central Canada: (1) to establish a detailed regional climatology (1814-21) as an historical benchmark; (2) to obtain year-by-year (1800-1900) estimates of monthly temperature anomalies (reference 1941-70) and wetness indices; (3) to produce schematic daily weather maps (east/central Canada, 1816-18).

The results of these studies indicate: (1) from 1800-10, May-October temperatures along the east coast of the Bay were akin to those today. The seasons then became cooler, with mean temperatures falling spectacularly in 1816 and 1817 to values below the modern record; from 1811-20, they averaged about 1.6°C below the 1941-70 normal; (2) the low temperatures were accompanied by greater-than-normal snowfall, and in 1816 and 1817, they essentially precluded the growth of many plants. Some areas of snowcover probably remained through the season in 1816, and offshore, the Bay was barely free of ice at the end of the season; (3) flow patterns over east/central Canada from 1 June to mid-July 1816 suggest that spring had been delayed or protracted by as much as six weeks; (4) although there was some recovery in 1818, seasonal temperatures below the 1941-70 energy level remained characteristic until the 1870s.

The period of unrelieved record cold from October 1815 to March 1818 may have been influenced by Tambora - a relatively short-term volcanic eruption exacerbating a longer-term (60-year) lowering of temperature already underway. But there can be no doubt that the exceptionally heavy Bay ice lingering through the summers combined with the high frequency of onshore north and west winds was a major factor in reducing summer temperatures on the east coast of the Bay.

Introduction

Until well into this century, climate loomed large in the daily living and even survival of the those inhabiting the Canadian Shield and Prairies, and the Hudson's Bay Company (HBC) Post Journals, Correspondence and Annual Reports provide a rich variety of information directly or indirectly pertaining to weather.

---

[1] 90 Holmside, Gillingham, Kent ME7 4BE, U.K.

The climatic information is of three kinds: (1) Meteorological Registers, often meticulously kept in accordance with the accepted practices of the time, but with one or two important exceptions on the east side of Hudson Bay, the periods of record are relatively short; (2) descriptive entries in the Post Journals encapsulating the day's weather, with occasional seasonal comment (the latter is also found in the Correspondence and Annual Reports); (3) proxy weather information, the impact of weather on the natural environment, and on the property, activities and well-being of the inhabitants (in all three sources).

In using this remarkable record as a database to extend the modern climatic series into the past, and to study past climatic anomalies, a major problem is that of fragmentation. This has resulted from accidents of history, Company policies and activities, and from the nature and interests of the individuals recording the events. The strength of the archive is its diversity, permitting cross-checking of results and the convergence of evidence. This paper describes briefly three approaches that I have used to integrate the material, so as to overcome the fragmentation and take full advantage of the diversity in reconstructing (Figure 1):

1. A regional climatology[1], as a climatic benchmark in the historical record. (The east coast of Hudson/James Bay, 1814-21).

2. Extended seasonal time series[1] - temperature and wetness indices. (The east coast of Hudson/James Bay, 1800-1900).

3. Schematic daily weather maps[2]. (East/central Canada, summers 1816-18. This study is still in its early stages).

Some aspects of the reconstructed climate in Canada from 1809 to 1820 have been selected to illustrate the rich potential of the HBC archives as an historical database.

The overall approach to the historical material was traditional, in which the researcher does the abstracting so as not to lose vital information offered by the context and subtext. With this in mind, weather and proxy data were abstracted in context. Climatically, the historical data were approached as far as possible in physical terms, from the standpoint of small-scale climatology, the approach of Landsberg (1967) and Geiger (1965). Even with the synoptic mapping, this approach was helpful in evaluating and interpreting the individual point data. In this, personal experience of several summers in the field at Great Whale[3], observing the weather, measuring surface energy exchanges and keeping weather journals, has played an integral part.

Owing to the detailed nature of this kind of work, I do not have space here to substantiate the methods or to discuss the assumptions and confidence limits. This information is available, together with a full account of the results, in four reports distributed by the Canadian Climate Centre (Wilson 1982, 1983a, 1985a, 1988) and three papers published by the National Museum of Natural Sciences in *Syllogeus* (Wilson 1983b, 1985b, 1985c); see also Wilson 1985d.

---

[1] Studies 1 and 2 were carried out under contract to the Canadian Climate Centre, Atmospheric Environment Service, Downsview, Ontario.

[2] The developmental stages of study 3 have been funded by the National Museum of Natural Sciences, Ottawa, as part of the Museum's *Climatic Change in Canada Project*.

[3] I am grateful to the Centre d'études nordiques, Université Laval, Québec, the National Research Council of Canada and the Canadian Atmospheric Environment Service for the logistical support and research funds which made this possible.

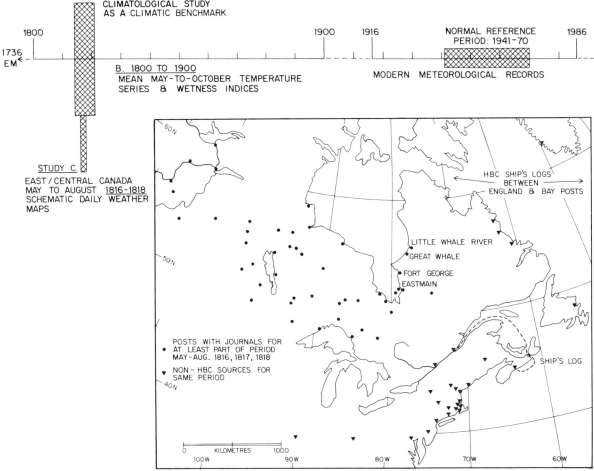

Figure 1: Studies of past climate in Canada: three approaches to the Hudson's Bay Company archives as an historical database.

Place and Time of Study

Considering the amount of work involved, the choice of region and period for study is critical. For the regional climatology and the construction of the time series, the east coast of Hudson/James Bay (HBC Posts: Eastmain, Fort George, Great Whale, Little Whale River) and the active season (May-October) were selected for reasons that follow.

The Hudson Bay region appears to be particularly sensitive to climatic fluctuations, and the eastern windward coast of the Bay provides an excellent laboratory. It is a marginal area with respect to the fluctuating arctic/subarctic boundary and the northern limit of tree growth. This vast sea, with its seasonal ice cover, open to arctic waters and arctic ice, extends the influence of polar climate into the heart of the continent (south of Latitude 52°N) in spring, and remains a cold sink in summer; in late autumn the presence of open water creates a snowbelt on this windward east coast. To the east lies the plateau of New Quebec/Labrador, a former centre of the Laurentian Ice Sheet. All forms of life are so finely tuned to climate along these marginal coastlands that any unusually severe or prolonged anomaly can soon disturb the ecological balance, and human life and activity - and incidentally make good copy for Journal writers. Other reasons for the choice of region include the availability of adequate modern weather and environmental records, and my first-hand knowledge of the area.

The earliest HBC Post Journal for this region is for Eastmain in 1736, but I began the time series with the nineteenth century, when better coverage was available for the warm season. With few exceptions, there was at least one Post reporting through each season during the century. In 1814, the Hudson's Bay Company gave top priority to a carefully defined program of weather and weather-related observations at their Posts in Canada. With wars closing markets in America and Europe, fiery competition in the field from the North West Company and a worsening economic and social climate at home, the aim was to study and develop the local agricultural potential at each Post, to cut the high costs of sending out European food. At a number of Posts, including Great Whale, Fort George and Eastmain, fixed-hour temperature and weather data were also recorded regularly in Meteorological Registers. The directive fell into abeyance after the amalgamation with the North West Company in 1821. Although the Company may never have applied its hard-won information, the archives from 1814 to 1821 remain a rich data source for detailed regional climatological studies of a period of unusual climatic interest, and were used here.

Again, by implementing this programme in 1814, the Hudson's Bay Company through its network of Posts and lines of communication in central, western and northern Canada provided a system of synoptic weather observation unique at this time, both in the discipline imposed, the consistency of purpose and of manner of observing and recording, and in the extent of its coverage. These Company records, together with the logs of the annual supply ships from England, coastal shipping and of canoe journeys, offer a basis for synoptic weather mapping.

From the results of the climatological study for the east coast of the Bay, and given the interest in the atmospheric circulation in the years around the eruption of Mount Tambora, the summers 1816 to 1818 were chosen for initial mapping and investigation. Figure 1 locates all HBC Posts with Journals for at least part of the period May-August 1816-18. The density of the network and type of weather information available varies through the season and from summer-to-summer, depending in part on the regular seasonal operations and needs of the fur trade itself, but to a greater extent on the conflict between the Hudson's Bay and the North West companies. Sadly, the battle for the Athabasca trade curtailed weather information from the Red River Valley westward from June 1816 through 1817. For the three summers, no alternative historical weather sources have been found for the west. For eastern Canada and northeastern United States a number of weather records, personal diaries, Mission reports and newspaper articles are available for this period to extend coverage (see Figure 1). Regular weather observations were also recorded at Godthaab (now Nuuk), Greenland. The search for additional information continues.

The Three Approaches

A Regional Climatology, 1814-21

The prime weather data sources in the HBC archives are the Meteorological Registers, even where they were kept for only a few years. To try to make full use of them, one approach is to analyze and integrate all aspects of the local or regional weather from all HBC sources for those years - temperature, winds, cloud, precipitation, extreme events and, rarely, pressure, together with all proxy-weather indicators - to gain as clear a picture as possible of the climate of that time in terms of the present-day climate: that is, to set up a climatic benchmark.

The dangers of this approach are only too well-known: the differences between the historical and modern instrumentation, exposure, observing practices and so on. Happily, a tradition of careful meteorological observation and reporting had become established on the west side of the Bay in the second half of the eighteenth century with the collaboration of the Royal Society, which advised on instruments and procedures. This tradition continued into the early nineteenth century. Provided that basic assumptions are made explicit and their import is clearly stated, and that every effort is made to compare like with like, I believe the results to be worth the time and effort required.

With the historical temperature readings at Great Whale/Fort George and Eastmain, I tackled the calibration from several directions, hoping in this way to approach a consensus and to avoid circular arguments. The three main lines of attack were:

1. *Historical* - the reconstruction of the early observing sites, and social context, and of the meteorological instrumentation and procedures accepted at the time. A study was also made of the history and homogeneity of the respective modern temperature records.

2. *Physical* - examining the systematic temperature differences that might arise from changes in site, instruments and their exposure, and observing practices, given the distinctive qualities of the subarctic surface conditions and regional and local weather.

3. *Statistical* - an application and extension of current Canadian quality-control procedures, the fitting of simple regression models, and the analysis of the fields of error.

Although corrections were made to the daily maximum and noon temperatures, and also to the daily minimum where the start of the climatological day differed, I was impressed by the consistency of the historical temperature record in the context of the instruments and procedures of the time.

Extended Seasonal Time Series, 1800-1900

The basic HBC sources of continuous and consistent weather information over extended periods are the descriptive entries in the Post Journals. A reading of the Journals often leaves a strong impression as to the relative heat or cold, dryness or wetness of the different seasons, through the subtle integration of the many different weather and environmental factors and their impact. Thus a second approach is to try to integrate daily and seasonal weather remarks and all forms of proxy-weather information, to obtain monthly estimates of the temperature anomaly with respect to a modern reference period (in this case 1941-70); also, to obtain monthly wetness indices with respect to the modern precipitation record. The benchmark set up in the first study acts as a useful reference for the early part of the century.

The approach is similar to that taken by Pfister (1980) in Switzerland. For the thermal series, a first approximation to the monthly temperature anomaly is obtained from the direct weather remarks, then the proxy indicators are applied to try to obtain an order of magnitude (timing, intensity, duration), or at least supporting or modifying evidence. One major difference here is that a greater variety of information must be used to compensate for the fragmentation of individual data series, to give a convergence of evidence.

As the method must accommodate so many different kinds, fragments and combinations of data, some firm climatological structure is required. A secure, yet flexible modern frame of reference was provided by the nine modern daily temperature curves for Great Whale, Fort George and Eastmain, respectively - comprising for each day of the May-October season the reference period daily mean, the highest and lowest daily mean, the daily mean maximum and the highest and lowest maximum, the daily mean minimum and the highest and lowest minimum. Other aids included a wide variety of modern temperature and temperature-related information, including analogues for warm and cold months.

The proxy-weather sources (Figure 2) can be grouped into three: snow and ice, phenological events (plants and animals), and human activities. These data have been approached from two points of view:

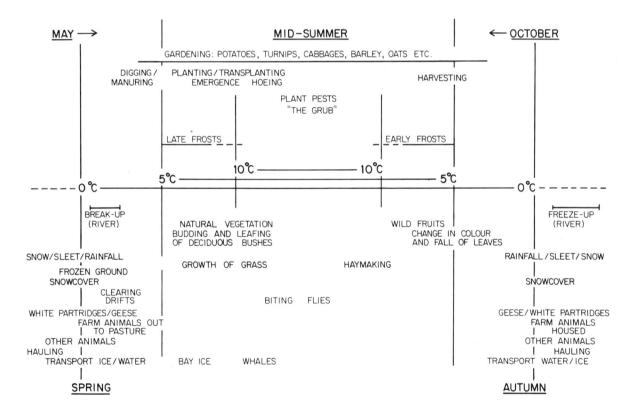

Figure 2: Summary of environmental indicators.

1. To obtain an indication of the timing and magnitude of seasonal and unseasonable events, and to compare where they intersect the modern daily or weekly temperature curves. One avenue of attack is the heat-unit concept, with thresholds 0°C, 5°C and 10°C; another, the cardinal points with respect to the different crops; a third, the agroclimatic capability classes, which delimit in climatic terms the crop potential of the region - and so on.

2. To try for a statistical link between the proxy data and the monthly temperature anomaly during the modern period of record, to provide guidelines or rules of thumb.

All the information was integrated for each season and each Post in direct comparison with the modern reference period to give the temperature anomaly month-by-month from May to October. Where more than one Post was reporting, the anomalies were then compared and combined. The period of greatest confidence is 1815-20.

The wetness index is a five-point scale based on the number of days with reported precipitation, together with supporting remarks and proxy evidence indicating its intensity or duration, and the degree and duration of dry periods. Each month and each season, for each Post, was assessed directly against the modern record of precipitation at the respective weather stations. Following Pfister, the upper and lower quartiles were chosen as limits between wet and dry months; also, the octiles define very wet and very dry classes. Adjustments were made to account for those days when rain fell only at night and may not have been reported, and to account for discrepancy in the modern reporting of the number of days with snowfall between 24-hourly observing stations and climatological stations (cf. Ashmore 1952; Manley 1978).

Schematic Daily Weather Maps, Summers 1816-18

A third approach permitting the integration of all available kinds and fragments of weather information is synoptic mapping, although the Meteorological Registers with their detailed timed weather observations act as linch pins in the historical analysis.

Since the historical data include few records of atmospheric pressure at this time, these weather maps for east/central Canada are primarily based on surface wind data - more readily and frequently recorded. Thus they offer schematic representations of the flow patterns, rather than the refinement of Kington's (1988; this volume) classic series of daily weather maps for western Europe and the northeastern Atlantic in the 1780s, and in 1816, which are firmly based on a network of barometer readings. That surface winds can be so used has already been elegantly demonstrated by Lamb (in association with Douglas) for the period of the Spanish Armada, May-October 1588 (cf. Lamb 1988). Caution is required. Regional wind direction and speed can be modified at some sites by local topography, the geometry of forest and other obstructions, or masked by the influence of local sea or lake breezes and valley winds. But marked local effects can usually be detected and allowed for if the network is reasonably dense and given some knowledge of the terrain.

As a first step in the map analysis, all the direct and proxy-weather information are plotted on daily base maps. Using transparent overlays, the data for morning and evening hours are transposed to separate charts. Each chart is then analyzed over the base map, which provides the necessary background information as well as intermediate history, and in conjunction with previous maps. Areas of cloud and precipitation are shaded in, the temperature and wind fields studied, the zones of maximum gradient, wind shear and the pressure tendencies noted. Frontal zones are tentatively indicated. Then an attempt is made to sketch in the pressure pattern, bearing in mind the wind speed, the nature of the surface and the most likely direction and speed of movement of the frontal systems. As an aid to analysis, where Registers exist, daily temperature curves have been drawn up by month, with winds, cloud and precipitation added, to give a visual display of the sequence of weather. All available pressure traces have also been plotted.

Here, the question of data calibration is partly resolved by the space and time smoothing implicit in the map analysis. With respect to temperature, it is the relative differences and changes that are important, and any significant errors would be expected to stand out. The results of the earlier calibration study of the HBC data suggested that observing practices at this time were consistent throughout the network; also that temperature readings were most reliable in the morning and evening and near freezing. The wind-force scale in use on the Bay was quite similar to the Beaufort Scale, and the latter has been used to convert all indicators to approximate speeds. Wind direction was generally assumed to be with respect to true north. The barometer readings have been reduced where possible to sea level.

Concerning the synchronization of data across Canada, the Hudson Bay region is considered as the reference time zone; the western and eastern extremities of the map are then within about $\pm$ two hours, which can be born in mind. The timing and frequency of the observations are seminal to the analysis. Problems here can often be overcome when maps are sketched in for both morning and evening. In drawing up early weather maps, "historical continuity" becomes a prime tool in grappling with the many difficulties, including that of sparse data. There is also the advantage of knowing (if only in part) the future as well as the past.

Notes on Climate in Canada, 1809-20

What can be gleaned from these studies about climate in Canada from 1809 to 1820? Are there any clues following the major volcanic eruption of Mount Tambora in April 1815 as to the possible influence of such massive atmospheric loading on regional climate? To what extent was the atmospheric circulation over east/central Canada anomalous at this time? The following notes illustrate some of the kinds of climatic information that can be reconstructed using the Hudson's Bay Company archives as an historical database.

Temperature

To set the climatic events of the period 1809-20 in context for the east coast of Hudson/James Bay, Figure 3 shows the reconstructed series of May-October mean temperature anomalies through the nineteenth century together with the modern twentieth century record. The reference period is 1941-70. During the first decade of the last century, mean temperatures were similar to those today, but then the seasons quite rapidly became cooler, tumbling in a spectacular fall in 1816 and 1817 to values far below the modern records. Seasons well below the 1941-70 energy level remained characteristic of the first half-century with a cluster of cold years from 1835 to 1840; decadal averages for 1811-20 and 1831-40 were about 1.5°C below. A certain mid-century amelioration was followed by some very cold weather in the 1860s, before a remarkably sudden change occurred in the early 1870s to a new higher energy level akin both to that at the beginning of the nineteenth century and to that today. Although not as warm as the 1870s, the last two decades sustained these milder conditions, and variability remained well within that for 1941-70. The upward trend from 1811-40 through 1870-1900 shows clearly in the overlapping 30-year means. This sequence of events has no analogue in this century. The volcanic eruptions of Tambora and Coseguina were followed by extreme cold, but the decline had set in beforehand. For Krakatau and Agung, any similar signal (if that is what it is) is weaker, and for St. Helens and El Chichón, absent.

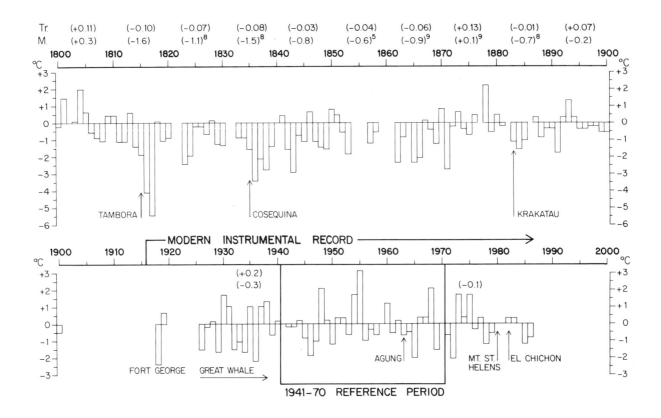

Figure 3: East coast, Hudson/James Bay. Reconstructed mean temperature anomalies (°C) summer seasons (May-October) 1800 to 1900. M, decadal means; superscripts, number of years. Tr, maximum tree-ring density, decadal mean anomaly (Parker et al. 1981). Anomalies with reference to the 1941-70 period.

Within the May-October season, the most striking feature of the nineteenth century is the coldness of spring (May/June) and autumn (September/October), and the effective shortening of the active season. The curve for the spring months closely parallels that for the season as a whole, but in autumn, cold seasons persist through the last two decades, although not below the modern record low of 1974. In complete contrast, midsummers (July/August) were in general not so very different from those today, although 1816 and 1817 with anomalies of about -5.5°C and -4.0°C were in a class by themselves; cold summers also occurred in 1836 (-2.5°C) and 1871 (-3.0°C). By far the coldest summer on modern record was 1965 (-3.5°C), two years after Agung.

Focusing on the second decade of last century, with its remarkably sudden lowering of temperature, Table 1 gives the absolute mean temperatures for 1808-20 (estimated for Great Whale from the reconstructed regional anomalies) together with the 1941-70 normal values. The first strike in the change to a colder mode was winter 1808-09, heralded by a very cold autumn in 1808. Gladman, Master at Eastmain (a keen observer and a veteran of these shores), found it the coldest winter he had ever experienced, with temperatures frequently below -40°C and scarcely any mild weather. The next event was the extremely early and very cold autumn of 1811, advent of the even colder longer winter of 1811-12; the two spring months May/June 1812 probably averaged below freezing at Great Whale (Table 1). These were considered extraordinary times. In 1811, the HBC annual supply ship from England did not arrive at Moose Factory until 25 September, which was so late in the season and unprecedented at that time that there had been great alarm at Eastmain lest no ship should arrive. As Gladman wrote: "these circumstances are altogether so new and unfortunate" (HBC.B59/b/30). When the ship sailed for England, it could no longer leave the Bay for ice, and wintered in Strutton Sound off Eastmain. An HBC supply ship had last wintered over (though storm damage) in 1715 (Cooke and Holland 1978). It was to happen again, as a result of ice, in 1815, 1816, 1817, 1819.

The most persistent and intensely cold period, continuously below the reference normal, began in autumn 1815 and lasted until late winter 1818. The degree of cold reached its nadir in January/February 1818, to be followed by a remarkable flip to an early, warm spring and a benign season. Following this, cold springs and autumns continued. For 1815-20, Figure 4 shows the monthly temperature anomalies for May-October at Great Whale, Fort George and Eastmain - adjusted values based on temperature observations in the Meteorological Registers (cf. Wilson 1983b). Looking more closely at these seasons in Table 1 and Figure 4, several features are outstanding.

1. The degree of the anomalous cold in 1816 and 1817, with many of the months below the modern record - Alder, Master at Great Whale, then at Fort George: "if summers I may call them" (HBC.B77/e/1a). At Great Whale in 1816, July appears to have been nearly 6°C below normal, that is 2°C below the lowest on record 1965. The modern standard deviation is 1.2°C. At Fort George in 1817, the season as a whole was some 5°C below normal. As a season 1817 was more severe than 1816, but the greater severity was in spring and autumn rather than midsummer.

2. From autumn 1815 through April 1818, the greater part of this coast experienced arctic conditions following Köppen's definition. During summers 1816 and 1817, the arctic/subarctic boundary lay close to Eastmain, some 3.5° latitude south of the present average position near Richmond Gulf. The closest modern analogue is probably the 1965 season.

3. In 1815 and 1816, the rise of the daily mean temperature through 0°C, the start of the active season, was two to three weeks later than the 1941-70 normal, akin to 1972. In 1817, it was some four to five weeks late and the return through 0°C three weeks early in autumn, hence the period above 0°C was some two months shorter. This was reflected in the break-up and freeze-up of river ice. It is almost certain that seasonal ice remained in the ground in many areas throughout the 1816 and 1817 seasons. It was also reflected in the snowcover, and type of precipitation. The implications with respect to plant growth can be clearly seen in Table 1.

Precipitation

Again, to set the 1809-20 period in context, Figure 5 shows the series of May-to-October wetness indices for the east coast of Hudson/James Bay through the nineteenth century, and the modern record expressed in the same form. In general, the first half of last century was not only colder but wetter than today, while the second half became warmer and drier. By far the wettest decade was 1811-20, with a run of wet seasons from 1814 to 1820. The three wettest seasons of the century 1816, 1817, 1820, at least matched the record in 1944. The three driest seasons were 1809 (which probably equalled the record 1920 season), 1807 and 1878. The rapid change from the warmer/drier mode of the first decade to the very cold/wet regime of the second is remarkable.

Table 1: Great Whale, Reconstructed Mean Temperatures (absolute values in °C) 1808-20.[1]

PERIOD	1941-1970	1808	1809	1810	1811	1812	1813	1814	1815	1816	1817	1818	1819	1820
May to October	6.3	5.3	6.8	6.8	5.3	5.3	7.0	5.0	4.5	_2.3_	_1.0_	6.5	5.3	5.5
May/June	3.6	3.1	2.6	3.6	5.1	_-0.9_	5.1	2.6	_-0.4_	_0.1_	_-2.9_	4.6	1.6	3.1
July/August	10.5	11.0	12.5	11.5	10.5	11.5	11.5	10.0	12.0	_5.0_	_6.5_	10.5	12.0	11.0
Sept./October	4.9	_1.9_	5.4	5.4	_0.4_	5.4	4.4	2.4	_1.9_	_1.9_	_-0.6_	4.4	2.4	2.4
June/August	9.1	9.8	10.4	10.4	9.4	8.4	10.4	8.8	8.8	_4.1_	_4.4_	8.8	9.1	9.8
August	10.3	9.3	12.3	11.3	10.3	11.3	11.3	9.3	12.3	_5.3_	_6.3_	10.3	11.3	10.3

[1] Unusually cold periods are underlined.

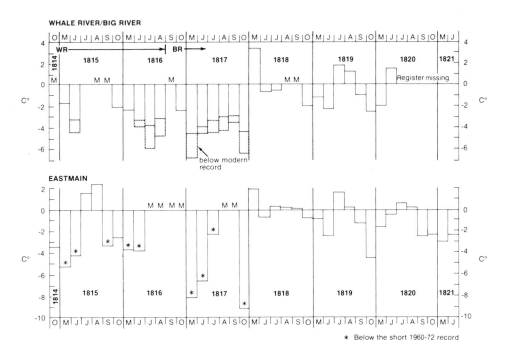

Figure 4: Whale River (Great Whale), Big River (Fort George), Eastmain, 1814-21: mean daily temperature (adjusted values) expressed as differences from the 1941-70 normals (WR, BR) or 1960-72 averages (EM). The shading and asterisks indicate where the historical mean was below the extreme monthly mean on modern record. M, data missing. These monthly anomalies, together with the absolute values, are tabulated in the Appendix.

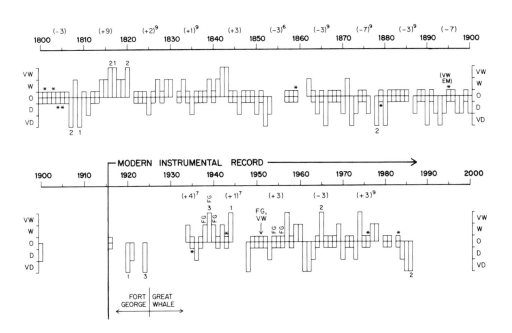

Figure 5: East coast, Hudson/James Bay. Wetness indices, summer seasons (May-October) 1800 to 1900. Index from +2, very wet to -2, very dry. In parenthesis, decadal sums of the index; superscripts, number of years. Asterisks indicate borderline cases, wet or dry; 1,2,3 wettest or driest season. Reference periods: Great Whale, 1926-76; Fort George, 1916-69; Eastmain, 1960-76.

Looking at the detail in the Meteorological Registers and Post Journals for the seasons 1815-20, the effect of such low temperatures on precipitation is evident. Figure 6 illustrates the greater number of days with snowfall from May through October, and the shortening of the snow-free season, contrasted with the modern period. The effect is especially noticeable on James Bay, suggesting southward extension of the autumn snowbelt. Of particular interest is the summer snowfall in 1816 at Great Whale; not only was there more snow in July than today, but even more fell in August - a month that has no modern record of snow having fallen. Summer 1816 provides a marginal case for a residual snowcover on the east side of the Bay. In summer 1817, snow conditions at Great Whale were most likely even more extreme. At Fort George, snowfall was extraordinarily frequent and often heavy in May and June 1817, but no snow fell in July and August, and the heavy rains of August probably washed away any snow remaining at the coast.

Thus the May-October seasons of 1816 and 1817 were such that had these conditions persisted, they might have resulted in the formation of permanent snowfields in parts of New Quebec/Labrador. From the impact of these seasons recorded in the Post Journals, Correspondence and Annual Reports, it was indeed possible to see the southward expansion of snow and ice forcing back the northern margins of habitation along this coast. The association of volcanic activity and incipient glaciation is an old idea in the literature of climatic change.

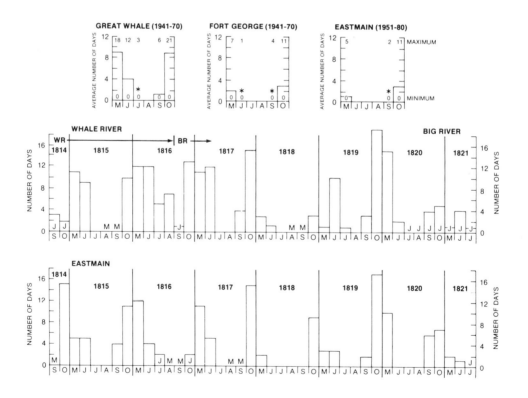

Figure 6: Whale River (Great Whale), Big River (Fort George), Eastmain, 1814-21: number of days with snowfall, together with modern reference values for Great Whale, Fort George and Eastmain. (The climatological day beginning at 8 a.m. at Whale River and Big River, 6 a.m. at Eastmain.) M, data missing; J, Journal entries, no Register; *, less than 1 day.

Regional Climate and Tambora

Circumstantial evidence for the east coast of Hudson/James Bay from autumn 1815 to later winter 1818 suggests a possible case for regional climatic cooling through the intervention of Tambora. That the material from the equatorial eruption in April 1815 should have entered the polar stratosphere by autumn of that year, with a residence time of more than one year, is in keeping both with the structure and behaviour of the atmospheric circulation and with studies of radioactive fallout in the 1950s and early 1960s. Moreover, empirical studies and certain theoretical considerations suggest that any resultant lowering of air temperature near the surface in higher latitudes might be expected to be most apparent in the warm season, and of greater magnitude than in lower latitudes. But in the case of Tambora (and of Coseguina), this appears to be at most a short-term feature superimposed on longer-scale climatic changes. The onset of cooling in this region occurred before the major event of Tambora (cf. Figure 3), with the return to a warmer mode some 50 years later. While a number of smaller eruptions did take place in the years preceding Tambora, there is also the concurrent event of the double cycle of abnormally low sunspot number (the lowest since the Maunder Minimum), which spanned the first two decades of the nineteenth century (Eddy 1976, p. 1191; this volume). A further consideration is the sudden swing to near-record warmth in April/May 1818 following hard upon the coldest weather recorded; this almost suggests an "over-compensation" in redressing the balance. Had there been a significant scavenging of the aerosol by the end of the very wet 1817 season?

For a different perspective on the 1816-17 seasons, it is useful to consider the regional climatic controls and energy exchanges - in so far as clues are offered in the HBC archives - without directly invoking Tambora.

Energy Exchanges: Advection

The thermostatic effect of the Bay on summer temperatures is a critical, if complex, influence along this windward coast, where background air temperature level appears to be closely related to the temperature of the Bay surface. Cold seasons do tend to have a higher proportion of Bay winds, and this was the case in 1816 and 1817.

The arctic summers of 1816 and 1817 were marked by heavy ice and late break-up and melt in the eastern and southern parts of the Bay. In both years, the last heavy ice was compacted in southeastern Hudson Bay, extending into northern James Bay - a pattern similar to modern maximum ice/water limits for mid-August to mid-September. In 1817, the timing was perhaps a week later. But in 1816, these stages occurred some four weeks later between mid-September and mid-October. In mid-September 1816, the ice in James Bay was akin to the normal for mid-July. Given the cold autumn of 1816, the season provides a marginal case for the carry-over of ice from one season to the next - a situation exceptional for the period itself. In contrast, the clearing of the ice for the 1818-20 seasons was relatively early in this part of the Bay.

The windroses for Great Whale, June to August 1816, are shown in Figure 7, together with those from the modern record. North winds are normally frequent here in May and early June, associated with a series of anticyclones which cross from the Arctic into southeastern or eastern Canada; this is then superseded by prevailing upper westerly flow with travelling depressions, which is characteristic of summer. Figure 7 shows how the spring pattern continues through July in 1816 with an unusually high frequency of north and west winds. In August the pattern changes, but winds are overwhelmingly from the west. These months are dominated, then, by the advection of cold air either from the Arctic or from passage over the Bay ice, and by the

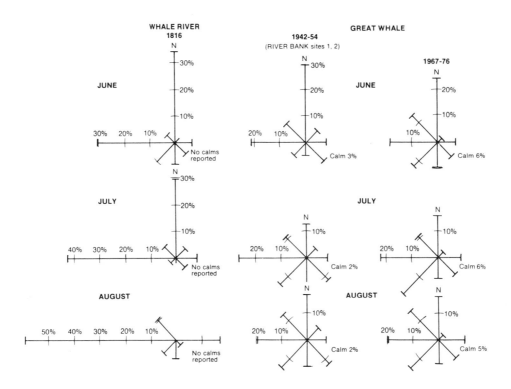

Figure 7: Windroses for Whale River (Great Whale) 1816, and Great Whale 1942-54, 1967-76. (The recent river bank sites are comparable with that of 1816.)

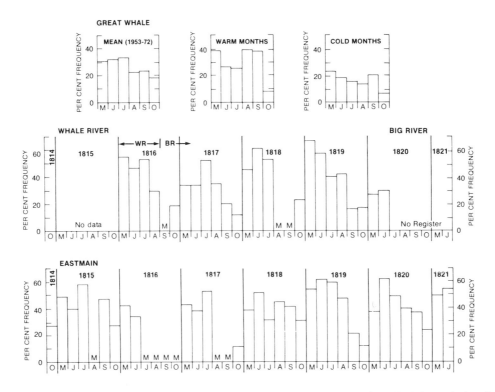

Figure 8: Whale River (Great Whale), Big River (Fort George), Eastmain, 1814-21: relative frequency of "clear" hours, together with reference values for average, warm and cold months at Great Whale. Morning, noon and evening hours. (A "clear" hour is defined by zero to five-tenths cloud cover.)

almost total absence of the warmer southerly or land components. In 1816, as today, snow in spring and early summer was brought by northerly and westerly winds. The persistence of onshore winds in turn served to pack the ice in along this coast all summer, and further depress coastal temperatures. At Fort George and Eastmain in 1817, the prevailing onshore winds from June through August indicated the frequent passage of depressions. This suggests that the exceptional cold of these two seasons was associated with different circulation systems.

Radiative Energy Exchanges
Today, low cloud is dominant in this region during the average summer season, and in exceptionally cold summer months even more pronounced (Figure 8). In sharp contrast, a striking feature of the seasons 1815-20 in general is the greater frequency of clear weather (zero to five-tenths cloud cover) from May to August. In 1816 at Great Whale, the frequency was double what might be expected today in very cold months, and particularly noticeable in July. Respective listings of clear hours against wind direction and damp, cloudy hours with Bay winds suggest, when compared with modern analogues, that the clear weather in 1816 was the result of: (1) the prolonged influence of arctic airmasses at this period; (2) the greater frequency, persistence and intensity of spring anticyclones over Hudson Bay, probably extending through July; and (3) the late break-up and unusual persistence of heavy ice in the Bay through the summer.

Still leaving aside Tambora, the clearer skies in 1816 and 1817 compared with very cold summer months today imply a larger receipt of incoming solar radiation at the surface, although the full potential may have been reduced, particularly in the region, as a consequence of the unusually "quiet" sun[1]. Considering the short-wave radiation balance, any increase in incoming radiation at Great Whale in 1816 could have been more than offset by increased losses resulting from the exceptional clarity of the air and through reflection from late snowcover and ice, enhanced into July and from the third week in August by fresh snowfall. In the case of the long-wave radiation balance, the dryness and clarity of the air and low sky temperatures would have encouraged loss from any more favoured sites or surfaces, while the net radiation through the summer would have been used primarily in melting snow and ice, and thawing and drying out the soil. These conditions together with the low-level advection of cold air could go some way to account for the very low air temperature at screen level at Great Whale in summer 1816.

Reintroducing Tambora, measurements of solar radiation following the eruption of El Chichón in 1982 suggested a reduction in the total short-wave radiation reaching the surface as a result of the stratospheric loading of dust and sulphur; while there is satellite evidence of enhanced infrared emission from the cloud, the effect of the volcanic material on the long-wave balance at the surface is not known.

To speculate, the evidence so far suggests that the cause of the extreme cold along this coast at this time was most likely multiple: a combination of unusual external factors converging on the years 1816 and 1817 (the general decrease in solar power, coupled perhaps with a high frequency of heavy volcanic aerosol in this particular region of the stratosphere), whose climatic effects in "summer" were magnified along this subarctic/arctic margin through the massive presence of

---

[1] Given the apparent connection between auroral/geomagnetic activity and sunspots, it is worth noting that compared with the high frequency of auroral activity observed in recent years, the only reports of auroral sightings in the HBC Journals for this coast during the nineteenth century were in 1878, 1879 and 1880 (cf. Figure 3).

unusually late ice[1] and snow, and the complexity of the ensuing surface - atmosphere interactions.

Atmospheric Circulation, June-July 1816 - Preliminary Remarks

The maps from the sequence 1 June to 13 July 1816 are first approximations to illustrate the work in progress (e.g., Figure 9). The results of the pilot study for 1-17 June (Wilson 1985c, 1985d) had indicated that useful daily schematic flow patterns for east/central Canada can be drawn for this early period, with the HBC archives providing the core database, supplemented where possible by other historical weather sources. Two sample maps are reproduced here (Figures 9a, b); additional information obtained more recently for the east coast of the United States and Canada (cf. Figure 1) is now serving both as a check on the original analysis and to refine the patterns. Although the HBC data are less complete in July, the coverage is still adequate when the weather patterns are well-articulated, which was generally the case in the first two weeks studied to date. At this early stage of the study, one or two preliminary remarks can be made concerning the atmospheric circulation during the first half of summer 1816.

During much of this period, flow over east/central Canada and the northeastern United States was predominantly meridional, interspersed by short periods of more zonal flow (notably the first week in July) with rapidly moving depressions, and brief northward extensions of the Subtropical High. The synoptic situation for the period 5-10 June points to blocking in the vicinity of Hudson Bay; from 6 July until the end of the present analysis on 13 July, there is some evidence, provided by the approaching HBC ships, of a blocking high east of Greenland.

The two exceptionally cold events over eastern Canada and the United States (5-10 June, 6-11 July) were apparently associated with these periods of blocking. In each case, a depression passed across the Great Lakes/northern Ontario (cf. Figures 9a, 10), lost speed abruptly over Québec and developed into a large system, gradually drifting eastward. Behind the depression, high pressure extended from the Arctic down over Hudson Bay, and very cold air was pulled unusually far south in the rear of the storm. In June, the bitter northwest winds brought frost and snow to the St. Lawrence Valley and New England (Baron, this volume). In July, the clear dry air brought very low temperatures, especially at night, at least as far south as Philadelphia, where it was as cool as late September - and mornings and evenings uncomfortably so[2].

The intensity and size of some of the systems, as well as the highly variable and contrasting extremes of temperature experienced throughout the region, bear witness to the vigour of the north-south energy exchanges, and suggest a much stronger mid-latitude temperature gradient than is usual today at this season. The storm tracks in early July were unseasonably far south. A strong temperature gradient was present at times to the south of Hudson/James Bay between the forested/spaghnum Shield country north of the Great Lakes, which on occasion became extremely warm, and the unusually complete and compacted ice-covered surface of the Bay; here, lows seemed to regenerate or develop.

---

[1] The possibility of submarine seismic activity in the Arctic as a source of kinetic and heat energy, easing the breaking up and outflow of previously compacted arctic ice, has still to be ruled out (cf. Wilson and MacFarlane 1986).

[2] Deborah Norris Logan's diary, Historical Society of Pennsylvania.

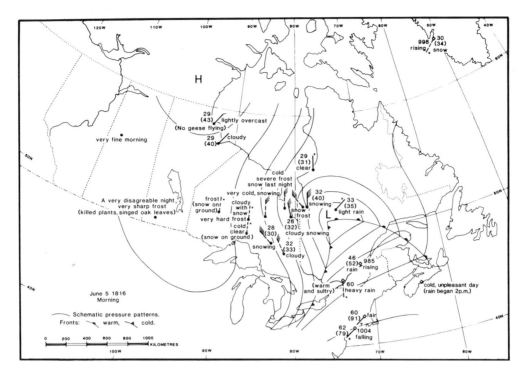

Figure 9a: Surface weather map, 5 June 1816, morning. Temperatures in degrees Fahrenheit; in parenthesis, mid-day values. Winds, short barb five knots, long barb 10 knots; broken arrow, one observation a day, time unknown; asterisk, speed unknown. Pressure in millibars; recent work has suggested that an adjustment of about +9 mb is required to reduce the station pressure at Québec City to sea level. (Reproduced with kind permission from *Weather* 10, p. 137.)

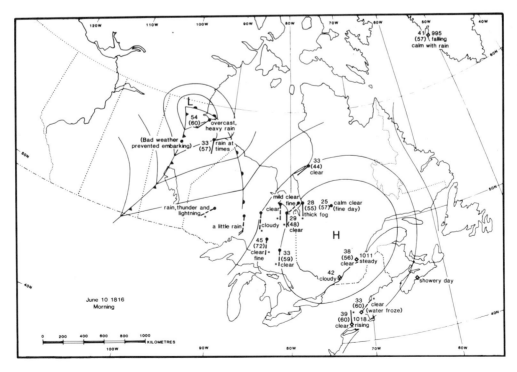

Figure 9b: Surface weather map, 10 June 1816, morning. For legend, see Figure 9a. (Reproduced with kind permission from *Weather* 10, p. 137.)

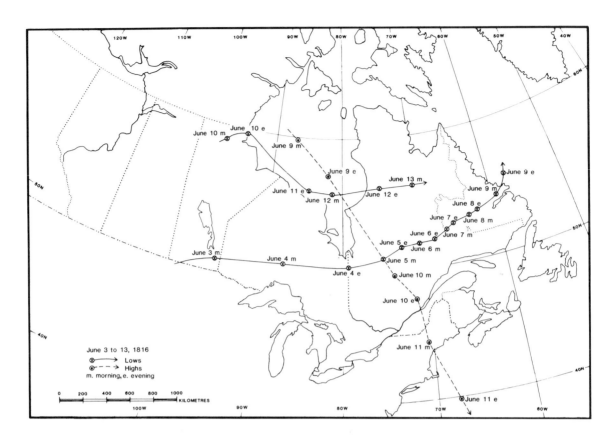

Figure 10: Trajectories of the surface high- and low-pressure systems, 3-13 June 1816. (Reproduced with kind permission from *Weather* 10, p. 138.)

It appears that the stalled situation of 5-9 June was associated with the intensification of an anticyclone over the Bay ice. On 9 June, the high began to move out to the southeast (cf. Figures 9b and 10) and was centred off the coast of New England by the evening of 11 June. Although less persistent, this anticyclonic pattern of flow was repeated in the third week of June - the trajectory now more southerly, over Lake Erie. The analysis is not yet complete, but a rather similar situation may also have occurred at the time of the stalled circulation in July. Today, the Hudson Bay High is characteristic of spring, (March-May) when anticyclones are more frequent over eastern Canada. Although some of these spring anticyclones remain cold-surface features, Johnson (1948) found that typically the Hudson Bay High is a deep system associated with a warm ridge aloft in the vicinity of 90°W. This upper ridge is, in turn, related to a trough down over western Canada and the United States near 112°W, while the trough off the east coast lies near its normal spring position (about 55°W over northern Newfoundland curving gently southwestwards). Such situations, which can persist up to five days or more, are preceded by blocking over the North Atlantic and western Europe (see also Treidl *et al.* 1981). In addition, Johnson noted that the trajectories of the highs shifted during the course of the season from southeastward over New England in March and April to a more southerly route in May.

Thus the map evidence strongly suggests that the atmospheric circulation over eastern Canada and the northeastern United States in June and the first half of July was abnormal in its timing rather than in kind - that spring had been delayed or protracted by as much as six weeks. This is in keeping with the evidence of the regional climate on Hudson/James Bay, where the normally brief summer was essentially obliterated. It points yet again to the importance, climatically, of the extraordinary ice and snow conditions over the Bay and northeastern Canada in summer 1816, and more generally to the key role of Hudson Bay at this season in the climate of this part of North America.

Concluding Remarks

For the east coast of Hudson/James Bay during 1809-20, the evidence associating the extreme cold with the eruption of Tambora remains circumstantial and intriguing. It is tempting to see the period of unrelieved record cold from October 1815 to March 1818 as influenced by high levels of stratospheric dust and acid - a relatively short-term feature exacerbating a longer-term (60-year) lowering of temperature, which had already begun. There can be no doubt that in 1816 and 1817 the combination of exceptionally heavy ice lingering through the summer in this part of the Bay and the high frequency of onshore north and west winds was a major factor in reducing coastal air temperature through the "warm" season.

From a different perspective, the daily surface weather maps reconstructed for east/central Canada, from 1 June to mid-July, indicate circulation patterns and sequences normally related to ice and snowcover over the Bay and northeastern Canada, and suggest that spring was running some six weeks late in 1816.

The riches offered by the HBC archives in terms of Canada's historical climate have scarcely been tapped. A major inhibiting factor is the labour-intensive, time-consuming nature of the work. The great challenge is to reduce the time and labour required without sacrificing information, quality control or physical reality - truly a worthy challenge to modern computer techniques.

Acknowledgements

Warmest thanks to Gordon McKay, Howard Ferguson and Mal Berry of the Canadian Climate Centre for long-term support since 1977 through a series of contracts, which have made the bulk of this work possible; also to my Scientific Officers Bruce Findlay and Joan Masterton, to Valerie Moore for processing all the manuscripts, to draftsman Brian Taylor (cf. Figures 4,6,7,8), and to the many others at the Atmospheric Environment Service who have helped me over these years.

I am grateful to Dick Harington of the National Museum of Natural Sciences (now Canadian Museum of Nature) for bringing me into the Museum's *Climate Change in Canada Project*, and for his encouragement particularly with the synoptic-mapping study, for which the Museum provided seed moneys. I appreciate too the financial aid and the other help that he has provided, with Gail Rice, in publishing my papers in *Syllogeus*, and the work of Edward Hearn (Ottawa University) who has drafted nearly all my figures for *Syllogeus*, most under contract to the National Museum.

I also thank the Hudson's Bay Company for permission to use the Company archives, HBC archivists Joan Craig and Shirlee Smith and their staff, and Alan Cooke, who in 1965 as a colleague at the Centre d'études nordiques, Université Laval introduced me to the climatic material contained in these archives, thereby opening up a new world.

References

Ashmore, S.E. 1952. Records of snowfall in Britain. *Quarterly Journal of the Royal Meteorological Society* 78:629-632.

Cooke, A. and C. Holland. 1978. *The Exploration of Northern Canada, 500 to 1920, A Chronology.* The Arctic History Press, Toronto. 549 pp.

Eddy, J.A. 1976. The Maunder Minimum. *Science* 192:1189-1202.

Geiger, R. 1965. *The Climate Near the Ground.* Harvard University Press, Cambridge, Massachusetts. 611 pp.

Johnson, C.B. 1948. Anticyclogenesis in eastern Canada during spring. *Bulletin of the American Meteorological Society* 29:47-55.

Kington, J. 1988. *The Weather of the 1780s over Europe.* Cambridge University Press. 166 pp.

Lamb, H.H. 1988. The weather of 1588 and the Spanish Armada. *Weather* 43:386-395.

Landsberg, H. 1967. *Physical Climatology.* Third edition. Gray Printing Company. Dubois, Pennsylvania. 446 pp.

Manley, G. 1978. Variations in the frequency of snowfall in east-central Scotland, 1708-1975. *Meteorological Magazine* 107:1-16.

Parker, M.L., L.A. Jozsa, S.G. Johnson and P.A. Bramhall. 1981. Dendrochronological studies on the coasts of James Bay and Hudson Bay. *In: Climatic Change in Canada 2.* C.R. Harington (ed.). *Syllogeus* 33:129-188.

Pfister, C. 1980. The Little Ice Age: thermal and wetness indices for central Europe. *Journal of Interdisciplinary History* 10:665-696.

Treidl, R.A., E.C. Birch, and P. Sajecki. 1981. Blocking action in the northern hemisphere: a climatological study. *Atmosphere-Ocean* 19:1-23.

Wilson, C. 1982. The summer season along the east coast of Hudson Bay during the nineteenth century. Part I. General introduction; climatic controls; calibration of the instrumental temperature data, 1814 to 1821. *Canadian Climate Centre Report* No. 82-4:1-223.

_____. 1983a. Part II. The Little Ice Age on eastern Hudson Bay; summers at Great Whale, Fort George, Eastmain, 1814-1821. *Canadian Climate Centre Report* No. 83-9:1-145.

_____. 1983b. Some aspects of the calibration of early Canadian temperature records in the Hudson's Bay Company Archives: a case study for the summer season, eastern Hudson/James Bay, 1814 to 1821. *In: Climatic Change in Canada 3*. C.R. Harington (ed.). *Syllogeus* 49:144-202.

_____.1985a. The summer season along the east coast of Hudson Bay during the nineteenth century. Part III. Summer thermal and wetness indices. A. Methodology. *Canadian Climate Centre Report* No. 85-3:1-38.

_____. 1985b. The Little Ice Age on eastern Hudson/James Bay: the summer weather and climate at Great Whale, Fort George and Eastmain, 1814-1821, as derived from the Hudson's Bay Company Records. *In: Climatic Change in Canada 5*. C.R. Harington (ed.). *Syllogeus* 55:147-190.

_____. 1985c. Daily weather maps for Canada, summer 1816 to 1818 - a pilot study. *In: Climatic Change in Canada 5*. C.R. Harington (ed.). *Syllogeus* 55:191-218.

_____. 1985d. Daily weather maps for Canada, summer 1816 to 1818. *Weather* 40:134-140.

_____. 1988. The summer season along the east coast of Hudson Bay during the nineteenth century. Part III. Summer thermal and wetness indices. B. The indices 1800 to 1900. *Canadian Climate Centre Report* No. 88-3:1-42.

Wilson, C. and M.A. MacFarlane. 1986. The break-up of Arctic pack ice in 1816 and 1817. *Weather* 41:30-31.

Appendix: Corrected Mean Daily Temperature (1814-21) Whale River, Big River, Eastmain.[1,2] Great Whale Record (1925-76); Fort George (1915-69).

	Whale River/Big River						Year	Eastmain[3]						
	M	J	J	A	S	O		M	J	J	A	S	O	
Mean Daily Temperature °C (Corrected)	-0.7	1.9	10.6	-	-	-	1814	-	-	-	-	-	1.2	Mean Daily Temperature °C (Corrected)
	-1.05	2.4	4.7	-	-	(0.2)	1815	-1.2	6.4	(14.8)	(14.6)	6.2	2.1	
	-4.3	4.2	7.9	(5.6)	/	(0.5)	1816	0.3	6.7	-	-	-	-	
	5.9	8.1	11.8	6.9	4.6	-3.4	1817	-4.1	3.9	10.9	-	-	(-4.4)	
	1.3	6.5	14.2	-	-	0.9	1818	6.0	9.9	13.5	12.4	9.7	3.9	
	0.5	10.4	-	12.4	7.1	0.3	1819	3.2	8.0	14.8	12.4	8.4	0.2	
	-	-	-	-	-	-	1820	2.5	10.3	13.8	12.4	7.1	2.3	
							1821	1.1	8.3	-	-	-	-	
Difference from 1941-70 Normal	-1.6	-4.4	0.0	-	-	(-2.1)	1814	-	-	-	-	-	-3.5	Difference from 1960-72 Average
	-2.4	-3.9	-5.9	-	-	(-2.4)	1815	-5.3	-4.2	(+1.6)	(+2.4)	-3.4	-2.6	
	-6.8	-4.6	-4.5	(-4.7)	-3.5	-6.3	1816	-3.8	-3.9	-	-	-	-	
	+3.4	-0.7	-0.6	-4.3	-	-2.0	1817	-8.2	-6.7	-2.3	-	-	(-9.1)	
	-1.2	-2.3	+1.8	-	-	-2.6	1818	+1.9	-0.7	+0.3	+0.2	+0.1	-0.8	
	-2.0	+1.6	-	+1.2	-1.0	-	1819	-0.9	-2.6	+1.6	+0.2	-1.2	-4.5	
	-	-	-	-	-	-	1820	-1.6	-0.3	+0.6	+0.2	-2.5	-2.4	
							1821	-3.0	-2.3	-	-	-	-	
Difference from Lowest on Record	-	-1.1	-	-	-	-	1815	*	*	-	-	*	-	* Below the Short 1960-72 Record
	-	-0.6	-2.1	(-1.6)	/	-	1816	*	*	-	-	-	-	
	-2.2	-0.7	-1.2	-1.3	-0.6	-1.9	1817	*	*	*	-	-	(*)	

[1] The HBC Meteorological Registers for Whale River (Great Whale), Big River (Fort George) and Eastmain, 1814-21, have been annotated and reproduced in Wilson 1983a, Appendix III.

[2] Values in parenthesis indicate more than five days are missing.

[3] At Eastmain the record was too short to give a useful anomaly.

Climatic Change, Droughts and Their Social Impact: Central Canada, 1811-20, a Classic Example

Dr. Timothy Ball[1]

Abstract

Changes in the climate of central Canada from 1760 to 1800 were marked by extreme fluctuations as the region began to emerge from the nadir of the Little Ice Age. The harshness of climate, particularly along the northern limit of trees, created severe ecological conditions. Evidence from the historical and meteorological records, maintained primarily by the Hudson's Bay Company, provides clear indications of the extremes and the impact climatic changes had on the socio-economic infrastructure of the region. Between 1800 and 1810 the climate was relatively benign, holding promise of better conditions and times in the nineteenth century. The promise was short-lived as temperatures began to decline in 1811, a trend that was to continue through to 1818. Most research on the period has stressed the temperatures, but detailed studies of the historical documents show that the period from 1815 to 1819 was one of severe drought as well as cold. The combination suggests that the mechanisms causing the drought were probably different than those that created the hot droughts of the 1930s.

Droughts in the Canadian prairies are usually attributed to a northward extension of the Pacific High (Subtropical). Droughts in the boreal forest region are usually associated with a southern position of the Arctic High. Both regions indicate a 22-year cycle of droughts that seems to coincide with sunspot cycles. With a northward shift of the mean summer position of the Arctic (Polar) Front there is a hot drought on the prairies. With a southerly location of the Front there is a cold drought. The 1815-17 period was a classic example of the latter, and is associated with an extreme degree of meridionality in the zonal index.

Variations in precipitation in the early nineteenth century had a significant effect upon the wildlife of the region. This resulted in a decrease in the food supply for Europeans and Indians, with the concomitant social and health stresses. A decline in the fur-bearing animals created declines in income that led to significant socio-economic adjustments.

Introduction

A great deal of attention has been paid to the exceptionally cold summer of 1816. A study of the eventful summer has been documented by Hoyt (1958) with a description of weather, food supplies, prices, and even population movements. Work focused upon the northeastern United States and Canada. Attention spread to what Post (1977) called, "the last great subsistence crisis of the western world". The most extensive and analytical study was the book *Volcano Weather* (Stommel and Stommel 1983). This work included a chapter on conditions reported in Europe. A brief reference at the end of the chapter suggests limits to the extent of the area influenced by cold conditions, "It would appear, however, that the truly exceptional character of 1816 weather was limited to a small portion of northeastern America, Canada and the extreme western parts of Europe" (Stommel and Stommel 1983, p. 51).

---

[1] Department of Geography, University of Winnipeg, 515 Portage Avenue, Winnipeg, Manitoba R3B 2E9, Canada.

The severe cold conditions have more recently been detailed in the Hudson Bay region of northern Canada (Catchpole 1985; Catchpole and Faurer 1985; Skinner 1985; and Wilson 1985). Such work has produced not only valuable information on the extent and intensity of the cold conditions in 1816, but it has also contributed to ideas concerning the possible causes of such exceptional conditions.

The debate centres on whether the cold conditions were caused by: (1) the intensity of the dust veil emitted by the equatorial eruption of Mount Tambora in April 1815; or (2) the influence of variable sunspot activity; or (3) natural variation caused by some atmospheric, or atmosphere/ocean phenomenon. This controversy remains unresolved. While each effect may exert an influence on large-scale atmospheric circulation, by reducing the Earth's radiation balance, they may also exert a simultaneous effect.

Documentary evidence for the summer of 1816 appears to agree on one aspect of the cold summer - that a ridge of high pressure extended south over eastern North America and western Europe bringing cold arctic air well south of its normal latitudes for the time of year. These systems are relatively common in the fall, winter and early spring, but are unusual in the summer. Their impact on the socio-economic conditions of that period were severe.

Here, I intend to show that the pattern of weather in 1816 can be generally defined from climatic information in Hudson's Bay Company records. The pattern indicates that cold conditions did not include the entire prairie region. Southern Alberta had normal conditions, while the north had an exceptionally wet summer. Overall weather conditions began to deteriorate in 1809, and continued to decline until 1816. That year, apart from being cold, was the first year of a severe drought that lasted until 1819. Comparison of conditions with modern synoptic charts suggest that this was a cold drought within the 22-year cycle of droughts experienced in the Great Plains.

The 22-year cycle of droughts correlating with sunspot activity has generally been established and accepted (Herman and Goldberg 1978). Very little detailed analysis of the nature of each drought period has been completed. It is generally accepted that droughts are coincident with hot weather, and the 1930s are cited as the classic example. Undoubtedly hot dry weather is especially damaging to modern agriculture, but lack of precipitation under any temperature regime is serious. The cold drought from 1816 to 1819 was especially damaging, as journals and diaries record.

These years of severe weather had a considerable impact upon wildlife, indigenous peoples and Europeans. Later, I will suggest that it served as a catalyst for an already volatile situation: the Seven Oaks massacre in 1817 at the Red River settlement.

The fur trade had been suffering from over-trapping and competition between the North West Company and the Hudson's Bay Company. Tensions between Indians, Métis and fur traders were somewhat overshadowed by the growing talk of permanent European settlement. The first group to arrive, the Selkirk Settlers, came from Scotland in 1811. Generally, they were unwelcome because they threatened the fur trade and the traditional ways of native people. Ironically, they had left Scotland because of severe weather. Now they had moved into a land that was suffering for the same reason. Diminished wildlife populations meant reduced food supply, with associated hunger and disease. Residents already felt threatened and unsure, thus they saw the settlers as an even greater threat. This situation heightened tensions and began a long period of conflict.

The Hudson's Bay Company Post Journals often provide a brief but daily summary of activities and weather conditions. Some Journals are incomplete in this period because of feuds between the Hudson's Bay and the North West companies, or absences from the Posts on expeditions between Inland Posts and Bay Posts to exchange furs for supplies. Sometimes the mere struggle for survival precluded maintenance of the records. Although the Journals are fragmented, the entries provide a series of proxy data that give some indication of the activities and weather conditions that occurred within these years.

Data Sources

Proxy data from the Journals include: (1) comments on garden-crop preparation, growth and damage; (2) remarks of frost or ice formation, movement and decay; (3) descriptive terms for winds or precipitation restricting outdoor activities or travel. Even limited comments of phenological data for animal appearances or migrations serve as an indicator of weather anomalies.

The Hudson's Bay Company Posts located in the west-central region of Canada are shown in Figure 1. Brandon House was an important Post adjacent to the Assiniboine River. Peter Fidler, the Factor in charge, provided much of the proxy data and insights into the weather through his records. Carlton House, although shifted several times, was transferred in 1810 to a site near "a crossing place" on the south bank of the North Saskatchewan River. The daily accounts written for each Post between May and October 1810 to 1820, were analyzed for proxy data that might be attributable to adverse weather.

Figure 1: Location of Hudson's Bay Company posts.

Gardening and crop production were a major part of the general way of life. The Company encouraged each Post to obtain much of its sustenance from local sources in order to be as self-sufficient as possible. Gardens were maintained, and hunting and fishing supplemented the diet with a fresh supply of meat. Cutting firewood for the approaching winter also took up much of the time (Ball 1987). Frequent canoe trips to Posts along Hudson Bay were embarked upon in the short summer season. The observations recorded during these activities are evidence of the severe cold experienced in the early to late summers of 1816 and 1817. Warmer weather returned in 1818, but the drought continued to 1819.

Peter Fidler's daily records for Brandon House yield valuable information. In 1816, the ice must have dispersed toward the end of April or early May as Indians were fencing in the Assiniboine River on 21 May "...half of mile above the House to kill sturgeon" (HBCA, PAM B22/a/19). Fidler mentioned that the presence of sturgeon usually gives an indication of when the ice goes out "...they annually come up every spring in great numbers when the ice goes away and they appear here about 10 to 12 days after it clears away..." (HBCA, PAM B22/a/19)[1].

Gardening and crop preparation began on 3 May, and an indication of the spring runoff levels and weather were observed by Fidler on 9 May as water levels were "...falling daily 1¾ inch - cold weather and strong wind these two days" (HBCA, PAM B22/a/19). The remainder of the crops were sown by 27 May, but it was not until 5 June that a severe cold spell occurred. "A very sharp frost at night and killed all the Barley, Wheat, Oats and garden stuff above the ground except lettuce and onions - the Oak leaves just coming out are as if they are singed by fire and dead" (HBCA, PAM B22/a/19). With a severe frost early in June the growth of crops and natural vegetation would certainly be curtailed, as this period is essential for their development to mature plants.

An interruption in the Journal occurs after this period due to the battle between the Hudson's Bay Company and the North West Company[2]. In the spring of 1817, Brandon House was subjected to severe weather, as on 18 June "...thin snow fell 2 inches deep" (HBCA, PAM B22/a/20). Visitors were late arriving at the Post due to the backward spring, and Fidler explained the cause of late arrival; "...he was detained long by the ice in the Little Winnipeg" (HBCA, PAM B22/a/20). Little Winnipeg refers to Lake Winnipegosis. The summer of 1817 was reported to be backward due to the unseasonable cold, but drought also had a direct effect on agriculture. "The crops exceedingly backwards - some potatoes only 4 inches above ground - whereas in other seasons there were new ones bigger than Walnuts, the grass is also remarkably short and ground dry - all the little runs of water now dry - so there is every reason to expect a bad crop on account of the great want of rain - the season has been colder than usual" (HBCA, PAM B22/a/20).

The summer continued to be dry, as the small saline lakes began to evaporate. The apparent migrations of buffalo southward also give an indication of the dry summer and unusual cold. Fidler recorded buffalo movements near the Post in the spring, and by 11 August they were "...very numerous - even extending so low down as the Forks" (HBCA, PAM B22/a/20). The cold weather continued. Fidler wrote in August of frosts that occurred on 17 and 23 July at Brandon which killed all the potato tops. The autumn season seems to end on 23 October when

---

[1] Hudson's Bay Company Archives, (HBCA) Provincial Archives of Manitoba (PAM), Journal Number.
[2] The conflict between the two companies was resolved in 1821 when the smaller Hudson's Bay Company incorporated the North West Company.

the Assiniboine River froze over. This occurrence is seen by Fidler as being "...very early in the season, about 20 days sooner than usual - and it set in early last fall" (HBCA, PAM, B22/a/20).

The spring of 1818 apparently began without mention of adverse conditions, as Fidler reports on 26 May "...the ice drove by about 5 weeks ago ..." (HBCA, PAM, B22/a/21). Despite a break in the daily reports for Brandon House, Fidler continued to write on his journey from Red River to Martins Falls near Albany Factory on James Bay.

Returning to Brandon House, Fidler recorded the late summer conditions: "Water very low in the river and a very dry season scarce a single shower of rain all summer, all the potatoes and garden stuff quite burnt out as also 2½ bushels of Barley sown there - when 3 inches high all killed by the great drought - these 3 summer past remarkably little rain ... quite different from what it used to be" (HBCA, PAM, B22/a/21).

Summers at Carlton House are also recorded in fragments due to continuing battles between Hudson's Bay and North West companies, and canoe trips to other Posts. However, direct information recorded in the Journals still provides an indication of summer conditions for 1816-18.

The spring of 1815 appeared to have a positive beginning: crops were planted as early as 29 April. However, conditions changed, and on 13 May John Pruden recorded the bleakness of the weather; "...hard frosts every night retards vegetation very much, none of the seeds that have been sown make their appearance above ground except the cabbage seed" (HBCA, PAM, B27/a/4).

By June, weather continued poor, but there was a different problem. "Wind SW blowing fresh part-clear and part cloudy weather, it has been remarkable dry wind weather all this month which keeps the garden stuff very backward" (HBCA, PAM, B27/a/5,2d). Things had not improved a month later: "The insects have eaten all our cabbage and turnips owing I suppose to the dry season, ..." (HBCA, PAM, B27/a/5,4). This was the first indication of a drought that was to grip the eastern half of the prairies for three years. The impact was to be quite severe.

The drought conditions are best summarized in Peter Fidler's General Report of the Red River District for 1819.

> The spring months have sometimes storms of wind and thunder even so early as March within these last years the Climate seems to be greatly changed the summers so backward with very little rain and even snow in winter much less than usual and the ground parched up that all summer have entirely dried up, for these several years loaded craft could ascend up as high as the Elbow or Carlton House but these last 3 summers it was necessary to convey all the goods from the Forks by land in Carts... (HBCA, PAM, B22/e/1,6).

We can discover the extent of the drought by noting which rivers are reported to be low. The North Saskatchewan, Assiniboine, Red, Hayes, Nelson and Steel rivers all receive attention in the journals. This means that the drought covered the drainage basins of all of these rivers, thus encompassing a large part of central North America.

The degree of the drought can be determined by the impact that it had on the environment, wildlife and subsequently the people. James Sutherland reports that water routes connecting the

Hayes and Nelson rivers were only made passable by the construction of dams (HBCA, PAM, B154/e/1,2). Peter Fidler notes that:

> ...as the country wherever I have been and by the invariable information of the different Tribes I have enquired at agree the country is becoming much drier than formerly and numbers of small lakes become good firm land will be covered with Timber of various kinds...(HBCA, PAM, B22/e/1,8d).

Fidler implies that he expects these conditions to persist in the future, although he does not specify for how long.

The value of his comments lie in putting the individual events into a larger and longer climatic framework. It is important to note that all seasons suffered from the lack of precipitation. "These 3 summers past remarkably little rain - as also very little snow in winter quite different from what it used to be" (HBCA, PAM, B22/a/21,29d). We also know that conditions were good prior to the drought; "...since 1812 there was always good crops of everything until 1816 when the dry summers commenced..." (HBCA, PAM, B22/e/1,8).

Prairie droughts are usually accompanied by the appearance of insects, particularly grasshoppers, that exacerbate the problems. Fidler makes some interesting comments when talking about the grasshopper infestation. He notes that, "...They first made their appearance the third week of August 1818 at 2 O'clock in the afternoon and came from the southwest" (HCBA, PAM, B22/e/1,20). The direction is significant because it indicates wind direction during the period. This is confirmed by John Pruden's observation at Carlton House that, "Wind SW blowing fresh part-clear and part cloudy weather, it has been remarkable dry wind weather all this month..." (HBCA, PAM, B27/a/5,2d). Then Fidler writes: "These insects (*grass-hoppers*) make their appearances in great numbers about every 18 years..." (HBCA, PAM, B22/e/1,6d). This implies cycles of infestation and possibly of climate.

Atmospheric Circulation

So far we have established that 1811-20 had below normal temperatures, especially in the years 1816 and 1817. Between 1812 and 1816, conditions were cool but generally good for crops and vegetables. In 1816 drought began in a large region including the central and eastern prairies. The drought ended in 1820 as temperatures and precipitation patterns returned to long-term normals.

How did the circulation pattern for these years differ from the long-term normal, and was the drought typical of those that occur regularly on the prairies?

The climate of central Canada is generally determined by the position of the Arctic Front[1]. In summer the mean position of the Front approximates the northern boreal forest limit. In winter it curves south in a great arc toward the centre of the continent to an approximate mean position

---

[1] There appears to be some confusion over the use of the term Arctic Front. Bryson and others have used the term Arctic Front to describe the major division between Arctic and Temperate air in North America. The term Polar Front is used by others, particularly in Europe, presumably to indicate that there is a similar front in the southern hemisphere. I have used Arctic Front because the paper is examining conditions in North America.

of 40°N Latitude. This curve occurs because the Rocky Mountains act as a barrier, and create a standing wave in the westerly flow of the general circulation.

Polar air north of the Arctic Front tends to be cold and dry, while subtropical air to the south tends to be warm and dry. Generally, moisture is brought to the region by cyclonic storms that move along the front; these usually occur in spring and autumn as the Front moves through the region in its annual migration. Most summer precipitation is convectional as instability develops in the warm subtropical air.

Dey (1973) analyzed synoptic conditions occurring during summer dry spells in the Canadian Prairies. He showed that the most severe droughts occurred when the Pacific High (subtropical) extended northward into the southern prairie region. Blocking occurs with the extreme meridional pattern that is formed. This configuration is commonly called an 'omega block'. Low pressure zones on the Pacific coast and in the region to the west of Lake Superior lie on each side of a large high pressure region. On the weather map this creates a pattern similar to the Greek letter omega - hence the name.

It is also possible to have dry conditions in summer if the Arctic Front extends southward over the region. This would produce cool, dry conditions under a predominantly northerly flow. The summers of 1815-17 are good examples, being marked by very cool dry conditions in the early summer as the Arctic Front remains well south of its normal position. When the Front finally retreats northward, the prairie sites experience the associated precipitation. For example, at Carlton House in July 1815, the Journal reads: "Wind easterly cloudy weather had a heavy shower of rain last night, the only one I may say since summer commenced..." (HBCA, PAM, B27/a/5,3). In June 1817, Swan River experienced three days of continual rain under cyclonic conditions as the Front migrated northward.

Further evidence to support this hypothesis is provided by the weather patterns at Fort Chipewyan. This post is ideally located to determine the latitudinal and longitudinal shifts of the Arctic Front in summer. Spring was late each year from 1815 to 1818 inclusive. The summers were cool and short, as autumn came early. This was especially true in 1816 as the entry for 30 September indicates: "One of the several days that I have witnessed at this season of the year, the ground covered with snow 3½ inches deep, blowing very fresh and extremely cold" (HBCA, PAM, B39/a/9,10d). 1818 saw the return to a longer summer, with the first snow falling on 13 October, and a comment that there was "...mild weather with wind from the south" on 22 October (HBCA, PAM, B39/a/14,8d).

Île-à-la-Crosse, south of Fort Chipewyan, has a limited record, but it does indicate normal conditions. For example, the earliest date of the water being clear of ice in the modern record is 12 May. In 1816 the ice was gone by 18 May. It is reasonable to use this station as the western limit of the outbreak of cold arctic air in the spring and early summer of that year.

In summary, it appears that 1815 saw the beginning of generally cooler conditions. The cold was more notable in 1816 and 1817, especially for the spring and early summer in the eastern half of the Prairies. The Jetstream and Arctic Front swung south so that the eastern half of the Prairies was cold and dry, under arctic air. Conditions changed significantly in 1818. The Arctic Front moved north, and an omega block system set in that appeared to dominate through 1819. Thus the *cold drought* that existed in 1816 and 1817 was replaced by a *warm drought* in 1818 and 1819.

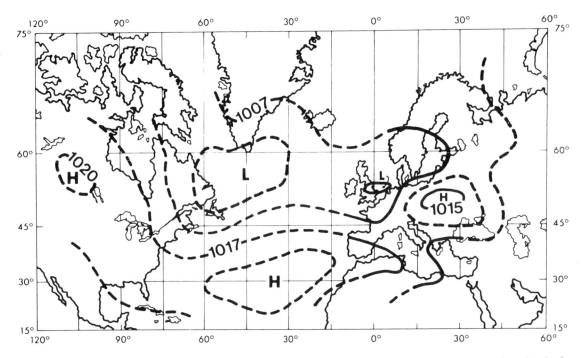

Figure 2: General reconstruction of the pressure patterns for North America and the North Atlantic for July of 1816 (after Catchpole 1985; Lamb and Johnson 1966).

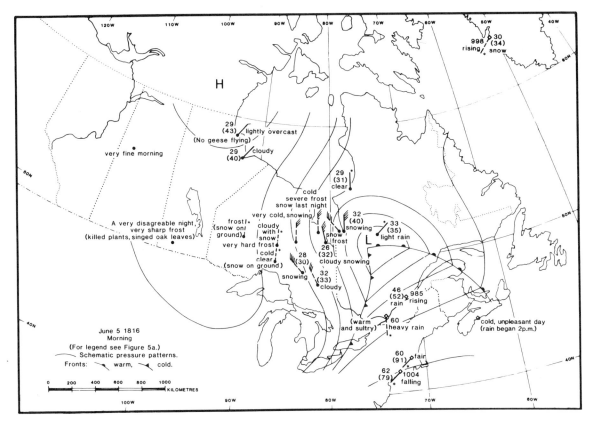

Figure 3: Surface weather map: morning, 5 June 1816 (after Wilson 1985).

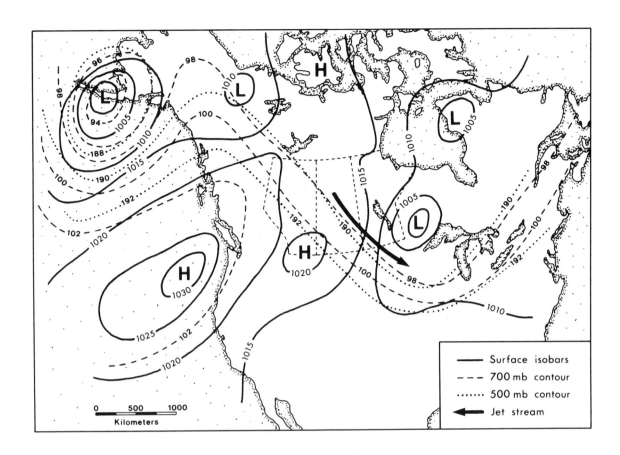

Figure 4a: Synoptic weather pattern for drought conditions in Western Canada (after Dey 1973).

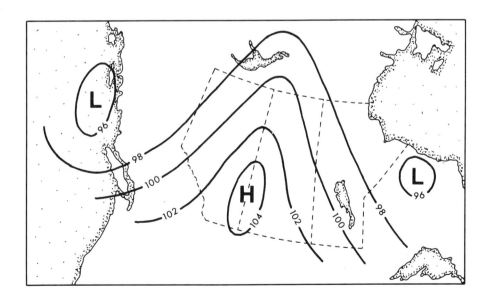

Figure 4b: Synoptic weather pattern for drought conditions in Western Canada (after Dey 1973).

This pattern is consistent with the general reconstruction suggested by Catchpole (1985) (Figure 2). It also demonstrates that Wilson's (1985) diligently drawn synoptic maps for three months of the summer of 1816 are valid. Wilson's map for 5 June 1816 (Figure 3) shows the surface conditions with a southern expansion of the Arctic High. Synoptic conditions as reconstructed by Dey (1973) would have been the general situation in 1818 and 1819 (Figure 4).

Climatic Impact

Regardless of the climatic mechanisms, there is no doubt that climatic conditions seriously affected wildlife and people's ability to grow food. The lack of precipitation reduced the planted crops, but it also affected the wild harvest of berries and other fruits. Lack of snow is devastating to many wildlife species, especially with colder-than-usual temperatures.

Climate also affected people's ability to travel. Early ice, late ice, too little snow, as well as shallow rivers and lakes all hampered movement for trade and hunting. Buffalo migration also indicates unusual conditions. Normally these animals would move westward or southward, but with changing snow patterns and dry conditions to the west they altered their behaviour. Fidler writes in 1817: "There are plenty of Buffalo not 15 miles off and all last winter and this spring they have been very numerous - extending even so low down as the Forks" (HBCA, PAM, B22/a/20,9d). Again in 1818: "The Catfish during these summers have also been very scarce...but fortunately vast numbers of Buffalo have kept pretty near all summer" (HBCA, PAM, B22/a/21,2).

Life has always been a struggle in this part of the world. Food supply varies dramatically with climate. It is truly a region of plenty or dearth. However, the 1811-20 period was one of special severity. The 1780s and 1790s had been periods of severe climatic conditions, with weather oscillating from one extreme to another. The author has argued elsewhere that this period created the pressure that forced the Company away from its complacent position on the Bay. A brief respite from 1800 to 1810 was then shattered by severe cold and drought. Starvation and hardship returned, as animals disappeared or changed their routines. Fur traders saw their industry threatened. Indians saw the fur trade threatened and their traditional way of life dashed. Conflict for the trade was at a peak as the Hudson's Bay and North West companies literally battled over the spoils. The Selkirk Settlers entered the scene unaware of the tensions and problems. Since the settlers posed a threat to all, it is not surprising that a series of confrontations occurred culminating in the massacre at Seven Oaks in 1817, when 20 men were killed. The stress of the impending social and cultural confrontation was underlain by exceptionally severe weather. Starvation and malnutrition made rational behaviour less likely.

Conclusion

There appear to be two types of drought on the Great Plains of North America; hot droughts and cold droughts. The former coincide with the omega blocking system that dominated the region in 1988. In this system the Pacific High extends northward over southern Alberta, southern and central Saskatchewan, and southern Manitoba. This creates very hot and dry conditions similar to those seen in the 1930s and again in the 1980s. The latter occur when the Arctic Front dips south across these same regions so that they are now dominated by clear, cool and equally dry conditions.

The 1816-19 period was one in which cold drought predominated. Journals of the Hudson's Bay Company provide much information about the extent and intensity of the conditions. They also allow estimation of the impact that these conditions had upon the environment, and therefore upon wildlife and people.

The heat and drought of the 1980s have led to current predictions of global warming and impending doom as droughts increase in frequency and severity in North America. My brief study suggests that this will not be the case. Perhaps the pattern of hot or cold droughts will change. A more northerly location of the Arctic Front might result in less southerly incursions of Arctic air.

References

Ball, T.F. 1987. Timber! *Beaver* 67(2):45-56.

Catchpole, A.J.W. 1985. Evidence from Hudson Bay Region of severe cold in the summer of 1816. *In: Climatic Change in Canada 5*. C.R. Harington (ed.). *Syllogeus* 55:121-146.

Catchpole, A.J.W. and M.-A. Faurer. 1985. Ships' Log-Books, sea ice and the cold summer of 1816 in Hudson Bay and its approaches. *Arctic* 38:(2)121-128.

Dey, B. 1973. Synoptic climatological aspects of summer dry spells in the Canadian Prairies. Unpublished Ph.D. thesis. University of Saskatchewan. Saskatoon. 180 pp.

Herman, J.R. and R.A. Goldberg. 1978. *Sun, Weather and Climate*. Scientific and Technical Information Office, NASA, Washington, D.C. Sp-426. 360 pp.

Hoyt, J.B. 1958. The cold summer of 1816. *Annals of the American Association of Geographers* 48:118-131.

Lamb, H.H. and A.I. Johnson. 1966. Secular variations of the atmospheric circulation since 1750. *Geophysical Memoirs* 110. H.M.S.O., London. 125 pp.

Post, J.D. 1977. *The Last Great Subsistence Crisis in the Western World*. Johns Hopkins University Press, Baltimore. 240 pp.

Skinner, W.R. 1985. The effects of major volcanic eruptions on Canadian climate. *In: Climatic Change in Canada 5*. C.R. Harington (ed.). *Syllogeus* 55:75-106.

Stommel, H. and E. Stommel. 1983. *Volcano Weather*. Seven Seas Press Inc., Newport, Rhode Island. 177 pp.

Wilson, C.V. 1985. Daily weather maps for Canada, summers 1816 to 1818 - a pilot study. *In: Climatic Change in Canada 5*. C.R. Harington (ed.). *Syllogeus* 55:191-218.

The Year without a Summer: Its Impact on the Fur Trade and History of Western Canada

Timothy F. Ball[1, 2]

Abstract

Edward Umfreville referred to the Hudson's Bay Company as "being asleep by the frozen sea". He was talking about the fact that the Company had established its trading posts along the shores of Hudson Bay and made no attempt to build permanent posts inland. Arthur Dobbes used this as evidence in his charge of monopoly against the Company. He claimed that the Company was deliberately protecting and hiding the potential of the interior of North America to ensure the dominance of the Hudson's Bay Company.

At the end of the eighteenth century the Company established its first inland post at Cumberland House on the Saskatchewan River. It has always been argued that the sole reason for this move was to counteract the expansionism of the North West Company. Increasing evidence suggests that climatic change brought about a dramatic decline in the ecology of the northern region and this was a major cause of the move inland. By 1810 the expansion created increasing conflict between the two companies. In 1812 a third component, the Selkirk Settlers, arrived and the turmoil continued to build.

Severe weather affected all three groups through their dependence upon the land for sustenance and economic profit. Two events, the Seven Oaks Massacre and the amalgamation of the Hudson's Bay and North West companies, followed the period of most severe weather in 1816-17. There is little doubt that the summer of 1816 was one of the worst in the historic record. It was referred to as "the year with no summer" and, more recently, as "the last great subsistence crisis in the western world". The effects on the living conditions in Western Canada were well-documented, and clearly placed a great deal of stress on native people and the European traders and colonists. Friction between the groups was exacerbated by the uncertainties of food supply. Probably the hardships created by the extreme weather were a significant catalyst for the events that occurred.

Introduction

The effect of climate on human behaviour has been a contentious issue in the twentieth century. The concepts that evolved from Friedrich Ratzel's *Anthropogeography*, published at the end of the nineteenth century, were transported and transposed by various people until they came to a distorted rest in Adolf Hitler's *Mein Kampf*. Since then the concept of climate influencing people or history has been anathema in the academic world. Unfortunately, it is evident from even a cursory glance at the patterns of climate and the sequence of history that we 'threw the baby out with the bathwater'.

---

[1] Department of Geography, University of Winnipeg, 515 Portage Avenue, Winnipeg, Manitoba R3B 2E9, Canada.

[2] The following article is a précis of a public lecture given during the conference *The Year Without a Summer? Climate in 1816* at the National Museum of Natural Sciences.

The purpose of this presentation is not to pursue the idea of climatic determinism, but rather to examine the pattern of the fur trade in the context of climatic conditions. The argument is presented that plants and subsequently all animals are limited in their options and reactions by climatic conditions. History must be examined in the context of climate because of its control over the fundamentals of life. I prefer to think that geography and history are inseparable; history is the play and geography the stage on which it is enacted.

It is interesting that anthropologists have little problem with the idea that primitive societies are essentially controlled by climate, but somehow historians reject the idea. What is the difference between the two? It is partly the fact that, until relatively recently, we have known little about the climate of this historic period. There is also a great deal of conceit in the belief that humans are not as affected by climate as other animals. This conceit has reached its highest levels in North America in the twentieth century where technology is believed to have the answers to all problems. Despite the fact that climate dictates over 80% of the yield on any farm there are no compulsory courses in climate or meteorology at Canadian schools of agriculture. The drought of 1988 brought the realities of the dominance of climate to the fore once again. It should have reminded us that man's mastery over the environment is a figment of his conceited imagination. I hope that, as we increase the amount of knowledge about past climates, we include it as a significant factor in the pattern of human actions; both past, present and future.

The pioneering work of people like Hubert Lamb delving into historical diaries and journals to reveal very different past climatic conditions has only occurred since the Second World War. Historical climatology has shown that climate has varied a great deal in time and space, thus altering the prosperity of different regions. The primary alteration is in the ability to produce food. However, climate also affects commerce, especially if it depends upon a natural product that is weather dependent.

One cannot examine the impact of the period from 1789 to 1820 upon the fur trade without considering the broader context. Rarely do singular climatic periods or events create direct change. Invariably a system is put under increasing pressure until certain climatic conditions become a catalyst to change.

The fur trade in North America is a good example of an enterprise almost totally dependant upon climate for its survival and success. Climate dictates: the number and quality of furs; conditions for the trappers and their families; the ease of transport through snow conditions or water levels in rivers and lakes; the ease of shipment across the oceans; the dependency of Europeans upon food supply from the land, to name a few items.

There is no point in blaming historians for ignoring climate as a factor in such change because the information has not been available. As reconstruction of climatic patterns continues, it is essential this be included as a major factor in the mosaic of variables that direct the human condition.

The period from 950 to 1200 is variously referred to as the medieval warm epoch or the Little Climatic Optimum. Regardless of the term, it was a period of much warmer conditions than at present. Oats and barley were grown in Iceland; the Domesday Book records commercial vineyards flourishing in England; while the eleventh and twelfth centuries later became known as the golden age in Scotland.

It is important to note what was happening in North America because of the parallels with current predictions of global warming. The warmer conditions resulted in northward migration of the agricultural people of the lower Mississippi Valley into Wisconsin and Minnesota. However, it also resulted in increasing aridity in the Midwest - that is the area west of the Mississippi. Analyses of Holocene pollen from the northern plains of Iowa indicate increasing aridity and a change from deciduous forest to grasslands. In Canada the northern limit of trees expanded northward up to 100 km in some regions, as warmer conditions brought a longer growing season.

After 1200, global climate began to cool. The circumpolar vortex expanded and the zone of cyclonic storm activity shifted south. Cultures that had benefited from the warmer conditions now saw a decline, but as with any climatic shift, others gained. For example, the Old Norse colony in Greenland collapsed as crop failures increased and permafrost returned. The settlements in Iceland and Norway experienced a decline in population as agricultural conditions deteriorated. In North America the increased strength of the westerlies resulted in a greater rainshadow effect in the lee of the Rocky Mountains and increased dryness on the Great Plains.

The problems in Europe were a litany of woes for people who had experienced the warm conditions of the Little Climatic Optimum. The woes included: increasing storm severity; harvest failures; abandonment of croplands and villages in higher elevations; and an increase in disease and mortality rates. In Scotland it has been estimated that the elevation at which agriculture could be practised lowered by 200 m between 1450 and 1600. The greatest loss was in the Highlands because the vertical loss converts into a substantial horizontal loss, which is devastating in a country with little level or arable land.

Martin Parry has estimated that harvest failures occurred one year in 20 in the thirteenth century. By the late seventeenth century this had been reduced to one year in two. Consecutive years of failure led to consumption of seed grain, thus accentuating the situation. The implications are that the initial Highland clearances were caused by climate, not by land-hungry landlords. With Highland clans forced to lower ground, the clan wars began. This was to be the beginning of many decades of extreme social upheaval. In 1675 the Philosophical Transactions of the Royal Society reported that a lake in Strathglass had "...ice on it in the middle, even in the hottest summer." It is also reported by others that there was permanent snow on the tops of the Cairngorms. We rarely stop to think that curling, which originated on lake ice, could not be played in many winters in the twentieth century.

It has been estimated that by 1691 over 100,000 Scots had been transplanted to Ulster, driven by such conditions as occurred in March 1674 when excessively heavy snows, severe frosts and 13 days of drifting snow resulted in the deaths of hundreds of sheep. Unfortunately the 1690s brought even worse conditions; in the eight years from 1693 to 1700 there were seven failures of the essential oats harvest. An excellent measure of the degree of cold during this general period was the winter of 1683, known as the year of the great frost, when 2 feet (0.6 m) of ice formed on the Thames River in London.

Hubert Lamb called 1450 to 1850 the Little Ice Age. The coldest portion of this period was from 1645 to 1715, with the nadir occurring in the 1690s. (Ironically this period coincides closely with the lifespan of the astronomer, Edmund Halley). The impact of these climatic conditions on Europe are receiving growing attention. A very important point is that this 70-year period coincides with a period known as the Maunder Minimum. During this time there were virtually no sunspots. There is increasing evidence that when there are high numbers of sunspots, as in the 1980s, the Earth is warm, and when there are few it is cold.

Although the initial hardships of the Little Ice Age were caused by deteriorating weather, the reaction of the landowners was, in most cases, reprehensible, ranging from absolute cruelty to benign neglect. One who did attempt to alleviate the problems faced by some of his tenants was Thomas Douglas, 5th Earl of Selkirk. It is essential to understand that even his motives are suspect. That is they were not as altruistic as people have proposed. His objective was to ensure colonization of the North American continent to halt expansion of the American revolutionary nation. He detested revolutionaries, but especially Americans. The family house had been attacked by John Paul Jones, Scottish-born naval hero of the American War of Independence. Young Selkirk - then seven years old - was so frightened and angered that he held a lifelong grudge against Americans.

Climate and Fur Trade in Western Canada

I will return to Lord Selkirk later. It is necessary now to look at the evolution of the fur trade. I believe it is significant that the Hudson's Bay Company[1] received its charter in 1670, just 10 years before one of the coldest decades in the last several hundred years. The demand for furs would have been much different in the warmth of the Little Climatic Optimum. The Company prospered as the demand for furs increased, and they expanded their operation accordingly. Interestingly they did not move inland from the shores of the Bay remaining, as Edward Umfreville described it, "asleep by the frozen sea." This was to place increasing pressure on the wildlife in the northern regions, which will be discussed later.

The first half of the eighteenth century saw a gradual growth of the fur trade. A widening region yielded more and more furs, but now there was a growing confrontation. The pedlars or Canadians (as the Hudson's Bay Company called the fur traders of the North West Company operating from Quebec) expanded westward across the prairies. Debate began within the Company about the need to open inland posts to offset the threat. Historians argue that the decision to move inland was totally due to this competitive factor. I contend that competition was a factor, but more significant was the impact of climate, especially in the northern regions around and west of the Bay.

While the debate was occurring, global climate was changing. The weather records for Churchill and York Factory show a significant shift in the pattern of winds, precipitation and other variables. Prior to 1760 the mean summer position of the Polar Front was south of Churchill and York Factory. This meant that both sites experienced subarctic climatic conditions. After 1760 the Front shifted north, so that Churchill continued with a subarctic climate but York Factory now had a more temperate climate capable of supporting the boreal forest. By the 1780s another shift in climate was occurring. Conditions deteriorated and the record becomes replete with comments on the lack of game, hard times and starvation among the Indians. At the same time there was a continued reduction in the number of furs being taken.

A study of the period from 1780 to 1800 by Stephen Wilkerson quantified the deteriorating conditions. Content analysis (a frequency count of the number of references to starvation and other key words) clearly shows a system under extreme stress. In some years, such as 1792 the number of comments about lack of food, starving people, malnutrition, and death are eight to 10 times above previous periods. An entry for 17 January 1792 reads, "Indeed this winter has been

---

[1] The body of water is named Hudson Bay: the company is correctly called "The Hudson's Bay Company" or sometimes just "the Company".

so far the most remarkable for scarcity of provisions for neither Englishman or Indians can find anything to kill."

These conditions are coincident with unusual patterns of animal behaviour. Joseph Colen recorded the following in the Journal for 22 October 1787, "Game of all kinds scarce but that White Bears are so numerous and trouble some as to attack them and their stages where their provisions is deposited." Later in the same year an entry for 23 December reads, "Late in the evening large herd of wolves surrounded the Factory." Both these events are unusual for the period. Colen was to become a victim of the conditions.

During this period the number of furs taken was significantly reduced. Colen wrote to London arguing that the cause was overtrapping. The Company accused him of mismanagement and removed him from his post. Actually he was right, but for the wrong reason. The number of furs taken would not normally have been a problem except that now the climatic changes had altered the thresholds. Wide variations in temperature created great stress on the plants and animals. For example, an entry for 5 February illustrates the unusual nature of the situation, "They (Indians) also inform me that the winter set in so early upwards that many Swans and other waterfowl were froze in the Lakes and they found many of the former not fledged, they likewise say that the snow is remarkably deep."

However variations in precipitation caused the greatest difficulties. An entry for 14 November 1783 reads, "Never remember the snow so deep at this season of the year." The next year an entry for 24 April informs that "Snow at least 10 feet deep." Too much snow creates great difficulty, especially for larger animals including man. Too little snow is devastating. Ptarmigan, lemmings and many other species that form a major portion of the base of the food chain die off without the insulating effects of snow. Low temperatures that would not have been a problem with deeper snow became deadly.

Expansion to the interior continued apace, and by the turn of the century competition between the two companies was placing even greater pressure on the resources. The Indians were caught in the middle of the conflict. They watched the battle and felt the effects as the land and its resources were hard hit. For 20 years the struggle continued, finally being resolved with amalgamation in 1821.

Climate improved briefly in the first decade of the nineteenth century, but by 1809 a cooling trend was beginning. Much has been written about 1816, the year with no summer. It was originally thought that the eruption of Tambora in 1815 was the cause of this dramatic, history-altering year. However, climatic records show that cooling had begun in 1809, and the volcanic eruption occurred at the nadir of the cool period 1809-20. It is interesting to speculate on the impact of the volcano if global temperatures had been increasing at the time.

Conditions in Europe had been very similar through the latter part of the eighteenth century, as the work of many scholars attests. Harsh conditions seriously reduced the food supply and placed populations under increasing stress. People responded in their traditional ways, starvation and/or migration. Interestingly, migration is not the choice of the majority. Lord Selkirk's offer of transport to new opportunities and better conditions made the decision a little easier; but still it was not everyone's choice.

The first groups who came from Scotland under his auspices went to Prince Edward Island and Ontario. The best known group was the Selkirk Settlers: under the leadership of Miles Macdonell they arrived at York Factory in the autumn of 1811. Too late to travel south, they wintered at a place known as the Nelson Encampment and experienced the harshness of conditions of that part of the world. Things were not much better when they arrived at the Red River settlement at the junction of the Red and Assiniboine rivers in 1812. The cooler conditions discussed above had already begun, and made the early years very difficult. The harsh conditions of the "year with no summer" were just the beginning. In his district report for 1819 Peter Fidler writes that from 1816 to 1819 a severe drought affected everything and everyone:

> The spring months have sometimes storms of wind and thunder even so early as March within these last three years the Climate seems to be greatly changed the summers so backward with very little rain and even snow in winter much less than usual and the ground parched up that all small creeks that flowed with plentiful streams all summer have entirely dried up, for these several years loaded craft could ascend up as high as the Elbow or Carlton House but these last 3 summers it was necessary to convey all the goods from the Forks by land in Carts...

The latter comments refer to the shift from river traffic on the Assiniboine River to the use of Red River Carts. This climate-induced shift is reflected in the pattern of roads and settlement across the prairies today.

Consider the situation that has developed by 1817. The Selkirk Settlers have been thrust into a new harsh landscape. They are almost immediately confronted with severe weather that made things as difficult as they had been in their native Scotland. In addition, they were not welcome, either by the Indians, the Métis, or the fur traders. The Indians saw them as a threat to their traditional way of life and usurpers of their land. They were particularly concerned because they were suffering severely by the harsh weather and consequent lack of food. The Métis were already the 'in-between' group and had nothing to gain from the addition of another faction. Besides, they also benefited as a key part of the fur trade. Fur traders saw the settlers as a threat to their freewheeling monopolistic style. They also recognized that agriculture and fur trapping were potentially mutually exclusive. Amalgamation between the two companies had not occurred yet, and the settlers were unwitting pawns in the conflict.

The entire situation was extremely volatile. It is not surprising that the Indians and Métis attacked the settlers or that the fur traders (particularly members of the North West Company) did little to assist them. The culmination of these conflicts was the Seven Oaks Massacre in 1817 when a group of Métis led by Cuthbert Grant killed 21 people. I do not think that the climate was the cause of this event, however, I suggest that the extreme climatic conditions and their impact on the economy and food supply created untenable and volatile situations. The fur trade was to continue for some decades as there was little useable land in northern Canada. However, the southern regions were irretrievably changed as the land was cleared at an increasing rate.

Conclusion

Borisenkov, the Soviet climatologist, has carried out extensive research using historical sources to reconstruct climate over the last several centuries. He writes that, "In the climatic sense the Little Ice Age was highly variable both spatially and temporally. The main feature of that period was the frequent recurrence of climatic extremes, during which Russia suffered 350 "hungry years" as a result of unfavourable climatic conditions." He makes links between the climatic disasters, such as drought, rainy and cool summers or severe winters and the pattern of peasant life. The correlation between particularly prolonged harsh conditions and peasant revolts cannot be ignored.

It is easy to blame historians for not considering climate as a major factor influencing important social events. They have not had the information - although some clues should have been apparent. Paintings such as those by Breughel showing winter conditions very different than today or Jan Griffier's frosty painting of the River Thames with 2 feet (0.6 cm) of ice in the great frost of 1683, cannot be accused of artistic licence. As the picture of historic climate is reconstructed we should be able to reach more precise conclusions about the relationship between climate and history.

The concept of climate influencing the pattern of history has suffered from the extreme distortions of fascism and the lack of information about actual climatic conditions. Climatic determinism is not the issue here, especially as it relates to human characteristics. My point is that climate affects the environment which has direct impact upon the food supply and economy, and therefore the people.

The fur trade of North America provides an excellent opportunity to study the relationship between climate and the pattern of history. Records maintained by the Hudson's Bay Company provide detailed evidence of the climate and its impact on the economy and lives of the people.

The Ecology of a Famine: Northwestern Ontario in 1815-17

Roger Suffling[1] and Ron Fritz[1]

Abstract

The extraordinary summer weather of 1816 has been blamed on the eruption of the Tambora volcano in 1815, and has been associated with various social and economic disruptions around the globe. In northwestern Ontario there were famines in the winters of 1815-16 and 1816-17 among native Ojibwa people and Hudson's Bay Company traders who relied on seven basic resources: moose, caribou, wildrice, potatoes, fish, wildfowl and furbearers. The first of these two famines was extremely severe. It resulted from an initial, non-climatically induced reduction in moose and caribou, and a natural cyclic crash in the snowshoe hare population. The early summer drought of 1815 reduced the potato crop, and late summer rain and cold ruined the wildrice harvest and fishing. These, combined with an early fall goose migration left both the Ojibwa and Hudson's Bay Company employees starving. In contrast, the dry cold summer of 1816 fostered production of a large potato crop, though of indifferent quality, and normal wildrice and fish harvests. The winter 1816-17 famine resulted primarily from deep snow conditions, but the Hudson's Bay Company was able to feed many starving natives. Thus the cold, droughty summer of 1816, the "year with no summer", may have done as much to ameliorate famine among the Ojibwa people as to create it.

Introduction

Northwestern Ontario (Figure 1) is a harsh land of subarctic forests. Even now, the population is sparse and, in the early nineteenth century, it probably never exceeded a few thousand (Bishop 1974). Though severe epidemics of smallpox and other diseases periodically devastated the Ojibwa and Cree peoples (e.g., Hearne 1791), starvation was often the most effective controlling factor, as the following report from Osnaburgh House (51° 90'N 90° 15'W) illustrates:

> "An Indian woman of the Crows gang came in to [sic] day with her four young children all much starved and with a very miserable report. Says that her husband starved to death two winters back since which she has been in wretchedness and want with four children to support. Her friends take but little notice of her being under the impression that she eat her husband when under one of the greatest of all miseries extreme starvation." (1 November 1814)[2]

The above is part of a daily record kept by Hudson's Bay Company (HBC) Post Masters, and known as a Journal of Occurrences. It spans the years 1786 to 1911 with few interruptions. The journals confirm the generality of the appalling conditions described above (though this is a severe case indeed). Of 126 years of records, 60 include allusions to people starving (Figure 2).

---

[1] Faculty of Environmental Studies, University of Waterloo, Waterloo, Ontario N2L 3G1, Canada.
[2] Unless otherwise noted, dates quoted are from the Osnaburgh House Journals of Occurrences, originals of which are kept at the Hudson's Bay Company Archives in Winnipeg, Manitoba, Canada.

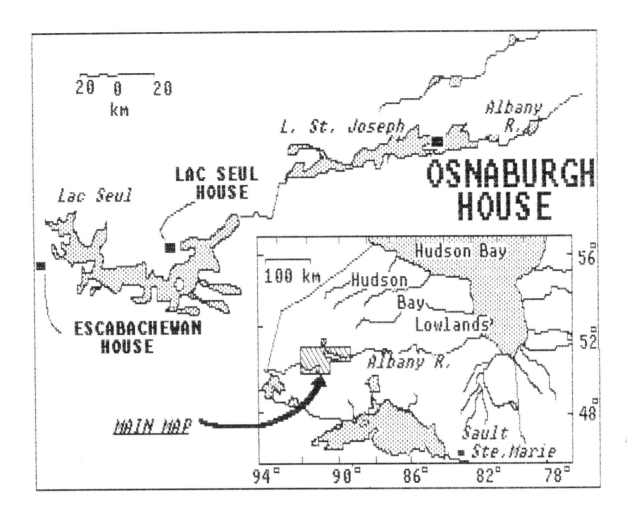

Figure 1: Locations mentioned in the text.

In using the word, "starving", we intend the same meaning attached by the Post Masters: that somebody was unable to procure food for days at a time. Sometimes this situation would be brief or intermittent. On occasion it continued until death ensued. Often, however, it is apparent that starvation contributed to death by other means. The woman cited above struggled on in the same pathetic condition for two more years before succumbing to an illness: she had killed her husband to survive a famine.

We have used the word "famine" to mean general starvation among people that was sufficiently prolonged to become life-threatening.

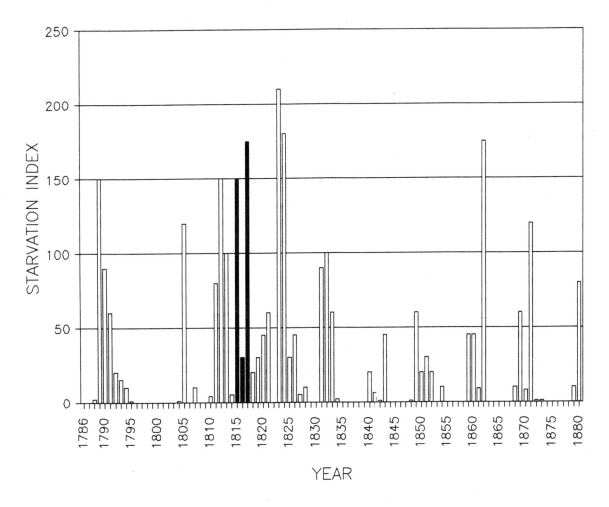

Figure 2: Incidence of starvation at Osnaburgh House. The starvation index is the product of the number of references to starvation in the Journal of Occurrence and the maximum number of people recorded as starving.

In the fall and winter of 1815-16 there was a particularly horrendous incident of famine in northwestern Ontario that is well illustrated by the Osnaburgh House records. It involved not only the native population, but also (and unusually) the better prepared HBC employees and their families. The 1815-16 famine was one of eight recorded that involved more than 30 people (out of a total population of about 200). In terms of numbers referenced in the journals, it was one of the worst eight incidents, and in deaths it probably ranks only second to the 1823-25 incident.

There was another famine in the 1816-17 winter but, though it was severe, and though it may have involved just as many people, its consequences were not as grave as those of the 1815-16 famine. Both of these famines are part of a prolonged series of incidents from 1810 to 1825 that broke the spirit and culture of the Ojibwa people (Bishop 1974). The famine series is associated with depletion of big game and an exceptionally cold climatic fluctuation.

The 1815-16 famine is particularly striking because it corresponds with climatic fluctuations evidently caused by the eruption of the Tambora volcano in what is now Indonesia, early in 1815. The question that we asked ourselves, therefore, was whether the 1815-16 famine was caused by unusual weather conditions. To find an answer, we examined the ecology of the Ojibwa people around Osnaburgh House to see which climatic or other conditions normally contributed to famine, and to see which of these pertained immediately before or during the 1815-16 incident, and the lesser famine of 1816-17.

Ojibwa Ecology and Food Sources in Early Nineteenth Century Northwestern Ontario

Big Game

Bishop (1974) has postulated that the Ojibwa people who lived near Osnaburgh House in the early 1800s had moved there from around Sault Ste. Marie.

Initially, they had been primarily big game hunters, subsisting on moose (*Alces alces*) and woodland caribou (*Rangifer tarandus*). Bishop believes that, after first European contact, the people began to make forays into northwestern Ontario in search of furbearers to use in the new commercial fur trade. Rival companies soon began to challenge the HBC's hegemony over this and other areas. First the French and then Scots from Montreal, and American traders moved into the area to trade furs at their source. This induced the Ojibwa to remain on the summering grounds year round, but forced a radical reorganization of their hunting strategy. Originally, they had hunted in large groups. Now, with the need to spread out to trap beaver, they broke into family groups, and they used firearms to kill moose and caribou. When they were available in sufficient quantity, moose and caribou meat and skins were traded to the HBC, putting further pressure on the herds. By 1815, both species were already somewhat depleted (Figures 3, 4), and this was beginning to wreak hardship, not only directly in terms of food availability, but also because leather for mocassins and snowshoes was becoming scarce. The people's very existence in this land of thinly-spread resources, was predicated on nomadic foraging, so a lack of leather hampered many food gathering activities, as well as in fur trapping. In a typical instance, a native arrived at Lac Seul asking to purchase a summer bear (*Ursus americanus*) skin, there being no caribou or moose leather: "The bear skin is for making his shoes without which he cannot leave his tent" (Lac Seul 28 April 1828).

Moose (Figure 3) and caribou (Figure 4) were stalked at all times of the year, and herded into the water for slaughter in the summer. Deep snow slowed the animals down in winter, making them easier to approach, but they could easily outrun hunters on thin snow - so the latter condition is associated with hardship. Extremely deep snow made both hunter and hunted less mobile and sometimes prevented the people from traveling between various moose and caribou wintering grounds.

Crusted snow gives human or other predators a marked advantage (J. Theberge 1988, personal communication). It must have occurred with greater frequency in the early nineteenth century as uncontrolled forest fires increased the proportion of open country where crusting occurs easily. Thus, even as the herds were reduced, the pursuit of the remaining animals may have become more efficient, ensuring further big-game depletion.

Moose and caribou meat were eaten fresh, or preserved by drying in strips over a fire, or in pemmican (a preserved mixture of fat, berries and shredded meat pounded together).

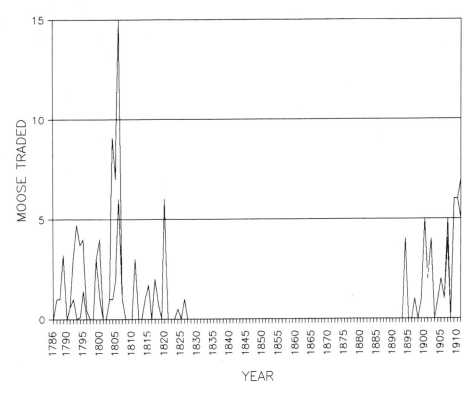

Figure 3: The number of moose involved in trading of meat and skins from natives to the HBC at Osnaburgh House. The upper line is a maximum estimate, and the lower line a minimum. Derivation of the data is given in Fritz (1988).

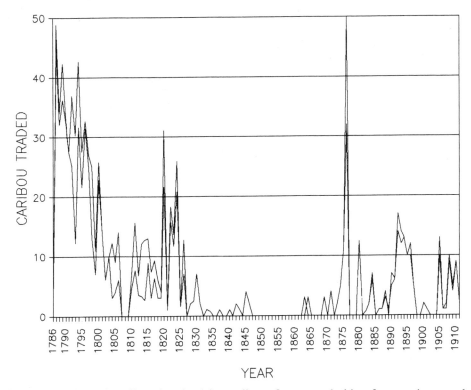

Figure 4: The number of caribou involved in trading of meat and skins from natives to the HBC at Osnaburgh House. The upper line is a maximum estimate, and the lower line a minimum. Derivation of the data is given in Fritz (1988). The maximum figure (102) for 1876 is off-scale.

Fish

The third major food source was fish. They were hooked, speared or netted, depending on species and season. Several species were used including: whitefish (*Coregonus clupeiformis*), sucker (*Catostomus* spp.), pickerel (*Stizostedion vitreum*), sturgeon (*Acipenser fulvescens*) and pike (*Esox lucius*). At Osnaburgh House, sturgeon appear to have been particularly critical to human welfare. They could be readily speared and netted when spawning in the early spring - an otherwise lean time of year. Spring and summer were employed in catching mostly pickerel, pike and whitefish; and the fishery continued until the water became warm and the eating quality of the fish declined. Fishing resumed in the fall as water temperatures fell, when a number of species came to spawn in the shallows and rapids of the rivers. Fishing continued until freeze-up and occasionally afterwards, under the ice, but the early nineteenth century natives do not seem to have mastered the art of ice fishing with nets as the HBC people had.

High water in the lakes generally signalled a failure of the fishery, especially in the fall. The high water could be caused by unusually heavy rainfall, cool weather, or a combination of both. Early freeze-up also hurt the fishery as it cut short the spawning seasons of the fish, and they withdrew to deeper water.

Fish were vitally important at northwestern Ontario HBC posts during winter - especially whitefish and, as soon as the weather became cold enough to store fish, they were netted intensively. In times of native starvation these fish were distributed to Ojibwa begging at the posts, as long as the HBC's own supply of stored or fresh fish remained assured. The motivation was partly charitable but hinged too on the economic need to preserve the lives and health of the beaver trappers who were the lifeblood of the Company's activities in these parts.

Wildfowl

The fourth major native food resource was wildfowl -- primarily geese. Both Canada Geese (*Branta canadensis*) and Snow Geese (*Chen caerulescens*) were shot, as well as a variety of ducks. At Osnaburgh House, wildfowl first appeared in April, moving north to the Hudson Bay Lowlands and beyond on the turbulent edge of the retreating Arctic air mass (Ball 1983). If snow lies on the coastal marshes of Hudson Bay at the time when goose eggs should be laid, a lower proportion of females than normal actually lays eggs. In addition, average clutch-size is reduced. Thus the cohort of young geese produced is small, as happened in 1967. Fall migrants then prove relatively sparse, as do those birds returning the following spring. If fall came early, sending the birds south too soon for native needs, then people had a longer time to wait between fall and spring migrations. The people were often starving in late winter, so that the return of the geese was awaited with eagerness by both natives and HBC men.

Wildrice

Wildrice was the only staple vegetable of the largely carnivorous Ojibwa. (The same cannot be said of the HBC men who also grew potatoes and some lesser crops). Wildrice is an annual aquatic grass found in slow-flowing rivers and shallow lakes (Dore 1969, Suffling and Schreiner 1979). It sets seed in late summer and is harvested in late August or early September in northwestern Ontario. The seeds, which were fermented, hulled, dried and stored for winter use, were a good hedge against starvation. High water in mid- to late-summer - especially rising high water - is disastrous to the crop. Also, windstorms can scatter the grain before it is harvested.

Hares

Snowshoe hares (*Lepus americanus*), usually called rabbits in the journals, were also an important food item. They never appear to have been a preferred food (Bishop 1974), but were snared in hard times when other victuals were lacking. Hare pelts that were the by-product of this activity were traded, but only commanded a minimal price at the HBC posts. Alternate freezing and thawing in winter made rabbit snaring impossible (Lac Seul, 3 February 1825).

The snowshoe hare exhibits a remarkable eight- to nine-year cycle of population density (MacLulich 1937, Elton and Nicholson 1942). It is notable that most of the peaks in human starvation at Osnaburgh House appear in the year after the crash of the hare population (Figure 5), a relationship which is statistically significant ($X2$, $P<0.01$). Thus snowshoe hare scarcity could precipitate famine.

Furbearers

Furbearers were the means by which the Ojibwa obtained non-local commodities such as iron knives, hatchets, guns, blankets and rum. The species trapped or hunted include marten (*Martes americana*), otter (*Lutra canadensis*), fisher (*Martes pennanti*) and lynx (*Lynx canadensis*), but the most important was beaver (*Castor canadensis*). Where they were available in large numbers, as at Lac Seul, muskrats (*Ondatra zibethicus*) were also very important in total, and as beavers were depleted, muskrats assumed an increased economic significance. Although beavers and muskrats had the added advantage that the carcasses were edible, generally furbearers did not contribute greatly to human nutrition. Their significance in this context is in how they influenced the pattern of trapping and hunting of other animals.

Potatoes

As a rule, natives in early nineteenth century northwestern Ontario did not grow potatoes, though there were a few individual attempts. This vegetable was, however, a staple of the HBC posts. Potatoes were planted in early May at Osnaburgh, and harvested in mid- to late-October.

The potato crop was highly variable. It appears to have suffered after hot, dry summers, and was of low quality if an early fall frost damaged the tubers. Though spring frosts were damaging to the top growth and may have reduced the yield, they do not seem to have been as serious a problem. Every few years, there were also damaging epidemics of "grubs" (so far unidentified). As with fish, potatoes were given to starving Ojibwa coming to the post for assistance - at least for as long as there was no threat of starvation to HBC employees themselves.

The Annual Cycle of Ojibwa Subsistence

The annual cycle of Ojibwa subsistence in the early nineteenth century (Figure 5) is not totally dissimilar to the modern pattern described by Sieciechowicz (1977). The Ojibwa year can be thought of as beginning in late September to October when natives arrived at the HBC post to obtain their outfit for the coming winter. At this time, goods were normally obtained on credit, a process described as "taking debt" or "outfitting". The fall fishing and goose harvest more or less coincided with taking debt. Then, as the weather hardened, and the furbearers came into prime pelage, there was a concentrated effort to trap - and especially after the first snows. The amorous bull moose could be readily killed at this time as they could be called in by a hunter using a birch bark trumpet to initiate a rival or a cow moose.

When the large lakes froze, usually in early November around Osnaburgh, winter began in earnest. By now only big game, hares and occasional grouse (*Canachites canadensis*, *Bonasa umbellus*, and possibly *Pedioecetes phasianellus*) were available. Since stored fish and wildrice gave out, often about the beginning of January, people had to since rely entirely on meat and fat. If starvation arose, it became apparent in the journal entries at this time, and it would become more severe after the onset of really bitter weather. The starving time of winter could be alleviated or avoided if large game abounded, if hares were present in large numbers, or if spring came early. The converse was also true.

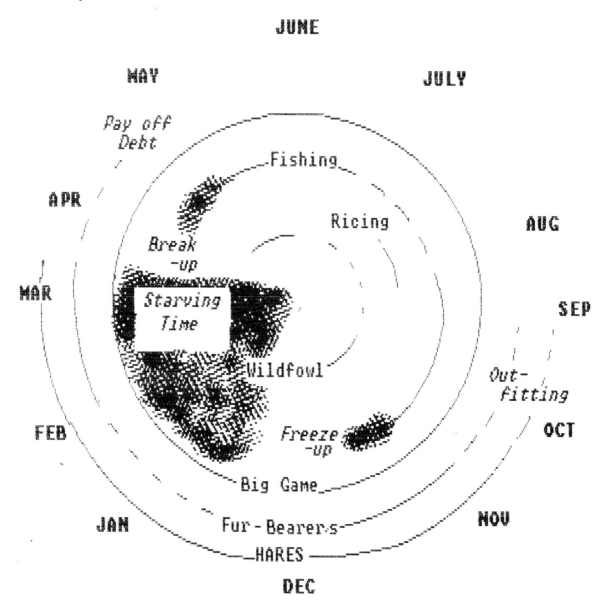

Figure 5: The annual cycle of early nineteenth century Ojibwa people living around Osnaburgh House.

If people were not starving severely, there was a second burst of fur-trapping activity in the relatively mild weather of late winter and early spring.

Normally in mid-April the first geese arrived, and sometimes set off what can only be described as a hunting frenzy at the HBC posts! At Lac Seul in 1828, for instance, the Post Master gave all his people three days off to hunt geese,"... *in the hopes of their setting to work afterwards*". Geese not only provided relief from starvation, but for the HBC men in particular, brought a welcome rest from the six-month monotony of potatoes and whitefish.

In late winter Ojibwa appeared at the posts to redeem their debts, to socialize, and to drink. It was a time when the HBC Post Master anxiously awaited the fur harvest, and when the extent of any lethal starvation became apparent through the non-arrival of families from the forest.

With the break-up of ice on the lakes and freshets of meltwater in the rivers came the spearing of sturgeon and pickerel as they spawned at the base of rapids. Bears (*Ursus americanus*) too came for the fish, and could be readily trapped then, if they had not been found in their hibernation dens and speared. They were generally too lean in the spring to provide much meat or fat. Fishing continued throughout the spring, supplemented by game hunting, as well as by collecting a variety of fruits and berries, and possibly birds' eggs.

In August the water became too warm for profitable fishing, but at the end of the month the wildrice harvest began. This was another time that brought people together, and it was closely followed by fall fishing and acquiring new outfits at the Posts.

The pattern described above is typical, but each year presented a slightly different situation, and the resources available around each trading post differed slightly. Osnaburgh had more sturgeon, Lac Seul had more muskrats and wildrice, etc. The trading policies of the HBC and its rivals, the weather, as well as availability of food and furs all varied enormously over time, and have been discussed at length by Bishop (1974).

Factors that Precipitated Starvation

The factors causing or excacerbating starvation are summarized in Table 1. Severe starvation might be avoided if only one or two factors were unfavourable in a given year, but if several coincided, then people would suffer accordingly. Most of the factors have been discussed above, but the incapacitation of hunters needs comment.

Injury, sickness, or death of menfolk was a constant peril to family groups - it could deny them access to big game. If the women and children's occupation of snaring hare was unavailable because of a hare population crash, then starvation was bound to follow. Repeated freeze-thaw cycles that prevented hare snaring sometimes had the same effect.

The 1815-16 Famine

January and February 1815 were fairly typical for the time of year - dry and cold. This weather persisted into April which, coupled with north winds, kept most of the geese and ducks from arriving. A few came, however, on 15 April - about the usual time.

May, too, proved very cold at first so that the Post Master remarked that the weather had "more the appearance of March than of May" (8 May 1815). The main body of geese arrived only on 14 May, a month late. The keeper of the journal considered that the latter half of May was warm for the time of year, and the lake ice broke only a little late.

Early June was judged to be warm for the season, but the latter half was rainy and cold. This weather persisted into early July until it became very hot during 9-13 June, and then again from 24 July to 10 August. The warm spell was sufficiently drying that the writer states:

> "Hookamarshish informs me that all his furs were burned, he says that he was going to move to another place and he forgot to put out his fire and so it set fire to the woods and burned all his furs." (15 August 1815).

Table 1: Factors Contributing to Starvation among the Early Nineteenth Century Ojibwa around Osnaburgh House.

	Causal Factor	Resource Affected
Climatic Contributors		
	High water in summer	Wildrice, fish
	Rising water in summer	Wildrice
	Thin snow	Moose, caribou
	Very deep snow	Moose, caribou
	Cold, damp spring on Hudson Bay	Geese
	Droughty summer	Potatoes
	Freeze/thaw in winter	Snowshoe hare
Non-Climatic Contributors		
	Increased forest fires due to fur trade	Moose, caribou
	Overhunting due to fur trade	Moose, caribou
	Cyclic population crash	Snowshoe hare
	Injury, sickness or death of menfolk	Moose, caribou

On 22 August came a sudden cooling with rainy, stormy weather accompanied by NW and E winds. These conditions persisted unabated until 17 September. By late September it was apparent that both the fall fishing and the wildrice harvest had failed on account of high water in the lakes and rivers:

> "The water being so remarkably high at this place the Indians is not made any rice worth while so that I have only got 64 gal in all. So that I am much afraid of starving in the winter as there is no fish to be got here when the water is high in the fall. Am sorry to inform you that this is a very poor place for most everything. There is no beaver nor moose to the indians to hunt and most of them were starving when I seed them but are all off now to hunt." (Letter from James Slatter at Escabachewan 23 September 1815 to the Master of Osnaburgh House).

At Osnaburgh, the problems were compounded by a lack of fishing twine and of available labour. Such fall starvation was unusual, but the people were probably cheered by the early arrival of the bulk of the fall geese on 1 October. In reality, this worsened matters for, with the early passage of wildfowl, the impending winter starvation was to last longer.

The potato harvest at Onsaburgh House was 77 kegs, down 20 from the previous year, so that the HBC people entered the winter with very little food to spare for visiting Ojibwa.

The first snow came on 22 October and the lake froze on 7 November, a trifle early. The subsequent ice fishery failed as miserably as had the fall netting. The first half of December was very cold and the the latter half mild.

Starvation is first mentioned again in the journals in December, and by late January 1816 it was general among the natives, even appearing at the HBC fishing outposts: *"The men are already feeling the iron hand of want."* (Osnaburgh House, 30 January 1816).

In February, which was cold even for that time of year, natives arrived at the Post both frozen and starved; but others coming from the north were heavily laden with furs and apparently well fed (according to Bishop (1974) there were still moose to be had in that quarter).

By 23 March, the potato ration for HBC people had been cut to two gallons per week (instead of the usual three three gallons), and by 25 March everybody was sent out to hunt or fish since the daily ration was, by that time, one small fish. The men at the marsh outpost of Osnaburgh were now too weak even to go to the House for food and one fellow, reduced to eating fish offal, became very sick.

Very few natives had visited the Osnaburgh House during the winter, either because they were too weak to travel or possibly because word was out that there was no food to be had there.

April was very cold, and the digging of the potato garden at Osnaburgh House began two weeks late (on 30 April) as a direct consequence. The famine finally broke with the arrival (three weeks late) of the first geese on 4 May. Sturgeon did not begin to spawn until 29 May - two weeks later than usual.

June 1816, likewise, was very cold with a hard frost on 4-6 June and another on the 23-27 June. On both occasions the gardens were badly frosted. On the latter, there was one-quarter inch of ice in the bottom of the canoes pulled up on the shore. This must have been a dry month as the lake fell six inches in three weeks.

July was cool and rainy with mostly NW winds, and this weather continued into the first half of August. In spite of this, the water remained low in the Albany River, suggesting perhaps that there had been little snow in the previous winter, and that water in the marshes must have been low all summer.

On 18 August, there was snow - an unheard-of event in this month, and with continuing cold weather the geese were already flying thick by 15 September - a month early. During 25-30 September there was a gale, remarkable not so much for its ferocity but for its duration. Its winds tracked from E to S, to SW. In contrast with 1815, there is no indication that the wildrice harvest or the fall fishing were other than normal at Osnaburgh House.

The ground froze by the 3 October (about three weeks early), so the HBC people were caught unprepared and the potatoes were frozen in the ground. In spite of this, they harvested 190 kegs, a large crop, though evidently of indifferent quality on account of the frost. The rye plants were six-feet tall but the grain was still green, and never had a chance to ripen. The whole crop was lost.

The lake froze a little early on 9 November and the weather continued cold until the second half of December which proved mild and snowy.

January 1817, and the rest of the winter, were cold with heavy snowfall, which was in marked contrast with the previous cold, and apparently dry winter of 1815-16.

There are almost as many citations of starvation in the 1816-17 journal as there had been for 1815-16, but there is little indication of the grinding life-threatening severity of famine which had overtaken people in the previous year. Apparently the 1816-17 starvation touched only the native people, and many of them were visiting the post for handouts.

Discussion and Conclusions

It would be easy to rush to the conclusion that the Tambora eruption of 1815 explains any unusual weather patterns in the subsequent couple of years. As the discussions of the "Year without a summer? Climate in 1816" conference demonstrated, it is difficult to unequivocally establish causal connections, even though there is a suspicious conjunction of climatic dislocations around the globe. At Osnaburgh House, the unusual conditions were: the sudden cool, wet end to the hot, droughty summer; the long, dry, cold winter of 1815-16; and the dry cold summer of 1816.

Likewise, the mere existence of famine in 1815-16 is not proof, *per se*, of the human ecological consequences of the Tambora eruption, or even of the effects of the harsh weather of these years. In reality, several factors contributed to the famine at Osnaburgh House and elsewhere in northern Ontario. They include the following ecological factors:

1. Depletion of moose and caribou herds by overhunting (Bishop 1974), or possibly by habitat change through forest fires in the late 1700s. Both of these are associated with the expansion of the European fur trade.

2. A cyclic crash in the hare population between spring 1814 and spring 1815 fur returns.

Neither of these two factors are climatic, but they are primary causes in the sense that they set the stage for the other events. The fate of many natives was sealed by other phenomena that were indeed climatic, namely:

1. Failure of the wildrice harvest in 1815 due to high, rising water levels in late summer.

2. Failure of the 1815 fall fishery due to high water, and failure of the ice fishery, possibly for the same reason.

3. The small 1815 potato harvest resulting from the early summer drought and hot weather. (The potatoes were grown in a dry, "hungry", sandy soil that warmed quickly in spring but was very vulnerable to drying.) Thus the HBC had few or no potatoes to spare for the natives during the famine.

4. The early-fall goose migration in 1815, and the late-spring migration of 1816. The former ensured that the natives entered the winter in poor nutritional condition, and the latter prolonged their suffering in spring.

One last factor was the lack of HBC labour for fishing, and lack of fishing twine in the fall 1815. These are minor economic or logistic causes.

The summer of 1816 was in marked contrast to that of 1815. Though there were late spring frosts and summer snow, the cool weather actually appears to have helped the potato crop. Likewise the droughty conditions kept water levels low and assured at least a normal fish and wildrice harvest. Spring frosts had few harmful effects, and the fall frosts did some damage to potatoes, but not enough to be serious.

It is as impossible to say that climate alone caused the 1815-16 famine as to claim successfully that non-climatic factors were responsible. People relied on seven major resources. Moose and caribou had already been depleted by 1810, and the worsening climate of 1810 to 1817 was the trigger for a series of famines that were only alleviated by the hare "high" of 1814. The subsequent crash of hares reduced the resources available to four: wildfowl, wildrice, fish and potatoes. The high water of 1815 knocked out two of these - fish and wildrice - leaving only potatoes and wildfowl - and even the potatoes were reduced by early drought. A catastrophe then became inevitable.

The fall of 1816 was different. Big game and hares were still scarce, but the water remained low, ensuring a wildrice and fish harvest. The cool summer evidently favoured the HBC with a large potato crop, although possibly more had been planted as a reaction to the previous famine. On the other hand, the October frost impaired the quality of the crop. The two frosts of June had evidently had little effect, although one might easily jump to the conclusion that they had caused the smaller famine of 1816-17. In reality the potatoes were the saving grace for a native population that probably was still reeling from the physical and psychological impact of the 1815-16 famine. The winter or 1816-17 was a year of starvation, but was not as serious as its predecessor. If anything, it must have been deep snow that limited travel to new hunting grounds or to the HBC for charity, that caused most hardship.

We conclude that the extraordinary cold, dry weather in the summer of 1816 may actually have done more to prevent famine than to create it. The cold, wet end to the summer of 1815 was, however, the proximal cause, but only the proximal cause, of the 1815-16 suffering.

The postscript of the famine is as interesting as the event itself. Between 1819 and 1820, the number of both moose and caribou traded at Osnaburgh House rose dramatically (Figures 3, 4). Perhaps enough hunters perished that the predation pressure on the herds was reduced and they started to increase. Deep snow in 1816-17 may have reduced hunting pressure with the same effect. Whatever the causes, moose and caribou then persisted until the mid-1820s before succumbing to hunting or other pressures. Thus one effect of the 1815-16 incident had been to prolong the survival of big game that were so important to the people, even as it helped to destroy their will and culture.

Acknowledgements

The research for our paper was conducted while one of us (R.F.) held a Natural Sciences and Engineering Research Council Undergraduate Internship. We thank John Theberge and Harold Lumsden for their advice concerning ungulate and goose ecology.

References

Ball, T. 1983. The migration of geese as an indicator of climate change in the southern Hudson Bay region between 1715 and 1851. *Climatic Change* 5:85-93.

Bishop, C.A. 1974. *The Northern Ojibwa and the Fur Trade, An Historical and Ecological Study*. Cultures and Communities Series. S.M. Weaver (general ed.). Holt, Rinehart and Winston, Toronto.

Dore, W.G. 1969. *Wildrice*. Canada Department of Agriculture Research Branch, Plant Research Institute Publication 1393. Ottawa. 84 pp.

Elton, C.S. and A.J. Nicholson. 1942. The ten year cycle in the numbers of the lynx in Canada. *Journal of Animal Ecology*. 1:215-244.

Fritz, R. 1988. Moose and caribou population decline in N.W. Ontario boreal forests of the Osnaburgh House (HBC) trade area: 1786-1911. Senior Honours Essay. Department of Geography, University of Waterloo, Ontario.

Hearne, S. 1791. *A journal of observations made on the journey inland from Prince of Wales Fort in latitude 58°50' North to latitude 72°00' Beginning 7th Decr. 1770, ending June 30th, 1772 by Samuel Hearne*. Manuscript in British Museum Library, London, U.K.

MacLulich, D.A. 1937. Fluctuations in the numbers of the varying hare (*Lepus americanus*). *University of Toronto Studies, Biology Series* 43:1-136.

Sieciechowicz, K. 1977. People and land are one: an introduction to the way of life north of 50°. *Bulletin of the Canadian Association in Support of Native Peoples* 18(2):16-20.

Suffling, R. and C. Schreiner. 1979. *A Bibliography of Wildrice (Zizania species) Including Biological, Anthropological and Socio-economic Aspects*. University of Waterloo School of Urban and Regional Planning, Working Paper 5, Waterloo, Ontario.

The Development and Testing of a Methodology for Extracting Sea-Ice Data from Ships' Log-Books

Marcia Faurer[1]

Abstract

The severity of the weather in 1816 in the Hudson Strait and Hudson Bay regions became apparent through the reconstruction of sea-ice conditions for the period 1751-1870. The sea-ice data were derived from an exceptionally large collection of ships' log-books. Current research is focusing on the development of a reliable methodology for use in further environmental reconstructions covering this period using historical documents.

This research applies a methodology called content analysis which was developed by the Social Sciences for extracting meanings from human communications in an objective manner. This technique has the ability to reduce the subjectivity inherent in the interpretation of historical documents by testing the level of reliability of the procedure that is used to extract sea-ice data from log-book descriptions. In this study, tests have been applied throughout the development of the methodology with the goal of devising an objective procedure. These tests also reveal the degree of detail that a particular source can reliably provide, as well as helping to reduce the difficulties associated with calibrating the historical terminology against the contemporary sea-ice vocabulary.

Introduction

Although content analysis (CA) has been used widely in the application of historical documents as proxy sources for climatic reconstructions, this methodology has not been applied to its fullest potential. This is primarily due to the general omission of its strongest attribute, which is the ability to test the reliability of the methodology that is used. This aspect of CA is not merely an option that may or may not be applied, it is actually an integral part of the CA process. Without this means of evaluation, the interpretation of historical texts may be guided by predetermined decisions about the information required for the reconstruction instead of by the information that the documents can objectively provide.

This case study was conducted to test the applicability of an objective methodology for extraction of environmental data from historical documents. The format of CA was followed closely by the repeated application of reliability tests. Whereas this reduced the information obtained, it insured that the derived data were obtained to a measured and acceptable level of reliability.

Data Sources and Background Information

The eighteenth and nineteenth century log-books of the Hudson's Bay Company are a potential source for a wide variety of environmental data. Although temperature readings were entered in the log-books, they appeared sporadically throughout the period of record (1751 to 1870). Wind directions were given on a fairly regular basis as well as other meteorological phenomena,

---

[1] Department of Geography, University of Manitoba, Winnipeg, Manitoba R3T 2N2, Canada.

however sea ice was chosen to be the focus of this study. While sea ice is not a meteorological element *per se*, it is a visible expression of several environmental factors. It was also selected because it posed a clear and present danger to the success of the voyage and to the lives of the crew. Therefore, it was anticipated that any event or observation related to sea ice would be faithfully recorded in the log-books.

Sea ice was encountered by the ships in Hudson Strait and Hudson Bay during the westward portion of a voyage between England and the Hudson's Bay Company's bayside posts. These locations and the routes of the ships are shown in Figure 1. Each year, the Company dispatched a small convoy of ships to supply these remote posts and to bring trade goods back to England. This collection of log-books provides an unbroken record of sea-ice descriptions that can be cross-checked because each of the ships in the convoy kept at least one log-book. Figure 2 is a reproduction of a log-book page showing the meticulous way in which the environmental observations were recorded. Fortunately, this format and the vocabulary used in the log-books remained virtually unchanged throughout the entire period of record.

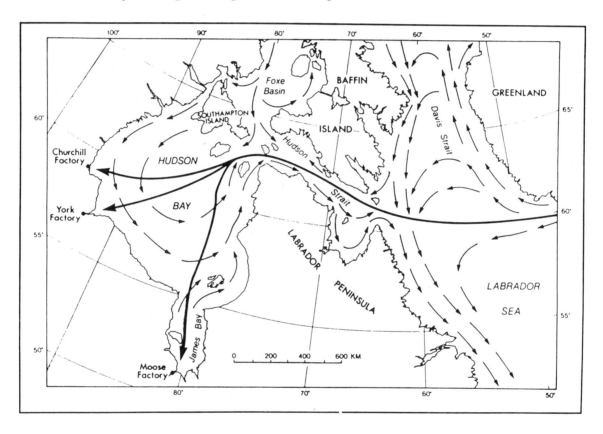

Figure 1: Sea currents and Hudson's Bay Company sailing route.

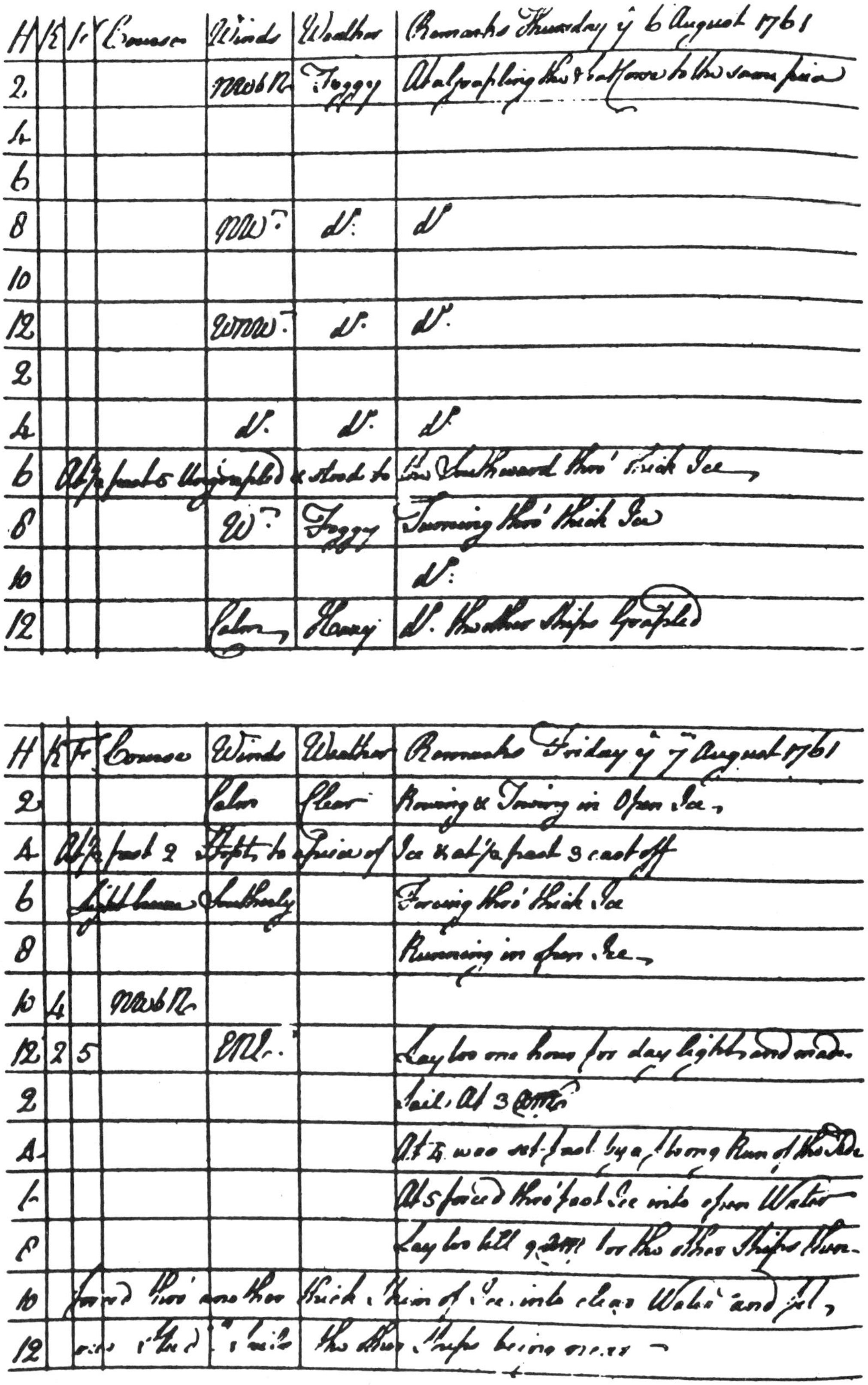

Figure 2: Sample log-book page.

Problems of Interpretation

The problems arising from this type of data source are twofold. The first aspect of the problem lies in the interpretation of the descriptive accounts that used a vocabulary which was different from the current terminology. Secondly there is the difficulty that arises in conversion of qualitative or verbal descriptions into quantitative or numerical data that can be compared with contemporary records. When properly applied, CA serves to resolve these problems reliably.

Although these historical sea-ice descriptions used the same terminology throughout the period of record, their conversion into a sea-ice index was complicated by two factors. The first is that the Hudson's Bay Company did not provide a dictionary of the terms since they were not originated by the Company but were passed down through two generations of ships' captains. Therefore it is not possible to translate the terms directly from an historical lexicon to their current counterparts. Secondly, even though the individual words may have been consistent through time, they were not used in isolation, but rather they were used in phrases and sentences. As a result, their meanings changed in relation to the context in which they were used. An initial survey of the log-books resulted in a compilation of 90 representative phrases, an example of which is given in Table 1.

Consequently, the system of classifying these phrases had to allow for any number of combinations of terms. In cases where there is a finite set of terms or phrases, it is possible to establish specific rules and guidelines for their classification. In this case however, this approach would not have been sufficiently flexible. As a result, it was obvious that a certain degree of subjectivity would be an unavoidable component of the analysis. The goal therefore, was to devise an objective system of categories.

Methodology

Figure 3 illustrates the CA procedure in the role of a translator from the written descriptions into numerical ice data. Even though the specifics of CA are usually designed to suit the needs of each individual research project, there is an established general format (Figure 4).

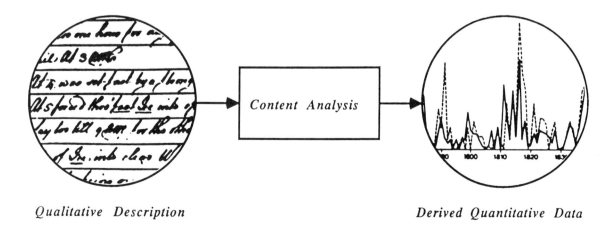

Qualitative Description   *Derived Quantitative Data*

Figure 3: Conversion of qualitative data to quantitative data through content analysis.

CONTENT ANALYSIS PLAN

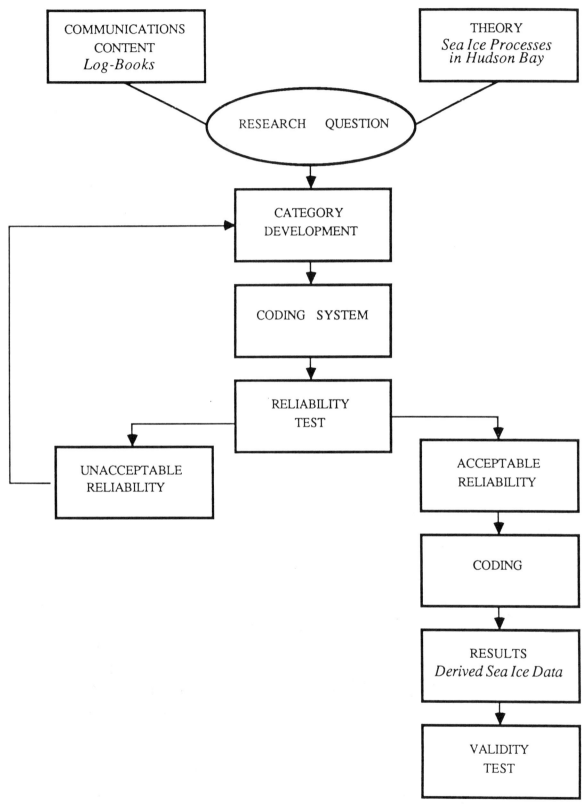

Figure 4: Content analysis plan.

Table 1: Representative Eighteenth and Nineteenth Century Sea-Ice Phrases.

1. Ice open and heavy
2. Ice close but much smaller
3. Sailing among heavy straggling ice
4. Pieces of ice
5. Passing thro' a deal of sailing ice
6. Passing thro' close ice
7. Heavy close packed ice
8. Saw some ice
9. Fast beset among close small ice can't move

The formulation of the research question is the first of a series of decisions that are made throughout the procedure. The crucial nature of this decision is due to the fact that CA is a linear process in which each step is the logical outcome of the previous step. Should an error be made anywhere along the line, then it will be carried throughout the entire analysis and may even be intensified in the process. The research question is based on two sources of information: the communications content and a body of theory. In the case of this study, the ships' log-books of the Hudson's Bay Company provided the communications content, and the Hudson Bay sea-ice processes provided the background theory.

The set of procedures that follow from the identification of the question (Figure 4) comprises the core of this research. The steps followed here are repeated until an acceptable degree of reliability has been attained. This means that a group of independent researchers who apply the same methodology using the same data will consistently produce the same results. Basically, this is a process of redefining categories with the goal of reducing the level of subjectivity. Once this has been accomplished using a sample of the data, the categories can be applied to the entire body of descriptions, and the reconstruction can be made and tested for validity.

This research followed the plan presented in Figure 5, and was divided into three phases. The goal was to objectively develop a set of categories into which the log-book descriptions could be grouped.

Phase I

The first phase involved 56 randomly selected log-book pages that included samples from the 120-year period, and five coders who all had considerable experience with CA and the log-books. This phase was essentially experimental. A set of five categories and codes (Table 2) was intuitively derived, and was based on the maximum amount of sea-ice information that was

RESEARCH PLAN

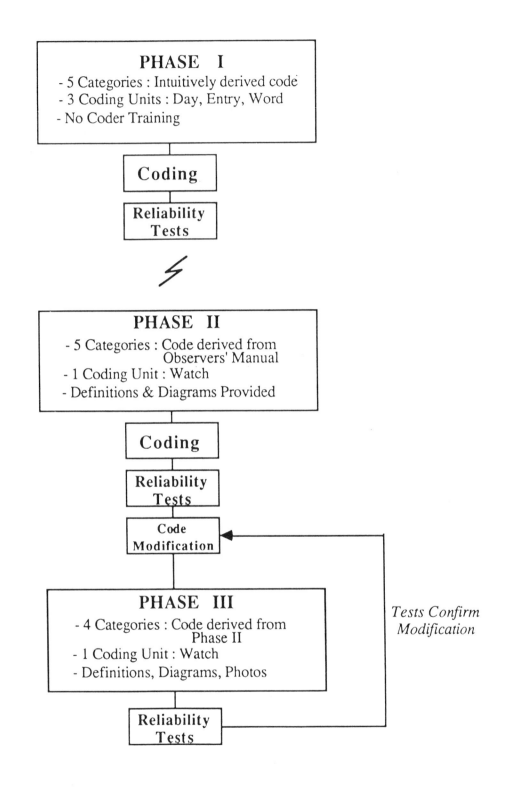

Figure 5: Research plan.

desired for the reconstruction. Each of the five major headings required the coders to make a dichotomous decision about the meaning of the log-book transcription. Besides the 10 codes, the coders were given the option of indicating those transcriptions that did not provide enough information to make a decision regarding the particular category. This was an important aspect of the code insuring that each decision made by the coder was done with some degree of certainty and was not a *forced* response. It also allowed decisions to be made later about the type of information that could be obtained from the log-book descriptions.

The first phase of coding was actually divided into three sections based on the unit of transcribed information that was coded (coding unit). In the first section, the coders were required to provide a five-digit code (one from each of the five categories - see Table 2) for each of the 56 days. In this way, a day was treated as one block of information or coding unit. This process was repeated by each coder five times so that their level of consistency could be determined. The second section involved the coding of each individual hourly entry for the same 56-day sample (a total of 261 entries), and again this process was repeated five times. The third coding unit was comprised of a list of 81 individual words, 24 of which were direct descriptions of sea ice, and 57 described the navigational activities employed to deal with the ice (e.g., grappling, tacking, rounding). These words were only coded twice.

To evaluate the reliability of the code, the coding units, and each coder, percentage agreements were calculated. This is the most common and rudimentary method of calculating reliability. It is a considerably less-than-ideal approach because it is biased in terms of the number of categories and coders, in such a way that the fewer of each the higher the percentage agreement is likely to be. The highest average intercoder agreement (among the coders) was 53.8% which was achieved when the *day* coding unit was used. The highest intracoder agreement (consistency level for each coder) was *80%* for the *entry* coding unit, although all three showed high levels of consistency (*day*=68% *word*=70%). It is important to note however, that a large proportion of the agreements was due to agreements that there was not enough information.

Phase II

The second phase was derived only slightly from the findings of the previous phase. Again there were no compulsory categories so that a coder could judge that the information was insufficient to make a coding decision. As a result of this option in the first phase, it was possible to conclude that sea-ice concentration was the only category in which a decision was possible as much as 70% of the time.

Phase II introduced a new coding unit: the seamans' watch, which was more in context with the log-book format since the entries were actually summaries of the six four-hour watches listed in Table 3. Therefore the coders were required to provide a code for each watch per day (whether there was an entry or not). Another difference between the two phases was that the second code was not intuitively derived nor did it evolve from the first coding system. Instead, it was based on the terms and definitions found in *The Ice Observer's Training Manual* (Environment Canada 1984). Since the final goal was to create a sea-ice reconstruction that could be compared with current records, the logical approach was to use modern definitions in developing the categories and codes, and in the coding process itself. The second set of categories and codes is given in Table 4, and with this the coders were also given definitions and diagrams (from the *Observer's*

Table 2: Phase I Code.[1]

Presence

0 = Ice not present in vicinity of ship
1 = Ice present in vicinity of ship

Concentration

2 = Small area covered by ice (<50%)
3 = Large area covered by ice (>50%)

Fragmentation

4 = Ice cover highly fragmented
5 = Ice cover **not** highly fragmented

Thickness

6 = Thin layer of ice
7 = Thick layer of ice

Motion

8 = Ice in motion
9 = Ice **not** in motion

[1] (No compulsory codes). Coding units: Day (5x), Entry (5x), Word (2x).

Table 3: Phase II.[1]

Watch	Time
1. Afternoon	Noon - 4:00 p.m.
2. Dog	4:00 p.m. - 8:00 p.m.
3. First	8:00 p.m. - Midnight
4. Middle	Midnight - 4:00 a.m.
5. Morning	4:00 a.m. - 8:00 a.m.
6. Forenoon	8:00 a.m. - Noon

[1] Coding unit: seaman's watch (3x).

Table 4: Phase II Code.[1]

A.	Concentration

1. Ice Free
2. Open Water
3. Very Open Ice
4. Open Ice
5. Close Ice
6. Very Close Ice
7. Consolidated/Compact Ice

B.	Floe Size

1. Giant Floe
2. Vast Floe
3. Big Floe
4. Medium Floe
5. Small Floe
6. Ice Cake
7. Small Ice Cake

C.	Openings

1. Crack
2. Open Lead
3. Blind Lead
4. Shore Lead
5. Flaw Lead

D.	Arrangement

1. Ice Field
2. Belt
3. Tongue
4. Strip
5. Ice Edge (compacted)
6. Ice Edge (diffuse)
7. Concentration Boundary

E.	Motion

1. Diverging
2. Converging
3. Shearing

[1] No compulsory codes. Coding units: seaman's watch (3x).

Manual) for the terms. Each coder applied this system to all of the watches on three separate occasions so that their consistency could be determined.

The analysis of these sessions followed a more complex process that eliminated biases inherent in the use of percent agreements. This was appropriate here because the development and application of the categories was more structured than in Phase I. In this case, Krippendorff's agreement coefficient was calculated by using the following equations.

$$\alpha = 1 - \frac{D_o}{D_e}$$

Where: α = agreement coefficient

D_o = observed disagreements

D_e = expected disagreements

and

$$D_o = \sum_b \sum_c \frac{x_{bc}}{x_{..}}$$

Where: x_{bc} = number of disagreements in a matrix of category codes (or coders)

$x_{..}$ = total of the marginal entries

and

$$D_e = \sum_b \sum_c \frac{x_{b.} x_{.c}}{x_{..}(x_{..}-m+1)}$$

Where: $x_{b.}x_{.c}$ = the products of all possible marginal entries of the matrix

$x_{..}$ = the marginal total

m = number of category codes (or coders)

The resulting coefficient is a number between 0 and 1 which, when multiplied by 100, gives the percentage by which the agreements are better than chance. Therefore, when the coefficient is 0, then any agreement is completely by chance. When the coefficient is 1, the agreements are based entirely on the coders' judgements with no degree of chance.

When this was applied to the intercoder agreements (for category A - Concentration), the coefficient was 0.468 or 47% better than chance. The average intracoder agreement was 0.591. It was then decided that these figures could be improved by modifying the categories since the problem was not due to unskilled coders. One of the many advantages of this agreement coefficient is that it can also be used as a diagnostic device in the restructuring of the categories (or reselection of coders, if necessary).

When the cause of low coefficient values is due to a problem with the categories, it is usually because the distinctions between the codes are not sufficient. This can be remedied by combining those codes that are most frequently confused with one another. Figures 6a-d provide an example of this testing procedure. Figure 6a is the basic matrix for the seven codes in the concentration category. The numbers in the cells indicate the frequencies with which each was used, so that all of the diagonal entries are the numbers of agreements among all five coders for each code, and the off-diagonals are the disagreements. The coefficient for this matrix was calculated to be 0.468. Figure 6b shows the matrix if it was collapsed into two codes: no ice (1) and ice (2-7). Intuitively, this would be expected to substantially increase the agreement coefficient. However, because there is no longer a bias in favour of fewer categories the value was increased by only 0.086. Figure 6c shows another regrouping into three codes: no ice (1), general ice descriptions (2-6), and complete ice coverage (7). This raised the coefficient by only 0.016. Finally, the codes were regrouped (Figure 6d) into four codes: no ice (1), open ice (2-3), close ice (4-5), and consolidated ice (6-7) and this increased the coefficient by 0.203 so the value became 0.689 (almost 70% better than chance). It should be stressed here that the coefficients in Figures 6b, c, and d were all calculated from the original matrix and not by recoding. This process was also applied to categories B (floe size) and D (arrangement) with increases in the coefficient of 0.333 and 0.479 respectively.

Phase III

The coding system for this phase resulted directly from the regroupings discussed above and is presented in Table 5. Category C (openings) was omitted due to infrequent usage by the coders, and the other four categories were regrouped as illustrated by comparing Tables 4 and 5. This coding session was only repeated twice because the coders' consistencies had been sufficiently tested by this point. The same definitions and diagrams were used here as in Phase II, the major difference being that category A (concentration) was compulsory. That is, a code designation was required for this category for every watch of every day. The agreements were analyzed as in Phase II. The coefficients for category A are given in Table 6. Because the other three categories were so rarely used, agreement coefficients were calculated only for category A.

Two observations are clear from Table 6. First, regrouping raised the agreement level by 21%. Secondly, although the averages of the coefficients for Phase III were lower than for the Phase II regrouped figures, they differed by only 2%. Therefore it is possible to use the calculated regroupings as a prediction for the Phase III coding agreements, and the third phase of coding could actually be eliminated.

It was concluded that the Phase III coding system produced an acceptable level of reliability since an average of only 37% of the agreements were made by chance and the remaining 63% were reliable agreements. As a result, the seaman's watch and the concentration category were adopted as the basis for the sea-ice reconstruction.

a COINCIDENCE MATRIX
PHASE 2 : CATEGORY A - CONCENTRATION

	1	2	3	4	5	6	7	
1	168	105	2	0	0	0	0	275
2	105	138	86	10	0	0	0	339
3	2	86	434	47	1	0	0	570
4	0	10	47	218	107	16	6	404
5	0	0	1	107	208	43	11	370
6	0	0	0	16	43	12	11	82
7	0	0	0	6	11	11	0	28
								2068

b COINCIDENCE MATRIX
PHASE 2 : CATEGORY A - CONCENTRATION
Agreement Coefficient = .554
Categories : No Ice & Ice

	1	2	3	4	5	6	7	
1	168	105	2	0	0	0	0	275
2	105	138	86	10	0	0	0	339
3	2	86	434	47	1	0	0	570
4	0	10	47	218	107	16	6	404
5	0	0	1	107	208	43	11	370
6	0	0	0	16	43	12	11	82
7	0	0	0	6	11	11	0	28
								2068

c COINCIDENCE MATRIX
PHASE 2 : CATEGORY A - CONCENTRATION
Agreement Coefficient = .484
Categories : No Ice, General Ice, Consolidated Ice

	1	2	3	4	5	6	7	
1	168	105	2	0	0	0	0	275
2	105	138	86	10	0	0	0	339
3	2	86	434	47	1	0	0	570
4	0	10	47	218	107	16	6	404
5	0	0	1	107	208	43	11	370
6	0	0	0	16	43	12	11	82
7	0	0	0	6	11	11	0	28
								2068

d COINCIDENCE MATRIX
PHASE 2 : CATEGORY A - CONCENTRATION
Agreement Coefficient = .689
Categories : No Ice, Open Ice, Close Ice, Consolidated

	1	2	3	4	5	6	7	
1	168	105	2	0	0	0	0	275
2	105	138	86	10	0	0	0	339
3	2	86	434	47	1	0	0	570
4	0	10	47	218	107	16	6	404
5	0	0	1	107	208	43	11	370
6	0	0	0	16	43	12	11	82
7	0	0	0	6	11	11	0	28
								2068

Figure 6: Coincidence matrices and agreement coefficients. (a) Basic matrix; (b) Two-category matrix; (c) Three-category matrix; (d) Four-category matrix.

Table 5: Phase III Code.[1]

A. Concentration
1. Ice Free
2. Open Water/Very Open Ice
3. Open Ice/Close Ice
4. Very Close/Consolidated/Compact Ice

B. Floe Size
1. Small Ice Cake
2. Ice Cake
3. Medium/Small Floe
4. Big Floe

C. Arrangement
1. Strip/Diffuse Ice Edge/Concentration Boundary
2. Belt
3. Tongue
4. Ice Field/Compacted Ice Edge

D. Motion
1. Diverging
2. Compacting

[1] Category A compulsory. Coding units: seaman's watch (2x).

Table 6: Intercoder Agreement Coefficients. Category A - Concentration.

Coding Session	Phase II	Regrouped Phase II	Phase III
1	.422	.653	.666
2	.463	.653	.603
3	.468	.689	[1]
Average Differences		+.214	−.019

[1] Phase III was only repeated twice.

Concluding Remarks

Although this case study was not directed specifically to the climatic anomaly of 1816, its relevance pertains to the entire time frame in which this event occurred. There is a potentially large volume of climatic information relating to key volcanic episodes that is in a descriptive format. This study provides a solution to the problem of interpreting this type of information reliably, an approach that in the past has been superficially addressed. Reconstructions based on categories that are developed from thoroughly-tested evolutionary process provide information that describes the reliability with which the original documents were interpreted. Furthermore, the results of these reliability tests must accompany the reconstruction so that the question of an acceptable level of reliability is relegated to the user of the reconstruction to a certain degree. This does not mean that any agreement coefficient should be accepted by the researcher. In the search for an acceptable level of reliability, an attempt should be made to balance the amount of information obtained against the degree of objectivity by which it was derived. In this study, a considerable amount of information was discarded throughout the testing procedure from the first phase to the final code so that sea-ice concentration was the only category to be used in the final reconstruction.

River Ice and Sea Ice in the Hudson Bay Region during the Second Decade of the Nineteenth Century

A.J.W. Catchpole[1]

Abstract

Analysis of documentary sources in the Hudson's Bay Company Archives has provided records of river- and sea-ice conditions in the Hudson Bay region during the eighteenth and nineteenth centuries. These include six records of dates of first-breaking and first-freezing of routes to the bayside trading posts. Several different validity tests have been applied to these data, and the results of these tests generally indicate the data are valid measures. The values of river- and sea-ice data in each year from 1810 through 1820 are compared with their values during the whole period of record. This enables the identification of years with anomalously early or late dates of breaking and freezing, and years with severe summer sea ice. Evidently, exceptionally cold summer weather occurred in the second decade of the nineteenth century. This was not initiated after the eruption of Tambora in April 1815, but was first apparent in 1811 and 1812. However, the most severe summer cold in the decade occurred in the two years following the eruption.

Introduction

In subarctic regions the dispersal of ice in spring and freezing of water bodies in fall are intimately linked to weather and climatic conditions. Anomalous weather in a particular year may cause exceptionally early or late breaking and freezing of rivers, lakes and seas. Ice observations therefore figure prominently among the routine climatologic and oceanographic observations made by the nations fringing the poles. Another property of ice is that it occurs in forms vividly apparent to casual observers, and under circumstances where it can severely restrict their physical activities. For these reasons, informal descriptions of ice also occur prominently in written historical sources that contribute to the reconstruction of climates in the recent past. So it was that the daily journals kept by servants of the Hudson's Bay Company graphically described the long anticipated breaking of the rivers in spring and their equally vital refreezing in fall. Likewise, the log-books of the Company's supply ships that annually sailed the ice-congested waters of Hudson Strait and Hudson Bay gave frequent, detailed descriptions of the ice which imperilled their passage.

The Ice Records

These sources have yielded six records of dates of first-breaking and first-freezing of the estuaries of rivers draining into Hudson Bay and three records of summer sea-ice severity encountered along portions of the sailing routes to the bayside trading posts. The various records commenced in the early or mid-eighteenth century, and in most cases ended in the latter part of the nineteenth century. Table 1 lists for each record its location, the year when the record commenced and ended, the total number of years in which ice data have been reconstructed and the source in which the reconstruction was originally published. Three stages in the seasonal development of river ice are dated:

---

[1] Department of Geography, The University of Manitoba, Winnipeg, Manitoba R3T 2N2, Canada.

1. *date of first-breaking* - the first day on which any evidence of breaking was observed, irrespective of whether or not the river remained broken thereafter;

2. *date of first partial freezing* - the first day on which the river was observed to become partially frozen, irrespective of the spatial extent of the ice cover or its continuity thereafter;

3. *date of first complete freezing* - the first day on which the entire surface of the river was frozen, irrespective of whether or not it remained completely frozen thereafter.

Each of these historical records was derived from daily journals written at trading posts located in the estuaries of rivers draining into Hudson Bay (Figure 1). In several of these estuaries the locations of the posts changed from time to time, but the records derived in the Severn, Albany, Moose and Eastmain estuaries are each based on single post locations. The Churchill journals were written both at the Old Fort, located inside the estuary, and at Fort Prince of Wales situated on an exposed promontory where the north shore of the estuary protrudes into Hudson Bay. The Churchill first-freezing data used in this paper were those reconstructed from the Old Fort journal, and the first-breaking data were those derived at Fort Prince of Wales. In the Hayes River estuary the location of York Factory was changed in 1791 to a site very close to the former on the north shore of the estuary.

Table 1: Historical Records of River Ice and Sea Ice Derived from Hudson's Bay Company Archives.

		Dates of First-Breaking of River Estuaries		
River	Location of Record	Limits of Record	Number of Years of Record	Sources
Churchill	Churchill Old Fort	1720-1866	110	Moodie and Catchpole (1975)
	Fort Prince of Wales	1731-1861	107	
Hayes	York Factory 1	1715-1790	74	Moodie and Catchpole (1975)
	York Factory 2	1791-1851	45	Moodie and Catchpole (1975)
Severn	Fort Severn	1763-1939	104	Magne (1981)
Albany	Fort Albany	1722-1939	190	1722-1866 (Moodie & Catchpole 1975); 1872-1939 (Magne 1981)
Moose	Moose Factory	1736-1871	133	Moodie and Catchpole (1975)
Eastmain	Eastmain House	1743-1939	109	Magne (1981)

Table 1: (cont'd)

Dates of First Partial Freezing of River Estuaries

River	Location of Record	Limits of Record	Number of Years of Record	Sources
Churchill	Churchill Old Fort	1718-1866	69	Moodie and Catchpole (1975)
	Fort Prince of Wales	1731-1845	42	Moodie and Catchpole (1975)
Hayes	York Factory 1	1714-1790	73	Moodie and Catchpole (1975)
	York Factory 2	1791-1850	44	Moodie and Catchpole (1975)
Severn	Fort Severn	1761-1940	100	Magne (1981)
Albany	Fort Albany	1721-1938	180	1721-1867 (Moodie & Catchpole 1975); 1872-1938 (Magne 1981)
Moose	Moose Factory	1736-1870	132	Moodie and Catchpole (1975)
Eastmain	Eastmain House	1743-1940	98	Magne (1981)

Dates of First Complete Freezing of River Estuaries

River	Location of Record	Limits of Record	Number of Years of Record	Sources
Churchill	Churchill Old Fort	1718-1865	95	Moodie and Catchpole (1975)
	Fort Prince of Wales	1722-1852	55	Moodie and Catchpole (1975)
Hayes	York Factory 1	1714-1792	76	Moodie and Catchpole (1975)
	York Factory 2	1793-1851	39	Moodie and Catchpole (1975)
Severn	Fort Severn	1760-1940	91	Magne (1981)
Albany	Fort Albany	1721-1921	178	1721-1864 (Moodie & Catchpole 1975); 1872-1921 (Magne 1981)
Moose	Moose Factory	1739-1861	117	Moodie and Catchpole (1975)
Eastmain	Eastmain House	1743-1940	97	Magne (1981)

Table 1: (cont'd)

Region	Summer Sea-Ice Severity Indices		
	Limits of Record	Number of Years of Record	Sources
Hudson Strait	1751-1889	137	1751-1870 (Catchpole and Faurer 1983); 1871-1889 (Catchpole and Hanuta 1989)
Eastern Hudson Bay	1751-1870	108	Catchpole and Halpin (1987)
Western Hudson Bay	1751-1869	111	Catchpole and Hanuta (1989)

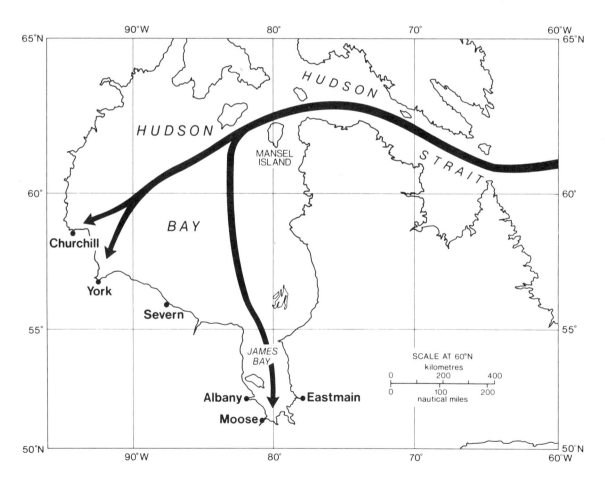

Figure 1: Location map showing sailing routes through Hudson Strait, across eastern Hudson Bay to Moose, and across western Hudson Bay to York and Churchill.

The three sea-ice severity records refer not to point locations but to the three portions of the sailing-ship route (Figure 1). These records were reconstructed from descriptions of ice given in the supply ships' log-books. These ice-severity indices are numerical in form but they function as ordinal not interval data. As such, the indices rank the years on the basis of summer-ice severity, but they are not numerical measures of the quantities of ice present in each summer. The frequency distributions of the ice indices are highly skewed, with very high proportions of small values and a few very large values. This property implies that the indices discriminate more accurately between the ranking of the few severe ice years than between that of the large number of moderate and light ice years.

Quality of Ice Records

The objective of this paper is to use the records listed in Table 1 to determine whether the river- and sea-ice conditions in the second decade of the nineteenth century were in any respects anomalous when compared with the ice conditions observed throughout the periods of record. In view of this objective it is pertinent to comment briefly on the quality of these historical data. Two aspects of the quality of climatic data derived from historical sources should be considered. These are the *reliability of the method* of derivation and the *validity of the data* derived. The reliability of the method determines the degree to which similar results will be obtained when the same method is applied to the same sources by the same person, or by different people with similar training. The validity of the data determines the degree to which the results are true measures of what they are intended to measure. There has been no fully comprehensive testing of the quality of these historical river- and sea-ice data. However, several studies have yielded information that bears upon their reliability and validity.

The derivation of the breaking and freezing dates of the river estuaries (Moodie and Catchpole 1975) included reliability testing as one of its major aspects. The test results showed that high degrees of reliability were obtained when dates based on *direct dating categories* were derived for *places where the journals were kept*. Much lower levels of reliability were obtained for dates based on less direct information. Marcia Faurer (this volume) is developing and applying an innovative approach to testing the reliability with which sea-ice data can be derived from sailing ships' log-books.

The validity of river-ice dates has been tested *internally* by examining the spatial homogeneity between similar dates derived at adjacent estuaries. These tests found high correlations between the dates of first-breaking at Fort Albany and Moose Factory and supported the conclusion that these are, therefore, true measures of the actual breaking dates in these river estuaries (Moodie and Catchpole 1976). Some studies have compared selected river-ice dates with tree-ring data derived from trees growing in the vicinity of the river estuaries. These studies were not designed as tests of the validity of the ice data but they do detect similarities between the trends revealed by tree-ring and ice data. In so doing they provide rudimentary indications of the validity of the ice data tested against *external* criteria. This approach is exemplified by a study of ice conditions in the Churchill River estuary conducted by Jacoby and Ulan (1982). This study used tree-ring data from near Churchill. It found a multiple correlation coefficient of 0.69 between tree growth and the date of complete freezing at Fort Prince of Wales during 1741-64. Jacoby and Ulan (1982) used this relationship to derive dates of complete freezing from tree-ring data in the period 1680-1977. In his reconstruction of temperatures in the Hudson Bay region during the past three centuries, Guiot (1986; this volume) assembled a database including early instrumental temperature observations, tree-ring data and river-ice dates.

Significant correlations were found between several of the records of the first-breaking and freezing of river estuaries and other records in this database (Guiot 1986, pp. 13, 19). Dates of first partial freezing and first complete freezing were generally found to be positively correlated with autumn temperatures measured at York and Churchill, whereas dates of first-breaking were generally negatively correlated with spring temperatures. Some of the tree-growth records were negatively correlated with the date of first complete freezing, and this finding is consistent with the results obtained by Jacoby and Ulan (1982). Lough and Fritts (1987; Lough this volume) used North American tree-ring data to assess the possible effects of volcanic eruptions on North American climate during 1602-1900. In this study they employed the mean dates of first-breaking and first complete freezing of the James Bay estuaries as "independent temperature records outside the area covered by the arid site tree-ring reconstructions." Using superposed epoch analysis, Lough and Fritts detected changes in ice dates following major volcanic eruptions that were consistent with the observed changes in tree growth.

Wilson (1988; this volume) has derived summer thermal indices for the southeast coast of Hudson Bay in the nineteenth century, using a miscellany of historical evidence in the Hudson's Bay Company Archives. A preliminary study of these indices shows that they may afford an indirect means of testing the validity of sea-ice data, in so far as anomalous summer cold in this region may be a result, or a cause, of severe summer ice on adjacent seas. This study involved a comparison between the incidence of severe ice years and negative anomalies in the thermal indices for May to June (Figure 2A) and May to October (Figure 2B). The May to June data were selected for this comparison because the ships' log-books were not among the historical sources used to derive these indices. The May to October data were selected because Wilson (1988, p. 13) considered that the index is most accurate over the entire summer season. However, the sea-ice indices and May to October thermal indices are not entirely independent because the ships' log-books did play a minor role as sources in the derivation of the mid-summer thermal conditions.

Figure 2 comprises graphs of Wilson's thermal indices upon which are superimposed vertical bars identifying *severe ice years*. In this context a severe ice year is defined as one of the years having the 10 highest ice indices in each of the three ice records derived for Hudson Strait, eastern Hudson Bay and western Hudson Bay. The 10 highest indices are based on the entire period of the sea-ice records, not the period 1800-70. A vertical bar on Figure 2 indicates that severe ice occurred in that year in Hudson Strait or in eastern or western Hudson Bay. It is judged to be appropriate to consider these three records together in this way and not separately. Severe ice in Hudson Strait could retard the entry of ships into the bay to such a degree that they would not encounter the bay ice in July and August, but rather in September. At this time even severe late summer ice is generally cleared from the bay. Furthermore, the eastern and western parts of the bay are not separate entities in the context of ice clearing, but rather the lateral limits of the waters in which the last remnants of ice tend to congregate under the influence of prevailing winds and currents (Danielson 1971). In years with zonal atmospheric circulation these remnants tend to be driven towards the east and accumulate in the sailing route to James Bay. A meridional atmospheric circulation permits late ice to remain in the west in the path of ships sailing to Churchill or York Factory.

Figure 2 reveals a tendency for severe ice years to concur with periods having negative thermal indices. This is most apparent in the middle of the second decade of the century, in the late 1830s and in the early to mid-1840s. It is noteworthy that these are generally periods in which Wilson

(1988, pp. 7, 8) noted the quality of the thermal indices as good to excellent. It is not appropriate to numerically evaluate the correlation between these data because the ice indices are ordinal not interval data.

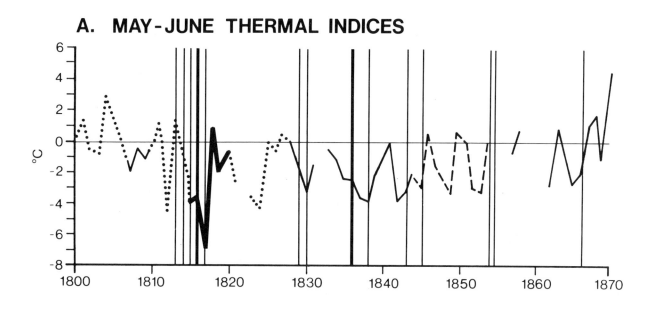

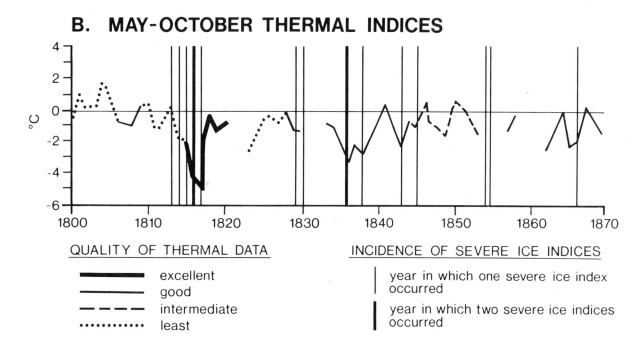

Figure 2: Thermal indices for the southeastern coast of Hudson Bay (from Wilson 1988), and years with severe summer ice in Hudson Strait and Hudson Bay, 1800-70. The thermal indices are estimates of departures from the 1941-70 normals of temperatures in the May to June (A) and May to October (B) periods. The quality of these indices in different time intervals was assessed by Wilson (1988, p. 7-8). A severe ice year is defined as a year having one of the 10 highest ice indices, in the period 1751 to 1870, within each of the three ice records.

Ice Conditions, 1810-20

River- and sea-ice conditions in each year from 1810 through 1820 are evaluated in Tables 2A-C and 3. The data given in these tables compare the ice condition in each year with the range of values of that condition reconstructed over the whole period of record. In the case of the river-ice dates (Tables 2A-C) the comparison is made by the calculation of the parameter Z.[1] This enables the identification of years with anomalously early or late dates of breaking and freezing. Table 4 lists these years and distinguishes between anomalies having less than 1, 2.5 and 5% probabilities of occurring by chance. Table 3 gives the rank order of occurrence of each sea-ice severity index among the indices reconstructed for the whole period of record. Table 4 identifies the years in which the sea-ice index was ranked among the upper 10 values in each record.

Fourteen of the river-ice records are designated anomalous in Table 4, and all of these are indicative of summer cold with significantly late-breaking and early-freezing. During this decade there was no occasion of early-breaking or late-freezing that produced a Z value so large that there was only a 5% probability of its occurring by chance. The greatest anomalies, in frequency and amount, were those of retarded first-breaking in 1817 and 1812. In 1817 the date of first-breaking was anomalously late in all of the river estuaries from which a date could be obtained in that year. The record was interrupted in 1817 at York and Severn (Table 2A). In 1812 this date was anomalously late in four estuaries but not at Moose or Eastmain. Furthermore, at York and Severn, the 1812 anomalies exhibited 5% probabilities of occurrence by chance, whereas all of the 1817 anomalies exceeded this level of significance. First partial freezing was significantly early at Eastmain in 1817 and at Churchill in 1811. First complete freezing was early at York and Eastmain in 1811 and at York in 1817.

This decade was marked by severe late-summer ice in eastern Hudson Bay in 1813 and by a cluster of high sea-ice indices in 1815 to 1817. This cluster included the highest ice index derived in Hudson Strait (1816), as well as severe ice in western Hudson Bay in 1815 and in eastern Hudson Bay in 1816 and 1817. 1816 provides a case in which the passage of the ships through Hudson Strait was so greatly delayed that they apparently entered the bay so late as to reduce their ability to monitor a mass of ice that persisted late in the summer within the bay. This ice was located in the east across the sailing route to James Bay. The ship in question (the *Emerald*) rounded Mansell Island and entered Hudson Bay on 7 September. This was 25 days later (standard deviation 9.7) than the mean date on which ships bound for Moose Factory entered the bay in the period 1751-1870. During this delayed passage to Moose in 1816, the *Emerald* encountered ice which yielded the seventh largest index (Table 3). Probably the 1816 ice in eastern Hudson Bay would have ranked even higher if the *Emerald* had sailed these waters closer to the normal sailing date. In 1816, the *Prince of Wales* sailed to York Factory. This ship also entered the bay on 7 September. However, it encountered no ice on its passage to the west coast and there is, therefore, no evidence that this ship would have encountered exceptionally late ice if it had sailed into the bay earlier than this late date.

---

[1] $Z = \dfrac{x - \mu}{\sigma}$

where:
x = date of breaking (first partial freezing, first complete freezing) in a particular year;
μ = mean date for whole period of record;
σ = standard deviation from this mean for whole period of record.

Table 2A: Dates of First-Breaking of River Estuaries, Standard Units Z.[1]

	Estuary (n=number of years of record)					
	Churchill[2] n=107	York[3] n=45	Severn n=104	Albany n=190	Moose n=133	Eastmain n=109
1810	+0.31	+0.05	+0.71	-1.49	-1.61	0
1811	+0.72	+0.37	-	-0.28	-0.39	-0.79
1812	+2.67	+1.98	+2.17	+2.38	+1.49	+1.76
1813	+0.45	-1.12	-	-0.72	-0.75	-
1814	0	-	-	+1.27	+1.07	-0.01
1815	+1.84	+0.60	+1.44	+1.60	+2.05	+1.53
1816	0	-1.87	-	+0.72	+0.83	+1.41
1817	+2.40	-	-	+2.27	+2.30	+3.03
1818	-0.95	-	-	+0.72	-0.39	-0.21
1819	+0.58	-0.91	-1.48	+0.06	+0.10	-0.44
1820	-0.81	-1.12	-1.69	-1.49	-1.49	-1.13

[1] $Z = \frac{x-\mu}{\sigma}$.
[2] Estuary of Churchill River at Fort Prince of Wales.
[3] Estuary of Hayes River at York Factory 2.

Table 2B: Dates of First Partial Freezing of River Estuaries, Standard Units Z.[1]

	Estuary (n=number of years of record)				
	Churchill[2] n=69	York[3] n=44	Albany n=180	Moose n=132	Eastmain n=98
1810	-	-	-	-0.77	-0.12
1811	-2.20	-1.29	-1.32	-1.12	-1.67
1812	-0.18	-1.15	-	+0.16	0
1813	-0.94	-	-1.07	-1.35	-0.83
1814	+0.96	-1.73	+0.39	+0.63	+1.31
1815	-	+1.29	+0.02	-0.30	-0.24
1816	-	-	+1.37	-	+1.19
1817	-0.43	-	-0.58	-0.88	-2.26
1818	-0.43	+0.71	+1.24	+0.86	+1.67
1819	-0.30	+1.15	-0.46	-0.77	-0.48
1820	-	-0.14	+0.15	-0.42	-0.48

[1] $Z = \frac{x-\mu}{\sigma}$.
[2] Estuary of Churchill River at the Old Fort.
[3] Estuary of Hayes River at York Factory 2.

Table 2C: Dates of First Complete Freezing of River Estuaries, Standard Units Z.[1]

	Estuary (n=number of years of record)				
	Churchill[2] n=95	York[3] n=39	Albany n=178	Moose n=117	Eastmain n=97
1810	-	-	-	-	-
1811	-	-2.20	-1.94	-1.82	-2.15
1812	-0.94	-2.13	-0.83	-1.15	-
1813	-0.68	-	-	-	-
1814	+0.60	+0.46	+1.02	+0.84	-
1815	-	-0.47	-0.83	-1.49	-0.52
1816	-	-	-	-0.24	+0.56
1817	-	-	-1.20	0	-1.70
1818	+0.74	+1.68	+1.30	+1.25	-
1819	-	-	-1.11	-	-1.34
1820	-	-	-0.83	-0.16	-0.34

[1] $Z = \frac{x-\mu}{\sigma}$.
[2] Estuary of Churchill River at the Old Fort.
[3] Estuary of Hayes River at York Factory 2.

Table 3: Summer Sea-Ice Severity Indices, Annual Ranking.

	Location (n=number of years of record)		
	Hudson Strait n=137	Eastern Hudson Bay n=108	Western Hudson Bay n=111
1810	88	39	32
1811	36	66	-
1812	36	27	32
1813	57	2	101
1814	12	78	101
1815	44	44	6
1816	1	7	101
1817	27	8	-
1818	40	95	41
1819	48	95	101
1820	135	81	15

Table 4: Incidence of Anomalous River-Ice Dates and Severe Sea Ice During 1810-20. The Probabilities of River-Ice Anomalies are Based on the Standard Units Z (Tables 2A-C). The sea-ice anomalies are the years having one of the 10 highest ice-severity indices in each of the three records. The fractions given compare the rank with the number of years in the record.

		1810	1811	1812	1813	1814	1815	1816	1817	1818	1819	1820
RIVER ESTUARY ICE DATES	First Breaking			LATE AT: CHURCHILL ALBANY York Severn			LATE AT: Moose		LATE AT: EASTMAIN CHURCHILL ALBANY MOOSE			
	First Partial Freezing		EARLY AT: Churchill						EARLY AT: EASTMAIN			
	First Complete Freezing		EARLY AT: York Eastmain	EARLY AT: York								
SUMMER SEA-ICE SEVERITY	Hudson Strait							SEVERE: 1/137				
	Hudson Bay (East)				SEVERE: 2/108			SEVERE: 7/108	SEVERE: 8/108			
	Hudson Bay (West)						SEVERE: 6/111					

PROBABILITY THAT THIS ANOMALY OCCURRED BY CHANCE

CHURCHILL: Less than 1%   **ALBANY**: Less than 2.5%   **York**: Less than 5%

Conclusions

The river- and sea-ice data presented here indicate that in the second decade of the nineteenth century cold summer weather was not initiated after the eruption of Tambora in April 1815, but was first apparent in 1811 and 1812. However, this evidence does show that the most severe summer cold in that decade occurred in the two years following the eruption.

The first of these cold episodes commenced in 1811 with early first partial freezing at Churchill and early first complete freezing at York and Eastmain. This was followed in the spring of 1812 with late first-breaking at Churchill and Albany and with late breaking, though less delayed, at York and Severn. In the fall of 1812 first complete freezing occurred early at York.

An isolated case of severe sea-ice occurred in eastern Hudson Bay in 1813, and this was followed by a cluster of years with severe ice in 1815 (western Hudson Bay), 1816 (Hudson Strait and eastern Hudson Bay) and 1817 (eastern Hudson Bay). This period culminated in late first-breaking in 1817 at Eastmain, Churchill, Albany and Moose. In the fall of 1817 early first partial freezing occurred at Eastmain. There were gaps in the historical record during both of these cold-summer periods, and these were most prominent in 1816 and 1817. In particular, data on first partial freezing and first complete freezing are unavailable for Churchill and York in 1816, and no river-ice data are available for York in 1817.

References

Catchpole, A.J.W. and M.A. Faurer. 1983. Summer sea-ice severity in Hudson Strait, 1751-1870. *Climatic Change* 5:115-139.

Catchpole, A.J.W. and J. Halpin. 1987. Measuring summer sea-ice severity in eastern Hudson Bay 1751-1870. *Canadian Geographer* 31:233-244.

Catchpole, A.J.W. and I. Hanuta. 1989. Severe summer ice in Hudson Strait and Hudson Bay following major volcanic eruptions, 1751 to 1889 A.D. *Climatic Change* 14:61-79.

Danielson, E.W. 1971. Hudson Bay ice conditions. *Arctic* 24:90-107.

Guiot, J. 1986. Reconstruction of temperature and pressure for the Hudson Bay Region from 1700 to the present. *Canadian Climate Centre Report* No. 86-11:1-106.

Jacoby, G.C. and L.D. Ulan. 1982. Reconstruction of past ice conditions in a Hudson Bay estuary using tree rings. *Nature* 298:637-639.

Lough, J.M. and H.C. Fritts. 1987. An assessment of the possible effects of volcanic eruptions on North American climate using tree-ring data, 1602 to 1900 A.D. *Climatic Change* 10:219-239.

Magne, M.A. 1981. Two centuries of river ice dates in Hudson Bay region from historical sources. MA. thesis, University of Manitoba, Winnipeg. 78 pp.

Moodie, D.W. and A.J.W. Catchpole. 1975. Environmental data from historical documents by content analysis: freeze-up and break-up of estuaries on Hudson Bay 1714-1871. *Manitoba Geographical Studies* 5:1-119.

_____. 1976. Valid climatological data from historical sources by content analysis. *Science* 193:51-53.

Wilson, C.V. 1988. The summer season along the east coast of Hudson Bay during the nineteenth century. Part III. Summer thermal and wetness indices. B. The indices, 1800 to 1900. *Canadian Climate Centre Report* No. 88-3:1-42.

The Climate of the Labrador Sea in the Spring and Summer of 1816, and Comparisons with Modern Analogues

John P. Newell[1]

Abstract

The wide range of natural variability in climatic conditions at the local and regional scales makes it necessary to examine data from as large an area as possible in order to determine the significance of past departures from present-day conditions. Many authors have demonstrated that the spring and summer of 1816 were among the coldest ever recorded in the region extending from the northeastern United States to Hudson Bay. Recent research on tree rings indicates that climate may not have been as severe in the western United States and Canada. This study examines proxy-climatic data for northeastern North America, extending from southeastern Newfoundland to Hudson Strait and including the waters of the Labrador Sea, in an effort to develop a more continental view of climate during this critical period.

The sources investigated include: weather narratives from both Newfoundland and Labrador; a daily weather diary from eastern Newfoundland; and sea-ice records for the waters adjacent to Newfoundland and Labrador. The study demonstrates that, during the spring and summer of 1816, climatic and sea-ice conditions in northern Labrador were among the most severe ever recorded; however, farther south in Newfoundland, conditions were by no means as severe, and may have been near nineteenth century normals.

The 1816 patterns of climatic and sea-ice conditions in Newfoundland and Labrador are compared with recent (post-1950) patterns of temperature, precipitation and sea-ice conditions in eastern North America to determine if modern analogues exist. This comparison indicates that conditions in 1816 have no clear analogues in the recent climatic record. However, there are patterns that, while not as severe, do provide some indications of the nature of the circulation in 1816. These patterns indicate that the circulation during the summer of 1816 was similar to the present normals for March and April. This agrees with the July circulation pattern for 1816 presented by Lamb and Johnson (1966).

Introduction

The unusual character of the summer of 1816 in the Labrador Sea is demonstrated by the following report from the records of the Moravian Church which operated several missions along the Labrador coast: "The *Jemima* [the moravian mission ship] arrived in the river [Thames] from Labrador, after one of the most dangerous and fatiguing passages ever known. As in almost every part of Europe, so in Labrador, the elements seem to have undergone some revolution during the course of last summer" (Periodical Accounts, Vol. VI, p. 263).

Modern research has demonstrated that the summer of 1816 was unusually cold in Europe (Manley 1974; Kelly *et al.* 1984; Briffa *et al.* 1988), eastern United States (Stommel and Stommel 1979; Ludlum 1966); and Hudson Bay (Wilson 1983; Catchpole 1985). Other authors

---

[1] 34 Cornwall Crescent, St. John's, Newfoundland A1E 1Z5, Canada.

have demonstrated that sea-ice conditions in both Hudson Strait (Catchpole and Faurer 1985) and the Labrador coast (Newell 1983) were extremely severe during the summer of 1816. By comparison, sea-ice conditions in the East Greenland Sea (Scoresby 1820) and near Iceland (Lamb 1977; Ogilvie, this volume), while more severe than normal, did not reach the record conditions experienced in eastern North America.

The only previous study (Lamb and Johnson 1966; Lamb, this volume) that directly considers climatic conditions in the Labrador Sea during 1816 is an analysis of January and July global sea-level pressure patterns for the years 1750 to 1962. It includes a map of July 1816 circulation over the North Atlantic indicating that a 1002 mb low-pressure centre was situated over the Labrador Sea, giving a northerly flow along the Labrador coast. This circulation pattern is more representative of conditions in April than of the normal circulation in July. It should be noted that Lamb and Johnson provide maps showing that the Labrador Sea was outside the limits of reliable isobars until the 1870s. While the exact data used to construct their map for 1816 are not given in the report, other sources (Lamb and Johnson 1959, 1961) indicate that it was likely based on wind data from New England and possibly Greenland.

This paper presents the results of an analysis of proxy-climatic records from areas surrounding the Labrador Sea (Newfoundland, Labrador, Hudson Strait and southwestern Greenland) and an attempt to reconstruct the atmospheric circulation pattern in this region for June 1816. In addition, temperature patterns over the area in June 1816 are used to select modern analogues for the 1816 circulation pattern. These modern analogues are then compared with the reconstructed circulation pattern. The study area and locations noted in the text are shown in Figure 1.

Analysis of Historical Data

The following brief review of the history of the study area in 1816 provides an indication of types of proxy-climatic data sources available. At the start of 1816, Newfoundland was in the midst of a financial crisis caused by the fall in fish prices after the end of the War of 1812, and in February 1816 a major fire struck St. John's, the capital of the island. At this time the main economic activity in Newfoundland was the inshore cod fishery. The only other significant economic activity was the seal "fishery" carried out off the northeastern coast each spring. Farther north along the Labrador coast, the Moravian Church operated missions it had established during the late eighteenth century. These missions were supplied each spring by a mission ship that sailed directly to Labrador from England. At the same time the Moravians also operated a number of missions in southwestern Greenland that were supplied by Danish ships sailing from Denmark to the Greenland settlements. Whaling ships from Britain also operated off the west coast of Greenland each spring. The only other significant shipping activity in the study area at this time involved Hudson's Bay Company ships that sailed from England to Hudson Bay each spring and returned in the fall.

A review of material available in the Newfoundland Archives revealed that government correspondence from this period is rather limited. This is partly due to the fact that prior to 1818 the governor was only resident in Newfoundland during the summer. The only pertinent remark was: "The weather during the greater part of the season [summer 1816] has been particularly unfavourable for the curing [the cod was dried in the sun] of fish" (Report of Fishery, December 1816, Government Letter Book, Newfoundland Archives). This situation could result from either damp weather or calm weather with clear skies. Analysis of historical catch statistics for cod in Newfoundland waters (Forsey and Lear 1987) indicate that 1816 was a relatively good year.

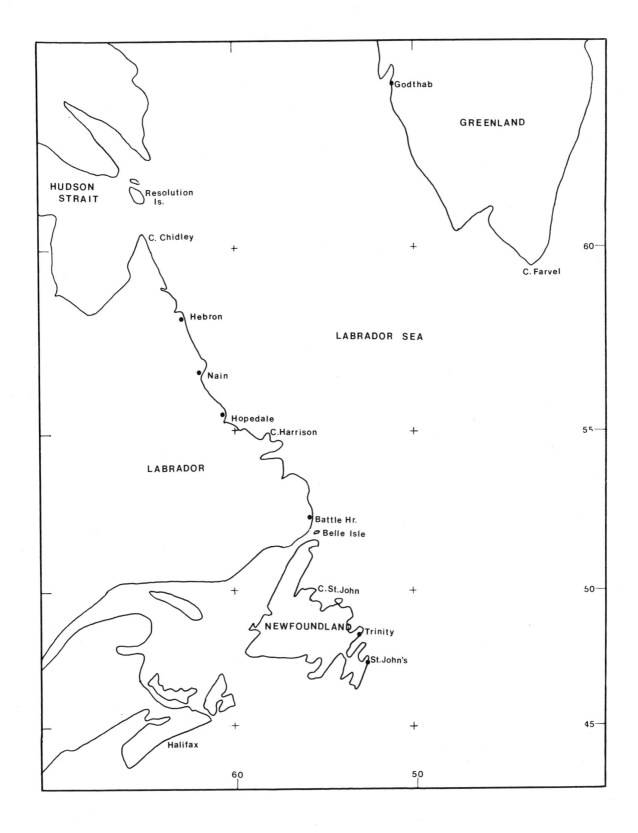

Figure 1: Study area.

While no statistics on the seal catch are available for 1816, available data do not point to a bad year. The records indicate however a very low catch in 1817, which was attributed to severe ice conditions.

A weather diary kept at Trinity, Newfoundland by the firm of Slade and Kelson provides the best information on the climate of Newfoundland in 1816. A review of the weather remarks and rain/snow frequencies given in the diary do not provide any evidence for cold conditions during the spring or summer of 1816. Analysis of the daily reports of wind for June 1816 indicate a high frequency from the southwest (55%) compared to present day normals for Bonavista, Newfoundland (less than 30%) and compared to Trinity in 1817 (48%) and 1818 (36%). A comparison of air temperature versus wind direction for St. John's, based on modern data, indicates that the two parameters are closely linked (Figure 2). Southwest winds are clearly warm winds, so it was likely that southeastern Newfoundland experienced normal to above normal temperatures in June 1816. The typical synoptic situation giving southwest winds over Newfoundland in June is a ridge of high pressure pushing northward from the Bermuda High.

In northern Labrador and Hudson Strait, Moravian records (Newell 1983) and Hudson's Bay Company records (Catchpole and Faurer 1985; Teillet 1988) indicate severe ice conditions with considerably delayed clearing dates. Newell (1983) states that in 1816 it was " likely that the sea ice had not completely cleared the coast by the start of the next [ice] season". Analysis of sea-ice clearing in this region based on satellite imagery for 1964-74 (Crane 1978) demonstrates that late clearing dates are associated with an increased frequency of northerly winds. Catchpole and Faurer (1985), investigating sea-ice conditions in Hudson Strait during 1816 using logs from Hudson's Bay Company ships, also found evidence for an increased frequency of northerly winds during the summer.

Besides providing valuable information regarding the offshore ice conditions, the Moravian mission reports also provide some indication of the weather experienced at the stations. The following remarks regarding the summer of 1816 at Okkak follow a description of the severity of the winter: "In spring, the frost continued so severe, that we could not work in our gardens at the proper time, and consequently expect but a poor crop of vegetables this year, for the whole summer season has been cold and dry" (Periodical Accounts, Vol. VI, p. 265). The following reports from the Moravian missions in southwestern Greenland suggest different conditions on the other side of the Labrador Sea: "It rains almost incessantly, and if it even ceases for a day, yet the heavens are overcast...I must say that for these four months past, we have not had one day on which the sun has shone throughout the whole day" (Periodical Accounts, Vol. VI, p. 452). An analysis of conventional meteorological data collected at the Labrador Moravian stations in the 1880s and 1890s suggests that in June cold dry conditions are associated with north or northwest winds and lower air pressures; both of which would occur with a mean low-pressure centre to the east and lows tracking well south of the area. The wet conditions in southwestern Greenland indicate that this area was near or just east of the main low-pressure centre.

In summary, the data presented indicate that during the spring and summer of 1816 a mean centre of low pressure was situated in the Labrador Sea with a trough extending north into Davis Strait (Figure 3). At the same time the main track of low-pressure systems was across southern Labrador and into the Labrador Sea. This pattern would give the north to northwest winds and cold/dry conditions in Labrador and the wet conditions in southwestern Greenland. South of the storm track, southeastern Newfoundland was under the influence of the Bermuda High. The temperature pattern for June 1816 has very cold conditions in northern Labrador and normal to above normal temperatures in Newfoundland.

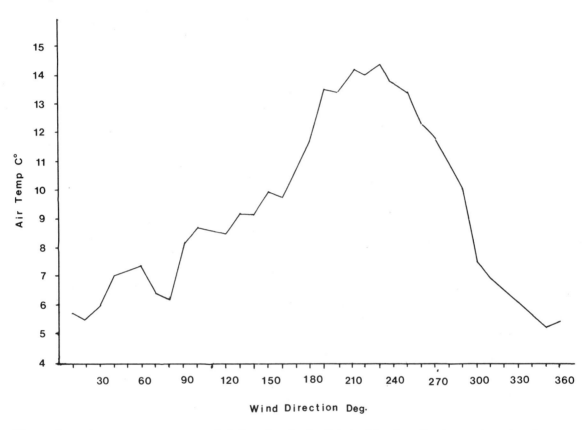

Figure 2: Air temperature versus wind direction for St. John's, Newfoundland. Based on data for June 1971-87, supplied by the Atmospheric Environment Service, Scientific Service Unit, St. John's.

Modern Analogues

To provide a check on the proposed circulation pattern for 1816 and to give more detail on the nature of the circulation, modern analogues for the temperature pattern observed in June 1816 were selected, and their circulation patterns compared to that proposed. The criteria used were below-normal temperatures in northern Labrador and normal or above-normal temperatures in southeastern Newfoundland in June. Monthly temperature patterns were obtained from maps in Environment Canada publications (*Climatic Perspectives* and *Monthly Record*). During the 30-year period 1958-87, five years had June temperatures that met the criteria (1969, 1971, 1972, 1978 and 1986; Figure 4).

All of the years selected as analogues had below-normal June temperatures at Churchill, Manitoba, on the west coast of Hudson Bay. This pattern agrees well with conditions in 1816 when temperatures at Churchill were considerably below normal (Catchpole 1985). All of these years except 1986 had cool to very cold conditions in central England; in fact, June 1971 and 1972 were colder than June 1816 (Manley 1974). The opposition of temperatures in Newfoundland and England agrees with the Burroughs' (1979) finding of an inverse relationship between temperatures in the two areas. The agreement between conditions in the five years mentioned above and 1816 is not as strong when conditions in New England are considered. Only two of the five years (1972 and 1986) had below-normal June temperatures at Boston: however, in all five years below-normal temperatures reached some part of New England.

In all but one of the five years considered, the mean centre of low pressure in the North Atlantic was near its normal position, over or near the Labrador Sea. The exceptional year was 1972, when the low was south of Iceland. However, in all cases mentioned, the circulation was more intense than normal. Of the five years considered, the circulation patterns in 1969 and 1971 seem most unlike that for 1816. In these cases the surface winds in northern Labrador had a strong southerly component - totally unlike 1816. Perhaps the surface temperatures indicated for northern Labrador during these years (based on data from surrounding stations) are in error, and the true temperatures were higher. In the case of 1972, while the temperature pattern matches that for June 1816 in England, New England, Hudson Bay, Newfoundland and Labrador, the nature of the circulation does not fit the proposed pattern for the eastern Atlantic/western Europe sector as proposed by Kelly *et al.* (1984). In fact the circulation for June 1972 is totally different from the conditions that usually produce below-normal June temperatures over England (Perry 1972).

The two remaining years (1978 and 1986) both have deeper-than-normal low-pressure centres over the Labrador Sea; however, in 1978 a high-pressure centre occurring over Hudson Bay was absent in 1986. Wilson (1985) proposed that such a high was centred over Hudson Bay during the summer of 1816. The 700-mb circulations in both of these years have some similarities and some major differences. Both years have above normal 700-mb heights southeast of Newfoundland. This feature also occurs during the three other years selected, and is likely related to above-normal temperatures in Newfoundland. Farther north the 700-mb patterns for the two years are different. In June 1986 there is a trough along the Labrador coast with the largest negative height departures over the mid-Labrador coast, whereas in June 1978 the trough is farther west, the greatest negative height departures being over Foxe Basin.

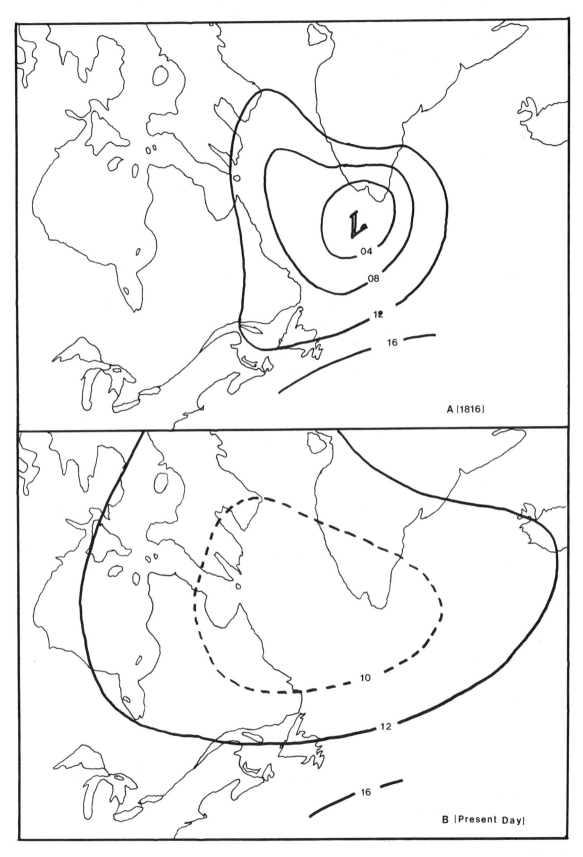

Figure 3: Mean June sea-level pressure: (A) estimated for 1816 and (B) present-day normals.

251

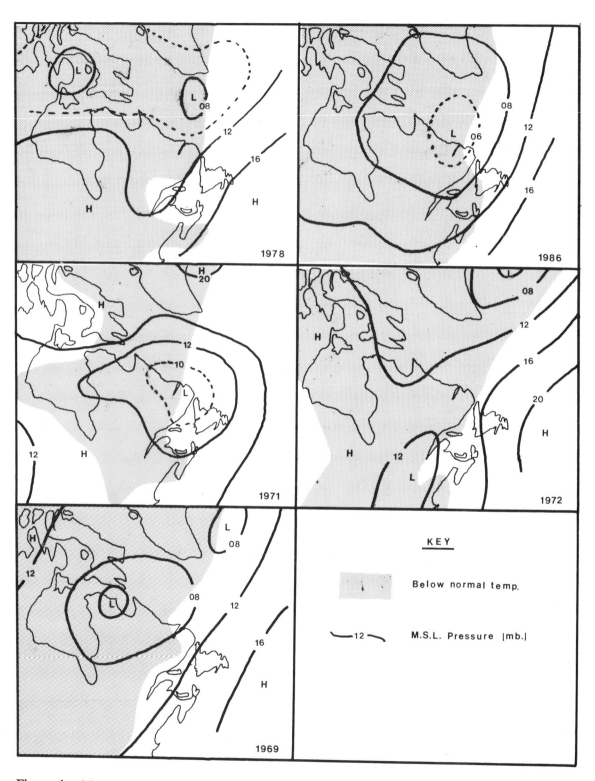

Figure 4: Mean sea-level pressure (mb) and regions with below-normal air temperature for June: 1969, 1971, 1972, 1978 and 1986.

Analysis of Environment Canada (Atmospheric Environment Service) ice charts indicate that clearing dates for the Labrador coast were later than normal in both 1978 and 1986; however, neither year represented record conditions. Ice conditions at the end of June 1978 were more severe than at the end of June 1986, but the rate of retreat during the month of June was greater in 1978 than in 1986. Since this analysis only considered conditions in June, it is not surprising that ice conditions were not as severe as in 1816. Conditions earlier in the spring, and the strong northerly flow in July 1816 indicated by Lamb and Johnson (1966), likely played an important role in the exceptional 1816 ice conditions.

Summary

Comparison of circulation and temperature patterns for June 1978 and 1986 with the proposed pattern for June 1816 (Figure 3) indicates that they are in general agreement. For example, all three maps have a deep low-pressure centre in the Labrador Sea. However, apparently the northerly circulation in 1816 must have been more vigorous than in 1978 or 1986 to give the lower temperatures reported. This would require that the low-pressure centre in the Labrador Sea be deeper than in either of those years. The actual pattern for June 1816 likely combined features of both June 1978 and 1986. This pattern also agrees with the North Atlantic circulation for July 1816 proposed by Lamb and Johnson (1966).

The occurrence of a circulation pattern such as the one proposed for June 1816 without outside forcing (such as volcanic cooling) does not seem unrealistic in light of the variability demonstrated in the five analogues considered in this study. Perhaps such an occurrence is especially likely considering that in 1816 the northern hemisphere was experiencing the last stages of the Little Ice Age, a period when such circulation patterns would have been more common. The main difficulty with this argument is that data from other sources demonstrate that conditions during July and August 1816 were equally unusual. A long-term data set of sea-ice conditions for the Labrador Sea that I am currently developing may assist in determining how the summer of 1816 compares with modern conditions and with other summers in the nineteenth century.

References

Briffa, K.R., P.D. Jones and F.H. Schweingruber. 1988. Summer temperature patterns over Europe: a reconstruction from 1750 A.D. based on maximum latewood density indices of conifers. *Quaternary Research* 30:36-52.

Burroughs, W.J. 1979. An analysis of winter temperatures in central England and Newfoundland. *Weather* 34:19-23.

Catchpole, A.J.W. 1985. Evidence from Hudson Bay region of severe cold in the summer of 1816. *In: Critical Periods in the Quaternary Climatic History of Northern North America. Climatic Change in Canada 5*. C.R. Harington (ed.). *Syllogeus* 55:121-146.

Catchpole, A.J.W. and M.-A. Faurer. 1985. Ships' logbooks, sea ice and the cold summer of 1816 in Hudson Bay and its approaches. *Arctic* 38:121-128.

Crane, R.G. 1978. Seasonal variations of sea ice extent in the Davis Strait-Labrador Sea area and relationships with synoptic-scale atmospheric circulation. *Arctic* 31:434-447.

Forsey, R. and W.H. Lear. 1987. Historical catches and catch rates of Atlantic Cod at Newfoundland during 1677-1833. *Department of Fisheries and Oceans, Canadian Data Report of Fisheries and Aquatic Sciences* No. 662:1-52.

Kelly, P.M., T.M.L. Wigley and P.D. Jones. 1984. European pressure maps for 1815-16, the time of the eruption of Tambora. *Climate Monitor* 13:76-91.

Lamb, H.H. 1977. *Climate: Present, Past and Future. Vol. 2; Climatic History and the Future.* Methuen, London. 835 pp.

Lamb, H.H. and A.I. Johnson. 1959. Climatic variation and observed changes in the general circulation, Parts I and II. *Geografiska Annaler* 41:94-134.

_____. 1961. Climatic variation and observed changes in the general circulation, Part III. *Geografiska Annaler* 43:363-400.

_____. 1966. Secular variations of the atmospheric circulation since 1750. *Great Britain, Meteorological Office, Geophysical Memoirs* 110:1-57.

Ludlum, D.M. 1966. *Early American Winters 1604-1820.* American Meteorological Society, Boston.

Manley, G. 1974. Central England temperatures: 1659-1973. *Quarterly Journal of the Royal Meteorological Society* 100:389-405.

Newell, J.P. 1983. Preliminary analysis of sea-ice conditions in the Labrador Sea during the nineteenth century. *In: Climatic Change in Canada 3.* C.R. Harington (ed.). *Syllogeus* 49:108-129.

Perry, A.H. 1972. June 1972 - the coldest June of the century. *Weather* 27:418-422.

Scoresby, W., Jr. 1820. *An Account of the Arctic Regions with a History and Description of the Northern Whale-Fishery.* Reprinted in 1969 by Augustus M. Kelley, Publishers, New York.

Stommel, H. and E. Stommel. 1979. The year without a summer. *Scientific American* 240:176-186.

Teillet, J.V. 1988. A reconstruction of summer sea ice conditions in the Labrador Sea using Hudson's Bay Company ships' log-books, 1751-1870. Unpublished M.A. thesis, University of Manitoba, Winnipeg. 161 pp.

Wilson, C.V. 1983. The summer season along the east coast of Hudson Bay during the nineteenth century. Part II: The Little Ice Age on eastern Hudson Bay: summers at Great Whale, Fort George, Eastmain, 1814-1821. *Canadian Climate Centre, Downsview, Report No. 83-9.*

_____. 1985. The Little Ice Age on eastern Hudson/James Bay: the summer weather and climate at Great Whale, Fort George and Eastmain, 1814 to 1821, as derived from Hudson's Bay Company records. *In: Critical Periods in the Quaternary Climatic History of Northern North America. Climatic Change in Canada 5.* C.R. Harington (ed.). *Syllogeus* 55:147-190.

Spatial Patterns of Tree-Growth Anomalies from the North American Boreal Treeline in the Early 1800s, Including the Year 1816

Gordon C. Jacoby, Jr.[1] and Rosanne D'Arrigo[1]

Abstract

Tree-growth anomalies based on 24 temperature-sensitive white spruce chronologies from boreal treeline sites in North America are mapped and analyzed for the interval 1805-24. This interval includes the volcanic eruption of Tambora in 1815 and the unusual "year without a summer" in 1816. The first few decades of the 1800s also are concurrent with a series of low-amplitude cycles in sunspot number which have been suggested as contributing to unusually cooler conditions during this time. It is inferred from the tree-ring data that climatic changes following the Tambora eruption influenced the North American boreal forest in different areas at different times from 1816 to 1818, with the coldest regional temperatures appearing to have occurred in the year 1816 in easternmost Canada. The series of anomaly maps provided here help to clarify the spatial patterns of climatic changes in remote northern regions during this extreme cooling event.

Introduction

The climatically unusual "year of no summer" (1816) was primarily documented as such by observers in Europe and eastern North America (Landsberg and Albert 1974; Stommel and Stommel 1983; Stothers 1984; Briffa et al. 1988). Cold air masses invaded most areas of Europe and the settled, eastern regions of North America (Stommel and Stommel 1983; Stothers 1984; Briffa et al. 1988). There is little documentation on weather variations for western North America about the time of the Tambora eruption (1815), and much of the documentation on weather variations for other parts of the world is only recently being brought into full consideration (e.g., Legrand and Delmas 1987, concerning evidence of the Tambora eruption in Antarctic ice-core data; also other papers in this volume). Overall, little is known regarding the regional-scale climatic variations following many volcanic eruptions (including Tambora), and it is likely that hemispheric-scale studies demonstrating a general cooling effect may obscure warming in some areas (Lough and Fritts 1987) that may result from changes in large-scale atmospheric dynamics following major eruptions (Hansen et al. 1978; Schneider 1983).

It has been hypothesized that volcanism can strongly influence climate, causing cooler temperatures (e.g., Lamb 1970; Mass and Schneider 1977; Sear et al. 1987; Bradley 1988). The mechanism is not thoroughly understood but the common theory is that stratospheric sulphate particles partially reflect and absorb incoming radiation, heating the stratosphere. This heat does not reach the troposphere, which then becomes cooler (e.g., Hansen et al. 1978). Although empirical modeling studies (Hansen et al. 1978) and superposed epoch analyses (Mass and Schneider 1977; Sear et al. 1987; Skinner, this volume) indicate a cooling of a few tenths of an

---

[1] Tree-Ring Laboratory, Lamont-Doherty Geological Observatory of Columbia University, Palisades, New York 10964, U.S.A.

degree following major eruptions, this cooling is within the level of natural climatic variability. Hence a link cannot be unequivocally proven, but there is strong evidence for a cause and effect relationship (Sear et al. 1987).

The eruption of Tambora in April 1815 was one of the greatest volcanic events of recent centuries (Simkin et al. 1981) . Because studies indicate (Sear et al. 1987) that southern hemisphere eruptions would cool northern hemisphere temperatures after a lag of about six months to a year, a response to this event at northern mid- to high-latitudes would not be expected to occur until late 1815-16. An unusually cold summer in 1816 is attributed by many to the effects of this event but the overall climatic anomalies of the period are more complex than a single event-response phenomenon. One phenomenon to be considered in the complexity is that the early 1800s were notable for a reduction in solar sunspot amplitudes, possibly reflecting solar irradiance changes (Lean and Foukal 1988; Kuhn et al. 1988) that may have contributed to a cooling during the early decades of the 1800s (Eddy 1977, this volume).

Perspective of This Study

The method used here to examine the climate of 1816 is the study of old-aged trees growing along the northern boreal forest zone of North America. All of the data are from white spruce [*Picea glauca* (Moench) Voss] near the forest-tundra transition zone. These trees primarily respond to summer temperature, with a secondary response to fall and spring temperature-related conditions (Jacoby and D'Arrigo, in press; Scott et al. 1988; Jacoby and Ulan 1982; Cropper 1982; Jacoby and Cook 1981; Garfinkel and Brubaker 1980). The average position of the Polar Front in summer largely coincides with the location of the northern treeline (Bryson 1966). Tree growth in this region can therefore be expected to record frontal shifts, extensive outbreaks of polar air masses and circulation changes influencing thermal conditions (e.g., Scott et al. 1988). Thus tree-ring data can provide useful information on climatic response following the 1815 eruption (and the early 1800s in general) in the northern boreal region, including the western region of North America where other data sources are scarce.

We have examined 24 time-series of absolutely dated tree-ring width indices (chronologies) throughout the region. Each time-series is usually based on about 10 trees with multiple cores (radii) from each tree. Some of these chronologies contain low-frequency response to climate, whereas others only preserve higher-frequency response. For compatibility and intercomparison, all chronologies were prewhitened (to remove low-frequency variation) and normalized. This procedure is appropriate since, in this case, we are evaluating variations in year-to-year response over a period of a few decades. Maps (Figures 1, 2) display the departures from the mean for each chronology location during the early 1800s (1805-24).

Distribution of Anomalies

For 1816, Figure 1 shows substantial negative departures (reflecting reduced radial growth/colder temperatures) in eastern Canada. These are the greatest anomalies for the period under review (1805-24). However, the rest of Canada and Alaska shows no such severe cooling. Northern Alaska was fairly cold (as in some other years), but central and western Canada are close to average or above for the year. The colder eastern region agrees well with reports and records from eastern Canada and the United States. For example in the northern region, Catchpole and Faurer (1983) demonstrate that the duration of westward passage of Hudson's Bay Company ships was the longest (54 days compared to a mean of 17.7 days) of the entire 1751-1870 record, representing severe sea-ice conditions. The authors (see also Catchpole, this volume) suggest that these conditions could be explained by enhanced meridional flow of arctic air masses over eastern North America at this time. Records from the eastern United States, do not show a continuously cold summer (Stommel and Stommel 1983; Baron, this volume). There were three distinct outbreaks of extremely cold air from the North at different times during the summer. These outbreaks had serious negative effects on food crops during the growing season (Stommel and Stommel 1983). Such records appear to support the theory of increased Arctic air flow over this region in 1816. Schneider (1983) suggests that since the cooling (about 3°C) would not have been sufficient to explain the documented frosts that occurred, a dip in the Jetstream and blocking of the mid-latitude westerlies could have contributed to the adverse conditions. He suggests that conditions to the west (about one-half wavelength away) of eastern North America would have been unusually warm if this had been the case. Our results support this contention since conditions in central Canada, although not unusually warm, do not demonstrate the pronounced cooling found in the east (Figure 1).

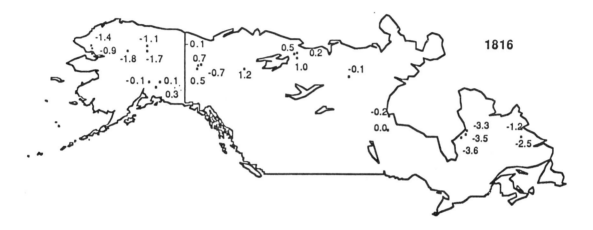

Figure 1: Tree-growth anomaly map for 1816. The growth departures are based on prewhitened and normalized tree-ring width indices for 24 white spruce chronologies from near the boreal treeline of North America.

To place this year in context, we review the years preceding and after 1816 beginning with 1805 (Figure 2). The three years of 1805-07 show little in the way of extreme cold temperatures. Except for southeastern Alaska, most other regions are near or above normal, and eastern Canada is substantially above normal. Then in 1808 colder temperatures prevailed in the Hudson Bay region, and Alaska was warm. In 1809 the Hudson Bay region was less cold but Alaska became cold. The distribution of regions of warmer or cooler temperatures correspond roughly to the configuration of the longwave pattern in the atmosphere. There is approximately one wavelength across the North American quadrant for a four-wave pattern, western Alaska to eastern Canada being slightly over 90° of longitude (Chang 1972).

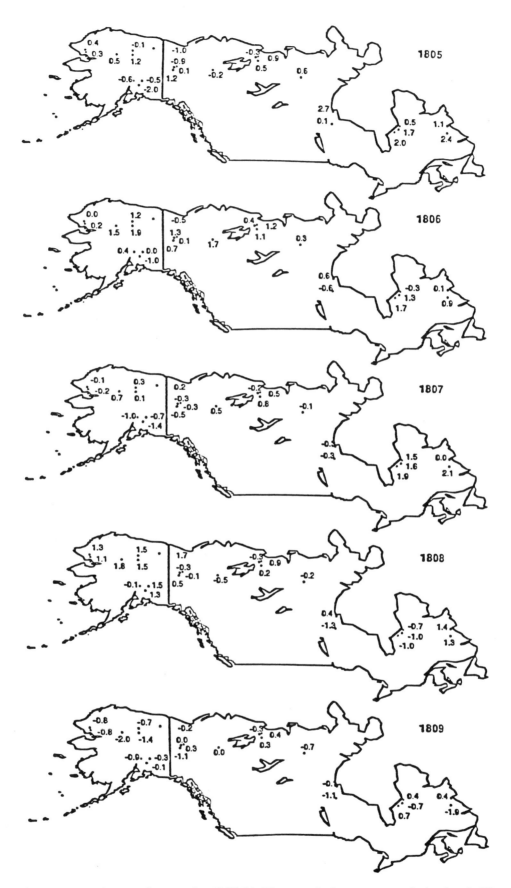

Figure 2: Tree-growth anomaly maps for 1805-24. The growth departures are derived as in Figure 1.

Figure 2 (cont'd):

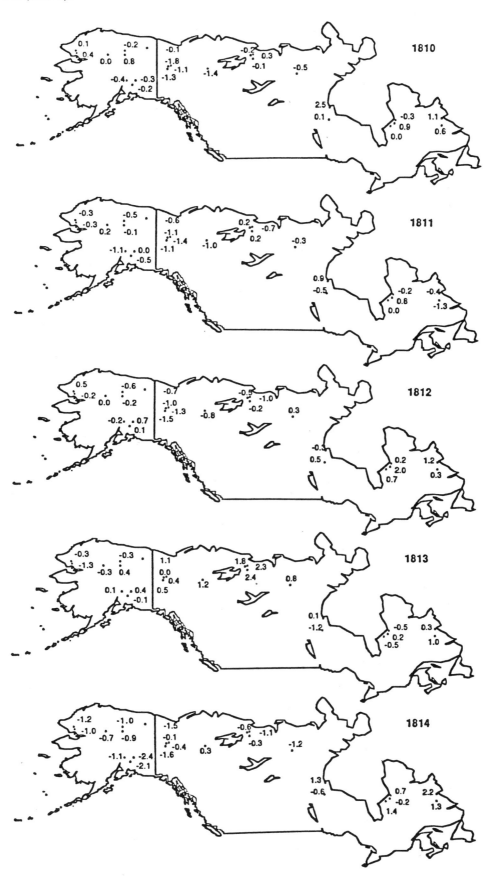

Figure 2 (cont'd):

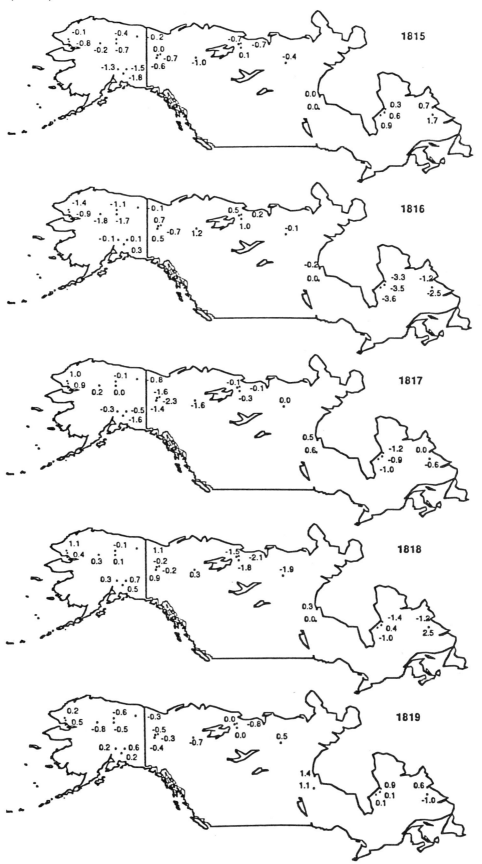

Figure 2 (cont'd):

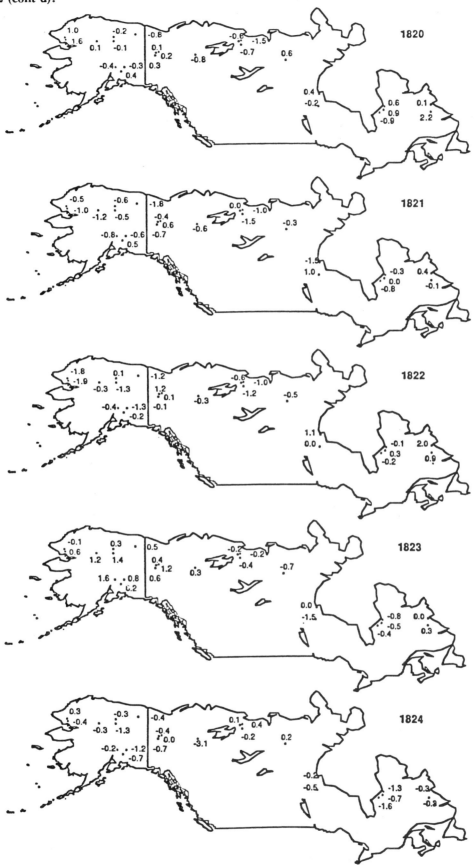

During 1810 through 1812 the main features of the maps are a cooling over eastern Alaska, the Yukon Territory and the western Northwest Territories, and in Labrador an alternating warm-cool-warm sequence. An indication of very warm conditions in 1813 over the Northwest Territories is followed by a reversal to quite cold conditions for Alaska, especially southeastern Alaska, and all of western Canada in 1814 accompanied by a warming in Labrador. The cooler conditions continue for the western region in 1815. Northern Alaska is fairly cold in 1816, but the most severe cold is restricted to eastern Canada (see also Figure 1). As noted above, central and western Canada are not unusually cold during 1816. More severe cold does not reach western Canada and eastern Alaska until 1817 when the anomalies are more negative than other years of the period, although 1809 is quite cold. The eastern region is still cold but recovering toward normal. In 1818, the coldest conditions occur in the Northwest Territories. Again these are the greatest negative anomalies for this area, although 1814 and 1821-22 are also cold. By 1819, almost all extreme negative anomalies are gone from the map region.

Extreme western Alaska and eastern Canada are warm in 1820 but cold pervades much of the entire map region during 1821 and 1822, except for Labrador in 1822. Alaska warms in 1823 but there is a return to cold temperatures in 1824 in both western Canada/Alaska and in eastern Canada.

In summary, there were significantly cold conditions in some northern areas before 1816. After the volcanic event, extreme cold affected all of the map region at different times until 1818. Cold temperatures pervaded some areas after 1818, and 1824 was fairly cool throughout most of the region.

Discussion and Conclusions

Our results show the spatial patterns of tree-growth anomalies from the North American boreal treeline during the anomalous period of the early 1800s, with an emphasis on the year 1816. The data provide added spatial coverage of western Canada and Alaska in relation to the climatic response following the Tambora (1815) event. In agreement with other studies, apparently the unusual cold in eastern North America may in part have resulted from meridional flow of cold Arctic air across this region. By contrast, conditions in western and central Canada (about a half wavelength to the west) were moderate in 1816 (Schneider 1983). This reflects a possible shift in atmospheric circulation which may or may not be directly linked to the volcanic event.

Records of climatic conditions in the early 1800s in other regions of the globe are rather fragmentary, as discussed, and the cooling in the early 1800s was probably not synchronous globally (see the Workshop section, this volume). Figure 3 shows a reconstruction of northern hemisphere annual temperatures (from Jacoby and D'Arrigo 1988) based on North American boreal tree-ring data, indicating a cooling during this interval that persisted for several decades. On a more regional scale, the cooling in Europe is well documented (Stommel and Stommel 1983; Briffa et al. 1988). Specifically, a sharp lowering of temperature is seen in England and central Europe from 1812-20 in reconstructed summer temperatures based on tree-ring density data (Briffa et al. 1988). Records from China and Japan (Stommel and Stommel 1983) do not indicate unusually cold conditions[1]. A detailed compilation of spatial temperature data during this time interval is clearly needed.

---

[1] But see Zhang et al. and Huang, this volume (editor).

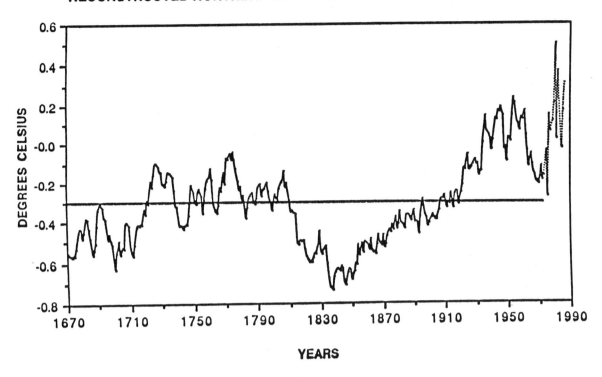

Figure 3: Reconstruction of annual northern hemisphere temperatures from 1671 to 1973 based on high latitude tree-ring data from North America. Temperature departures from 1974-87 from Hansen and Lebedeff, 1987, 1988 [see Jacoby and D'Arrigo (in press)]. Note the abrupt cooling in the early 1800s.

The detection of a direct cause and effect signal due to volcanism is difficult, in part due to the influence of other forcing functions on the climatic system. These include unusual (diminished) solar fluctuations that also occurred in the early 1800s (Eddy 1977). El Niño events occur on the same time scale as volcanic eruptions (largely high frequency) and can obscure their signal, as can random climatic variations (Robock 1981). Modeling studies (e.g., Gilliland and Schneider 1984; Robock 1981) show good agreement between model estimates (based on volcanic indices) and actual temperature data but other forcings must also be considered. Finally there are many complicating factors for individual eruptions [e.g., season and latitude of eruption, height and chemistry of ejecta, state of atmospheric circulation at time of eruption (Lamb 1970; Lough and Fritts 1987)] which complicate attempts to detect a common event-response pattern. Improvements in understanding volcanic forcings are necessary for isolating effects of other forcings such as CO_2.

The oft-applied term "year of no summer" for 1816 is obviously a misnomer in the context of the northern boreal forests of North America. To understand climatic change in Canada and the rest of North America, it is necessary to move away from this oversimplification and study spatial and temporal differences and dynamics of the early 1800s, as these decades are a time of substantial climatic variation. Here we have provided a series of maps from the early 1800s that help clarify the spatial patterns of climate at remote high-northern latitudes during this interval, and which may be useful in determining causes of and responses to such extreme climatic events.

Acknowledgements

This research was supported by the Climate Dynamics Division of the National Science Foundation, under grants ATM85-15290 and ATM87-16630. We thank J. Hayes and W. Ruddiman for helpful reviews, and the Canadian Forestry and Atmospheric Environment services for technical assistance. Lamont-Doherty Geological Observatory Contribution No. 4566.

References

Bradley, R.S. 1988. The explosive volcanic eruption signal in northern hemisphere continental temperature records. *Climatic Change* 12:221-243.

Briffa, K., P.D. Jones and F.H. Schweingruber. 1988. Summer temperature patterns over Europe: a reconstruction from 1750 A.D. based on maximum latewood density indices of conifers. *Quaternary Research* 30:36-52.

Bryson, R.A. 1966. Air masses, streamlines, and the boreal forest. *Geographical Bulletin* 8:228-269.

Catchpole, A.J.W. and M.A. Faurer. 1983. Summer sea ice severity in Hudson Strait, 1751-1870. *Climatic Change* 5:115-139.

Chang, J.H. 1972. *Atmospheric Circulation Systems and Climates*. Oriental Publishing Co., Honolulu. 326 pp.

Cropper, J.P. 1982. Climate reconstructions (1801 to 1938) inferred from tree-ring width chronologies from the North American Arctic. *Arctic and Alpine Research* 14:223-241.

Eddy, J.A. 1977. Climate and the changing sun. *Climatic Change* 1:173-190.

Garfinkel, H.L. and L.B. Brubaker. 1980. Modern climate-tree growth relationships and climatic reconstruction in subarctic Alaska. *Nature* 286:872-874.

Gilliland, R.L. and S.H. Schneider. 1984. Volcanic, CO_2 and solar forcing of northern and southern hemisphere surface temperatures. *Nature* 310:38-41.

Hansen, J.E., W.C. Wang and A.A. Lacis. 1978. Mount Agung eruption provides test of a global climatic perturbation. *Science* 199:1065-1068.

Jacoby, G.C. Jr. and E.R. Cook. 1981. Past temperature variations inferred from a 400-year tree-ring chronology from Yukon Territory, Canada. *Arctic and Alpine Research* 13:409-418.

Jacoby, G.C. Jr. and L.D. Ulan. 1982. Reconstruction of past ice conditions in a Hudson Bay estuary using tree-rings. *Nature* 298:637-639.

Jacoby, G.C. Jr. and R. D'Arrigo. 1988. Reconstructed northern hemisphere annual temperature since 1671 based on high latitude tree-ring data from North America. *Climatic Change* (in press).

Kuhn, J.R., K.G. Libbrecht and R.H. Dicke. 1988. The surface temperature of the sun and changes in the solar constant. *Science* 242:908-911.

Lamb, H.H. 1970. Volcanic dust in the atmosphere; with a chronology and assessment of its meteorological significance. *Philosophical Transactions of the Royal Society of London* 255:425-533.

Landsberg, H.E. and J.M. Albert. 1974. The summer of 1816 and volcanism. *Weatherwise* 27:63-66.

Lean, J. and P. Foukal. 1988. A model of solar luminosity modulation by magnetic activity between 1954 and 1984. *Science* 240:906-908.

Legrand, M. and R.J. Delmas. 1987. A 220-year continuous record of volcanic H_2SO_4 in the Antarctic Ice Sheet. *Nature* 327:671-676.

Lough, J.M. and H.C. Fritts. 1987. An assessment of the possible effects of volcanic eruptions on North American climate using tree-ring data, 1602 to 1900 A.D. *Climatic Change* 10:219-239.

Mass, C. and S. Schneider. 1977. Statistical evidence on the influence of sunspots and volcanic dust on long-term temperature records. *Journal of Atmospheric Science* 34:1995-2004.

Robock, A. 1981. A latitudinally dependent volcanic dust veil index and its effect on climate simulations. *Journal of Volcanology Geothermal Research* 11:67-80.

Schneider, S. 1983. Volcanic dust veils and climate: how clear is the connection? - an editorial. *Climatic Change* 5:111-113.

Scott, P.A., D.C.F. Fayle, C.V. Bentley and R.I.C. Hansell. 1988. Large-scale changes in atmospheric circulation interpreted from patterns of tree growth at Churchill, Manitoba, Canada. *Arctic and Alpine Research* 20:199-211.

Sear, C.B., P.M. Kelly, P.D. Jones and C.M. Goodess. 1987. Global surface-temperature responses to major volcanic eruptions. *Nature* 330:365-367.

Simkin, T., L. Seibert, L. McClelland, W.G. Melson, D. Bridge, C.G. Newhall and J. Latter. 1981. *Volcanoes of the World*. Smithsonian Institution, Washington, D.C.

Stommel, H. and E. Stommel. 1983. *Volcano Weather: The Story of 1816: the Year Without a Summer*. Seven Seas Press, Newport. 177 pp.

Stothers, R.B. 1984. The great Tambora eruption in 1815 and its aftermath. *Science* 224:1191-1198.

Early Nineteenth-Century Tree-Ring Series from Treeline Sites in the Middle Canadian Rockies

B.H. Luckman[1] and M.E. Colenutt[1]

Abstract

Preliminary data from tree-ring series at treeline sites in the Canadian Rockies are evaluated for evidence of anomalies associated with the 1815 eruption of Tambora. Three maximum density and nine ring-width chronologies (six *Picea engelmannii*, one each for *Abies lasiocarpa*, *Larix lyallii* and *Pinus albicaulis*) are presented, covering 1780-1860. The absence of narrow or light latewood marker-rings associated with 1816 or 1817 indicates that there is no distinctive tree-ring signal associated with the Tambora event at these sites. Most records do however contain a sharp decrease in ring widths during the 1810-20 decade, similar to that reported from latitudinal treeline sites elsewhere in North America, and which appears to be associated with an abrupt deterioration of climate initiated several years prior to the Tambora eruption.

Introduction

Evaluation of the spatial extent of climatic anomalies associated with the eruption of Tambora in 1815 requires assessment of proxy-data series throughout North America. The annual resolution of tree-ring series, combined with strong relationships between ring characteristics and climate, potentially provide a powerful tool to accomplish this goal. Here we present data from preliminary tree-ring chronologies for 1780-1860 at several treeline sites in and adjacent to Banff and Jasper national parks in the Canadian Rocky Mountains (Figure 1). These data are examined to see whether significant anomalies are present in the 1815-17 period that could be attributed to climatic effects associated with the Tambora eruption.

The principal direct climatic effect associated with volcanic eruptions is a reduction in solar radiation received at the surface because of stratospheric dust veils (Lamb 1970). This often results in cooler summers, and the spatial extent and severity of this effect depends on the magnitude, timing and nature of the eruption. LaMarche and Hirschboeck (1984) have demonstrated a strong relationship between the presence of frost rings in the Bristlecone pine chronology from treeline sites in the White Mountains of California and major volcanic eruptions. These data include a frost-ring date of 1626 B.C. for the eruption of Santorini in Greece (which destroyed the Minoan civilization of Crete, and probably had global climatic effects). Baillie and Munroe (1988) show that Irish bog oaks had very narrow rings in the 1620s (B.C.), which appear to confirm this result. In Canada, Filion *et al.* (1986) have shown that light latewood rings from black spruce chronologies in northern Quebec correspond with periods 0-2 years after major volcanic eruptions. In these records the 1816-17 rings have light latewood in 75% of the series studied, and 1784 (the year following the Laki eruption) is also a prominent marker ring. Parker (1985) and Jacoby *et al.* (1988) also note the exceptional nature of the 1816 and 1817 rings in white-spruce chronologies on the eastern shores of Hudson Bay. The severe climate of these two summers is amply demonstrated by several papers in this volume.

---

[1] Department of Geography, University of Western Ontario, London, Ontario N6A 5C2, Canada.

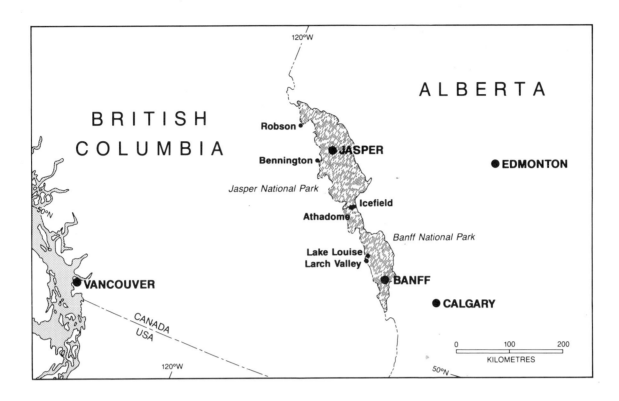

Figure 1: Location of the main study sites.

Parker (1985) evaluated selected tree-ring series from western and central Canada to determine whether he could detect a signal associated with the eruptions of Tambora or Krakatau (1888). He used ring-width and densitometric data from 135 trees (four different species) at 15 sites between Vancouver Island and Hudson Bay. The data were aggregated into six regional chronologies, and indexed data were used to compare the eruption year with groups of three years before and after the eruption. Only one site, Cri Lake on Hudson Bay, showed a significant growth reduction following the Tambora eruption.

These results suggest that, under certain conditions, a volcanic signal can be detected via its influence on climate, and thereby on tree-ring characteristics. Many authors have demonstrated strong relationships between tree-ring width or density series and summer temperatures - particularly at treeline sites (e.g., Parker and Henoch 1971; Luckman *et al.* 1985; Jacoby and Cook 1981; Jacoby *et al.* 1988). It would be anticipated, therefore, that trees at these sites would be particularly sensitive to reductions in summer insolation, and therefore most likely to record evidence of dust-veil-related volcanic effects. As most of the montane sites used by Parker (1985) are well below treeline, we decided to evaluate treeline records from the Rockies to see whether the 1815-17 record contained any distinctive signal that could be attributed to the effects of the Tambora eruption.

Sample Sites

Nine preliminary living-tree ring-width chronologies are available from our tree-ring studies in the Canadian Rockies (Table 1, Figure 1): eight are from treeline sites and five (Robson, Bennington and Icefields/Athabasca sites) are adjacent to Little Ice Age terminal moraines. Six of these chronologies utilize Engelmann spruce (*Picea engelmannii*) because it is the most ubiquitous, long-lived tree at treeline in this area. Single chronologies for alpine larch (*Larix lyallii*), alpine fir (*Abies lasiocarpa*) and whitebark pine (*Pinus albicaulis*) are also used. Tree-ring densitometric data are also available for three of these sites.

The Robson site (Figure 2) is an isolated stand of spruce on a low bedrock knoll overlooking an inactive outwash fan from Robson Glacier. At its Little Ice Age maximum position, Robson Glacier advanced against the upvalley side of the knoll and built a terminal moraine along its crest. Heusser (1956) estimated the date of formation of the three outermost moraines of Robson Glacier as 1787, 1801 and 1861 based on tree-ring sampling and allowing a 12-year ecesis interval. The oldest tree in the stand outside the moraine is just over 400 years old, i.e., it predates the maximum glacier advance by about 200 years. The trees were sampled in 1981 and 1983. Preliminary results are given by Watson (1983): the results presented here use both data sets.

Figure 2: The Robson Glacier site, view east from Adolphus Lake, Alberta (foreground) toward Rearguard Mountain and Mount Robson (snow-covered, top right). The sampled stand (a) is visible with the lighter-toned Little Ice Age moraine complex of the Robson Glacier extending from left to right across the middle ground behind the trees.

Table 1: Tree-Ring Series in the Canadian Rockies Used to Evaluate the 1810-20 Period.

Site	Lat. N	Long. W	Elevation M	Aspect	Species Used[1]	Series Used[2]	Chronology
Robson Glacier	53.10	119.10	1690	W	PE	RW	1568-1982
Bennington	52.41	118.20	1765	S	PE PA	RW RW, MXD	1661-1981 1276-1985
Athadome	52.12	117.25	2000	N	PE	RW	1646-1980
Icefields	52.12	117.25	2000	S	PE	RW, MXD	1316-1981
Lake Louise	51.14	116.15	1650	E	PE	RW, MXD	1705-1982
Larch Valley	51.20	116.13	2200	S	PE AL LL	RW RW RW	1568-1986 1725-1986 1349-1986

[1] PE = *Picea engelmannii*; PA = *Pinus albicaulis*; AL = *Abies lasiocarpa*; LL = *Larix byallii*.

[2] RW = Ring width; MXD = Maximum density.

The Bennington site is an open grown, almost pure stand of whitebark pine growing on a coarse talus and bedrock slope overlooking the lateral moraine of Bennington Glacier (Figure 3) which is dated to about 1700 and 1825 by dendrochronology (MacCarthy 1985). This stand contains a number of very old trees, the oldest of which has a pith date of 1112 A.D. at breast height, and is thought to be the oldest whitebark pine in Canada (Luckman *et al.* 1984). Chronology development at this site is incomplete, and data given are for the four trees for which ring width and densitometric data are presently available for the 1780-1860 interval. The spruce stand at this site is on the lower slope, slightly downvalley of the area shown in Figure 3. The tree-ring series from this site show high tree-to-tree variability due to rockfall disturbance of the site (Watson 1983). The results are included here solely for comparison with spruce chronologies at other sites.

The Athadome (Figure 4) and Icefield (Figure 5) sites are both adjacent to the Athabasca Glacier on opposite sides of the Sunwapta Valley, less than a kilometre apart. The Athadome site has a unique microclimate because it lies between the lateral moraines of Athabasca and Dome glaciers and was almost completely surrounded by and below the level of the adjacent ice surface about 1714 (Heusser 1956) and between approximately 1840 and 1920 (Luckman 1988). By contrast, the Icefield site is a well drained lower valley side slope just beyond the outer limits of the Athabasca Glacier. Both chronologies are Engelmann spruce, but sampling in 1980 and 1981 (Luckman 1982) indicated that some trees at the Icefield site were considerably older. Intensive sampling at this site in 1982 provided the present chronology (Jozsa *et al.* 1983), which is based on trees with a mean age of over 500 years (Table 2) and includes the oldest known Engelmann spruce (Luckman *et al.* 1984).

The Lake Louise site is the only non-treeline site presented here. It occurs in the lower subalpine forest about 300 m below treeline on a valley side bench overlooking Lake Louise townsite (Hamilton 1984). This site was the closest Engelmann spruce stand to the meteorological station at Lake Louise, and was used to explore climatic tree-ring relationships for this species (Luckman *et al.* 1985). The Larch Valley site (Figure 6) is at treeline, some 10 km south of Lake Louise. It is about 2 km from the Wenkchemna Glacier, and considerably above it on a broad valley side bench overlooking the main valley. Chronologies were developed for three species in the same stand at this site because of difficulties in crossdating the larch record which has several periods with very tight or missing rings (Colenutt 1988). This larch chronology is the best-replicated and most sensitive (mean sensitivity 0.38) of those discussed here. The Larch Valley and Lake Louise chronologies are also less likely to show local climatic effects from adjacent glaciers than the other chronologies reported here.

Chronologies for most sites were developed by standard methods using the Laboratory of Tree-Ring Research (Tucson) programs INDEX and SUMAC (Graybill 1982). However, chronologies for the Icefield and Lake Louise sites were developed by Forintek, Vancouver using a 99-year running mean to remove the growth trend (Parker *et al.* 1981). Therefore some of the longer-frequency trends in these chronologies have been removed resulting in a lower amplitude of response (Luckman *et al.* 1985).

Results

The results from the nine indexed ring-width chronologies are shown in Figure 7, and high-pass filter data (Fritts 1976) from these series are presented in Figure 8. The indexed values for 1810-20 are listed in Table 2 with some summary statistics for the chronologies used. These

Table 2: Indexed Ring-Width Values for the 1810-20 Period for Selected Treeline Chronologies in the Canadian Rockies.

Site	Species[1]	N[2]	Age[2]	1810	1811	1812	1813	1814	1815	1816	1817	1818	1819	1820
Robson	PE	15	113	118	128	109	107	**103**	109	112	121	97	91	110
Bennington	PE	17	61	99	96	90	92	**90**	**90**	99	102	102	95	101
Bennington	PA	4	364	138	119	96	115	**84**	86	100	108	91	85	97
Athadome	PE	14	75	146	163	128	136	**110**	123	132	151	120	126	131
Icefield	PE	20	319	98	101	101	103	**89**	90	97	115	89	89	97
Lake Louise	PE	16	78	107	108	102	98	**92**	93	**92**	103	88	90	99
Larch Valley	PE	11	95	101	121	92	96	71	70	**64**	79	60	63	73
Larch Valley	AL	17	39	70	87	65	77	**57**	63	58	69	59	66	75
Larch Valley	LL	29	112	94	113	101	87	90	72	**71**	121	76	87	92

[1] AL = *Abies lasiocarpa*, LL = *Larix lyallii*, PE = *Picea engelmannii*, PA = *Pinus albicaulis*.
[2] Number of cores and mean tree age in 1790 (mean age may be underestimated as pith is not always present). Highlighted values are local minima for the 1813-17 period.

Figure 3: The Bennington Pine site, view north across the north lateral of Bennington Glacier (July 1986); the main sampled area in the centre of the photo contains many standing snags and trees over 500 years old. The spruce and fir flanking the moraine are much younger in age. The ridgecrest is about 1000 m above the valley floor.

Figure 4: View south across the forefield of the Athabasca Glacier to the lateral moraines of Athabasca Glacier (left) and Dome Glacier (right). The Athadome site is near the small group of trees at (a), the Icefields site is visible in the foreground, north of the Icefields Parkway.

Figure 5: View west across the forefield of the Athabasca Glacier to the Icefields site (area b). Note the well-marked trimline denoting the Little Ice Age limit (about 1842 A.D.) of the Athabasca Glacier against the slope.

Figure 6: Part of the main stand sampled at Larch Valley. View north toward Sentinel Pass with Mount Temple (right).

chronologies show considerable differences in the amplitude and nature of tree-ring response at these sites. This may be attributed to a number of factors such as site-to-site differences in microclimate or other site factors, differences in response between species (e.g., the Larch Valley sites), differences in vigour between sites due to age (e.g., Athadome and Icefield), differences in standardization procedure and length of record used to derive indices (compare for example, Larch Valley fir and Athadome spruce). Despite this diversity, a number of common elements also occur. Except for the two most northerly spruce chronologies, 1799 and 1824 are readily identifiable (Figures 7 and 8) as conspicuous, narrow marker rings that bracket the period of particular interest here [in fact, the 1824 ring was missing in 27 of the 35 larch cores measured at Larch Valley, (Colenutt 1988)]. Generally the records (Figure 7) can be divided into three parts: a period of relatively high growth, particularly between about 1790 and 1810; a period of declining growth, usually from about 1810 to 1820 or 1830; and a period of low growth or general recovery thereafter. The relative intensity, timing and magnitude of the decline varies between sites but it is particularly marked in the Larch Valley, Lake Louise and Bennington pine chronologies. At Bennington the oldest pine is not included in this chronology because, following the sharp decline in ring width in the early nineteenth century, the post-1820 rings are too narrow to measure accurately with densitometry.

Several authors, particularly Jacoby and coworkers (Jacoby *et al.* 1985, 1988; Jacoby and D'Arrigo 1989; Ivanciu and Jacoby 1988) have reported an abrupt cooling in the early 1800s based on high-latitude North American tree-ring series and other data. This decline is shown to some extent by all of the alpine treeline chronologies reported here. At most sites 1816 occurs in the middle, or at the end of, this period, and is not a marked departure from the trend. Detailed examination of the index values of these chronologies (Table 2, Figure 8) shows that only one of the nine chronologies (Larch Valley spruce) has a significantly narrower ring in 1816. Two others have local minimum values in 1816, but these have similar values to preceding rings in 1814 (Lake Louise) and 1815 (Larch Valley larch). Based on these data, although 1816 is often represented by a narrow ring, the 1814 or 1815 rings are narrower - a fact that cannot be attributed to the Tambora event. It is not possible, therefore, to detect a marked decline in growth in 1816 from the ring-width records at these sites.

Filion *et al.* (1986), Parker (1985), Jacoby *et al.* (1988) in northern Quebec and Jones *et al.* (1988) in Europe report that the 1816 tree ring is distinctive because of its light latewood and low maximum-density values. Figure 9 shows the available (three) maximum-density indexed chronologies for the sites previously discussed. Although 1813 appears to be a significant marker ring, none of these three chronologies show light marker rings associated with 1815, 1816 or 1817. In fact, 1816 appears to have a greater maximum density than adjacent years at these sites suggesting that, if anything, conditions may have been a little warmer than adjacent years (Parker and Henoch 1971; Luckman *et al.* 1985).

Several of the papers in this volume draw attention to the possible effects of the 1783 Laki eruption and, in preparing this paper, the diagrams were extended to 1780 to include this period. The data (Figures 7, 8) show considerable variability in the 1780s but Table 3 indicates that 1784 or 1785 is the narrowest ring for the 1780-89 decade in seven of the nine chronologies (Larch Valley spruce and fir chronologies have slightly lower values in 1782). 1784 is narrower than 1783 in all chronologies and, except for the two northernmost spruce chronologies, this decrease is marked (7-44%). However, the relative widths of tree rings representing 1784 and 1785 are inconsistent: at four sites 1785 is much narrower; two sites have 1784 significantly narrower (including Larch Valley larch, which has a missing ring); and the indexed values are similar at

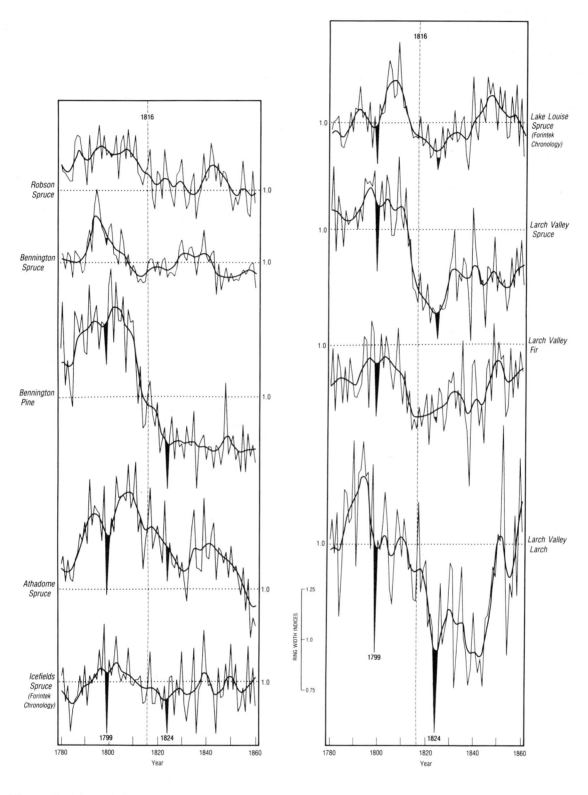

Figure 7: Ring-width chronologies (1780-1860) for nine sites in the Canadian Rockies. The ring-width series are standardized to a mean of 1.0 over the entire period of record (200-600 years; Table 1) and are plotted at the same scale. The lighter line is the chronology; the thicker line is a 13-year low-pass filter (see Fritts 1976).

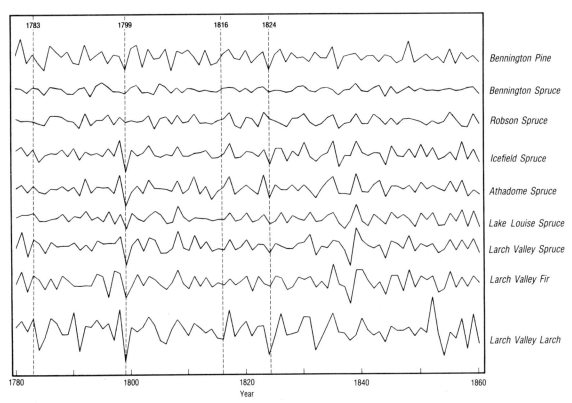

Figure 8: 13-year high-pass filter of ring-width chronologies for nine Rocky Mountain tree-ring sites. These data are standardized and plotted at the same scale. These data correspond to the deviations from the low-pass filter curve (Figure 7). 1799 and 1824 are significant narrow marker rings at most sites. 1783 is the date of eruption of Laki in Iceland.

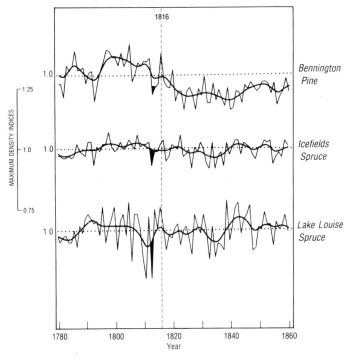

Figure 9: Standardized maximum density (MXD) chronologies for three sites in the Canadian Rockies. All are plotted at the same scale. The thin line is the annual indexed value; the thicker line is a 13-year low-pass filter (Fritts 1976). The shaded year is 1813.

Table 3: Indexed Ring-Width Values for the 1780-89 Period for Selected Treeline Chronologies in the Canadian Rockies.[1]

Site	Species	1780	1781	1782	1783	1784	1785	1786	1787	1788	1789
Robson	PE	118	110	109	107	**105**	**105**	125	128	127	119
Bennington	PE	105	104	97	105	103	**92**	106	104	102	95
Bennington	PA	114	147	105	118	103	**95**	143	138	132	141
Athadome	PE	110	116	123	114	105	**104**	114	118	128	124
Icefield	PE	100	107	90	98	**75**	86	91	93	103	93
Lake Louise	PE	94	100	104	103	88	**87**	91	95	105	95
Larch Valley	PE	105	132	**89**	117	109	92	110	103	105	106
Larch Valley	AL	74	98	**67**	95	88	70	89	71	93	86
Larch Valley	LL	99	112	95	116	**64**	89	127	125	110	113

[1] For abbreviations and further details see Table 2. The highlighted values are the minimum ring-width in each series for this decade.

the three other sites. Generally, these data indicate that the two years following the 1783 Laki eruption are considerably narrower, but the pattern is not consistent enough to use 1784 or 1785 as a marker ring (Figure 8).

Conclusions

In this paper we have examined nine ring-width chronologies and three maximum-density chronologies from four different species at treeline sites in the central Canadian Rocky Mountains. Although the 1810-20 decade showed a marked decrease in ring-width at all sites, probably as a result of climatic cooling, there is no indication of a marker (narrow) ring associated with 1816 or 1817 in the years following the Tambora eruption. Examination of the three maximum-density series available for these sites indicated no light latewood rings in 1815, 1816 or 1817. It would therefore appear that, unlike sites east of Hudson Bay, the treeline sites we have examined have no distinctive tree-ring signal to suggest significantly poorer growth conditions in 1816 or 1817. A major growth decline is identified at all sites during 1810-20 but, as that decade begins some years prior to the Tambora eruption, that eruption cannot be its principal cause. This significant period of declining ring-width has been identified elsewhere in North America, and reflects the most abrupt deterioration in climatic conditions during the last few centuries.

Acknowledgements

We thank: the Natural Sciences and Engineering Research Council of Canada for support of this research; Parks Canada and Mount Robson Provincial Park staff for permission to carry out research at these sites; L. Jozsa, Forintek Canada Corporation, for assistance in the field and in processing density data; F.F. Dalley (1980), G. Frazer (1981, 1983), S. Ulansky (1982), J. Hamilton (1983-84), D.C. Luckman (1985-87), D. McCarthy (1986), R. and S. Colenutt (1987) for coring assistance; and M.I. Johnson, H. Watson, K. Harding, B. Schaus, J. Hamilton and G. Frazer for ring-width measurements.

References

Baillie, M.G.L. and M.A.R. Munroe. 1988. Irish tree rings, Santorini and volcanic dust veils. *Nature* 332:344-346.

Colenutt, M.E. 1988. Dendrochronological studies in Larch Valley, Alberta. B.Sc. thesis, Geography Department, University of Western Ontario, London, Ontario. 123 p.

Filion, L., S. Payette, L. Gauthier and Y. Boutin. 1986. Light rings in subarctic conifers as a dendrochronological tool. *Quaternary Research* 26:272-279.

Fritts, H.C. 1976. *Tree Rings and Climate*. Academic Press, New York.

Graybill, D.A. 1982. Chronology development and analysis. *In: Climate from Tree Rings*, M.K. Hughes, P.M. Kelly, J.R. Pilcher and V.C. LaMarche Jr. (eds.). Cambridge University Press, Cambridge. pp. 21-28.

Hamilton, J.P. 1984. The use of densitometric tree-ring data as proxy for climate at Lake Louise, Alberta. B.A. thesis, Geography, University of Western Ontario, London, Ontario. 116 pp.

Heusser, C.J. 1956. Postglacial environments in the Canadian Rocky Mountains. *Ecological Monographs* 26:263-302.

Ivanciu, I.S. and G.C. Jacoby. 1988. An abrupt climatic cooling in the early 1800s as evidenced by high-latitude tree-ring data. *In: The year without a summer? Climate in 1816*. An International Meeting Sponsored by the National Museum of Natural Sciences, Ottawa, 1986. Abstracts. p. 29.

Jacoby, G.C. and E.R. Cook. 1981. Past temperature information inferred from a 400-year tree-ring chronology from Yukon Territory, Canada. *Arctic and Alpine Research* 13:409-418.

Jacoby, G.C. and R. D'Arrigo. 1989. Reconstructed northern hemisphere annual temperature since 1670 based on high-latitude tree-ring data from North America. *Climatic Change* 14:39-59.

Jacoby, G.C., E. Cook and L.D. Ulan. 1985. Reconstructed summer degree days in central Alaska and northwestern Canada since 1524. *Quaternary Research* 23:18-26

Jacoby, G.C., I.S. Ivanciu and L.D. Ulan. 1988. A 263-year record of summer temperature for northern Quebec reconstructed from tree-ring data and evidence of a major climatic shift in the early 1800s. *Palaeogeography, Palaeoclimatology, Palaeoecology* 64:69-78.

Jones, P.D., K.R. Briffa and T.M.L. Wigley. 1988. Climate over Europe during the summer of 1816. *In: The year without a summer? Climate in 1816*, An International Meeting Sponsored by the National Museum of Natural Sciences, Ottawa, 1988. Abstracts. p. 34.

Jozsa, L.A., E. Oguss, P.A. Bramhall and S.G.Johnson. 1983. Studies based on tree ring data. Report to Canadian Forestry Service, Forintek Canada Corporation, 33 pp.

LaMarche, V.C. Jr. and K. Hirschboeck. 1984. Frost rings in trees as records of major volcanic eruptions. *Nature* 307:121-126.

Lamb, H.H. 1970. Volcanic dust in the atmosphere: with a chronology and assessment of its meteorological significance. *Philosophical Transactions of the Royal Society of London* A266:425-533.

Luckman, B.H. 1982. Little Ice Age and oxygen isotope studies in the Middle Canadian Rockies. Report to Parks Canada, Ottawa. 31 pp.

_____. 1988. Dating the moraines and recession of Athabasca and Dome glaciers, Alberta. *Arctic and Alpine Research* 20:40-54.

Luckman, B.H., J.P. Hamilton, L.A. Jozsa and J. Gray. 1985. Proxy climatic data from tree rings at Lake Louise, Alberta: a preliminary report. *Geographie physique et Quaternaire* 39:127-140.

Luckman, B.H., L.A. Jozsa and P.J. Murphy. 1984. Living seven-hundred-year-old *Picea Engelmannii* and *Pinus albicaulis* in the Canadian Rockies. *Arctic and Alpine Research* 16:419-422.

McCarthy, D.P. 1985. Dating Holocene Geomorphic Activity of Selected Landforms in the Geikie Creek valley, Mount Robson Provincial Park. M.Sc. thesis, University of Western Ontario, London, Ontario. 304 pp.

Parker, M.L. 1985. Investigating the possibility of a relationship between volcanic eruptions and tree growth in Canada. *In: Climatic Change in Canada 5*. C.R. Harington (ed.). *Syllogeus* 55:249-264.

Parker, M.L. and W.E.S. Henoch. 1971. The use of Engelmann spruce latewood density for dendrochronological purposes. *Canadian Journal of Forest Research* 1:90-98.

Parker, M.L., L.A. Jozsa, S.G. Johnson and P.A. Bramhall. 1981. Dendrochronological studies of the coasts of James Bay and Hudson Bay. *In: Climatic Change in Canada 2*. C.R. Harington (ed.). *Syllogeus* 33:129-188.

Watson, H.M. 1983. A dendrochronological study of two sites in Mount Robson Provincial Park. B.A. thesis, Geography Department, University of Western Ontario, London, Ontario. 147 pp.

How Did Treeline White Spruce at Churchill, Manitoba Respond to Conditions around 1816?

David C.F. Fayle[1], Catherine V. Bentley[2] and Peter A. Scott[3]

Abstract

Annual radial increment throughout the stem, and height increment of individual white spruce trees at Churchill, Manitoba were reconstructed through measurement of ring widths on sections taken at close intervals throughout the stem. For the period around the eruption of Tambora in 1815, four trees each from open-forest and forest-tundra sites provided data. On each site, one tree was less than 1 m in height in 1816, the others ranging from 3 to 6 m. Growth of the larger trees, as indicated by height and radial increment, was generally declining over the two decades prior to 1816. In the upper stem, particularly of the forest-tundra trees, radial increment was least in 1818. Effects were less severe in the lower stem and recovery in open-forest trees had begun in 1818 after a low in 1817. Net-height gain of the forest-tundra trees during 1816-20 was one-third that of the previous five years, whereas in open forest trees it more than tripled relative to reduced growth in the previous five years. In combination with the radial-increment data, this suggests the occurrence of conditions in 1816, or possibly late summer of 1815, that led to damage of the terminal bud and upper crown with loss of foliage and (or) reduction of foliar efficiency and production of new foliage. Such effects were much less severe on open-forest trees. The decline in overall tree growth was statistically significant in 1817-18 compared with the variability in tree growth for 10 years prior to 1815. Comparisons made with the period around 1835 (eruption of Coseguina) show subsequent growth reductions were greater than after Tambora.

Introduction

The relationship between climatic variability and tree-ring widths is often difficult to establish and unclear at best. However, this relationship can be somewhat clarified by sampling climate-sensitive trees found in treeline areas where the annual energy deficit is restricting to growth (e.g., Jacoby and Ulan 1981); inclusion of other tree growth parameters may add considerable information from the outset (for a review see Fritts 1976).

A severe climatic anomaly that coincides with a large volcanic explosion, such as reported for the eruption of Tambora during 1815 (Rampino and Self 1982; Catchpole 1985; Parker 1985; Wilson 1985), offers an opportunity to identify anomalous patterns of tree growth that coincide with the event (Parker 1985; Filion *et al.* 1986; Lough and Fritts 1987). By examining representative samples of tree populations, damage to the forest can be assessed which not only indicates the climatic impact of such an eruption, but also reveals information on how the climatic conditions may influence the forest environment.

---

[1] Faculty of Forestry, University of Toronto, Toronto, Ontario M5S 1A1, Canada.
[2] R.R. 1, P.O. Box 22, Churchill, Ontario L0L 1K0, Canada.
[3] Department of Zoology, University of Toronto, Toronto, Ontario M5S 1A1, Canada.

Methods

The field methods and development of the subsequent tree-growth index have been documented elsewhere (Scott *et al.* 1988). Briefly, 11 white spruce [(*Picea glauca*) (Moench) Voss)] and two tamarack [*Larix laricina* (Du Roi) K. Koch] were harvested in 1982 from five sites near the treeline at Churchill, Manitoba (58°45'N, 94°04'W). The trees were all open grown and ranged in age from 88- to 347-years old near their bases. Where identifiable in the upper stem, the lengths of annual height increments were measured and cross sections cut from their mid-point. Elsewhere, sections were cut at 10-cm intervals throughout the stem, except where branches were present. The sections were air dried, sanded and the ring widths measured to 0.01 mm on the four cardinal directions with a Holman DIGIMIC (Fayle *et al.* 1983).

Specific volume increment (SVI) was used as a measure of the metabolic activity for a tree in each year (Shea and Armson 1972). This is the annual volume of wood produced relative to the surface area of the cambium that produced it (Duff and Nolan 1957); mathematically SVI is the average width of the growth layer. An advantage of SVI is that it is not a unidimensional parameter, such as ring width, because it integrates both diameter and height. Furthermore, since the reference point is a unit area of cambium, a common base is provided for comparison between trees.

To develop the tree-growth index, the SVI series for each tree was standardized with a robust estimator (Draper and Smith 1981) using a negative exponential or, in the case of negative indices, a straight line of negative slope or through the average. The standardized SVIs were converted to ratios of the individual growth curves and then averaged to produce the final growth index.

The final growth index was based on all trees sampled. However, only four of the white spruce from the open forest and four from the forest-tundra were present around 1816. Three from each type were greater than 3 m in height at that time and were used to analyze radial-longitudinal patterns of increment in relation to possible influences of climate. The fourth tree from each type was less than 1 m in height around 1816, and did not provide sufficient information for this particular purpose.

Results and Discussion

All of the trees show a decline in SVI of varying magnitude either during 1815 or in 1816 which persists for one to three years following (Figure 1, top and centre). The individual lags in response and magnitude do not allow for immediate conclusions regarding conditions during the summer of 1816. However, if we examine the overall status of the regional tree-growth index 10 years prior to 1815, the growth during 1817 and 1818 is below the 95% confidence interval from what would be expected (Figure 1, bottom). The inference that a volcanic eruption may influence tree growth is strengthened by repeating the confidence interval test for the 1835 period. The eruption during 1835 of Coseguina, which is much closer to Churchill than Tambora, may have had more potential for a stronger impact. In fact the 1835 period is the largest sudden decline in growth throughout the 1710-1982 period of the index.

The cumulative net-height growth patterns for the individual open-forest trees do not indicate any consistent deleterious effect subsequent to 1815 (Figure 2a). Indeed, height increment appeared to be slowing down during the previous decade and recovered shortly thereafter (Figure 2b). In contrast, net-height growth was affected in the forest-tundra trees, where it was reduced for several years before recovering in the 1820s. The greater loss of terminal growth on the forest-tundra than on the open forest trees is reinforced by the similarity in pattern after 1835 (Figure 2b); a substantial net gain in height did not occur on the forest-tundra trees for two decades.

The yearly longitudinal distribution of ring width throughout the tree stems for 1815-20 shows that changes did not occur uniformly (Figure 3). Reductions from 1815 to 1818 were greatest in the upper rather lower stem, and more severe on the forest-tundra than on open-forest trees. The occurrence in the upper part of the growth layers of a 'bulge' in ring width, such as in 1819 for A1 and W2, may be the influence of lateral-branch development and (or) an increase in foliar amounts following damage to the current terminal or existing foliage.

Minimum widths throughout the stem of the average forest-tundra tree occurred in 1818 with a 58%, 50% and 27% reduction compared to 1815 in the upper 0.5 m, upper 0.5-2.0 m, and basal 0.5-2.0 m respectively (Figure 4). In the average open-forest tree, the minimum occurred in 1818 in the upper stem, but recovery was underway in the lower stem with the minimum occurring in 1817; reductions during 1815-18 were 24%, 14% and 3% for the upper 0.5 m, upper 0.5-2.0 m and basal 0.5-2.0 m respectively.

The reductions in height growth and in ring width in the upper stem subsequent to 1815 and 1835 (Figure 3) indicate damage to, or loss of, the terminal buds and of foliage. The contribution of photosynthates and growth hormones by a branch to stem growth is related to the amount and proportion (by age) of the foliage it bears, the distance of this foliage from the stem and the amount of light it receives. A relatively short branch system with a high proportion of well-lit young foliage will make a high contribution to stem growth.

In white spruce, the number of new needles that will be produced in the current year, and the potential shoot elongation were determined when the bud was formed in the late summer of the previous year (Owens et al. 1977). The degree of elongation and production of photosynthate are determined by conditions in the current year. Needles can be retained for 10-15 years at Churchill but there is a loss of photosynthetic efficiency with age. Current, one- and two-year- old foliage may contribute as much as 60% of the total assimilation in white spruce (Clark 1961). The influence of a favourable or unfavourable part of or whole growing season will therefore not only have different, but also lag effects, on growth, which can be compounded if there is a physical loss of new needles or premature loss of old needles.

We have found that the loss of the terminal bud or shoot in treeline white spruce at Churchill is a common phenomenon often occurring at the same time as reduced width of the growth layer throughout the tree, but particularly in the upper stem. Poor growth occurs for several years while a lateral bud or branch establishes itself as the new terminal. Loss can come about through direct or indirect causes. An example of the latter could occur if the roots remain frozen while growth is under way (e.g., Scott et al. 1987), leading to desiccation and death of the needles and buds in the entire upper part of the tree (Sakai 1970; Kullman 1988). The proportion affected will determine the degree of growth reduction and, in combination with ongoing climatic conditions, the length of time to full recovery. The difference between open-forest and forest-tundra trees may be in the more exposed nature of the latter and the longer retention of older needles on the former, which provides a greater reserve. It is clear, from the slow recovery after the decline in growth during 1835, that many trees were apparently damaged this way, although there is little evidence of this occurring subsequent to 1815.

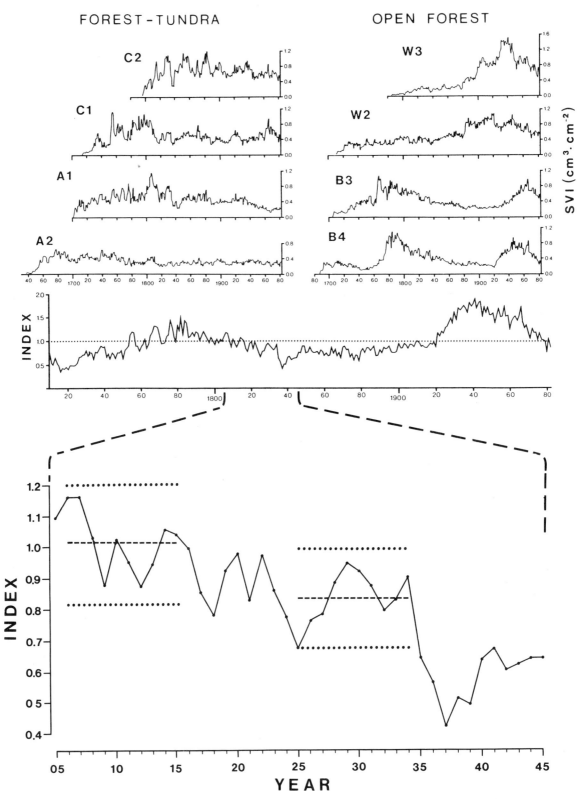

Figure 1: The specific volume increments (SVI) of the four open-forest and forest-tundra white spruce (top) that compose the growth index during the period around 1816. (This index, shown in the centre, is based on the SVIs of 13 trees.) Enlargement of the 1805-45 years (bottom) includes the mean (dashed line) and upper and lower (dotted line) 95% confidence limits for the 10-year periods prior to the 1815 eruption of Tambora and the 1835 eruption of Coseguina, to show that the years following these eruptions exhibit unusually poor growth.

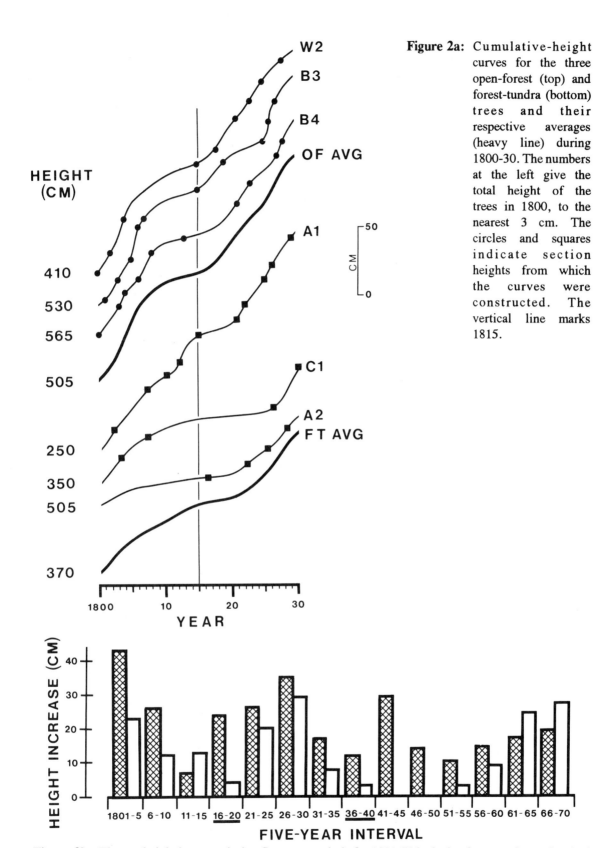

Figure 2a: Cumulative-height curves for the three open-forest (top) and forest-tundra (bottom) trees and their respective averages (heavy line) during 1800-30. The numbers at the left give the total height of the trees in 1800, to the nearest 3 cm. The circles and squares indicate section heights from which the curves were constructed. The vertical line marks 1815.

Figure 2b: The net-height increase during five-year periods for 1801-70 inclusive for open-forest (hatched bar) and forest-tundra (open bar) trees. The five-year periods immediately following the eruptions of Tambora and Coseguina are underlined.

285

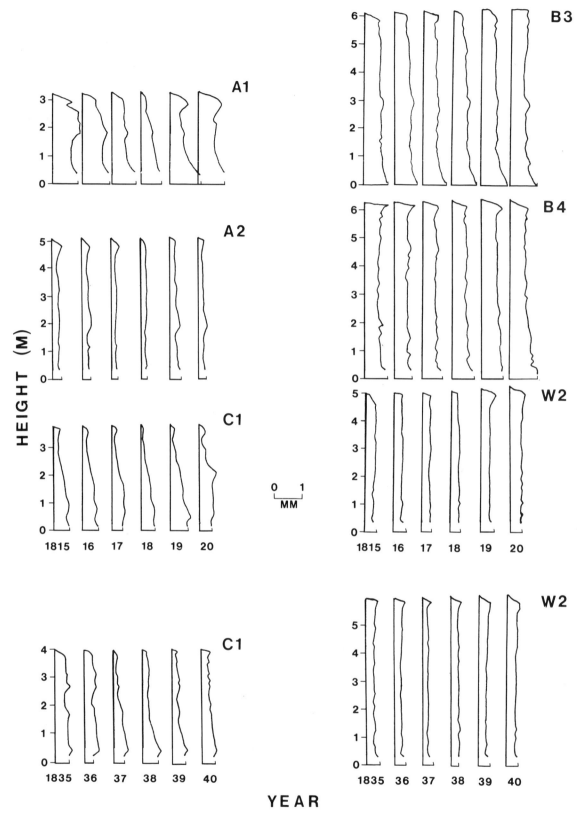

Figure 3: The width of the growth layer (average of four radii) for the three open-forest (OF) and three forest-tundra (FT) trees during 1815-20, and for one example of each for 1835-40.

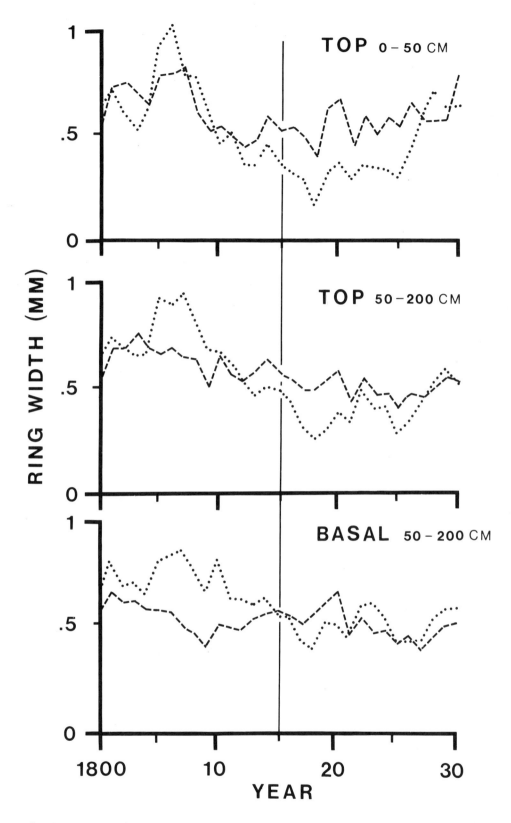

Figure 4: The average ring width for the 0-0.5 m and 0.5-2.0 m intervals from the contemporary apex, and for the 0.5 - 2.0 m interval above the stem base, for open-forest (dashed line) and forest-tundra (dotted line) trees. The vertical line marks 1815.

The fact that radial growth is least in 1818 rather than 1816 is probably the result of cumulative effects arising from adverse conditions in 1816 or late summer of 1815. We suggest that the occurrence of conditions at that time led to damage of the terminal bud and upper crown with loss of foliage and (or) overall reduction of foliar efficiency and production of new foliage. Adverse conditions in August, after height increment was completed in either or both years, would reduce the amount of foliage produced the following year. Recovery of the open-forest trees below the apical 0.5 m in 1817, and a lesser reduction from 1817 to 1818 than from 1816 to 1817 in the forest-tundra trees, suggests that growing conditions were improving in late 1817. The reduced growth of the forest-tundra trees in 1818 may have been due to the cumulative effects of reduced foliar area, particularly in younger age classes, rather than adverse growing conditions *per se*. All trees showed improved radial growth in 1819 suggesting favourable conditions for bud development existed in 1818 (Figure 5).

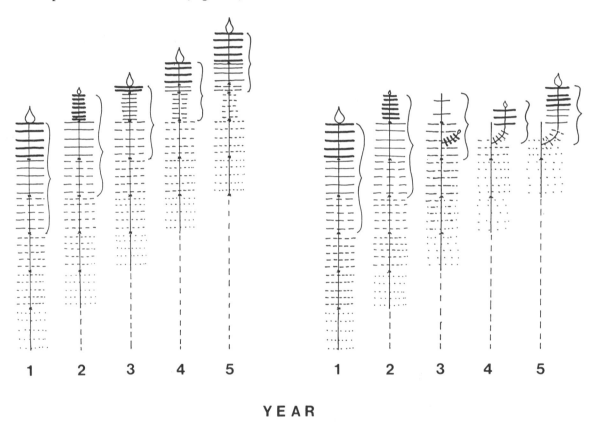

YEAR

Figure 5: Diagrammatic presentation of reduction in shoot growth due to adverse effects on bud development, shoot and needle elongation without (left) and with (right) damage to the terminal bud, shoot and needles. Horizontal lines represent needles of different age classes, from current year (heavy line) to five-years old (dotted line). Older years are not shown. The brackets indicate the needle classes that would normally contribute the bulk of photosynthate.

Year one (e.g., 1815) shows normal growth and bud development. In year two unfavourable conditions throughout the growing season restrict shoot and needle elongation and bud formation. Incipient damage to the bud and needles may occur. In year three, growing conditions are more normal and needle elongation and bud formation are not restricted, but the amount of new foliage is reduced due to previous adverse conditions. Where damage occurred, a lateral bud may begin to assume dominance, but its small size has restricted the amount of needles produced. In year four, growth and development are near normal. Where damage had occurred, the quantity of photosynthetically-efficient foliage is still low and ring width is minimal. Recovery occurs in year five here, whereas it was already underway in the undamaged shoot.

The above scenario is complemented by the observations of Filion *et al.* (1986) who reported a high occurrence of 'light rings' in 1816 and 1817 in krumholz black spruce in northern Quebec. We have not had the opportunity yet to determine their presence in our trees. If they do occur, which is likely, unfavourable conditions in the late part of the growing season and (or) a shortage of photosynthate, the result for example of needle loss, are suggested. In the 'light rings' illustrated by Filion *et al.* (1986), the last formed tracheids in the annual ring show a normal, narrow radial diameter but wall thickening is minimal. A supply of photosynthate is required to complete the process of wall thickening and environmental conditions must permit the completion of the maturation process for normal latewood formation.

From the examination of tree growth during 1815 and particularly 1835, it appears that a stochastic event, such as a volcanic eruption, occurring many thousands of kilometres away may have a widespread detrimental effect on forest productivity. Climatic conditions at Churchill following the eruption of Tambora in 1815 and Coseguina in 1835 did have adverse effects on growth of white spruce. Correspondingly, Wilson (1985) reports that, on the east side of Hudson Bay, conditions were poor during the late summer of 1815, during 1816 and possibly the first half of 1817. Similarly, while it is apparent that 1816 was not truly without a summer at Churchill, it may have been one of 5°C temperatures instead of the long-term average of 10°C.

Acknowledgements

We thank Ed Cook and Gordon Jacoby for helpful advice and supplying some analysis programs. Ring-width measurements were made using facilities of the Ontario Tree Improvement and Forest Biomass Institute, Ontario Ministry of Natural Resources, with grants supplied to the authors by Environment Canada and to P. Scott by Indian and Northern Affairs Canada. NSERC provided travel funds for D. Fayle. We also thank Roger Hansell for his help in the project and C.R. Harington for his support.

References

Catchpole, A.J.W. 1985. Evidence from Hudson Bay region of severe cold in the summer of 1816. *In: Climatic Change in Canada 5*. C.R. Harington (ed.). *Syllogeus* 55:121-146.

Clark, J. 1961. Photosynthesis and respiration in white spruce and balsam fir. *Syracuse University, State University College of Forestry Technical Publication* 85:1-72.

Draper, N. and H. Smith. 1981. *Applied Regression Analysis*. Second Edition. John Wiley & Sons, Inc., Toronto. 709 pp.

Duff, G.H. and N.J. Nolan. 1957. Growth and morphogenesis in the Canadian forest species. II. Specific increments and their relation to the quantity and activity of growth in *Pinus resinosa* Ait. *Canadian Journal of Botany* 35:527-572.

Fayle, D.C.F., D.C. MacIver and C.V. Bentley. 1983. Computer-graphing of annual ring widths during measurement. *The Forestry Chronicle* 59:291-293.

Filion, L., S. Payette, L. Gauthier and Y. Boutin. 1986. Light rings in subarctic conifers as a dendrochronological tool. *Quaternary Research* 26:272-279.

Fritts, H.C. 1976. *Tree Rings and Climate*. Academic Press Inc. New York, New York. 567 pp.

Jacoby, G.C. and L.D. Ulan. 1981. Review of dendroclimatology in the forest-tundra ecotone in Alaska and Canada. *In: Climatic Change in Canada 2*. C.R. Harington (ed.). *Syllogeus* 33:97-128.

Kullman, L. 1988. Subalpine *Picea abies* decline in the Swedish Scandes. *Mountain Research and Development* 8:33-42.

Lough, J.M. and H.C. Fritts. 1987. An assessment of the possible effects of volcanic eruptions on North American climate using tree-ring data, 1602 to 1900 A.D. *Climatic Change* 10:219-239.

Owens, J.N., M. Molder and H. Langer. 1977. Bud development in *Picea glauca*. I. Annual growth cycle of vegetative buds and shoot elongation as they relate to date and temperature sums. *Canadian Journal of Botany* 55:2728-2745.

Parker, M.L. 1985. Investigating the possibility of a relationship between volcanic eruptions and tree growth in Canada (1800-1899). *In: Climatic Change in Canada 5*. C.R. Harington (ed.). *Syllogeus* 55:249-264.

Rampino, M.R. and S. Self. 1982. Historic eruptions of Tambora (1815), Krakatau (1883), and Agung (1963), their stratospheric aerosols, and climatic impact. *Quaternary Research* 18:127-143.

Sakai, A. 1970. Mechanism of desiccation damage of conifers wintering in soil-frozen areas. *Ecology* 51:657-664.

Scott, P.A., C.V. Bentley, D.C.F. Fayle and R.I.C. Hansell. 1987. Crown forms and shoot elongation of white spruce at the treeline, Churchill, Manitoba, Canada. *Arctic and Alpine Research* 19:175-186.

Scott, P.A., D.C.F. Fayle, C.V. Bentley and R.I.C. Hamsell. 1988. Large scale changes in atmospheric circulation interpreted from patterns of tree growth at Churchill, Manitoba, Canada. *Arctic and Alpine Research* 20:199-211.

Shea, S.R. and K.A. Armson. 1972. Stem analysis of jack pine (*Pinus banksiana* Lamb.): techniques and concepts. *Canadian Journal of Forest Research* 2:392-406.

Wilson, C. 1985. The Little Ice Age on Eastern Hudson/James Bay: the summer weather and climate at Great Whale, Fort George and Eastmain, 1814 to 1821, as derived from Hudson's Bay Company records. *In: Climatic Change in Canada 5*. C.R. Harington (ed.). *Syllogeus* 55:147-190.

The Climate of Central Canada and Southwestern Europe Reconstructed by Combining Various Types of Proxy Data: a Detailed Analysis of the 1810-20 Period

J. Guiot[1]

Abstract

In this study, an attempt is made to synthesize various kind of proxy records and to reconstruct complete climatic series. In central Canada, tree-ring series and historical records from the Hudson's Bay Company (i.e., ice-condition and early instrumental data) have been assembled to reconstruct a seasonal temperature and sea-level pressure network back to 1700. The summer of 1816 was among the coldest in the period studied. The beginning of the nineteenth century was also cold, especially after 1807, but the main characteristic is great variability (e.g., 1818 was one of the warmest years since 1700). These results are compared with those obtained in a similar manner for southwestern Europe and northwestern Africa on the basis of tree-ring series, ^{18}O records, wine-harvest and other archival data. The 1810-20 period was also among the coldest of the last millennium, and 1816 was one of the four coldest years since the eleventh century. In the Mediterranean region, this period was far less cold. Some details are given on the method used for these reconstructions. As the proxy series are not homogeneous, particular devices are needed to estimate missing data and to reconstruct low-frequency components. The techniques are adapted from multiple regression, digital filtering, bootstrap analysis and principal component analysis.

The Data in Central Canada

The meteorological network is made up of 67 stations selected from the meteorological database of the Atmospheric Environment Service of Canada. The period of analysis is restricted to 1925-83 and the region is delimited by 61° - 105°W and 47° - 73°N. The monthly data are averaged into seasonal series.

The second Canadian data set is built from the proxy series (Figure 1) available at the date of the study (Guiot 1985a), including:

- freeze-up and break-up dates of rivers entering the western shore of Hudson and James bays: nine series (1714-1871 at maximum) derived by Catchpole and Moodie (1975) and extended to the modern period using recent data (Allen 1977);

- freeze-up and break-up dates of the Red River at Winnipeg, extending from 1798 to 1981: two series build by Rannie (1983);

- monthly temperature data for York Factory (1774-1910) and Churchill Factory (1768-69/1811-58) derived by Ball and Kingsley (1984), spatially and temporally averaged into four seasonal series for the York-Churchill region.

---

[1] CNRS UA 1152, Laboratoire de Botanique Historique et Palynologie, Faculté de St. Jérôme, 13397 Marseille Cedex 13, France.

These data are completed by tree-ring series. The trends of these series are modeled by negative exponentials, polynomials or filtered curves by the authors as proposed by Fritts (1976). Indexed series are obtained by dividing each ring width by its trend. The series used are the following:

- two white spruce ring-width indices series from Nain, Labrador (1769-1973) and Border Beacon, Labrador (1660-1976) from Cropper and Fritts (1981);

- one larch ring-width indices series from Fort Chimo (now Kuujjuaq), Québec (1650-1974) also from Cropper and Fritts (1981);

- two white spruce series from Cri Lake, near Kuujjuarapik, Québec, (1750-1979) (Parker *et al.* 1981), the first being ring-width indices and the second being ring maximum densities;

- two white spruce ring-width indices from Churchill, Manitoba (1691-1982) by P. Scott in Hansell (1984), the first was sampled in open forest and the second in forest-tundra.

Finally a total of 22 proxy series are available to reconstruct temperature in central Canada for 1700-1979.

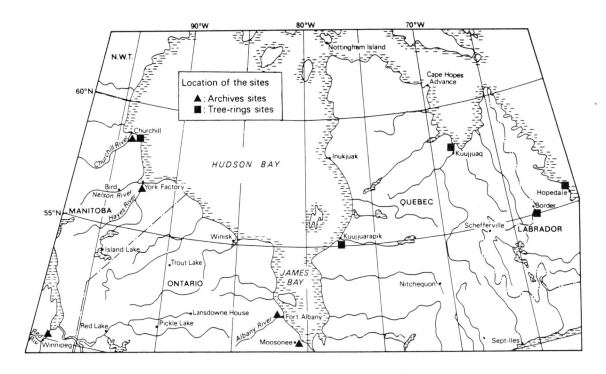

Figure 1: Location of the proxy-series sites in Canada.

The Data in Europe and North Africa

Meteorological data are the annual series gridded by Jones *et al.* (1985) extending from 35° to 55°N by steps of 5°, and from 10°W to 20°E by steps of 10°. So 20 series are available from 1851 to 1984, with missing data mainly before 1900.

The second data set, the proxy series, are collected from the longest proxy series existing for Europe and Morocco (Figure 2). A part of them consists of tree-ring chronologies of various species from various sites. They are also detrended as suggested by Fritts (1976). The set includes:

- oak ring-width series from west of the Rhine, near Trier, Germany (820 to 1964) collected and indexed by Hollstein (1965);

- oak ring-width series from the Spessart forest area (50°N, 9°30'E) in Germany (840 to 1949) collected and indexed by Huber and Giertz-Siebenlist (1969);

- oak ring-width series from Belfast, Northern Ireland (1001 to 1970) collected and indexed by Baillie (1977);

- oak ring-width series from southwestern Scotland (946 to 1975) collected and indexed by Pilcher and Baillie (1980);

- pine ring-width series from southern Italy (1148 to 1974) collected and indexed by Serre-Bachet (1985);

- larch ring-width series from Vallée des Merveilles, southern French Alps (1100 to 1974) collected and indexed by Serre (1978);

- fir ring-width series from Mont Ventoux, southern France (1660 to 1975) collected and indexed by Serre-Bachet (1986);

- pine ring-width series from northern Italy (925 to 1984) collected and indexed by Bebber (personal communication);

- larch ring-width series from Orgère, northern French Alps (1353 to 1973) collected and indexed by Tessier (1981);

- two larch ring-width series from Mercantour, southern French Alps (1701 to 1980 and 1732 to 1981) collected by Guibal (personal communication) and indexed for this study.

Another group of proxy series is composed of data derived from archives. These historical data have been compiled by various historians and/or climatologists:

- decadal temperature estimates of Bergthorsson (1969) for Iceland (1050 to 1550). These data were analyzed by Ogilvie (in Ingram *et al.* 1978), and those before 1170 and after 1450 were reported as unreliable - the unreliable decades are considered as missing;

- summer temperature index of Bray (1982) based on German and French wine-harvest data and central England (Manley) temperatures (1453 to 1973);

- the Pfister (1981) thermal indices in Switzerland, averaged on an annual basis from 1550 to 1829;

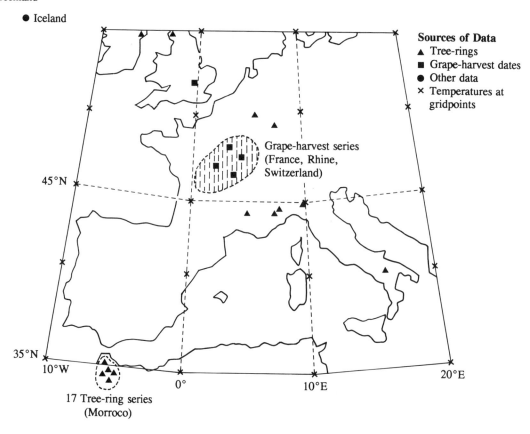

Figure 2: Location of the proxy-series sites in Europe and Morocco.

- the mean annual dates at the beginning of the grape harvest in northeastern France, French Switzerland, and southern Rhineland of Le Roy Ladurie and Baulant (1981) (1484 to 1879);

- the average annual dates at the beginning of the grape harvest in Switzerland reported by Legrand (1979) (1502 to 1979);

- frequency of southwesterly surface winds in England (1340-1978) from direct observations (1669 to 1978) in the London area and from historical proxy data before. These data are reconstructed by Lamb (1982);

A last category is provided by ^{18}O data in the Arctic ice. These isotopic series can be considered as good indicators of temperature, since the condensed vapour is enriched in heavy isotopes:

- Camp Century, Greenland, ^{18}O quasi-decadal values (1200 to 1970) collected and analyzed by Dansgaard *et al.* (1971);

- two isotopic series in central Greenland, 30-year running means of annual maxima of ^{18}O in ice cores compiled by Williams and Wigley (1983) (1180 to 1800).

Finally, to these data are added the first three principal components of the 17 longest cedar ring-width series in Morocco, sampled by A. Munaut and C. Till and analyzed by Till (1985). These series (1068 to 1979) represent nearly 40% of the total variance of the 17 raw series.

The period 1068-1979 is retained for a total of 23 series. To simplify matters, European data will include both European and northwestern African data.

Data Conditioning

The predictand matrix as well as the predictor contains missing data. Therefore it is fairly natural to estimate the gaps before beginning any detailed analysis. For the Canadian data, the method employed to estimate the missing data is explained in Guiot (1985a) and, with more details, in Guiot (1986). The general procedure is similar to that used for the management of the European series, described here.

The best analogues method is used to estimate the missing data of the proxy-series matrix. The main advantage of this method is that we have not to assume any linear relationship between the variables. This is particularly recommended when, like here, the series are highly heterogeneous. The estimate of a missing observation for a given series is provided by the most similar observations (analogues) of the same series within the 1200-1900 interval, the distance between observations being established on the m_{ik} observations of the series available.

$$d_{ik}^2 = \sum_{j=1}^{m_{ik}} (x_{ij} - x_{kj})^2 \tag{1}$$

The observations, denoted by k, available among the 20 best-fit analogues of observation i provide the wanted estimate

$$x_{ij} = \frac{\sum_k x_{kj}/d_{ik}^2}{\sum_k d_{ik}^2} \tag{2}$$

The correlation between estimates and actual values computed on the available data and averaged on the 23 series is 0.73 (ranging from 0.45 to 0.86), which is highly significant. For observations outside the 1200-1900 period, the mean correlation remains high, say 0.60 (ranging from 0.22 to 0.89). It must be noted that we have not estimated any coefficients so that the statistics computed on the calibration interval as well as on the verification one can be considered as independent. The mean and the standard deviations of the estimates are quite close to the actual ones, with discrepancies less than 15% of the mean standard deviation. Depending on the number of degrees of freedom, we can consider that the estimates are reliable.

For the meteorological data matrix, multiple regression was used. This method cannot be applied directly because the number of regressors is not constant on the total calibration interval (1851-1984). If m_i is the number of regressors available for observation i (i.e., with no missing data), the regression equation may be written as follows:

$$\hat{x}_{ij} = a_{oj} + \sum_{k=1}^{m_i} a_{kj} x_{ik} \qquad (3)$$

The correlation between estimates and actual values averages 0.76, with the highest values in the Northwest (0.90). These coefficients are highly significant, but the estimates must be considered as less reliable at the southern margin of the region analyzed (correlation around 0.70).

Extrapolation of the Temperature Series

When the predictor and predictand matrices, are fully determined, it is possible to extrapolate the annual-temperature series from the proxy series, using the common observations to calibrate a relationship. It is advisable first to transform the raw series into principal components (PCs). Indeed, a large proportion of the high order PCs represents extremely small proportions of variance, so that they can be assumed to be indistinguishable from statistical noise.

Reduction of the Number of Variables

For the European annual-temperature series, 10 PCs are used explaining together around 90% of the variance. For the European proxy series, 19 PCs are used explaining 95% of the variance. For the Canadian season temperature series, the first four PCs used explain between 82.5% (for summer) and 91% (for autumn) of the total variance. For the proxy series, the number of PCs depends on the season reconstructed.

Bootstrap Regression

In central Canada, a multiple regression has been employed to calibrate the relationship between climate and proxy series. In Europe, a more sophisticated approach, termed bootstrap regression, seemed more advantageous.

Bootstrapping is a recent technique devised by Efron (1979) to estimate statistics for unknown population distributions by Monte Carlo simulations. The idea is to resample the original observations in a suitable way to construct pseudo-data sets on which the estimates are made. In regression, this is particularly useful when the residuals are non-normal or autocorrelated, or when the data set is too small.

The bootstrap method is in fact a generalization of jackknife replication. The frame of the method can be summarized in a few lines. From the interval (1,n), where n is the size of the original data set, n pseudorandom numbers are randomly taken with replacement using a uniform distribution protocol. These n numbers are used to resample the actual observations. We should insist here on the fact that an observation is the vector of the m proxy data and p climatic parameters corresponding to the same year. The n observations selected in this way provide a pseudo-data set. This is repeated an arbitrary number NC times, and at each time, a regression is computed.

The reliability of a particular statistical model must be assessed by calculating a number of verification statistics measuring the degree of similarity between predictand observations and their estimates for time periods independent of the calibration. So a successful reconstruction is one for which it is demonstrated that independent estimates continue to be accurate at a level greater than would be expected solely by chance. *"The process used to optimize the coefficients of the model virtually ensures that the results will be more accurate for the calibration data than for any other observations to which it may be applied. It is why the decreasing of accuracy should be measured whenever possible."* (Fritts and Guiot 1988). Bootstrap regression enables one to integrate verification in the calibration process and to use the n observations both for the calibration and the independent verification:

- for each of the NC replications, the regression coefficients are computed and applied to proxy series to obtain the corresponding reconstruction;

- the reconstruction is compared to the actual climatic series both on the set of retained observations and on the others; thus verification statistics are calculated NC times;

- the mean and standard deviations of the verification statistics are obtained on the dependent and independent data set;

- the final reconstruction is given by the median of the NC replicated reconstructions, and a 90%-confidence interval is given by the 5th and 95th percentile.

Decomposition of the Spectra into Two Bands

Before computing a bootstrap regression, the predictors and the predictands are filtered, so that their spectra are decomposed into two bands (Guiot 1985b). Once more, the method is illustrated with European data. We use a nine-weights low-pass filter with a cut-off period of seven years. The effect of this filter is illustrated in Figure 3 with the first PC of the proxy series. The complementary high-pass filter enables us to retain the short-term fluctuations of the series. The raw series is the sum of both low-frequency and high-frequency components (Figure 3).

In the two frequency bands, bootstrap regressions are calibrated on the common period, 1851-1979. This method is particularly necessary for the low-frequency components dominated by large autocorrelations, which induce troubles in the interpretation of the fit quality. The "abnormality" of these smoothed data is compensated for by a lot of simulations.

Table 1 presents some statistics useful for the evaluation of the regressions. For each of the 50 simulations, the estimated means and standard deviations are compared to the actual ones on the randomly-drawn observations, as well as on the others. The deviations of these statistics are averaged over the 50 simulations (Table 1). Apparently the standard deviations are slightly underestimated, as expected, and the biases are not greater on the independent observations. Concerning the calibration data set, the correlations between estimated and actual observations are lower for the high-frequency components than for the low-frequency components. Nevertheless this must be appreciated regarding the reduced number of degrees of freedom of autocorrelated series. The most important feature is the lack of stability, appearing in the independent data set, of the high frequencies for components 3 to 6 and 8 to 10, while the low-frequency components estimates are quite stable. This justifies the spectral decomposition.

The regression coefficients are applied in each band to extrapolate the 10 temperature PC series back to 1068. The entire spectra are recomposed by adding the reconstructed high-frequency PC series to the low-frequency ones. Table 2 presents the effect of this addition for two periods in the calibration period. During the first period (1851-1900), temperature observations are less abundant and of lesser quality than during the second (1901-79). The means and standard deviations in the older period appear to be systematically more underestimated than in the more recent one by a factor of 2. These underestimates are negligible for the means: they represent between 0.7 and 2.5% of the total variance of the PCs (that is 2.10^4). The biases are higher for

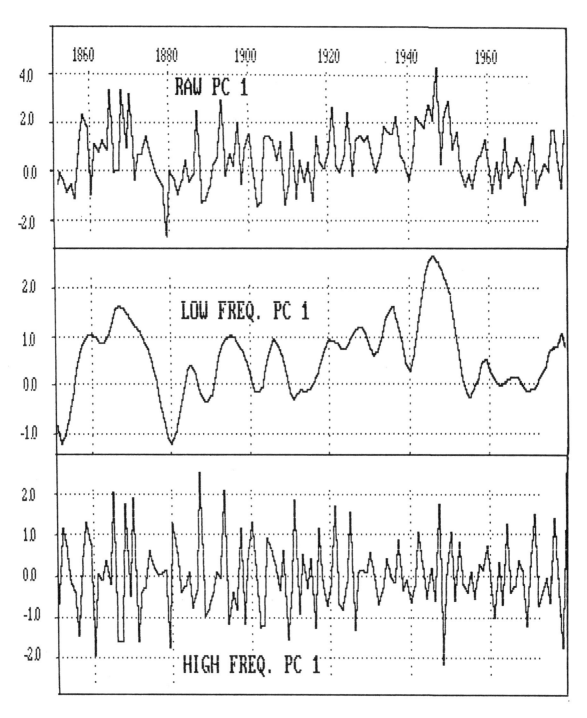

Figure 3: The first principal component of the European proxy series (1850-1979) and its spectral decomposition.

Table 1: Verification Statistics for the Reconstruction of the Low- and High-Frequencies Component of the First 10 PCs of the Annual Temperatures. [These statistics are averaged on the 50 replications: (a) on the calibration observations (randomly drawn); (b) on the others. dM = estimated mean minus actual mean; dS = estimated standard deviation minus actual standard deviation; R = correlation coefficient between estimates and actual variables $\pm$ 1 standard deviation.]

						Low Frequencies					
Var.		1	2	3	4	5	6	7	8	9	10
dM	(a)	0.00	0.00	0.00	0.00	0.00	0.00	0.00	0.00	0.00	0.00
	(b)	-0.01	0.01	-0.01	0.00	0.00	0.04	0.01	0.01	-0.01	-0.01
dD	(a)	-0.15	-0.28	-0.30	-0.13	-0.33	-0.21	-0.17	-0.22	-0.34	-0.17
	(b)	-0.15	-0.30	-0.29	-0.11	-0.33	-0.22	-0.17	-0.22	-0.33	-0.19
R	(a)	0.85	0.72	0.70	0.87	0.67	0.79	0.83	0.78	0.66	0.83
	$\pm$	0.03	0.03	0.03	0.02	0.04	0.03	0.03	0.03	0.04	0.03
	(b)	0.74	0.53	0.57	0.83	0.48	0.65	0.75	0.67	0.47	0.75
	$\pm$	0.09	0.09	0.09	0.04	0.10	0.09	0.05	0.07	0.09	0.07

						High Frequencies					
Var.		1	2	3	4	5	6	7	8	9	10
dM	(a)	0.00	0.00	0.00	0.00	0.00	0.00	0.00	0.00	0.00	0.00
	(b)	0.02	-0.01	0.04	-0.01	0.01	0.03	0.00	-0.01	-0.01	0.03
dD	(a)	-0.35	-0.50	-0.44	-0.54	-0.49	-0.62	-0.42	-0.48	-0.57	-0.51
	(b)	-0.29	-0.45	-0.41	-0.54	-0.43	-0.56	-0.40	-0.43	-0.53	-0.49
R	(a)	0.65	0.50	0.56	0.46	0.51	0.38	0.58	0.52	0.43	0.49
	$\pm$	0.04	0.05	0.05	0.06	0.04	0.06	0.04	0.04	0.07	0.05
	(b)	0.48	0.24	0.18	0.12	0.19	-0.10	0.31	0.15	-0.01	0.11
	$\pm$	0.09	0.09	0.11	0.15	0.10	0.11	0.09	0.09	0.11	0.11

the variability: between 10 and 20% of the total variance. These results are clearly better than those obtained without spectral decomposition. With this last method, the correlation gain is about 0.14, and the underestimating factor is divided by 2. We conclude that spectral decomposition increases the quality of fit, but we cannot infer that this best fit is warranted on independent periods. The example dealt with in Fritts and Guiot (1988) nevertheless confirms the stability of such extrapolations.

The reconstruction of the raw-temperature series is obtained by postmultiplying the 10 PC series matrix by the eigenvector matrix. The mean correlation between estimated and actual values is 0.63 (from 0.46 to 0.76) on the 1851-1979 period, with a maximum of more than 0.70 in the Northwest. These reconstructed series are provided with a 0.90-level confidence interval that is 0.3°C in mean.

Analysis of the Reconstructions as a Whole

The reconstructed series in Europe are analyzed from different points of view. The trend of the 16 series, at latitudes ranging from 40° to 55°N is plotted in Figure 4, on the basis of 20-year

Table 2: Verification of the Reconstructions of the 20 Annual-Temperature Series PCs (Multiplied by 100). The sum of the low and high frequencies are first verified, and then the regression, without separating low from high frequencies: (a) actual means; (b) estimated mean; (c) actual standard deviations; (d) estimated standard deviations; (e) correlation between actual and reconstructed PCs. SSD = sum of squared differences [sign (-) means underestimates].

	Low and High Frequencies									
	1901-1979					1851-1900				
	a	b	c	d	e	a	b	c	d	e
PC01	-41	-41	270	187	0.69	87	37	296	195	0.73
PC02	33	8	167	95	0.61	-55	-31	170	89	0.51
PC03	-17	-19	154	88	0.50	12	11	190	85	0.61
PC04	-15	-18	100	84	0.83	27	20	104	41	0.57
PC05	13	-11	69	48	0.37	-24	6	148	75	0.79
PC06	-34	-28	67	43	0.47	69	34	74	46	0.59
PC07	12	-2	58	42	0.68	-3	-1	69	41	0.74
PC08	6	6	62	41	0.62	-16	-17	82	43	0.71
PC09	-7	10	70	38	0.58	-3	-6	80	29	0.47
PC10	-8	-5	49	31	0.56	8	3	66	38	0.72
SSD	(-)1504		(-)20867			(-)5290		(-)43559		
mean					0.59					0.64

	Standard Regression									
	1901-1979					1851-1900				
	a	b	c	d	e	a	b	c	d	e
PC01	-41	1	270	144	.61	87	20	296	160	.65
PC02	33	-1	167	33	.35	-55	-1	170	40	.23
PC03	-17	-21	154	67	.37	12	19	190	52	.58
PC04	-15	-7	100	66	.74	27	14	104	35	.45
PC05	13	-5	69	31	.27	-24	5	148	37	.62
PC06	-34	-2	67	26	.33	69	18	74	36	.41
PC07	12	3	58	31	.62	-3	9	69	27	.58
PC08	6	4	62	28	.47	-16	-14	82	26	.51
PC09	-7	-3	70	21	.38	-3	-9	80	19	.43
PC10	-8	-1	49	18	.27	8	-4	66	18	.60
SSD	(-)4498		(-)53549			(-)11393		(-)83891		
mean					.45					.51

periods. A climate generally colder than now, results interrupted by a few warm periods. A spatial distribution of the anomalies is presented at some key 20-year periods for the 20 gridpoints.

Before 1200, the temperature was extremely low in the whole region. From 1200 to 1400, the analyzed region experienced a relatively warm climate mainly in the Southwest. This warm period is often called the "Little Climatic Optimum" of the Middle Ages. The results for the 1070-1420 period are verified by comparing the indices of Alexandre (1987) that are valuable for Mediterranean and non-Mediterranean western Europe. In order to make these indices representative of both winter and summer, we have used the difference "Winter severity index minus Summer precipitation index" to represent the annual temperature. It confirms that a warming is obvious from the beginning of the thirteenth century, with a mean of -1.8 before 1220 and 0.23 after (Figure 4).

From 1420 to 1460, conditions were very cold, except in North Africa. The warming at the end of the Middle Ages lasted from 1460 to about 1550. The spatial distribution was similar to that of 1200-1400, with a maximum to the West.

The period generally called the "Little Ice Age" seems to have begun about 1500. The first part of this period (1550-1610) was effectively cold mainly in areas along a diagonal extending from the British Isles to Tunisia. The seventeenth century was generally warm in the Southwest. In the Southeast, the cooling started around 1550. In fact, the Little Ice Age really started at the end of the seventeenth century. Temperature was low everywhere except in the Southwest. It lasted until 1860, with two particularly cold periods about 1700 and 1815. It had no equivalent in the Southwest, although the precipitation reconstructions of Till and Guiot (1988) indicate increasing moisture - particularly during these two extreme episodes. Richter (1988) confirms the climatic differences between the southwestern Mediterranean Basin and northwestern Europe. Its reconstruction of summer precipitation from pine tree-ring series shows that 1810-20 was wet in central Spain and its reconstruction of winter temperature shows that the same area was warm. As central Spain is located at the midpoint between Morocco and the rest of western Europe, these reconstructions are simultaneously, a confirmation of our temperature reconstructions and the precipitation reconstructions of Till and Guiot (1988).

The modern warm period began in the mid-nineteenth century, with a maximum between 1930 and 1950 - especially in the Northwest and in the Southeast.

Similar reconstructions for the four seasons have been obtained in central Canada, but only for the last three centuries. The reconstructions are detailed in Guiot (1985a). WI focus here on a comparison with Europe. Figure 5 shows three synchronous long periods on both continents: 1700-50; 1780-1820; and 1850-1920. After the beginning of the twentieth century, the general warming appeared in Canada some five years later than in Europe. The synchronism during the Little Ice Age could mean that it is forced by an external common phenomenon.

The Year 1816

If 1810-20 appears, as a whole, very cold in Europe, it is highly variable in Canada (Figure 5). For example, the summer was very cold in 1816 but it was very hot two years later. On the two continents, the cooling began at the beginning of the nineteenth century. 1816 is only a period where this cooling reached an extreme. If the volcanic eruption of Tambora in 1815 had an influence on the severe climate of the following year, it only accentuated a trend, and its eventual effects must be placed in the context of the cold period of the "Little Ice Age". This trend is noticeable as well in Canada as in Europe. Figure 5 also shows that the summer of 1816 was the coldest of the last three centuries in central Canada. In Europe, we have found three other years as cold as 1816: 1081, 1454 and 1703. The four coldest years of the millennium reached mean anomalies of -1.5°C (in the range 10°W - 20°E and 35° - 55°N).

Central Canada reconstructions are sufficiently precise for a chronology of the cooling in the region to be established. Figure 6 presents the distribution of temperature anomalies in relation to 1950-79. The temperature of the North is a blank because the proxy series used is not representative of latitudes higher than 65°N. Winter 1816 was nearly normal in the whole region studied except in the Southwest where the cooling was already perceptible (negative anomalies

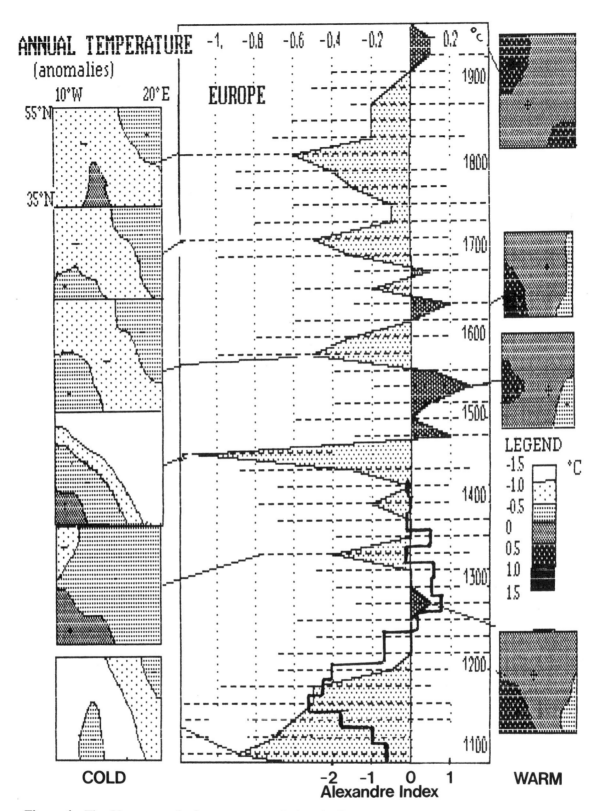

Figure 4: The 20-year trend of temperature variations in Europe (area restricted to latitudes 40°-55°N and longitudes 10°W-20°E). The distribution of the anomalies for some characteristic periods is shown for the total area (including 35°N). The broken horizontal lines represent the 90%-confidence intervals computed by bootstrapping. Between 1070 and 1410, are the smoothed Alexandre (1987) indices representing winter severity and summer precipitation.

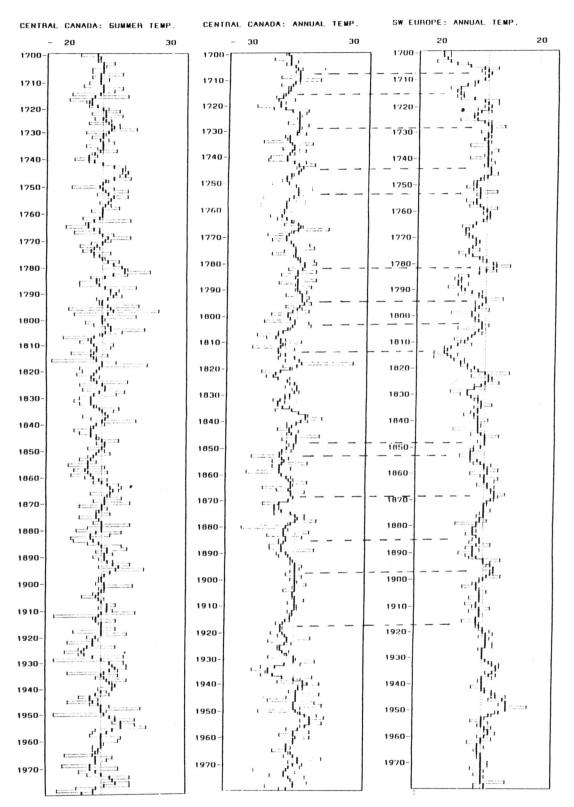

Figure 5: The annual temperature in southwestern Europe and central Canada (both being averages of the individual reconstructed series) and the summer temperature in central Canada. The series are smoothed with a digital filter (cut-off period = seven years).

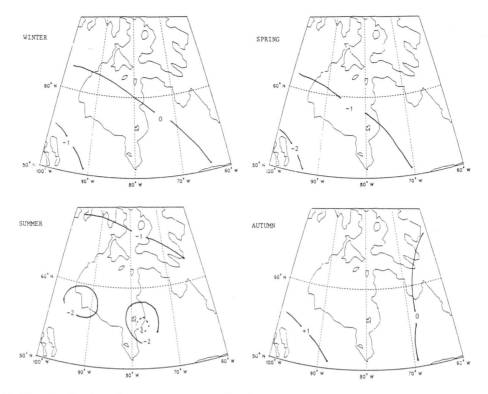

Figure 6: The distribution of temperature anomalies for the four seasons in central Canada.

greater than 1°C). The cooling affected the whole region in spring, with a maximum in the Southwest where negative anomalies of 2°C are reached. In summer, the temperature anomalies were globally -1°C with a minimum of almost -3°C in the region of Kuujjuarapik (southeastern Hudson Bay). A secondary minimum of -2°C occurs in the Churchill region (western Hudson Bay). The Southwest has already begun to warm, since spring anomalies are -2°C and summer ones -1°C. Autumn is nearly normal everywhere except in the Southwest where the positive anomalies are +1°C.

In Europe, because emphasis was laid on the ability to reconstruct temperature series over a millennium, it is impossible to collect a sufficient number of series to obtain a seasonal resolution. Figure 7 is nevertheless instructive respecting the spatial differences of the cooling. Briffa *et al.* (1988; this volume) have already shown that the summer of 1816 in central Europe was less cold than in western Europe. This is confirmed as far as the annual temperatures are concerned, and it is possible to more precisely judge the temperature of the western Mediterranean Basin. The maximum negative anomalies concern the British Isles and northern France (-3°C), but they extend far away to Africa - especially Tunisia. This teleconnection between northwestern Europe and North Africa is a classical synoptic configuration, which is well known nowadays in southern France during "Mistral" and "Tramontane" winds. Northwesterly air masses are canalized under the influence of an anticyclone located over Spain and a low pressure centre over the southern Alps and Gulf of Genova. The winds accelerate and become drier down the Rhone Valley (Mistral) and by invading the area between Pyrénées and Massif Central mountains (Tramontane) so that their influence (when they are exceptionally strong) is sometimes felt in Corsica, and even in North Africa. Perhaps this meteorological situation occurred very often during the summer of 1816. However, the coldness of this year was not general: apparently Morocco and southern Spain were largely influenced by southerly winds since the negative anomalies are lower than 1°C. Central Europe, with a more continental climate, was also less affected by this general cooling.

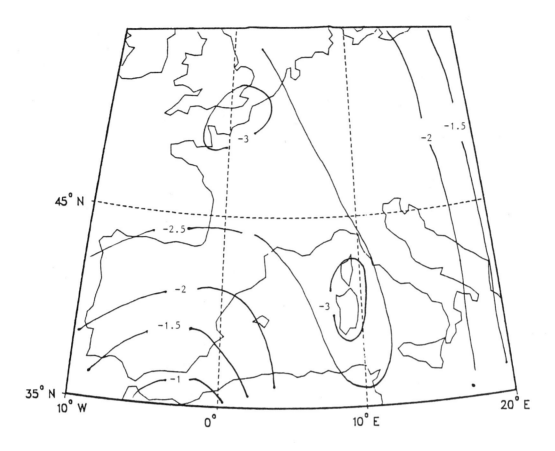

Figure 7: The distribution of the annual temperature anomalies in Europe and northwestern Africa in 1816.

Conclusions

The summer of 1816 was the coldest in the last three hundred years in central Canada. The other seasons have been about normal or slightly colder. The greatest negative anomalies concern the area southeast of Hudson Bay, and the smallest ones the region north of Hudson Bay. In Europe, 1816 was among the coldest years of the millennium with minimum temperatures (anomalies close to -3°C) extending from the British Isles to Tunisia. Northwesterly winds chilled western Europe: these cold air masses accelerated and dried between the Pyrénées and Massif Central mountains (Tramontane wind) and down the Rhone Valley (Mistral wind), crossing the Mediterranean Sea to Tunisia. At the same time, Morocco remained relatively warm. In central Canada, more details are available about 1816 from seasonal records. Apparently the cooling began in winter in the Southwest and ended by the close of summer, whereas it began a season later (in spring) in the East, also ending later (in autumn).

The severe climate characteristic of this year must be placed in context. The cooling began a few years before 1816, at the beginning of the decade. Then aerosols from the volcanic eruption of Tambora only exacerbated a trend already existing in Europe and central Canada. Comparison of results from the two continents shows strong coherency in the low-frequency variations of temperature on both sides of the Atlantic Ocean during this globally-cold period of the Little Ice Age. The coherency is weak in warmer periods.

This study shows how to synthesize various proxy series available to provide a better knowledge of past climatic changes in Europe. More records must be used in order to obtain maximum reliability of the gridded temperature reconstructions. It also appears that the Mediterranean climate, which is now very different from the northern European one, has been so for many centuries. Information concerning northern Europe cannot be directly extended to southern Europe. More proxy series related to the Mediterranean climate must be collected to achieve better reliability.

References

Alexandre, P. 1987. *Le climat en Europe au Moyen-Age*. Ecole des Hautes Etudes en Sciences Sociales. Paris. 825 pp.

Allen, W.T.R. 1977. Freeze-up, break-up and ice thickness in Canada. Fisheries and Environment Canada, *Atmospheric Environment Service* CLI-1-77:1-185.

Baillie, M.G.L. 1977. The Belfast oak chronology to 1001. *Tree-Ring Bulletin* 37:1-12.

Ball, T.F. and R.A. Kingsley. 1984. Instrumental temperature records at two sites in central Canada: 1768 to 1910. *Climatic Change* 6:39-56.

Bergthorsson, P. 1969. An estimate of drift ice and temperature in 1000 years. *Jökull* 19:94-101.

Bray, J.R. 1982. Alpine glacial advance in relation to proxy summer temperature index based mainly on wine harvest dates, 1453-1973. *Boreas* 11:1-10.

Briffa, K.R., P.D. Jones and F.H. Schweingruber. 1988. Summer temperature patterns over Europe: a reconstruction from 1750 based on maximum latewood density indices of conifers. *Quaternary Research* 30:36-52.

Catchpole, A.J.W. and D.W. Moodie. 1975. Changes in the Canadian definitions of break-up and freeze-up. *Atmosphere* 12:133-138.

Cropper, J.P. and H.C. Fritts. 1981. Tree-ring width chronologies from the North American Arctic. *Arctic and Alpine Research* 13:245-260.

Dansgaard, W., S.J. Johnson, H.B. Clausen and C.C. Langway Jr. 1971. Climatic record revealed by the Camp Century ice cores. *In: The Late Cenozoic Glacial Ages*. K.K. Turekian (ed.). Yale University Press, New Haven. pp. 37-56.

Efron, B. 1979. Bootstrap methods: another look at the jackknife. *The Annals of Statistics* 7:1-26.

Fritts, H.C. 1976. *Tree-Rings and Climate*. Academic Press, New York. 567 pp.

Fritts, H.C. and J. Guiot. 1988. Methods for calibration, verification and reconstruction. *In*: *Methods in Tree-Ring Analysis: Applications in the Environmental Sciences*. L. Kairiukstis and E. Cook (eds.). Reidel, Dordrecht. (in press).

Guiot, J. 1985a. Reconstruction of seasonal temperature and sea-level pressure in the Hudson Bay area back to 1700. *Climatological Bulletin* 19:11-59.

_____. 1985b. The extrapolation of recent climatological series with spectral canonical regression. *Journal of Climatology* 5:325-335.

_____. 1986. Reconstructions des champs thermiques et barométriques de la région de la Baie d'Hudson depuis 1700. Environment Canada, Atmospheric Environment Service, *Canadian Climate Centre Report* 86-11F (français) *Rapport* 86-11E (English): 111 pp.

Hansell, R.I.C. (Editor). 1984. Study on temporal development of subarctic ecosystems - determination of the relationship between tree-ring increments and climate. Report to Department of Supply and Services Canada, Atmospheric Environment Service, Downsview. Contract OSU84-00041.

Hollstein, E. 1965. Jahrringchronologische von Eichenholzern ohne Waldkande. *Bonner Jahrbücher* 165:12-27.

Huber, B. and V. Giertz-Siebenlist. 1969. Unsere tausendjährige Eichen-Jahringchronologie durchschnittlich 57 (10-150) fach belegt. *Sitzungsberichte der Osterreichischen Akadamie der Wissenschaften* 178:37-42.

Ingram, M.J., D.J. Underhill and T.M. Wigley. 1978. Historical climatology. *Nature* 276:329-334.

Jones, P.D., S.C.B. Raper, B. Santer, B.S.G. Cherry, C. Goodess, P.M. Kelly, T.M.L. Wigley, R.S. Bradley and H.F. Diaz. 1985. A grid point surface air temperature data set for the northern hemisphere. United States Department of Energy, Contract DE-AC02-79EV10098, Washington, TR022. 256 pp.

Lamb, H.H. 1982. *Climate, History and the Modern World*. Methuen, London. 387 pp.

Legrand, J.P. 1979. Les variations climatiques en Europe Occidentale depuis le Moyen-Age (suivi d'une note historique de LeRoy Ladurie). *La Météorologie, VIème série 1* 61:67-182.

LeRoy Ladurie, E. and M. Baulant. 1981. Grape harvests from the 15th through the 19th century. *In: Studies in Interdisciplinary History*. R.I. Rotberg and T.K. Rabb (eds.). Princeton University Press, New Haven. pp. 259-269.

Parker, M.L., L.A. Jozsa, S.G. Johnson and P.A. Bramhall. 1981. White spruce annual ring-width and density chronologies from near Great Whale River (Cri Lake), Quebec. *In: Climatic Change in Canada 2*. C.R. Harington, (ed.). *Syllogeus* 33:154-188.

Pfister, C. 1981. An analysis of the Little Ice Age climate in Switzerland and its consequences for agriculture production. *In: Climate and History*. T.M.L. Wigley, M.J. Ingram and G. Farmer (eds.). Cambridge University Press, Cambridge. pp. 214-248.

Pilcher, J.R. and M.G.L. Baillie. 1980. Eight modern oak chronologies from England and Scotland. *Tree-Ring Bulletin* 40:45-58.

Rannie, W.F. 1983. Break-up and freeze-up of the Red River at Winnipeg, Manitoba, Canada in the 19th century and some climatic implications. *Climatic Change* 5:283-296.

Richter, K. 1988. Dendrochronologische und Dendroklimatologische untersuchungen an Kiefern (*Pinus* sp.) in Spanien. Ph.D. thesis, Universität Hamburg, Hamburg.

Serre, F. 1978. The dendoclimatological value of the European larch (*Larix decidua* Mill.) in the French Maritime Alps. *Tree-Ring Bulletin* 38:25-33.

Serre-Bachet, F. 1985. Une chronologie pluriséculaire du sud de l'Italie. *Dendrochronologia* 3:45-66.

_____. 1986. Une chronologie maîtresse de sapin (*Abies alba* Mill.) du Mont Ventoux (France). *Dendrochronologia* 4:87-96.

Tessier, L. 1981. Contribution dendroclimatique à la croissance écologique du peuplement forestier des environs des chalets de l'Orgère. *Travaux Scientifiques du Parc de la Vanoise* 12:9-61.

Till, C. 1985. Recherches dendrochronologiques sur le cèdre de l'Atlas (*Cedrus atlantica* (Endl.) Carrière) au Maroc. Ph.D. thesis, UCL, Louvain-la-Neuve. 231 pp.

Till, C. and J. Guiot. 1988. Reconstruction of precipitation in Morocco since 1100 based on *Cedrus atlantica* tree-ring widths. *Quaternary Research*. (in press).

Williams, L.D. and T.M.L. Wigley. 1983. A comparison of evidence for late Holocene summer temperature variations in the northern hemisphere. *Quaternary Research* 20:286-307.

Climatic Conditions for the Period Surrounding the Tambora Signal in Ice Cores from the Canadian High Arctic Islands

Bea Taylor Alt[1], David A. Fisher[2], and Roy M. Koerner[1]

Abstract

The Tambora volcanic signal (acid layer) has been identified in ice cores taken from Agassiz and Devon ice caps in the Canadian High Arctic. Oxygen isotope values (representative of annual precipitation temperature) and melt percent values (representative of summer temperatures) from core sections surrounding the volcanic signal have been examined in detail and compared with present day conditions. The results suggest that the Tambora volcanic eruption did not produce significant cooling in the Canadian High Arctic.

On Agassiz Ice Cap the ice representing the year after the volcanic signal shows an increase (warming) of both oxygen isotope and melt percent values followed by a return to pre- volcanic conditions. On Devon Ice Cap the oxygen isotope values began to decrease (cool) prior to the Tambora signal and cool to a minimum 25 years later. Melt percent values on Devon Ice Cap had already reached a minimum by the time of the eruption and this persisted for 45 years.

Based on modern synoptic studies, the circulation pattern during the summer season containing the Tambora signal (1816) is best represented by the 1972 analogue. In this analogue a long, narrow vortex at 500mb (50kPa) extends from the Siberian side of the central Arctic Ocean across the Pole deep into Labrador-Ungava, and is held tight against Greenland by a strong ridge of high pressure in the Alaska-Beaufort Sea area. This pattern results in strong cold northwesterly flow, with frequent light precipitation and very little melt on the ice caps. The pattern is broken occasionally by the joining of the Alaska and Greenland ridge which brings clear skies and some melt to the islands along the northwestern edge of the archipelago.

Introduction

The records from deep ice cores extracted from ice caps in the Canadian Arctic Islands provide insight into the climatic conditions in the islands during the decade surrounding the eruption of Mount Tambora in Indonesia. It is also possible from these data to address the question of whether single volcanic events (such as the Tambora eruption) produce significant deviations in proxy annual temperatures and/or proxy summer temperatures in this area of the High Arctic. Using modern synoptic-climate analogues, inferences can be made about synoptic circulation conditions at the time of the eruption of Mount Tambora.

---

[1] Terrain Sciences Division, Geological Survey of Canada, 601 Booth Street, Ottawa, Ontario K1A 0E4, Canada.
[2] Department of Glaciology, Geophysical Institute, University of Copenhagen, Haroldsgade 6, DK-2200, Copenhagen N, Denmark.

Geological Survey of Canada Contribution No. 156789.

For this study the most complete data are available from a core drilled in 1984 at the top of a local dome on the Agassiz Ice Cap (A84) on northern Ellesmere Island (Figure 1). Results from another Agassiz Ice Cap core, drilled in 1977, 1.2 km down the flow line from the dome (A77) and from a combined time series of three cores taken from the top of Devon Ice Cap (Figure 1) in 1971, 72 and 73 (referred to as D123) are also examined.

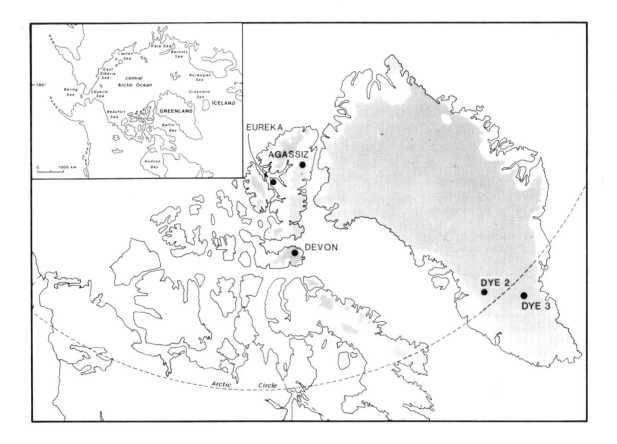

Figure 1: Location of ice cap deep core drill sites in the Canadian Arctic Islands and Greenland.

These cores have been extensively discussed elsewhere (Paterson *et al.* 1977; Koerner 1977; Koerner and Fisher 1981; Fisher *et al.* 1983; Koerner and Fisher 1985; Fisher *et al.* 1985; Alt 1985; Alt *et al.* 1985; Fisher and Koerner 1988). Here we will confine ourselves to the parameters and analyses which come closest to providing climatic data with an annual resolution. Every attempt has been made to provide an accurate annual time scale and to make these consistent for the period surrounding the Tambora eruption. It should be noted that northern hemisphere eruptions deposit acidic aerosols on the snow within months of the event, but southern hemisphere acid signatures first appear as much as one year after the southern eruptions.

Melt Features and Oxygen Isotopes

Melt layers (ice formed by the refreezing of meltwater) can be identified in ice cores by their relatively low concentration of air bubbles. In the upper reaches of an ice cap this melt is indicative of summer warmth (Koerner 1977). It is expressed as either melt-layer thickness (**m**) or as a percent of the total annual accumulation (**PC**). As melt (**m** or **PC**) can never be less than 0, possibly very severe summers are not adequately represented in the melt record. Table 1 gives the size of errors associated with time series of **PC**.

$\delta(^{18}O)$ is the $^{16}O/^{18}O$ ratio expressed as the fractional difference between the ratio in the sample and the ratio in "standard mean ocean water" (SMOW) measured in percent. In polar snow, δ is negative. Initially, δ was used as an indicator of mean annual temperature due to its dependence on the temperature at which condensation takes place. However up to seven non-temperature effects can alter the δ at a given site (Dansgaard *et al.* 1973; Fisher 1979; Fisher and Alt 1985; Johnsen *et al.* 1989). For the present discussions the δ values should be viewed as representing the mean annual precipitation temperature (temperature during precipitation events). Table 1 gives the size of the errors associated with time series of δ.

The Tambora Volcanic Signal

The volcanic peaks are identified by measuring the electrical conductivity (**ECM**) of core segments using brass electrodes with a 1250 DCV potential between them. The resulting values are plotted on a time scale derived from models and measurements of annual layer thickness (Koerner and Fisher 1985). Major acid layers are correlated with those in the absolutely dated Dye 3 core (Hammer 1980). The Tambora volcanic signal appears in the Dye 3 core in 1816.

In the A84 core the Tambora signal was identified in core 16 (Figure 2, bottom left). This core segment is in the firn at a depth of 27 m, is 146 cm long and represents approximately 16 years of accumulation. The core segment containing the peak signal from the eruption of Mount Laki in Iceland which occurs at 30 m depth in the ice is shown in Figure 2, bottom right. The peak **ECM** value for this Icelandic volcano is much higher than that from the Indonesian volcano, Tambora.

The melt-layer thickness values **m** for these cores have been plotted in a manner consistent with the **ECM** values. The melt and acid feature data are lined up within 5 cm. The Tambora event falls between two melt features whereas the Laki acid layer comes 40 cm above the big melt feature in core 18 (Figure 2, top).

Time Series of Acid, Melt and Oxygen Isotope Values for Agassiz

The annual average values of **ECM**, **PC** and δ have been plotted on the volcanic time scale for the Agassiz 84 core (Figure 3). This is probably the best time scale of the various Agassiz Ice Cap cores, and great care has been taken to align the three stratigraphies, however discrepancies of a year are possible. The Laki signal is absolutely dated as 1783 and the Tambora signal as 1816.

Both the δ and **PC** values (Figure 3) reach a maximum just following Tambora. The average annual δ has been calculated for the period of 1941-70, which is used as a meteorological normal. Compared to this, the 1816 δ value is slightly above the modern normals. The 1941-71 melt normals could not be calculated for A84. Instead the mean **PC** (dotted line, Figure 3) and the

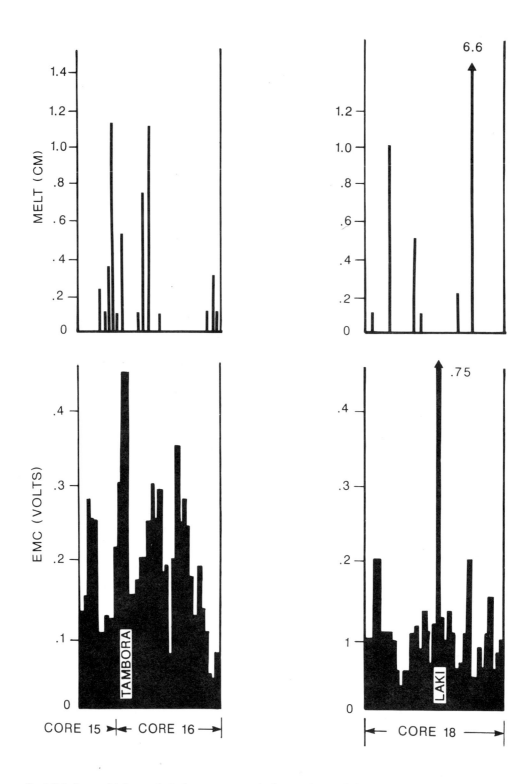

Figure 2: Melt-layer thickness (m) shown as actual observations of the thickness of individual layers of ice (top) and volcanic electrical conductivity measurements ECM values from the Agassiz 84 core segments containing the Tambora and Laki volcanic signals (bottom). Values are plotted on a depth scale with top of the core to the left. More than one melt layer can occur in a year. Core 16 is 146 cm long, so 5 cm is equivalent to the smallest horizontal increment on the ECM plot.

Table 1: Errors in Percent Melt PC and Oxygen Isotope δ Time Series.

Site	Interval years	Start AD	PC average %	SD PC noise 1 yr %	SD PC noise 5 yr %	δ average %	SD δ noise 1 yr %	SD δ noise 5 yr %	Accumulation (ice) cm/yr
A77	500	1946	2.8	>10	2.5	-31.5	0.48	0.35	17.5
A84	800	1961	4.1	>25	5.5	-28.5	0.32	0.23	9.8
Devon[1]	500	1956	7.0	> 8	1.6	-28.0	0.55	0.40	23.0

[1] Devon combined record; δ(73+72) and PC(71+72+73).
Note: SD is the standard deviation. The Devon SD(noise) data has been measured, but the A77 and A84 noise data is estimated.

Eureka mean July temperature (5.7°C) for the 1951-60 decade have been calculated. The 1941-70 Eureka July normal temperature (5.4°C) is slightly colder than the 1951-60 decade. The 1816 melt is also below the 1951-60 decade mean or probably near the modern normal, whereas the 1817 melt is considerably greater - comparable to the 1951-60 warm period.

When the A84 values are plotted as five-year averages (Figure 4) the δ profile might well be interpreted as showing a cooling immediately following Tambora. This is in direct contrast to the annual averages that show an immediate warming. Care must be used, therefore, in interpretation of averaged values in studying the short-term effects of single volcanic eruptions.

As mentioned, the A84 core has the most accurate time scale but it is at the top of a dome. Here the light winter snow is consistently scoured (i.e., blown away). This results in a δ record which is "warmer" than it would be if the winter snow was included. The A77 core lies sufficiently downslope from the dome to escape the scouring effect. A detailed plot of annual average δ values for the Tambora period from A84 and A77 (Figure 5) shows that the minimum preceding the Tambora signal is much colder in the unscoured core than in the scoured (A84) core. Based on the most recent time scale for the A77 core, this puts 1816 near the bottom of this minimum followed by a rise to 1970 normal values by 1819.

Time Series from Other High Arctic Cores

From the Devon blended 1971, 72 and 73 core records (D123) only five-year averages of δ and melt percent are available (Figure 6). The time scale is the vertical velocity time scale fine-tuned by analysis of annual layering as deduced by seasonal swings in microparticle concentrations and radiocarbon dating of gas bubbles in the ice (Paterson *et al.* 1977) and corrected for the location of the well-marked Laki eruption in the meltwater electrolytic conductivity records (Koerner and Fisher 1981). It is accurate to within a few years at the 1816 level. The 1941-70 normals are shown for both δ and melt percent (Figure 6).

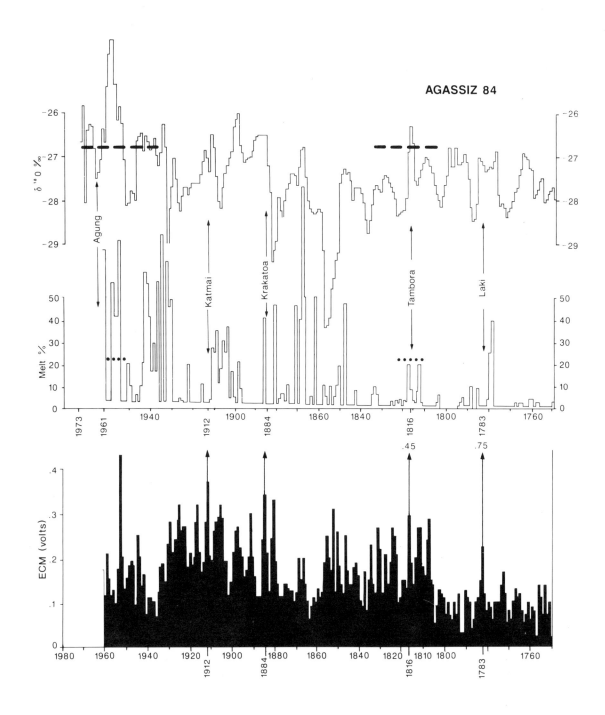

Figure 3: Annual averages from the Agassiz 84 core (A84): (a) oxygen isotope values δ; (b) melt expressed as percent of accumulation **PC**; and c) volcanic **ECM** values. The arrows on the **ECM** plot show the magnitude of the actual measured peak **ECM** value for various volcanic signals. The dotted line shows the 1951-60 **PC** decade average. The dashed line shows the 1941-70 δ 30-year average.

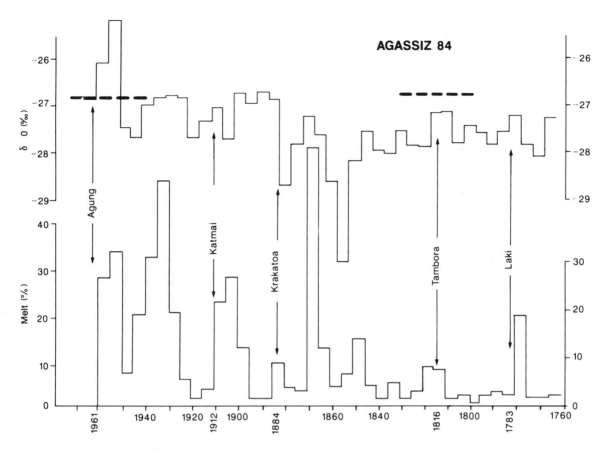

Figure 4: Five-year averages from A84 of oxygen isotope δ (dashed line 1941-70 average) and melt percent PC.

The most striking feature in the Devon Ice Cap blended record is the very cold summer period indicated by consistently low melt values during the whole period 1810-55. The eruption of Tambora occurred well after the beginning of this period. The Laki eruption, on the other hand, occurred during a period of increasing melt which reached almost to present normal values by 1800.

On the δ plot 1816 falls on a cooling trend which began before 1810 and reaches its lowest values for the 200-year period during the 1830s.

Comparing the 10-year melt averages for A77, D123 and Dye 2 on Greenland (Figure 7), we see that 1816 falls in a period of generally cold summers at all sites. In all cases the cold period began before the eruption of Tambora. At Dye 2 the lowest melt values occur in the 1820s and 30s but the long flat cold period of the D123 cores is not present.

The Effect of Single Volcanic Events on High Arctic Climate

None of the ice core records examined above shows definitive evidence of cooling resulting from the eruption of Mount Tambora. Those records which reach a minimum at some time following 1816 all show a cooling trend beginning before the Tambora volcanic signal. The same is true of the Laki eruption. Two other volcanic signals have been identified in the A84 core (Figures 3 and 4), dated by correlation with the Dye 3 volcanic record and also plotted on the D123

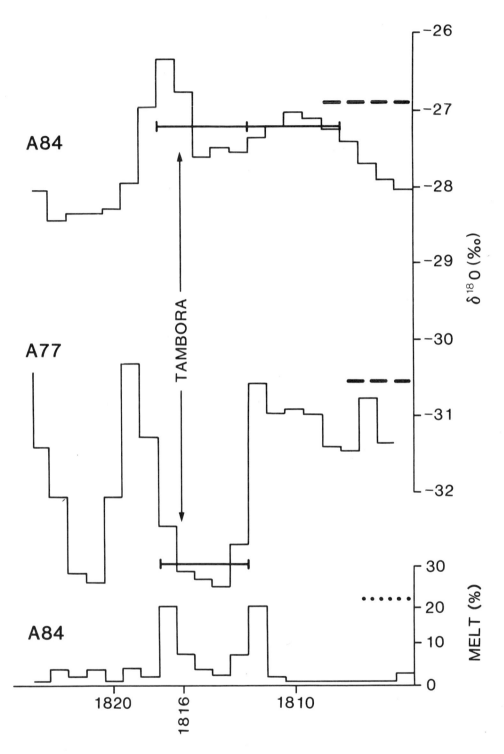

Figure 5: Detailed comparison of the annual mean A84 and A77 oxygen isotope values δ and the A84 melt percent values **PC** from around the Tambora volcanic signal. The five-year mean values (solid lines) and the 30-year δ normals from 1941-70 (dashed lines) are shown for the oxygen isotope values. The dotted line shows the average **PC** values for 1951-60.

record (Figure 6). The year of the eruption of Mount Agung in Indonesia is also shown on these figures. The Agung signal has not been positively identified in the Canadian cores as it was not sufficiently acidic. The five-year A84 **PC** averages (Figure 4) suggest cooling following Katmai, but close examination of the annual melt record (Figure 3) shows the season prior to Katmai was also cold. Both melt and δ values show high values following the Krakatau signal. On Devon Ice Cap the five-year averages for Krakatau drop sharply in summer melt but the δ values are already low. The Agung eruption appears to occur at the bottom of a δ minimum in the A84 core. Melt data are not available past 1961. In the Devon cores the eruption follows a δ minimum and is on a well-established downward melt trend.

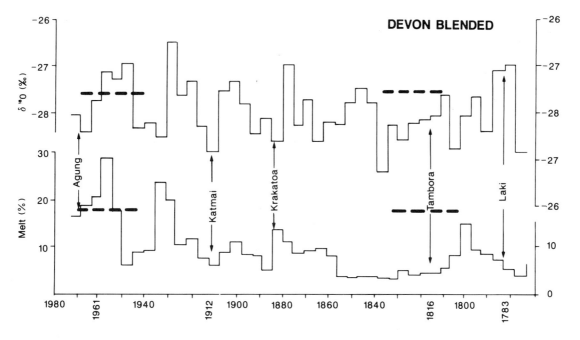

Figure 6: Five-year averages for D123, the Devon blended record (1971, 72 and 73 cores), of oxygen isotope δ and melt percent **PC**. The 30-year normals, 1941-70 are indicated (dashed lines).

The very cold summer of 1964 in the Canadian High Arctic, and the subsequent generally lower summer temperatures have been attributed to the effects of dust from Agung (Bradley and England 1978). Close examination of the hemispheric temperature plots of Dronia (1974) and Kelly *et al.* (1982), Figure 8, show that in both cases the hemispheric temperatures had begun to cool long before the eruption of Agung in 1963. The rather dramatic drop of July mean temperatures seen in the plots from the northern Canadian Arctic Island stations (Figure 9) is, in fact, a result of the record high temperatures in the 1962 season.

These results do not appear to indicate that single volcanic eruptions cause lower summer or annual temperatures in the northern Canadian Arctic Islands. This does not rule out the possibility that multiple eruptions in a period could have a cumulative effect on temperatures (Hammer *et al.* 1980) or that single volcanic events produce abrupt, short-lived temperature depressions on a hemispheric scale (Bradley 1988). Single events could also be responsible for significant anomalies in the atmospheric circulation regime in the Canadian Arctic Islands such as occurred in the summer of 1964 (Alt 1987).

Summary of Core Results

Now we can review what the ice core analyses reveal about climate in the area at the time of the Tambora volcanic signal. The results are expressed in Table 2 as simple estimates of the temperature anomalies with respect to the modern normals (1941-70).

On the Agassiz Ice Cap the summer melt conditions, and thus the summer temperatures, were near or slightly below the 1941-70 normals. There was a rise in the melt values immediately after the Tambora event to values similar to the relatively warm 1951-60 decade as experienced at Eureka.

The annual temperature (or more accurately the annual precipitation temperature) on Agassiz Ice Cap appears to have been lower than the 1941-70 normals. The scoured A84 core shows the Tambora signal to be part of a slight rise from below-normal conditions to above 1941-70 normals. The unscoured core A77 shows 1816 to be on a warming trend from a very cold period.

On Devon Ice Cap there was very little summer melt, indicating very cold conditions. These conditions began around 1810 and persisted until the late 1850s. This is the longest very cold period in the 800-year record.

On Devon Ice Cap the Tambora signal falls on a cooling trend of annual (or precipitation) temperature beginning about 1810, when the oxygen isotope values were very near the modern normals. This cooling trend could be viewed as part of a general decline beginning before the time of the Laki signal.

Table 2: Conditions on Canadian Arctic Island Ice Caps During the Period of the Tambora Volcanic Signal as Deduced From Ice-Core Records.

Season	Ice Cap	Temperature	Remarks
SUMMER (from melt percent records)	Agassiz	0 (normal)	then rising
	Devon	- - (very cold)	already very cold
ANNUAL (from oxygen isotope values)	Agassiz	- (cold)	on a warming trend
	Devon	- (cold)	on a cooling trend

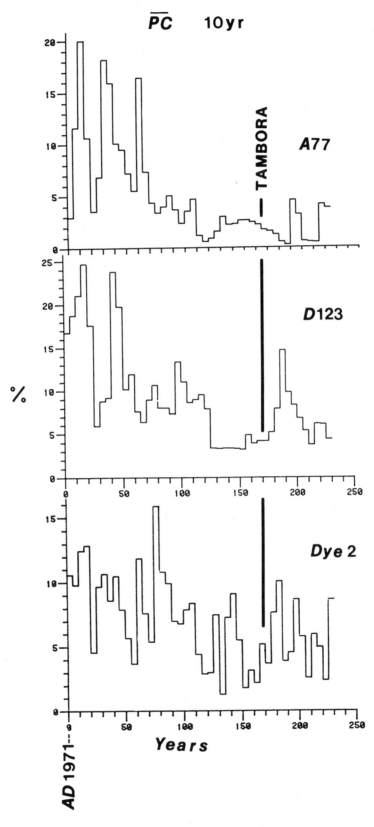

Figure 7: Comparison of 10-year averages of A77, Devon blended D123 and Dye 2 (Greenland) melt percent values.

Synoptic Conditions

Based on the core results for the period around the Tambora signal it is now possible to examine the synoptic circulation patterns which would be expected to produce these conditions on the two ice caps. Previous studies of synoptic analogues and ice-core results (Alt 1985; Alt *et al.* 1985; Alt 1987) have suggested that this period of the Little Ice Age was dominated by summers similar to the summer of 1972 (Figure 10). The most important feature of the 1972 circulation analogue for the study area is the persistence of a long deep 500mb (50kPa) vortex held against Greenland by a strong ridge over Alaska and western Canada. This produces persistent northwesterly flow into the Canadian Arctic Islands from the central Arctic Ocean and a deep layer of very cold air in the northern Baffin Bay area.

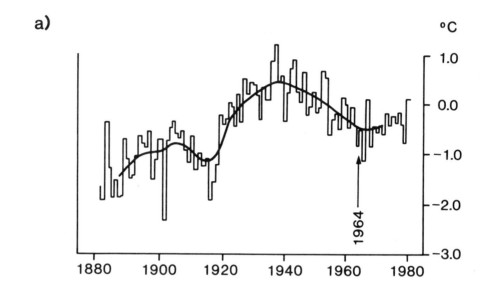

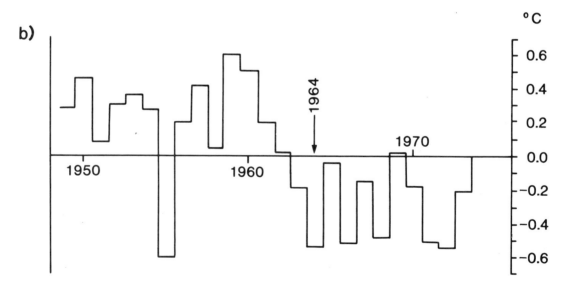

Figure 8: Two depictions of the annual temperature record for the arctic: (a) annual temperature departures from the 1946-60 reference period for 65-85°N (after Kelly *et al.* 1982); and (b) annual deviations from the 25-year mean 1949-73 of thickness of the 500/1,000mb layer for 65-90°N (after Dronia 1974).

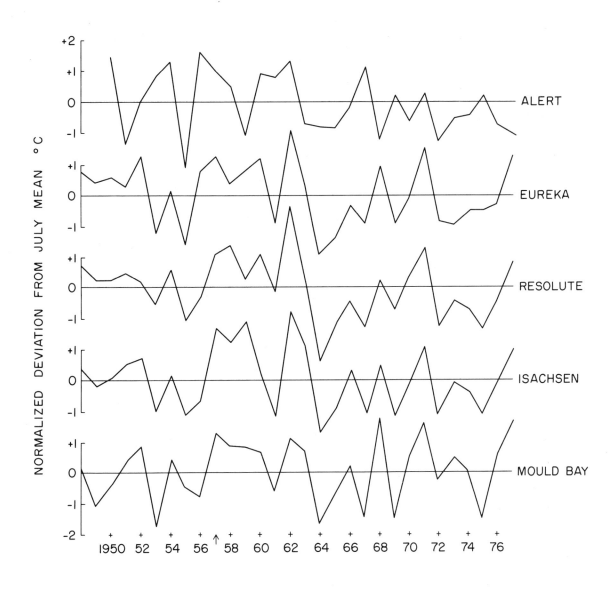

Figure 9: Normalized deviation from the mean of July temperature for Canadian Arctic Islands stations [(July mean - July normal)/July standard deviation] from P. Schofield (personal communication).

These features are evident from comparison of the mean July 500mb (50kPa) height contours for the period 1948-78 with those of 1972 (Figure 11a,b). We see that the 1972 vortex is deeper and narrower than the mean, and shifted eastward from the mean position by a ridge over the Beaufort Sea. The flow into the High Arctic Islands is stronger than normal as seen by the closer spacing of the contour lines. There is also a strong ridge over the Barents Sea. The mid-latitude circulation during the entire 1971-72 season was stronger than usual and distinctly meridional (i.e., with strong north-south components).

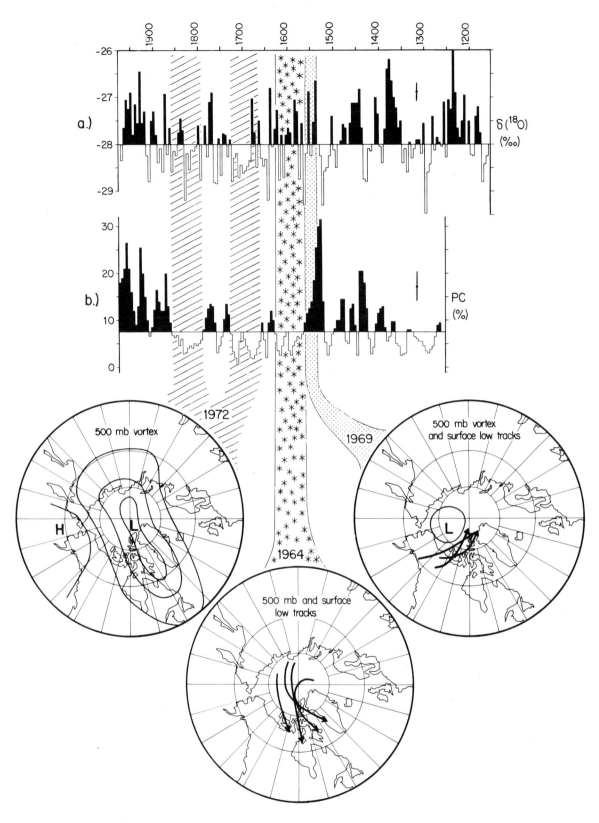

Figure 10: Schematics of summer synoptic characteristics for the Agassiz and Devon core-site area for various periods of the last 800 years. These were deduced by applying modern analogues to the ice-core results; represented here by the five-year averages of oxygen isotope δ and melt percent **PC** from Devon Ice Cap (Alt 1985).

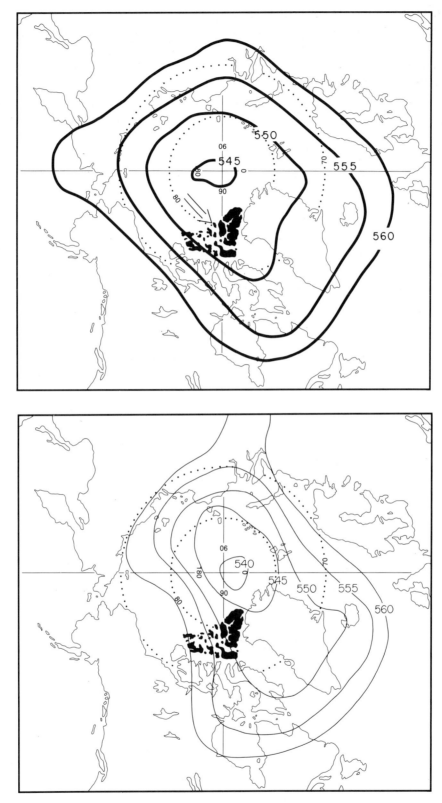

Figure 11: Mean July pattern of 500mb (50kPa) height contours in decameters (dam). In order to focus on the polar vortex, contours up to 560 dam only are shown: (a) for 1948-78 (after Harley 1980); and (b) for 1972.

In order to examine the surface weather conditions associated with the 1972 anomaly, the actual synoptic charts for 3 July 1972 are shown in Figure 12a,b. Here we see the long upper vortex (Figure 12a) with multiple centres, one over Labrador-Ungava, a second in northern Baffin Bay and a third over the Pole. This supports a surface trough (Figure 12b) across northern Ellesmere Island down the west coast of Greenland to Davis Strait. The strong northwesterly flow, indicated by the closely packed isolines, extends across the Canadian Arctic Islands into Keewatin, Hudson Bay and James Bay at all levels from the surface to 500mb (about 5,000m). Cold moist air from the central Arctic Ocean is pushed south into these areas. In the northern islands extensive low cloud, high humidity and frequent light precipitation accompany these conditions. Over Devon Ice Cap precipitation may be enhanced by the persistence of the northern Baffin Bay low, which picks up additional moisture from the open water.

Both the mean July 1972 (Figure 11b) and 3 July 1972 (Figure 12a and b) patterns appear to be consistent with conditions proposed for the Hudson Bay region during 1816. Wilson's (1983) studies show prevailing NW-NE winds in June and July 1816 and also suggest as a modern analogue the summer of 1972. High pressure west of Hudson Bay (in the case shown here, an extension of the Alaska ridge) is an important feature of her proposed 1816 circulation patterns.

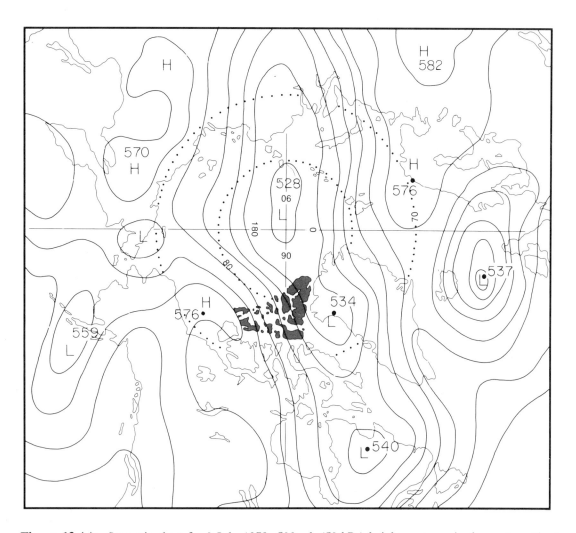

Figure 12 (a): Synoptic chart for 3 July 1972: 500 mb (50 kPa) height contours in decametres (dam).

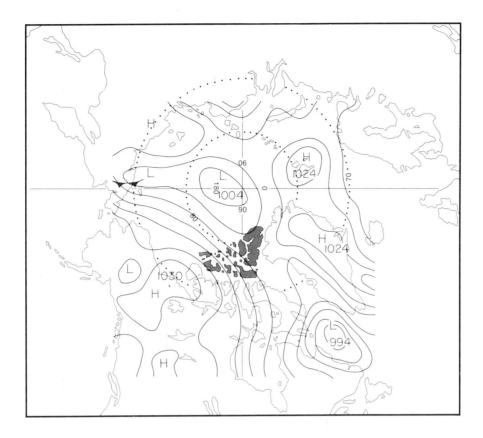

Figure 12 (b): Synoptic chart for 3 July 1972: surface pressures in mb.

Figure 12a,b also gives us an indication of the synoptic conditions that previous detailed synoptic studies of summers from 1960-78 have shown could produce melt on Agassiz Ice Cap and not on Devon Ice Cap. Close inspection shows that the ridge extending from the Beaufort Sea to southern Manitoba (which is responsible for keeping the trough or vortex tight against Greenland) can be seen pushing northeast across the northern Canadian Arctic Islands toward the north Greenland ridge. If these join (as they did later in July 1972), the vortex is cut off, becoming a closed low. The closed system can lie over Labrador-Ungava or farther west, as happened in July 1972. Along the northwestern edge of the islands the ridge results in subsidence through the whole troposphere, which dissipates the cloud and fog. Melt is produced by increased solar radiation (sometimes aided by warm-air advection) over the northern and western ice caps. Devon Ice Cap, however, is often under the influence of the cyclonic circulation in Baffin Bay and not as likely to experience melt. In fact, on Devon Island summer accumulation (snow) may occur under these conditions. These ridging conditions are often brief but they can produce significant melt on the ice caps and a touch of summer in the northwestern islands, as happened in mid-July 1972 (Alt 1987).

We can also say that this pattern resembles the mean winter conditions, and suggest that in years of this kind the winter circulation is never really broken down. The temperature gradients remain strong as do the mid-latitude westerlies. Summer comes only briefly to the ice caps if and when the blocking ridges join across the northern islands. These ridging conditions are more effective in the northwestern islands and may not produce any melt at the core site on Devon Ice Cap.

Conclusions

The Tambora signal can be identified as an acid layer in the Agassiz 84 core. The oxygen isotope δ and melt values **PC**, even allowing for a one-year discrepancy and other considerations such as noise and scouring, do not show evidence of cooling due to the eruption of Mount Tambora, although it may have occurred part way down a cooling trend. Nor is there definitive evidence of cooling in the northern Canadian Arctic Islands following the eruption of Laki, Krakatau, Katmai or Agung.

On Agassiz Ice Cap, conditions in the year dated as 1816 were near, or somewhat below, modern normals (1941-70) but rise to a secondary peak immediately following the Tambora signal. Care must be taken when interpreting average values for periods longer than a year as they can easily obscure the short-term variations. However, on the Devon Ice Cap blended five- year average plots, 1816 falls in a prolonged period of very low summer melt and below modern normal δ levels (annual or precipitation temperature); both of which began about 1800. Similarly the Dye 2 10-year average plot shows Tambora occurring on a well established cooling trend.

The climatic conditions suggested by the ice-core analyses around the Tambora eruption strongly resemble those of the summer of 1972, which has been identified as the modern analogue for melt suppression on High Arctic ice caps. This pattern, which features a long deep upper vortex extending from the Siberian side of the central Arctic Ocean across the eastern Canadian Arctic Islands to Labrador-Ungava and a strong ridge from the Beaufort Sea and Alaska into the prairies, appears to be compatible with synoptic interpretations from other parts of Canada. This pattern represents an intensification of the conditions which appear to have dominated the latter part of the Little Ice Age in the High Arctic islands.

References

Alt, B.T. 1985. A period of summer accumulation in the Queen Elizabeth Islands. *In: Critical Periods in the Quaternary Climatic History of Northern North America. Climatic Change in Canada 5.* C.R. Harington (ed.). *Syllogeus* 55:461-479.

_____. 1987. Developing synoptic analogues for extreme mass balance conditions on Queen Elizabeth Island ice caps. *Journal of Climate and Applied Meteorology* 26(12):1605-1623.

Alt, B.T., R.M. Koerner, D.A. Fisher and J.C. Bourgeois. 1985. Arctic climate during the Franklin Era as deduced from ice cores. *In: The Franklin Era in Canadian Arctic History.* Pat Sutherland (ed.). *National Museum of Man, Mercury Series, Archaeological Survey of Canada Paper* 131:69-92.

Bradley, R.S. 1988. The explosive volcanic eruption signal in northern hemisphere continental temperature records. *Climatic Change* 12:221-243.

Bradley, R.S. and J. England. 1978. Volcanic dust influence on glacier mass balance at high latitudes. *Nature* 271:736-738.

Dansgaard, W., S.J. Johnsen, H.B. Clausen and N. Gundestrup. 1973. Stable isotope glaciology. *Meddelelser om Grønland* 197(2):1-53.

Dronia, H. 1974. Uber Temperaturanderungen die frier Atmosphare auf der Nordhalbkugel in den letzten 25 Jharen. *Meteorologische Rundschau* 27:166-174.

Fisher, D.A. 1979. Comparison of 10^5 years of oxygen isotope and insoluble impurity profiles from Devon Island and Camp Century ice cores. *Quaternary Research* 11(3):299-305.

Fisher, D.A., R.M. Koerner, W.S.B. Paterson, W. Dansgaard, N. Gundestrup and N. Reeh. 1983. Effect of wind scouring on climatic records from ice-core oxygen-isotope profiles. *Nature* 301(5897):205-209.

Fisher, D.A., R.M. Koerner, N. Reeh and H.B. Clausen. 1985. Stratigraphic noise in time series derived from ice cores. *Annals of Glaciology* 7:76-83.

Fisher, D.A. and B.T. Alt. 1985. A global oxygen isotope model - semi empirical, zonally averaged. *Annals of Glaciology* 7:117-124.

Fisher, D.A. and R.M. Koerner. 1988. The effects of wind on $\delta(O^{18})$ and accumulation give an inferred record of seasonal δ amplitude from the Agassiz Ice Cap, Ellesmere Island, Canada. *Annals of Glaciology* 10:34-38.

Johnsen, S.J., W. Dansgaard and J.W.C. White. 1989. The origin of Arctic precipitation under present and glacial conditions. *Tellus* Series B, 41B(4):452-468.

Hammer, C.U. 1980. Acidity of polar ice cores in relation to absolute dating, past volcanism, radio-echoes. *Journal of Glaciology* 25(93):359-372.

Hammer, C.U., H.B. Clausen and W. Dansgaard. 1980. Greenland Ice Sheet evidence of post-glacial volcanism and its climatic impact. *Nature* 288(5788):230-235.

Harley, W.S. 1980. Northern hemisphere monthly mean 50 kPa and 100 kPa height charts. *Environment Canada, Atmospheric Environment Service CLI-80*:29.

Kelly, P.M., P.D. Jones, C.B. Sear, B.G. Cherry and R.K. Tavakol. 1982. Variations in surface air temperatures. Part II: Arctic regions, 1881-1980. *Monthly Weather Review* 110(2):71-83.

Koerner, R.M. 1977. Devon Island Ice Cap: core stratigraphy and paleoclimate. *Science* 196(4285):15-18.

Koerner, R.M. and D.A. Fisher. 1981. Studying climatic change from Canadian High Arctic ice cores. *In: Climatic Change in Canada 2.* C.R. Harington (ed.). *Syllogeus* 33:195-215.

_____. 1985. The Devon Island ice core and the glacial record. *In: Quaternary environments; eastern Canadian Arctic, Baffin Bay and western Greenland.* J.T. Andrews (ed.). Allen and Uwin, Boston. pp. 309-327.

Paterson, W.S.B. and seven others. 1977. An oxygen-isotope climate record from Devon Island Ice Cap, Arctic Canada. *Nature* 266(5602):508-511.

Wilson, C.V. 1983. The Little Ice Age on eastern Hudson Bay: summers at Great Whale, Fort George, Eastmain, 1814-1821. *Canadian Climate Centre Report* No. 83-9:1-145.

Europe (including Iceland)

1816 - a Year without a Summer in Iceland?

A.E.J. Ogilvie[1]

Abstract

There has been considerable speculation as to whether the eruption of Mount Tambora in April 1815 caused a world-wide lowering of temperatures and a "year without a summer" in the following year of 1816. In this paper, the weather during 1816 is detailed for one specific location: Iceland. The weather data used are taken from documentary accounts written at 10 different sites in Iceland. These suggest that the winter and spring of 1816 were very cold and unfavourable in most parts. The summer was mainly cold in the north, wet in the east and highly variable elsewhere. Many accounts of the autumn focus on the variability of the weather. Although it would seem that, on the whole, the summer weather was not sufficiently extreme for this year to be termed a "year without a summer," adverse weather did cause some impact on society. It seems very likely that there was direct climatic impact on important agricultural practices such as the hay harvest and the growing of vegetables. The Arctic sea ice, although not unusually heavy or prolonged in 1816, had a direct impact in northern Iceland, hindering fishing and sealing. Indirect impacts on society are less easy to establish. However, it seems likely that some social stress described in 1816 may be at least partly attributed to the climate.

Introduction

Although the precise nature of the effects of volcanic eruptions on the general circulation of the atmosphere are, as yet, unknown, there can be little doubt that major volcanic eruptions do affect the Earth's climate (e.g., Lamb 1970; Kelly and Sear 1984; Sear *et al.* 1987; Bradley 1988). The possible effects of one very large eruption - that of Mount Tambora in April 1815 - has excited particular interest. Although some researchers (e.g., Landsberg and Albert 1974) have concluded that this eruption did not have significant climatic effects, others have provided convincing evidence to show that the subsequent year, 1816, was anomalously cold in many places (Stothers 1984; Kelly *et al.* 1984). The year 1816 has even been termed the "year without a summer" (Stommel and Stommel 1979).

In this paper, the weather during 1816 is considered for one specific location - Iceland. In order to place the year in context, the general climate of Iceland is considered first, both for the twentieth century, and in terms of climatic variations in the past. Possible climatic impact in Iceland during 1816 will also be discussed.

The Present and Past Climate of Iceland

The Twentieth Century Context
Our knowledge of the climate of Iceland is derived from two main data sources. The principal of these is modern instrumental data. By the late nineteenth century, around 20 observing stations were in existence, and with the establishment of the Icelandic Meteorological Office (*Veðurstofa*

---

[1] Climatic Research Unit, School of Environmental Sciences, University of East Anglia, Norwich NR4 7TJ, U.K.

Íslands[1]) in 1920, the number of stations grew. By 1955, there were 66. From 1966 onwards, the number has varied between 120 to 130 (Einarsson 1976, pp. 12-13). Information from these, and other observing stations in the North Atlantic and Polar regions, plus oceanographic data, has enabled a general picture of key factors in the climate and weather of Iceland to be established. These are summarized below. For more detailed discussions on this topic, see Eythórsson and Sigtryggson (1971) and Einarsson (1976).

Main Features of the Climate of Iceland

The principal features of Iceland's climate are determined by its location at the frontier zone of two very different air masses; cold polar air from the north, and warmer maritime air from the Atlantic. Depressions moving toward Iceland from the western Atlantic often slow down as they near the southwestern corner of Iceland, thus maintaining a flow of mild Atlantic air over the country. This process causes thaws in winter, and rain and cool temperatures in summer. When these depressions cross Iceland and move toward Norway, a flow of polar air may take their place and bring much colder weather, especially in the northern part.

The alternating cold and milder air masses that Iceland experiences at varying intervals, and for different durations, are the prime cause of the variability of Iceland's climate. This variability is exacerbated by the two major ocean currents which flow around the island; the cold East Greenland polar current, and the warmer Irminger current. The Arctic drift ice also has considerable influence on the climate of Iceland. Most noticeably, when the ice is present off the coasts, both land and sea temperatures are lowered.

Although weather conditions in Iceland vary greatly, generally winters are mild compared with other northern continental locations, and summers tend to be cool. Typical temperature ranges during the winter months of December, January and February vary between -2 and 1°C. The warmest summer month is generally July, with a mean temperature varying from around 8 to 11°C, depending upon location.

The Past Climate of Iceland: Introduction

Information about the climate of Iceland derived from modern data is augmented and amplified by what is known of the past climate of Iceland. This is derived largely from documentary, historical evidence: the nature and use of such evidence is discussed briefly below. Although we cannot hope to gain as accurate a picture from documentary evidence of climate as from modern instrumental data, such evidence can act as a guide to what may have occurred in the past when no other data are available. To this end, proxy temperature variations based on the use of historical documentary evidence have been derived by Bergthórsson (1969) and by Ogilvie (1984a, 1986, 1990). Incidence of the sea ice off the coast of Iceland in the past has been estimated by these same authors, and by Koch (1945).

In the sections below, probable variations in the past climate of Iceland from medieval times to the early nineteenth century are outlined. Prior to this, the available data sources for this period are discussed.

---

[1] The Icelandic characters "þ" and "ð" (for "th") and all accents are retained wherever these are used in the original.

Data Sources
The accuracy of any proxy-temperature indicator will depend on the quality of the data used. To ensure high quality of documentary evidence, all sources must be analyzed carefully in order to establish their reliability. Key questions to ask here are whether the author was close in time and space to the events described; if this is the case, then a source is much more likely to be reliable than if he were not. For more detailed discussions on source analysis in general, see Bell and Ogilvie (1978) and Ingram et al. (1978). For discussions of the analysis of Icelandic sources see Vilmundarson (1972) and Ogilvie (1981, 1984a, 1990, 1991).

Iceland's climatic history may be traced back to early settlement times in Iceland (from about A.D. 870 onwards). However, the quality and availability of climatic and weather data vary considerably. For the period up to about 1170, there are no contemporary documentary sources, and only brief and sporadic comments on weather and climate may be found in existing sources. For the thirteenth century, a few reliable sources give some indication of possible changes in climate. Many more descriptions of weather and climate exist for the fourteenth century. The fifteenth century and the first half of the sixteenth century are very poorly documented. Typical sources for this period are certain sagas, the medieval annals and works of geographical descriptions.

From the early seventeenth century onward, many more reliable documents become available. For the early to mid-eighteenth century, there is extensive coverage from a variety of different sources including annals, travel accounts, government reports and weather diaries. These give information for most seasons in many different parts of Iceland. For the late eighteenth and early nineteenth centuries, sources of climatic and weather information are very full and detailed.

The earliest quantitative observations taken in Iceland date from the mid-eighteenth century (Eyþórsson 1956; Kington 1972). However, these, and subsequent late-eighteenth and nineteenth-century observations only cover a few months or years. Continuous temperature observations commence in 1846 (Sigfúsdóttir 1969). These were made at Stykkishólmur, in the west. For the period 1820-54, observations of temperature were taken in Reykjavík or the near vicinity by Jón Þorsteinsson (1794-1855). A part of this important series was subsequently lost for many years. However, the missing data were recently found by Trausti Jónsson of the Icelandic Meteorological Office, and he is engaged in their analysis (Jónsson, personal communication).

The Climate of Iceland from Settlement Times to about 1600
Iceland was settled, primarily from Norway, in the late ninth and early tenth centuries. Circumstantial evidence suggests a fairly mild climate around this time. A cold period may have occurred from about 1180 to 1210, while from about 1211-32 the climate may have become milder. An early geographical treatise written in approximately 1250 (*The King's Mirror*) mentions much sea ice between Iceland and Greenland at this time, and refers to Iceland's cold climate. However, it is difficult to draw firm conclusions from statements such as this. From about 1280 to 1300 the climate seems to have been fairly cold. During the early years of the fourteenth century, severe weather is mentioned only infrequently (in 1313, 1320, 1321 and 1323), so this period may have been mild. Milder weather may well have continued to past the mid-fourteenth century. The years 1360-80 are likely to have been colder. Little information is recorded from the 1380s. Only two severe years are noted for the 1390s. Evidently, 1412-70 was mild, and the 1480s or 1490s were years of dearth, possibly caused by severe weather. However,

very little information is available for 1430-1560. Likely the latter part of the sixteenth century was mainly severe. A detailed discussion of all medieval historical sources containing comments on the weather and climate, together with an analysis of their evidence, may be found in Ogilvie (1991).

The Climate of Iceland from 1601 to about 1850

The first and second decades of the seventeenth century were, overall, probably relatively mild. The years 1620-40 were cold, but 1641-70 was distinctly mild. From 1671-90, temperatures were colder. The 1690s were very cold. The early years of the eighteenth century were relatively mild, especially the first decade. The 1730s, 1740s and 1750s were cold, especially the two latter decades. The 1760s were somewhat milder, the 1770s cooler again. The period 1781 to 1820 was cold on the whole. The year 1816 must be assessed in the context of this prevailing background, with mainly cold conditions spanning most of the preceding four decades. From 1821 to 1841 the climate is likely to have been milder, while the 1840s were very mild. For a fuller account of climatic variations in Iceland during the seventeenth and eighteenth centuries, see Ogilvie (1981, 1984a, 1986, 1990).

The Weather in Iceland During 1816

Data Sources

In order to build up a clear picture of weather and climate during 1816, a number of sources were selected for detailed analysis. The main sources used are letters written by "Sheriffs" or government officials in the 20 or so different districts or *Sýsla* (plural *Sýsslur*) of Iceland. These letters contain information on such topics as grass growth and hay crop, trade, health and disease. They also report on the weather, sometimes in very great detail. One letter used here (from Snæfellsnessýsla, in the west) gives daily data, as well as seasonal summaries. The letters were sent at least annually, sometimes more frequently, to the Danish government in Copenhagen. Written in Danish, they are all unpublished, and are now kept in the National Archives in Reykjavík[1].

For this discussion of 1816 weather, letters were chosen from nine different sites in Iceland. Use was also made of one other source; an annal, called *Brandsstadaanáll*. This was written by Björn Bjarnason (1789-1859) at Brandsstaðir, Blöndudalur in Austur-Húnavatnssýsla in the north. This annal describes events that occurred in Iceland each year from 1783 to 1858, and includes detailed weather descriptions. Other available sources were not used. It was felt, however, that the sources used here were adequate to provide good regional coverage over the year. As Iceland's climate is regionally quite variable (Eythórsson and Sigtryggsson 1971; Ogilvie 1984a) it was essential to consider different parts of Iceland.

The sites at which these various sources were written are shown in Figure 1. The sources are from (in the order given on the map): (1) Ketilsstaðir in the district of Suður-Múlasýsla in the east; (2) Garður in Suður-Þingeyjarsýsla in the north; (3) Möðruvellir in Eyjafjarðarsýsla in the north; (4) Viðvík in Skagafjarðarsýsla in the north; (5) Brandsstaðir in Austur-Húnavatnssýsla in the north; (6) Gröf in Snæfellsnessýsla in the west; (7) Síðumúli in Mýrasýsla in the west; (8) Leirá in Borgarfjarðarsýsla in the west; (9) Reykjavík in Gullbringusýsla in the southwest; (10) Vík in Vestur-Skaftafellssýsla in the southeast. Some use was also made of a letter written at Grund, a site very close to Möðruvellir.

---

[1] All translations of sources are by the author.

1816 - The Evidence

Introduction

In Table 1, a brief synopsis is given of comments on weather from the sources described above. In the left-hand column, the place at which the letter was written is shown according to its number on the map. The columns in the centre show the main characteristics of the seasons. The term "winter" here refers to the period from mid-October of one year (1815) to mid-April of the next (1816). "Spring" covers mid-April to mid-June, "summer" is mid-June to August, and "autumn" is September to mid-October. The column on the right-hand side of Table 1 shows descriptions of sea ice.

Winter

From Table 1, it may be seen that letters from most districts report a severe winter. Two of the letters written in the north, at Mödruvellir in Eyjafjardarsýsla, and at Gardur in Adaldalur in Audur-þingeyjarsýsla, stated that the winter was, respectively "more than unusually severe" and "very severe." Two other sources, one from Vík in Vestur-Skaftafellssýsla in the southeast, and the other from Leirá in Borgarfjardarsýsla in the west, both noted that the winter was "very severe." The writer of the account from Reykjavík wrote that the winter was "severe with much snow and frost."

Some sources stress the variability of the weather this winter. Thus, the letter from Vidvík in Skagafjardarsýsla in the north reported that there was "much snow and alternating thaws and sharp frosts," and the letter from Sídumúli in Myrasýsla in the west gives a similar account. According to *Brandsstadaannáll*, the winter was also very severe, but there were some spells of calm and good weather in between, for example, from 25 November to 15 December (1815) and from 15 January to about 21 February. Interestingly, the letter from Ketilsstadir in Sudur-Múlasýsla in the eastern part of Iceland stated that the winter was merely average, although there was much snow.

Spring

The spring of 1816 was also relatively cold in most districts. The letters written at Vidvík and Mödruvellir in the north characterized the spring as "dry and cold" and "quite severe", respectively. According to the latter, the severity took the form of persistent northerly winds, frost and cold air. These the writer attributed to the presence of sea ice which lay off the northern coasts all spring. The letter from Gardur does not contain a description of spring weather as such, but does mention sea ice. This is stated to have been present from the beginning of March to mid-June. The spring was said to be "unusually cold" in the letter from Reykjavík.

According to *Brandsstadaannáll*, April was severe, but the weather improved at the end of the month. The account from Leirá, in the west, also notes severe cold in April, but says that from the end of the month to mid-May the weather was mild. From then it became cold again with northeasterly winds, sleet and night frost to about 24 June. The Sheriff of Skaftafellssýsla, writing at Vík, recorded a severe April and a mild May. The first 10 days of June were dry and frosty.

Summer

The weather during the summer of 1816 in Iceland was quite variable regionally. The northern sources used make it clear that, in most northern areas, the weather was very poor. Most accounts from the south and west report a mixture of both favourable and unfavourable weather. The eastern source used here states that the summer was wet. We may look at accounts from these regions in more detail, starting with the north. A summary of the data may also be found in Figure 2.

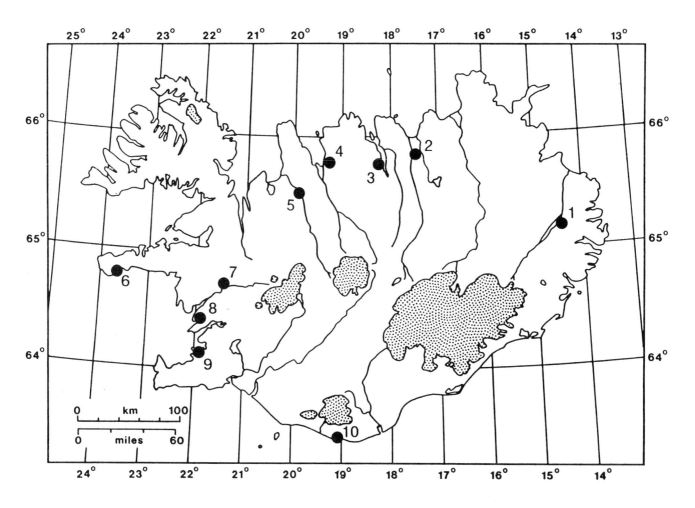

1. Ketilsstaðir, Suður-Mulasýsla
2. Garður, Suður-Þingeyjarsýsla
3. Möðruvellir, Eyjafjarðarsýsla
4. Viðvík, Skagafjarðarsýsla
5. Brandsstaðir, Austur-Húnavatnsýssla
6. Gröf, Snæfellsnessýsla
7. Síðumúli, Mýrasýsla
8. Leirá, Borgarfjarðarsýsla
9. Reykjavík, Gullbringusýsla
10. Vík, Vestur-Skaftafellssýsla

Figure 1: Sites of sources used for 1816 weather reconstruction.

Table 1: Main Characteristics of the Seasons in 1816.[1]

	WINTER	SPRING	SUMMER	AUTUMN	SEA ICE
1(E)	Average	—	Very damp and rainy	—	—
2(N)	Very severe	—	Unpleasant. Alternating snow, rain and frost	—	From beginning March to mid-June
3(N)	More than unusually severe	Quite severe due to northerly winds, frost and cold air	Very cold at Grund	Very wet September	Lay off northern coasts during spring
4(N)	Severe. May be compared to winters of 1784 and 1802	Dry and cold	Dry and cold then wet from mid-August	Wet. Frequent strong winds	—
5(N)	Mainly severe	Weather improved 25 April	Dry and good during harvest	Snow and frost late September. Calm and dry first half October	—
6(W)	Fairly severe	Variable	June mainly windy and wet. Good to 14 August then mainly wet	Stormy and rainy	—
7(W)	Severe	—	Averagely good	—	—
8(W)	Very severe	Mainly severe to about 24 June	Dry to end August	From September, damp with southerly winds	—
9(SW)	Severe	Unusually cold	Cold and inconstant on the whole	Stormy all autumn	—
10(SE)	Very severe	April severe. May mild. Dry and frosty early June. Then wet	Rain and storms 11 June to 11 July and good 12 July to 11 August Rain 12 August to 12 September	Mainly wet in September. Dry and good 3-9 October then unusually changeable	—

[1] The data are taken from sources written at the following sites: 1 Ketilsstaðir; 2 Garður; 3 Möðruvellir; 4 Víðvík; 5 Brandsstaðir; 6 Gröf; 7 Síðumúli; 8 Leirá; 9 Reykjavík; 10 Vík. N North; W West; E East; SW Southwest; SE Southeast.

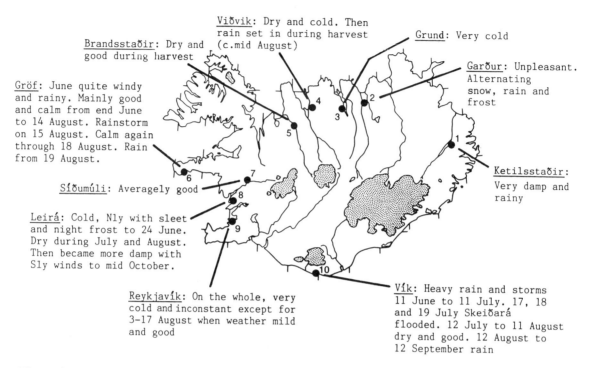

Figure 2: Summer weather reported in Iceland in 1816.

The Sheriff writing at Viðvík in Skagafjarðarsýsla stated that the summer was dry and cold except during the harvest around mid-August when there was rain. Stefán Þórarinsson at Möðruvellir also mentions long-lasting rain during the harvest. The report from Grund, near Möðruvellir, was of a very cold summer. In the letter from Garður, the summer is said to have been unpleasant with alternating snow, rain and frost. The other northern source used here, *Brandsstaðaannáll*, disagrees with these accounts. According to this source, the weather became good and calm after 25 April, and the summer as a whole was favourable. It should be noted that such variability between different areas in Iceland, even sites in close proximity, is not unusual. Furthermore, Brandsstaðir in Blöndudalur, where *Brandsstaðaannáll* was written, is in a fairly sheltered location.

In the east, Sheriff Páll Þorðarson Melsted, writing at Ketilsstaðir, commented on the damp and rainy summer. At Vík, in the southeast, the summer weather was quite variable. However, this variability took the form of quite long spells of fairly stable weather patterns, rather than short-term variation on a scale of days. From 11 June to 12 July there were storms and heavy rains. Then, from 12 July to 11 August, there occurred "the driest and best weather of the whole summer." On 12 August, when the hay harvest had just begun, a rainy period set in and lasted, with the exception of a very few days, to September. The rainy weather in the southeast in the early part of the summer may have been partially caused by the volcanic eruption that occurred in Skaftafellsjökull. This eruption, and the flood in the river Skeiðará, are discussed below.

Some western accounts of the summer weather show a pattern of variability not dissimilar to that noted above for Vík, although the timing and duration of cold, wet and dry spells is different in different locations. At Leirá in Borgarfjarðarsýsla, the Sheriff noted that the weather was cold, with northeasterly winds, sleet and night frost to about 24 June. Throughout July and August it was dry. At Reykjavík, just south of Leirá, the report stated that, on the whole, the weather was very cold and inconstant except for 3-17 August when the weather was mild and good. The account from Gröf, in Snæfellsnessýsla in the west (like Reykjavík, an exposed coastal site), contains daily weather data, and is much more detailed than most of the sources used here. It accords quite well with the Reykjavík report. The daily data given may be summarized as follows. On the whole, the month of June was mainly windy with snow, sleet or rain. Ten days were characterized as being calm. Only one day of rain is mentioned in July, but many days were described as breezy. No storms occurred in July. From 1-14 August, the weather was quite favourable. On 15 August and 19-24 August there were rain storms. On 25-27 August there were strong south-southwesterly winds with rainshowers, and on 28 August the wind was northeasterly with rain and fog. Northeasterly winds continued to the end of the month. One other western account, from the inland site of Síðumúli in Mýrasýsla, characterized the summer as "averagely good."

In spite of the reports from Síðumúli and Brandsstaðir, and some intervals of good weather at other sites, when the summer weather of 1816 is considered over Iceland as a whole, it must be classed as unfavourable. However, it was not extremely so, and the phrase "a year without a summer" does not, therefore, seem appropriate for this year in Iceland.

Autumn
Most sources characterize the autumn as mainly stormy and changeable. Thus, in the letter from Reykjavík, it is said to have been "stormy all autumn." The report from Gröf is of "stormy and inconstant rainy weather." At Viðvík, the weather is said to have been wet and inconstant, often with strong winds. At Leirá, southerly winds are noted. The weather was damp up to mid-October. At Vík, September is said to have begun with severe night frost. Subsequently, more rain than frost occurred. From 3 to 9 October, there was dry good weather with rime frost. From 10 October, there were mainly westerly winds with hail, snow, rain, sleet, frost and layers of ice on the ground. The autumn weather is said to have been, in general, "unusually changeable." According to *Brandsstaðaannáll*, there was snow and frost in late September. The first half of October was calm and dry, then snows fell.

Other Environmental Events
In 1816, a volcanic eruption also occurred in Iceland. This is known from the letter written by Sheriff Lyður Guðmundsson at Vík in Vestur-Skaftafellssýsla. According to him, the eruption began under Skaftafellsjökull (glacier) some time in May. In June, the eruption was visible over 16 miles (24 km) away, with an enormous column of rising vapour. "This later divided itself into clouds, and caused a bitingly sharp, cold drought until the clouds finally dispersed, and fell as a malignant, cold, severe and lasting heavy rain." The eruption does not appear to have had any serious effects on the populace, although the vegetable and hay crops were said to have been adversely affected.

Lyður Guðmundsson also reported flooding of the River Skeiðará on 17, 18 and 19 July. He described the river as "flowing out of the bowels of the Skaftafell glacier." Today, the river flows adjacent to the neighbouring Skeiðarárjökull. This discrepancy may be explained by the fact that the glaciers are undoubtedly smaller and of a different shape now than they were in the early nineteenth century, and the river is also likely to have changed its course. The Sheriff noted that the river flooded a large part of Skeiðarásandur (a stretch of sandy plain, washed out from the glaciers) and cut off all passage over a much greater distance. Probably the flood was largely caused by ice melting during the volcanic eruption.

Climatic Impact in Iceland in 1816

The Study of Climatic Impact: Methods and Approaches

In order to provide an analysis of past events that is as accurate as possible, recent research in the field of climatic impact has emphasized the need to adopt a rigorous methodology (Wigley *et al.* 1981; Kates *et al.* 1985). This is because of the difficulty in isolating and quantifying the effect that climate might have had on society, given all the other social, political and economic factors present. Such an exercise is difficult to carry out with present-day data. In the past, when fewer economic and climatic data were available, it becomes even more problematic. Although the difficulties arising from this can never be entirely eradicated, a number of measures may be adopted in order to provide a valid picture of possible climatic impact in the past. Important issues to consider before undertaking such a study are: (1) the location of the area to be studied; (2) the quality of the available data; (3) the time scale involved; (4) the economy and social structure of a given area (e.g., whether primitive or sophisticated); and (5) the strategy, or methodology to be employed. These points will be considered further below, with regard to Iceland.

Iceland's Location
Concerning the first of these points, location, Iceland occupies a marginal area on the borderline between environments that lend themselves easily to human habitation, and those that do not (e.g., Bergthórsson 1985). The cool climate of the area will obviously be a major factor in determining the growth of vegetation of all kinds. The many mountain and cold-desert regions mean that any attempts at agriculture will be limited, not only by the climate, but also by the amount of land available for such activities. Iceland's geographical situation thus makes it highly suitable for climatic-impact studies.

Data and Time Scales
Climatic data available for Iceland during the pre-instrumental era, and the specific sources used here, have been discussed above, and their quality established. Regarding the question of time scale, the data available make it possible to study climatic impacts in the long term (centuries), medium term (years to decades), and short term (a year or less) as it is here. The study of climatic impact in the short term has been criticized as giving undue attention to certain crisis years (Ingram *et al.* 1981). However, in this case, it is done within the context of previous studies of longer periods (Ogilvie 1981; 1984b).

Iceland's Economy and Society
Before about the mid-nineteenth century, no settlement large enough to be considered a town, or even a village, existed in Iceland. There were only isolated farmsteads and a few fishing stations on the coasts. The farms were scattered in order to make best use of the land available. Settlement was concentrated primarily in the coast and lowland areas of the southwestern, western and northern regions. The less hospitable areas of the northeast, southeast and northwest were even more sparsely populated.

Iceland's economy was based on animal husbandry: the main animals kept were sheep and cattle. These provided food in the form of meat and milk products, and also other useful items such as wool for clothing. Horses were used for transportation. Fishing was also important, but it was not until the twentieth century that this became a major industry.

During the short summer season, the major task for most Icelanders would be to bring in the annual hay harvest - still of great importance today. Hay was grown on the "homefields" (*tún*), near the farm, and on outlying pastures (*engi*). The hay was given to livestock during winter so that they could survive if there was little or no vegetation available. When the weather was favourable, certain of the livestock, particularly horses, sheep and gelded cattle, were expected to graze outside. These were known collectively as *útigangspeningur* or "outside livestock."

From 1380 to the Second World War, Iceland was ruled by Denmark. The trading monopoly enforced by Denmark for much of this time frequently worked in Iceland's disfavour as the Danish merchants controlled both prices and the goods available to the Icelanders.

Strategy
Throughout Iceland's recorded history, there are many "crisis-years." These are when the sources recount failure of the hay crop, livestock deaths, serious difficulties among the population such as the desertion of farms, begging, and even human mortality. Such events invariably occurred during very cold years or decades. Because people were so dependent on a successful hay harvest for supplementary winter fodder, it appears very likely that a poor harvest or a severe winter might have a considerable impact on the populace.

Rather than take this coincidence of events at face value, however, it is possible to adopt a strategy that will help to establish more clearly exactly what was occurring. To this end, this possible impact of climate may be divided into *direct* and *indirect* impact.

It is not difficult to demonstrate that climate had a considerable direct impact on biological and physical processes; (e.g., on grass growth, hay yield, and on other plants). This may be shown statistically (Ogilvie 1981, 1984b). For example, relationships between temperature, precipitation, grass growth and hay harvest may be tested by means of contingency tables (e.g., Table 2).

Indirect effects of climatic impact include deaths of livestock by starvation (although such effects may be compounded by direct impact in the form of cold and damp), and these may also be demonstrated by means of contingency tables (Table 3). In both of the tables shown here, the results are highly statistically significant.

Further indirect effects of climatic impact, such as the social problems mentioned above, are far harder to prove. Yet frequently much circumstantial evidence is available making it possible to show that such effects were very likely to have occurred (Ogilvie 1981). However, in all such studies, it is vital to take political, economic and social factors into account, as these invariably play a larger role than climate.

It is not possible to carry out the kinds of statistical tests mentioned above when considering data for one year only. However, as climate did have both direct and indirect impacts over longer time scales, clearly these would also be felt on an annual time scale. The reality of climatic impact has been demonstrated by several researchers using both modern and historical data (e.g., Bergthórsson 1966, 1985; Fridriksson 1969, 1972; Bergthórsson *et al.* 1988).

Table 2: Summer Temperature and Grass Growth in Iceland 1601-1780[1].

Grass growth	Summer Temperature			
	Cold	Average	Mild	Totals
Poor	52(24.6)	20(22.2)	20(45.2)	92(40.4%)
Average	7(17.1)	9(15.4)	48(31.4)	64(28.1%)
Good	2(19.3)	26(17.4)	44(35.4)	72(31.6%)
Totals	61(25.8%)	55(24.1%)	112(49.1%)	228(100%)

[1] $chi^2 = 84.0$

Table 3: Winter Temperature and Livestock Deaths in Northern Iceland 1601-1780[1].

Livestock	Winter Temperature			
	Cold	Average	Mild	Totals
Deaths	43(20.8)	2(6.5)	2(19.7)	47(28.1%)
Poor condition/disease	5(8.9)	6(2.8)	9(8.4)	20(12.0%)
Good condition	26(44.3)	15(13.8)	59(41.9)	100(59.9%)
Totals	74(44.3%)	23(13.8%)	70(41.9%)	167(100%)

[1] $chi^2 = 62.8$

In the section below, climatic impact during 1816 is considered. The direct impact of climate on grass growth and harvest, and on the vegetable crop, plus the direct physical impact of sea ice, is discussed first. Second, indirect climatic impact on domestic animals and humans is considered.

Direct Impacts of Climate
Grass Growth and Harvest
During 1816, both grass growth and hay harvesting varied considerably around the country in terms of quality and quantity. Only one source, *Brandsstaðaannáll*, from Blöndudalur in Húnavatnssýsla, gives an unqualified report that the grass grew well. Haymaking began on 25 July, and there was a successful hay harvest. Another source, that written at Síðumúli in Mýrasýsla, categorizes the grass growth as "quite good", and states that the harvest, and subsequent use of the hay, also went reasonably well. It is interesting to note that these are the only two sources which report a good, or, in the latter case, an "averagely good" summer as far as weather is concerned.

At Ketilsstaðir in the east, Sheriff Páll Þórdarson Melsteð judged that the grass growth was good "on the whole", although "lack of warmth" meant that the outlying pastures did not grow as well as the homefields. The harvest, however, was below average. This the Sheriff attributed to the damp and rainy summer which prevented the hay drying. The Sheriff of Snæfellsnessýsla in the west, Sigurður Guðlaugsson, who lived at Gröf, noted a similar situation. The grass seemed to grow well, but in the end turned out to be average. "The harvest from the homefields was very mediocre due to rain and damp weather." Writing about the autumn of 1816, he commented further: "On account of the autumn's stormy and inconstant rainy weather, the harvest was very poor in many parts of the district, especially on higher ground where some of the hay blew away and was washed away from the ground." The opposite situation is reported by Jónas Scheving, Sheriff of Borgarfjarðarsýsla at Leirá: "The grass growth, especially in the outlying pastures was average, but poorer from the homefields. However, the actual harvesting was excellent." The average to poor grass growth he attributed to cold weather from mid-May to about 24 June and, more particularly, the dry weather which followed this. The harvest "did not begin until the end of this month (July)." The dry weather, which lasted to the end of August, undoubtedly facilitated the harvest. The final state of the grass is also said to have been average in the account written at Reykjavík. However, grass growth was said to be very late due to cold spring weather. The harvest was "difficult." A letter of March 1817, states that in Árnes district, in the south, "the weather is supposed to have been not unfavourable to the harvesting of the hay." Furthermore, in spite of the difficult harvest, "with the exception of a few individual farms in Kjós district" there has not been a lack of hay up to this time. However, the severity of the winter 1816-17 meant that the upland farmers had to give outside livestock hay almost constantly. "It is thus feared that if the winter should remain severe during the present and next month, the lack of this item will be considerable."

In most northern districts, the situation regarding grass and hay during the summer and autumn of 1816 seems to have been more difficult than that of most other regions. Stefán Þórarinsson, writing from Möðruvellir in Eyjafjarðarsýsla, commented that, as a result of the cold spring, the grass growth was no more than average in most places in the north. This is echoed by other letters from the north. The account from Viðvík, for example, states that cold, dry weather prevented grass growth; and at Grund, Sheriff Gunnlaugur Briem noted that grass growth was unfavourable due to a very cold summer. All these northern letters mention that an epidemic, which affected people in many parts of Iceland this summer, served to hinder the hay harvest. The letter from Viðvík also commented on the rain that set in during the middle of the harvest. Stefán Þórarinsson also noted that long-lasting rain during September, together with storm winds that blew some of the hay away, caused a setback to the harvest of the outlying pastures, and resulted in this being, in his opinion, below average.

Comments on grass growth and the harvest in the different sources used here are summarized in Table 4. Also included is a summary of the characteristics of the winter, spring and summer seasons. In Table 5 the perceptions of the writers on how the weather affected the grass and harvest are shown. The main characteristics of the spring and summer seasons, plus grass growth and hay yield at each location, are shown in Table 6. Spring weather and grass growth are compared, and summer weather and the harvest. There can be little doubt that the summer weather directly affected the harvest. For example, if rain or snow or strong winds occurred, the harvest would be jeopardized. The exact effect of the spring weather on grass growth is much more complex, involving other variables such as soil condition, use of fertilizer, etc., but, from previous work (Bergthórsson 1966; Fridriksson 1972; Ogilvie 1981, 1984b; Bergthórsson *et al.* 1988) it is known that unfavourable weather (whether excessively cold, dry or wet) has a damaging effect on grass growth. It is interesting therefore to compare the incidence of favourable/unfavourable weather with favourable/unfavourable grass growth or harvest in the different locations (Table 6). Where these coincide a line is drawn between them. The harvest and summer weather agree in every case but one. However, it would be reasonable to assume agreement in this latter case also, as the Mödruvellir site, where the summer weather was not reported, lies only a few kilometres from Grund where the weather was said to be very cold. Grass and spring weather, as might be expected, do not agree as well, but the agreement (in six out of 11 cases) is nevertheless striking.

Vegetable Cultivation

From the latter part of the eighteenth century onwards, a serious attempt was made by the Danish authorities, and by enlightened individuals, to get ordinary people to supplement their diet by growing vegetables. The most commonly-planted species were potatoes, cabbage and turnips. These crops failed almost everywhere in Iceland in 1816. At Vidvík in the north, for example, Sheriff Jón Espólín noted that the number of gardens in use had increased greatly, but that they had not done well this year due to "the severe weather and storms" and also to the epidemic which affected people almost the whole summer, and prevented them from working. Early in 1817, he wrote again, commenting that gardening activity had ceased as the ground was frozen. He continued: "... one cannot think without sorrow of... the many years of dearth in most places in this district..."

Accounts from elsewhere for 1816 are similar to Jón Espólín's. Stefán Þórarinsson, writing from Mödruvellir, stated that some turnips and cabbage had grown, but that the potato harvest had failed completely. Sheriff Jónas Scheving, at Leirá, wrote that vegetables had done very badly over the past year. This he attributed to lack of sufficient seed, and also to cold spring weather, and dry weather in July. A poor vegetable crop also occurred in Vestur-Skaftafellssýsla.

However, the Sheriff there, Lydur Gudmundsson, mainly attributed their "pale and sickly appearance" to the effects of the volcanic eruption that occurred under Skaftafellsjökull in June 1816.

As with the grass growth and harvest, it seems reasonable to assume that, aside from the effects of this eruption, the weather of 1816 did play a considerable role in the failure of the vegetables. This is also suggested by previous work on crop/climatic relationships (e.g., Parry *et al.* 1988).

The Impact of Sea Ice

As noted in the early part of this paper, Iceland is close to the seasonal boundary of Arctic drift ice. When the ice reaches Iceland (most commonly, the northern, northwestern and eastern coasts) the most striking climatic effect is a lowering of temperatures in the areas affected (see also Wilson, this volume, regarding the cooling effect of sea ice lingering near the eastern coast of Hudson Bay). Rain and mist may be associated with the ice. The presence of the ice also has a direct physical impact. Because the ice prevents access to the open sea or makes it hazardous, activities such as fishing and sealing are prevented or hindered. This is no less true today than in past centuries, but, in the twentieth century, sea ice has not been common near Iceland. Other activities, such as gathering of shellfish from the shore, and the grazing by livestock of seaweed and marine plants, are also curtailed by land-fast ice. Such dietary supplements for humans and animals are of relatively little importance today, but played a vital role in the past.

The sea ice did bring some benefits, mainly in the form of driftwood and the occasional beached whale or other sea mammal, driven ashore by the encroaching ice. Wood was always in short supply, and a whale would greatly augment the food supply. For a more detailed discussion of the effects of the sea ice on flora and fauna, see Fridriksson (1969).

During the period 1809-20, heavy ice years occurred in 1811, 1812 and 1817. During these years, ice was present off the northern coasts and elsewhere from some time in January to July or August. During 1818 and 1819 very little ice appeared. The former year was very unusual in that the sea ice occurred in August, although not for long. In the latter year ice was seen briefly in April.

The year 1816 may be classed as a moderate ice year. During this year, sea ice affected the northern coast of Iceland from the beginning of March to the middle of July. Stefán Þórarinsson, at Möðruvellir, commented that the ice caused persistent northerly winds, frost and cold air. Briefly, ice prevented the arrival of the first trading ships at Eyjafjord. At Garður, in Suður-Þingeyjarsýsla, Sheriff Þorður Björnsson stated that the seal fishing had been very good until sea ice came and prevented this. The shark fishing was poor for the same reason. According to the account at Gröf, ice also prevented fishing in parts of Breidafjördur, in the west. But the layers of ice "far out to sea" reported by Sigurður Guðlaugsson, were caused by the sea itself being frozen, and not by actual sea ice. The Sheriff commented that in the 11 years he had been there, the fishing had never been as poor as this year.

Although sea ice undoubtedly caused some inconvenience during 1816, there is little evidence to suggest that it had a major impact on food supplies.

Indirect Impacts of Climate

Livestock

Most sources mention the severe winter this year, and the frequently frozen ground that prevented grazing. Nevertheless, there were no serious losses of livestock. Indeed, only Stefán Þórarinsson, writing at Möðruvellir in Eyjafjord district in the north, reported that some people lost a number of their outside livestock. He wrote:

Table 4: Summary of Seasons, Grass Growth and Harvest in 1816.

Place	Seasons[1]	Grass	Harvest
Ketilsstaðir	W - Average Sp - Cold, calm Sm - Wet	Good on homefields; not as good on outlying pastures	Below average
Garður	W - Very severe Sp - Sea ice present Sm - Rain, snow, frost	――	Poor
Möðruvellir	W - Very severe Sp - Quite severe Sm - ――	Average	Below average
Grund	W - ―― Sp - ―― Sm - Very cold	Unfavourable	Unfavourable
Viðvík	W - Severe Sp - Dry and cold Sm - Dry & cold then wet	Poor	Poor
Brandsstaðir	W - Mainly severe Sp - Weather improved Sm - Dry and good during harvest	Good	Good
Gröf	W - Fairly severe Sp - Variable Sm - Variable	Average	Very poor
Síðumúli	W - Severe Sp - ―― Sm - Averagely good	Quite good	Reasonably good
Leirá	W - Very severe Sp - Mainly severe Sm - Dry to end Aug.	Average on pastures; poorer on homefields	Very good
Reykjavík	W - Severe Sp - Unusually cold Sm - Mainly cold and inconstant	Average	Difficult
Vík	W - Very severe Sp - Severe Sm - Unfavourable	Poor	Meagre and spoilt

[1] W winter; Sp spring; Sm summer

Table 5: Contemporary Perceptions of Climatic Impact on Grass Growth and Harvest in 1816.

Ketilsstaðir
Cold spring meant that the outlying pastures did not grow as well as the homefields. Nevertheless, grass growth good on the whole. Harvest below average due to wet summer. Not possible to dry hay - therefore stacked up damp.

Garður
Harvest poor due to bad weather and epidemic.

Möðruvellir
In spite of the cold spring, the grass growth was about or almost average in most places here in the north. In the east it is said to have been poorer. The summer's harvest did not live up to the promise of the grass growth, however. This was due to the epidemic which occurred everywhere in the north at the beginning of the harvest. Then rains in September plus storm winds adversely affected the hay on outlying pastures. Thus, on the whole, harvest below average.

Grund
Grass growth unfavourable due to cold summer. Harvest also, primarily due to epidemic.

Viðvík
Grass did not grow well due to cold spring and summer weather. Harvest poor due to rains and epidemic.

Gröf
In most places the grass growth looked quite good to begin with, but turned out to be only average and, on account of wet weather, the harvest of the homefields was mediocre. Due to stormy and inconstant rainy weather, harvest very poor in many parts of the district, especially on higher ground where some of the hay blew away it was washed away from the ground.

Síðumúli
Dry weather from about 24 June meant that grass growth poorer than last years, so harvest did not begin until end July.

Reykjavík
As a result of the cold spring weather, the grass growth was only average and the harvest very difficult. Nevertheless, most people do not lack hay.

Vík
Poor grass growth due to volcanic eruption. Harvest spoiled by rains.

Table 6: A Comparison Between Spring Weather and Grass Growth, and Summer Weather and the Harvest in 1816.[1]

Location	Spring	Grass	Summer	Hay
Ketilsstaðir	Cold but calm ——	—— Good	Wet ——	—— Poor
Garður	Sea ice present	No comment	Rain, snow ——	—— Poor
Möðruvellir	Quite severe	Average	No comment	Below average
Grund	No comment	Unfavourable ——	—— Very cold ——	—— Unfavourable
Viðvík	Dry and cold ——	—— Poor	Dry and cold then wet ——	—— Poor
Brandsstaðir	Improved ——	—— Good	Good ——	—— Good
Gröf	Variable ——	—— Average	Good then wet ——	—— Poor
Síðumúli	No comment	Quite good	Averagely good ——	—— Reasonably good
Leirá	Mainly severe ——	—— Average to poor	Dry ——	—— Very good
Reykjavík	Very cold	Average	Mainly cold ——	—— Difficult
Vík	Severe ——	—— Poor	Unfavourable ——	—— Meagre and spoilt

[1] Lines between columns indicate coincidence of favourable weather with favourable grass growth or hay harvest and vice versa.

> ... the long lasting layers of ice in most places in this region... caused a good many farmers, here and there, to suffer a lack of fodder. They therefore lost a number of their so-called outside livestock, especially horses, due to emaciation. However, the latter loss (of the horses) only applied to some of the inhabitants of Skagafjord and Húnavatn districts who, to their own detriment, keep far too many horses. On the whole, the loss of outside livestock was neither general, nor of great importance.

Two other letters reported that lack of fodder meant that some livestock had to be slaughtered. These letters are from Garður, in Aðaldalur in the north, and Gröf in Snæfellsnessýsla, in the west. The account from the former stated that, although some people were forced to slaughter their livestock toward the spring, livestock deaths were not general. The latter source commented that in many places people had to slaughter their sheep as the usual winter grass failed in most places.

Other sources remark on the difficulties for livestock during 1816, but emphasize that, on the whole, they were kept alive. At Ketilsstaðir, in the east, the winter was said to be only average but, because of large amounts of snow in some places, the outside livestock had to be given fodder for a long time. However, "this did not last so long that the animals died of hunger." Sheriff Pétur Otteson, writing from Síðumúli in the west, stated that, because of the layers of ice and snow, virtually no grass was available for the livestock. He continued: "they would have died in great numbers if there had not been sufficient fodder after last year's good harvest.

The letter from Vík does not comment on the livestock during the winter, but says that, during heavy rain and storms from 11 June to 11 July, cows and ewes needed food and shelter. The Sheriff added: "After the severe winter, this could scarcely be spared," implying that, here too, the livestock needed extra fodder during the winter. According to this letter, the poor hay harvest this year caused livestock, especially cows, to be slaughtered in the autumn. From 9 October onwards, changeable weather with "hail, snow, layers of ice, rain, sleet and frost" meant that little grass was available. The Sheriff at Vík commented: "The horses and sheep have become emaciated and have sometimes needed to be given fodder, and this has had to be shared with the few remaining cows." The state of the livestock in the autumn and early winter is also noted by the Sheriff of Borgarfjarðarsýsla, at Leirá. After mid-October, "the winter set in with alternating frost and drifting snow, thaws and rain. This made it very difficult for the livestock, the horses and sheep who need to find their own food, as the frost caused the large quantities of snow and water which fell to form a frozen layer on the ground." We may conclude that, during 1816, conditions for livestock were, if not easy, not usually difficult either.

Social Stress
Research carried out for the period 1601 to 1780 (Ogilvie 1981) has shown that it is very likely that during this time climate did play a part in the occurrence of social stress, which manifested itself in such phenomena as the desertion of farms, begging and petty crime, plus hunger-related diseases and mortality among the people. During 1816, however, such problems were not widespread. Only one district reported general difficulties of this kind. This was Snæfellsnessýsla, in the west. Here, Sheriff Sigurður Guðlaugsson wrote:

> Great lack of food among inhabitants. People pressed by beggars from here and also from other districts. The majority of the district's populace have already got into debt at the trading places in previous years, and have scraped together all that they could in order to pay. So now they have to give all the best fish to the merchants and have little left for themselves except for flatfish and cod's heads. This is poor winter

provision, particularly on the coast among the poor fishermen who do not earn sufficient during the summer to buy other necessary foodstuffs from the farmers, and who therefore frequently live in the greatest misery.

The lack of food must be partly attributed to the fact that, as the Sheriff noted elsewhere in his letter, the trading places were very poorly supplied with corn wares and other imported foodstuffs. Furthermore, the fishing, of great importance in this district, largely failed this year. Clearly this was largely due to climate. The Sheriff describes how "although there should have been fishing in the latter part of the winter months, the severe frost and layers of ice far out to sea, frequently prevented the fisherman from getting out to sea for many days on end."

Because Snæfellsnes and nearby areas were important fishing centres, they attracted people whose inland sources of food had dwindled. Thus, although most other districts do not report social stresses this year, their silence on such matters may be partly attributable to the fact that the people in difficulties had already left to try their luck at the western and southern fishing stations.

Conclusions

During 1816, most districts in Iceland experienced a very severe winter. One source, the letter written by Sheriff Jón Espólín at Viðvík, compared it with two other very severe winters in recent times, 1784 (see Wood, this volume, regarding climatic effects of the Laki eruption) and 1802. The spring was also mainly severe in most places. It was a moderate sea-ice year, with ice present off the northern coasts from the beginning of March to mid-July. The summer was unfavourable, at least for part of the season, in most districts in Iceland with various combinations of excessive cold, wet or drought reported. In certain parts, the epithet "year without a summer" may have been appropriate, but if we consider the whole summer, over all Iceland, then it would not have been. The regional variability reflected in the sources used here is quite in accord with what is known of local climatic effects in Iceland (Eythórsson and Sigtryggsson 1971; Ogilvie 1984a).

If the summer of 1816 had been unfavourable in all parts of Iceland, as happened in true "years without summers" such as 1756 (Ogilvie 1981) and 1783 (Ogilvie 1986), then the climatic impact felt might have been greater. However, it might also have been greater if a favourable harvest had not occurred in 1815, thus boosting haystocks.

It is not difficult to demonstrate that direct impact, for example, on grass growth and hay yield did occur in 1816. The indirect role of climate on society this year is harder to define. While it is clear that there were difficulties amongst the populace, these were not widespread and were compounded by political and economic factors (e.g., by difficulties with trade). Several accounts this year report that supplementary foodstuffs received from Denmark were insufficient or of poor quality. There were also reports of poor fishing catches. It is true that fish are affected by climate, but the relationship is complex and, as yet, not fully documented. Certainly, poor fishing catches at sea are not directly linked to climate on land except in the case of heavy storms or when lowered temperatures cause ice to form on the sea, thus preventing fishing (as occurred off Snæfellsnes district this year). The presence of sea ice may also hinder fishing as happened off the north coast of Iceland this year.

In spite of the difficulty in allotting specific roles to economic, political and climatic factors in the general well-being of the Icelanders in 1816, there can be little doubt that some indirect climatic impact was felt this year. In the climatic context alone, 1816 was certainly an interesting year, if not a "year without a summer."

Acknowledgements

Dick Harington, Tim Ball and Cynthia Wilson deserve praise for their efforts in organizing the meeting " The Year Without a Summer? Climate in 1816" held in Ottawa June 1988. As always, I am grateful to many Icelanders for their help. Here I should like to acknowledge in particular Þórhallur Vilmundarson, Aðalgeir Kristjánsson and Trausti Jónsson. Part of the research for this paper was supported by grant GR3/7013 from the Natural Environment Research Council. This paper is dedicated to Valmore C. La Marche Jr. (1937-1988), who had been looking forward to joining in the debate on the climate of 1816.

> I had a dream, which was not all a dream
> The bright sun was extinguish'd, and the stars
> Did wander darkling in the eternal space,
> Rayless, and pathless, and the icy Earth
> Swung blind and blackening in the moonless air.

(From "Darkness" by Lord Byron. Written in 1816.)

References

Manuscript Sources
Þjóðskjalasafn (National Archives), Reykjavík.

Islands Journal 12
Stefan Þórarinsson. Möðruvellir. 26 September 1816, no. 2621.

Gunnlaugur Briem, Grund. 5 October 1816, no. 2650.

Jón Espólín, Viðvík. 26 August 1816, no. 2593 and 30 June 1817, no. 2818.

Sigurður Guðlaugsson, Gröf. 18 February 1817, no. 2811.

Pétur Ottesen, Síðumúli. 31 December 1816, no. 2808.

Johan Carl Thueracht v. Castenskiold, Reykjavík. 17 August 1816, no. 2519 and 5 March 1817, no. 2847.

Islands Journal 13
Pall Þórðarson Melsteð, Ketilsstaðir. 19 October 1816, no. 392 (formerly Islands Journal 12, no. 2645).

Jónas Scheving, Leirá. 31 July and 31 December 1816, no. 24.

Lýður Guðmundsson, Vík. 17 January 1817, no. 27.

Íslenzka Stjórnardeild 8
Þorður Björnsson, Garður. 23 September 1816, no. 505 (formerly Islands Journal 12, no. 2644).

Published Sources

Bell, W.T. and A.E.J. Ogilvie. 1978. Weather compilations as a source of data for the reconstruction of European climate during the medieval period. *Climatic Change* 1:331-348.

Bergthórsson, P. 1966. Hitafar og búsæld a Íslandi. *Veðrið* 11(1):15-20.

_____. 1969. An estimate of drift ice and temperature in Iceland in 1,000 years. *Jökull* 19:94-101.

_____. 1985. Sensitivity of Icelandic agriculture to climatic variations. *Climatic Change* 7:111-127.

Bergthórsson, P., H. Björnsson, O. Dýrmundsson, B. Guðmundsson, Á. Helgadóttir and J.V. Jónmundsson. 1988. The effects of climatic variations on agriculture in Iceland. *In: The Impact of Climatic Variations on Agriculture.* Volume 1: *Assessment in Cool Temperate and Cold Regions.* M.L. Parry, T.R. Carter and N.T. Konijn (eds.). Kluwer Academic Publishers, Dordrecht, Boston, London. pp. 383-509.

Bjarnason, B. 1941. *Brandsstaðaannáll.* Sögufélagið Húnvetningafélagið í Reykjavík, Reykjavík. 237 pp.

Bradley, R.S. 1988. The explosive volcanic eruption signal in northern hemisphere continental temperature records. *Climatic Change* 12:221-243.

Einarsson, M.Á. 1976. *Veðurfar á Íslandi.* Iðunn, Reykjavík. 150 pp.

Eyþórsson, J. 1956. Elztu Veðurathuganir með mælitækjum á Íslandi. *Veðrið* 1(1):27-28.

Eythórsson, J. and H. Sigtryggsson. 1971. The Climate and Weather of Iceland. *In: The Zoology of Iceland* 1(3). S.L. Tuxen (managing ed.). Ejnar Munksgaard, Copenhagen and Reykjavík. pp. 1-62.

Friðriksson, S. 1969. The effects of sea ice on flora, fauna and agriculture. *Jökull* 19:146-157.

_____. 1972. Grass and grass utilization in Iceland. *Ecology* 53:785-796.

Gunnlaugsson, G.Á., G.M. Guðbergsson, S. Þorarinsson, S. Rafnsson and Þ. Einarsson. (Editors). 1984. *Skaftáreldar 1783-1784 Ritgerðir og Heimildir.* Mál og Menning, Reykjavík. 442 pp.

Ingram, M.J., D.J. Underhill and T.M.L. Wigley. 1978. Historical climatology. *Nature* 276:329-334.

Ingram, M.J., G. Farmer and T.M.L. Wigley. 1981. Past climates and their impact on Man: a review. *In: Climate and History. Studies in Past Climates and Their Impact on Man.* T.M.L. Wigley, M.J. Ingram and G. Farmer (eds.). Cambridge University Press, Cambridge. pp. 3-50.

Kates, R.W., J.H. Ausubel and M. Berberian. (Editors). 1985. *Climatic Impact Assessment. Studies of the Interaction of Climate and Society.* SCOPE 27. Scientific Committee on Problems of the Environment (SCOPE) of the International Council of Scientific Unions (ICSU). John Wiley and Sons, Chichester, New York, Brisbane, Toronto, Singapore. 625 pp.

Kelly, P.M. and C.B. Sear. 1984. Climatic impact of explosive volcanic eruptions. *Nature* 311:740-743.

Kelly, P.M., T.M.L. Wigley and P.D. Jones. 1984. European pressure maps for 1815-16, the time of the eruption of Tambora. *Climate Monitor* 13:76-91.

The King's Mirror (Speculum Regale - Konungs Skuggsjá). 1917. Translated by L.M. Larson. *Scandinavian Monographs* 3. The American-Scandinavian Foundation, New York. 388 pp.

Kington, J.A. 1972. Meteorological observing in Scandinavia and Iceland during the eighteenth century. *Weather* 27:222-233.

Koch, L. 1945. The East Greenland ice. *Meddelelser om Grønland 130(3)*:1-373.

Lamb, H.H. 1970. Volcanic dust in the atmosphere; with a chronology and assessment of its meteorological significance. *Philosophical Transactions of the Royal Society of London* A266:425-533.

Landsberg, H.E. and J.M. Albert. 1974. The summer of 1816 and volcanism. *Weatherwise* 27:63-66.

Ogilvie, A.E.J. 1981. Climate and society in Iceland from the medieval period to the late eighteenth century. Unpublished Ph.D. dissertation. School of Environmental Sciences, University of East Anglia, Norwich. 504 pp.

_____. 1984a. The past climate and sea-ice record from Iceland, Part 1: data to A.D. 1780. *Climatic Change* 6:131-152.

_____. 1984b. The impact of climate on grass growth and hay yield in Iceland: A.D. 1601 to 1780. *In: Climatic Changes on a Yearly to Millenial Basis.* N.A. Mörner and W. Karlén (eds.). D. Reidel Publishing Company, Dordrecht. pp. 343-352.

_____. 1986. The climate of Iceland, 1701-1784. *Jökull* 36:57-73.

_____. 1990. Documentary evidence for changes in climate in Iceland A.D. 1500 to 1800. *In*: *Climate Since A.D. 1500*. R.S. Bradley and P.D. Jones (eds.). Harper Collins. (in press).

_____. 1991. Climatic changes in Iceland c.A.D. 865 to 1598. *Acta Archaeologica*. (in press).

Parry, M.L., T.R. Carter and N.T. Konijn. (Editors). 1988. *The Impact of Climatic Variations on Agriculture. Volume 1: Assessment in Cool Temperature and Cold Regions*. Kluwer Academic Publishers, Dordrecht, Boston, London. 876 pp.

Sear, C.B., P.M. Kelly, P.D. Jones and C.M. Goodess. 1987. Global surface-temperature responses to major volcanic eruptions. *Nature* 330:365-367.

Sigfúsdóttir, A.B. 1969. Hitabreytingar á Íslandi 1846-1968. *In*: *Hafísinn*. M.Á. Einarsson (ed.). Almenna Bókafélagid, Reykjavík. pp. 70-79.

Stommel, H. and E. Stommel. 1979. The year without a summer. *Scientific American* 240(6):134-140.

Stothers, R.B. 1984. The great Tambora eruption in 1815 and its aftermath. *Science* 242:1191-1198.

Vilmundarson, Þ. 1972. Evaluation of historical sources on sea ice near Iceland. *In*: *Sea Ice: Proceedings of an International Conference, 10-13 May 1971*. Þ. Karlsson (ed.). National Research Council, Reykjavík. pp. 159-169.

Wigley, T.M.L., M.J. Ingram and G. Farmer. (Editors). 1981. *Climate and History. Studies in Past Climates and Their Impact on Man*. Cambridge University Press, Cambridge. 530 pp.

First Essay at Reconstructing the General Atmospheric Circulation in 1816 and the Early Nineteenth Century

H.H. Lamb[1]

Reconstructions of the general atmospheric circulation in January and July year by year back to 1750, based on the best available network of monthly mean M.S.L. barometric pressure values over as much of the world as possible, from observation data in the archives and library of the United Kingdom Meteorological Office, were published by Lamb and Johnson (1959, 1961, 1966). The maps were all analyzed by me, and the analyses were tested by a simulation procedure: maps of the years 1919-39 were first analyzed using only restricted networks of data corresponding to the information available in the period 1786 to 1820, and these were then compared with maps for the same (inter-war) years analyzed with the use of full data. The distribution of errors was then studied. On this basis, it was decided that isobars on maps drawn for years in the late eighteenth and early nineteenth centuries could be considered satisfactorily reliable within regions where the standard error on the test maps was less than 1.0 mb in July (or less than 2.5 mb in January, this figure corresponding approximately to the ratio of the standard deviation of the observed values in January compared with those in July).

This meant in practice that isobars could only be presented with confidence over, or very close to, Europe between southern Scandinavia, Britain and the western Mediterranean on the maps for individual Januarys and Julys in the decade 1810-19. Decade and longer-term mean isobars could be reliable over a wider area, spanning most of the Atlantic Ocean between latitudes about 30 and 50 to 65°N. Isobars at 5-mb intervals were printed as unbroken lines over the areas established by the tests as reliable within the limits mentioned (and in regions of slack pressure gradients an intermediate isobar might be drawn in at a 2.5-mb interval).

On the maps for individual Januarys and Julys the isobars were extended, as broken lines, over regions where it seemed that the pattern must be broadly reliable, though the pressure values could not be relied upon.

In the case of July 1816 - as with some other seasons of historically dramatic weather - use could be made of a wealth of descriptive data on the weather experienced in many places so that it seemed reasonable to extend the isobar pattern, as broken lines, far beyond the limits of where the pressure values were known. This produced the map for the average conditions prevailing in July 1816 (Figure 1).

The coldness of that summer in eastern Canada, and in northeastern North America generally, appears here as attributable to prevalence of air drawn directly from the Canadian Arctic and the closeness of a focus of cyclonic activity to Labrador, Newfoundland and off-lying waters. The coldness of the summer in Britain, southern Scandinavia and the western part of continental Europe is seen to be due to the prevailing concentration of a low pressure region - unusually far south for summer - over the areas named, together with indraught of Arctic air from the source regions nearby. This is a similar explanation to that more tentatively shown for northeastern North America.

---

[1] Climatic Research Unit, University of East Anglia, Norwich NR4 7TJ, U.K.

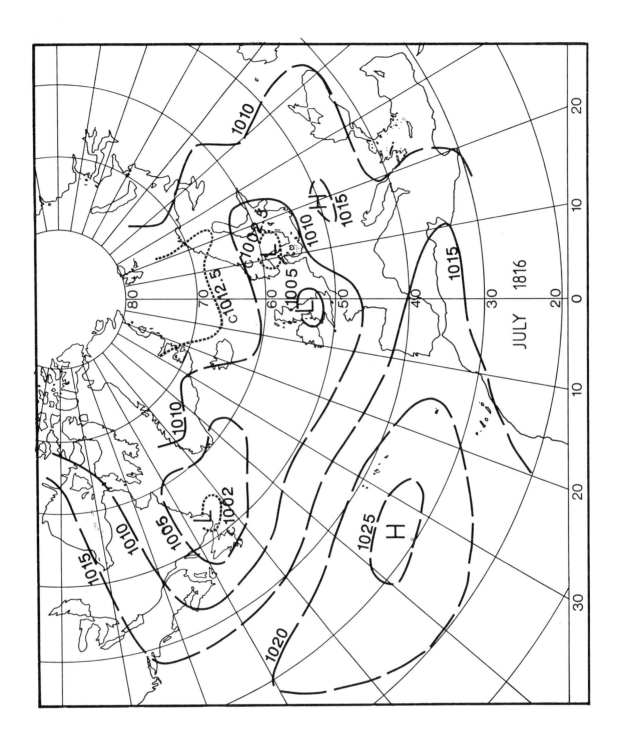

Figure 1: Average pressure conditions reconstructed for the area between eastern North America and western Europe (July 1816). See text for explanation.

The much better weather (and crops) experienced in Shetland - and to some extent all over the northern half of Scotland - and elsewhere in northern and also eastern Europe, extending south to the Crimea, is readily attributable to the higher pressures (and probably greater sunshine) over those areas.

References

Lamb, H.H. and A.I. Johnson. 1959. Climatic variation and observed changes in the general circulation: Parts I and II. *Geografiska Annaler* 41:94-134.

_____. 1961. Climatic variation and observed changes in the general circulation: Part III. *Geografiska Annaler* 43:363-400.

_____. 1966. Secular variations of the atmospheric circulation since 1750. *Geophysical Memoir* 110. (Her Majesty's Stationery Office, for Meteorological Office). London. 125 pp.

Weather Patterns over Europe in 1816

John Kington[1]

Abstract

An outline of the state of meteorology during the early nineteenth century is presented with particular reference to the introduction of the synoptic method of analyzing daily weather maps by Heinrich Brandes in 1816.

Links with the historical weather data made and collected in the 1780s are mentioned in relation to the series of daily weather maps for Europe that I am constructing from 1781.

The feasibility of undertaking a program of similar research for a period of years centred on 1816 is discussed. As an example, a run of daily charts for Europe in July 1816 is presented together with a preliminary analysis of the circulation patterns brought to light in the process.

Comparisons are made with events in the 1780s, in particular the cold summer of 1784 that followed the formation of the exceptional volcanic dust veil after the great eruptions in Iceland and Japan the preceding year.

Historical Weather Data: Comparison of 1816 with the 1780s

Writing in Breslau, Silesia towards the close of 1816, the German meteorologist Heinrich Brandes observed:

> ... If one could collect very accurate meteorological observations, even if only for the whole of Europe, it would surely yield very instructive results. If one could prepare weather maps of Europe for each of the 365 days of the year, then it would be possible to determine, for instance, the boundary of the great rain-bearing clouds, which in July [1816] covered the whole of Germany and France; it would show whether this limit gradually shifted farther towards the north or whether fresh thunderstorms suddenly formed over several degrees of longitude and latitude and spread over entire countries ... In order to initiate a representation according to this idea, one must have observations from 40 to 50 places scattered from the Pyrenees to the Urals. Although this would still leave very many points uncertain, yet by this procedure, something would be achieved, which up to now is completely new.

As a meteorological observation network did not then exist, Brandes was unable to examine the weather conditions of July 1816 but pursued his hypothesis by making use of data collected 30 years earlier by the *Societas Meteorologica Palatina*. Thus the first observations to be studied by means of the synoptic method devised by Brandes were those for 6 March 1783 (Figure 1), a day on which, like many of those in July 1816, stormy weather prevailed over western and central Europe.

---

[1] Climatic Research Unit, University of East Anglia, Norwich NR4 7TJ, U.K.

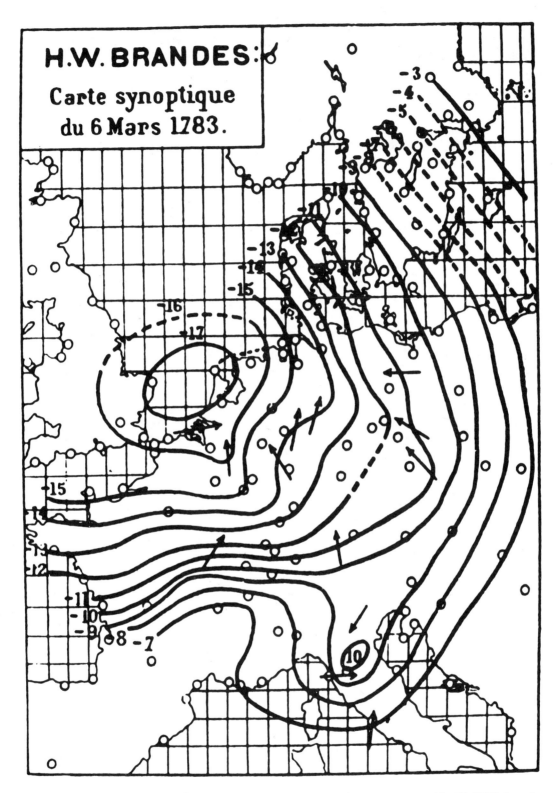

Figure 1: Synoptic weather map for 6 March 1783 by H.W. Brandes, reconstructed by H. Hildebrandsson. Surface wind directions are shown by arrows and the field of pressure by isopleths of equal departure of pressure from normal (e.g., -17, -16, -15, etc.). By overcoming the uncertainty about the height at which the barometer readings were made, the observations were successfully combined to allow the equivalent of isobars to be drawn at a constant level (from Ludlam 1966).

During the Enlightenment, hopes had been raised in scientific circles that a systematic study of meteorological observations would show that the seemingly disordered array of weather variations were subject to predictable forms of behaviour. Consequently, extensive networks of observing stations were established by two scientific societies in Europe during the early 1780s, namely, the *Société Royale de Médecine* and the *Societas Meteorologica Palatina* centred at Paris and Mannheim, respectively.

Unfortunately, tangible results proved to be elusive by the statistical approach then applied, which earlier had been so successful in predicting the motion of the stars and planets. However, the collections of reports from these two societies, together with further data from private individuals and ships' logs, means that a large array of daily instrumental and quantitative observational material became, and is still, available for the 1780s over Europe. These data provide a network of more than the "40 to 50 places" advocated by Brandes (Figure 2). This information is now being subjected to twentieth-century concepts in synoptic meteorology to "yield very instructive results" in the construction of daily historical weather maps as envisaged by Brandes 170 years ago. As an example of the series of charts now becoming available from 1781, the map for the same day as earlier constructed by Brandes (6 March 1783) is illustrated in Figure 3.

The early blossoming of meteorology in the late eighteenth century was brought to a halt by the political confusion and social unrest that followed the outbreak of the French Revolution. The two main scientific societies which had been promoting international cooperation in the exchange of weather information were disbanded in the mid-1790s. After a lapse of two decades, was it the "year without a summer" with its exceptionally cold wet weather and disastrous harvests that provided the stimulus for a revival of efforts to understand and predict weather changes? In any event, the idea of mapping over a large area simultaneous daily observations of meteorological elements such as pressure, wind and temperature (the concept upon which synoptic weather studies are based) was, as earlier stated, presented at this time by Brandes.

Having demonstrated recently that it is indeed possible to map historical weather data on a daily basis over Europe for the 1780s (Kington 1988), and knowing that comparable, albeit less well sorted and organized, observations are available for 1816, a pilot scheme was initiated for a monthly period in that year, following the kind invitation to attend this conference by Dr. C.R. Harington. July was chosen for several reasons, not least being the month first highlighted by Brandes in his letter of 1816, as quoted above.

In 1967 the German climatologist, Hans Von Rudloff, examined the weather patterns of 1816 in his study of the fluctuations and oscillations of European climate since the beginning of instrumental weather observing in the seventeenth century. His analysis showed that there was an abnormal distribution of pressure over Europe in the summer of 1816. The subtropical high pressure system, the "Azores High", which usually extends northeastwards over the region at times during the summer, appears to have been completely absent. Instead, systems of low pressure persisting over central Europe allowed polar air streams to be advected farther south than normal over the region (Von Rudloff 1967).

At about the same time as Von Rudloff's study, Hubert Lamb presented an investigation of secular variations in atmospheric circulation since 1750 by means of a series of maps showing mean pressure distribution for the months of January and July (Lamb 1967). The chart for July 1816 (Lamb, this volume) again shows an unusual distribution of pressure, with the "Icelandic Low" positioned well to the south of its normal summer latitude.

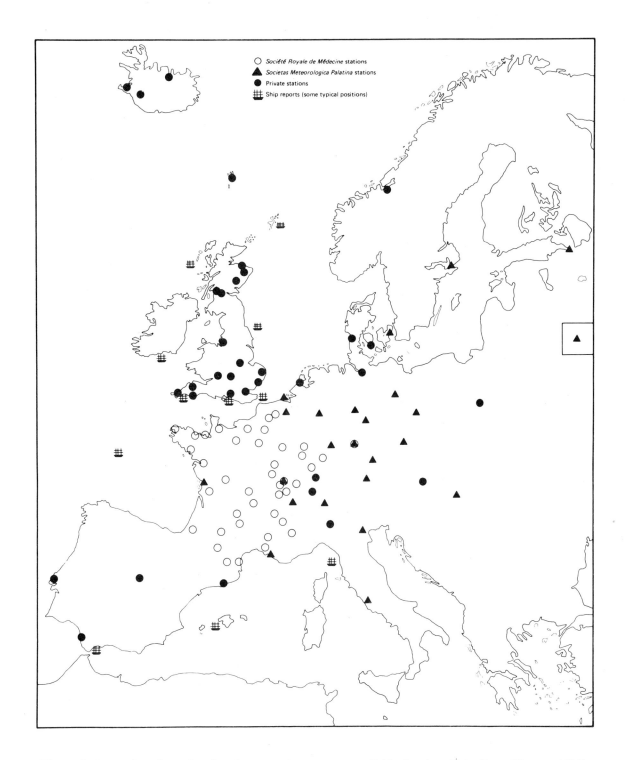

Figure 2: Map of stations showing the synoptic coverage available for the 1780s (from Kington 1988).

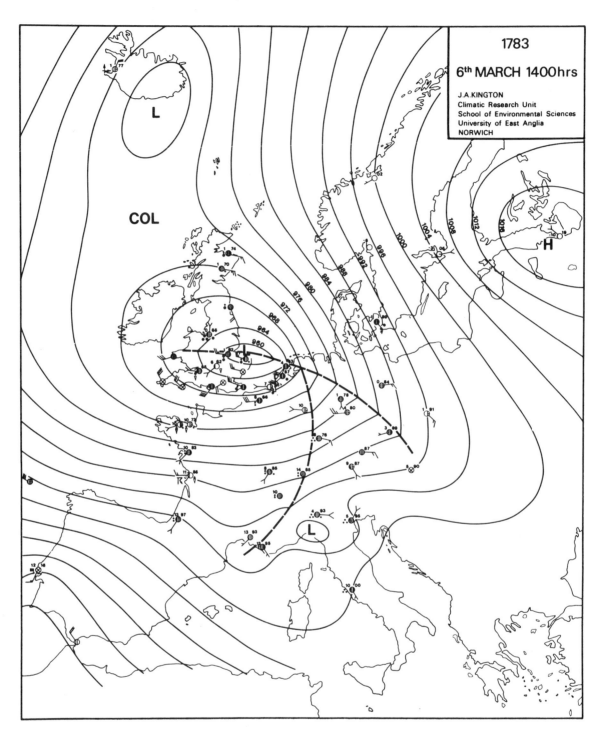

Figure 3: Synoptic weather map for 6 March 1783 (from Kington 1988).

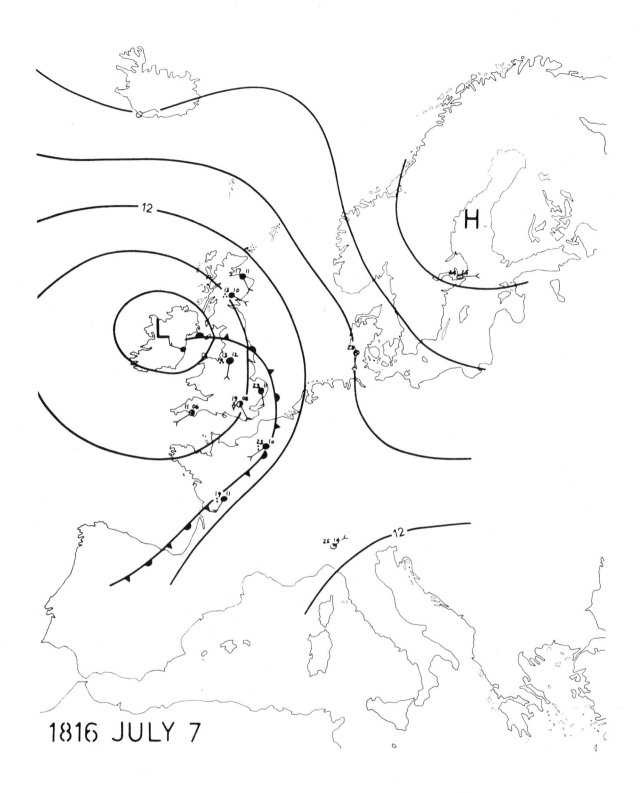

Figure 4: Synoptic weather map for 7 July 1816 illustrating the Lamb British Isles Cyclonic weather type.

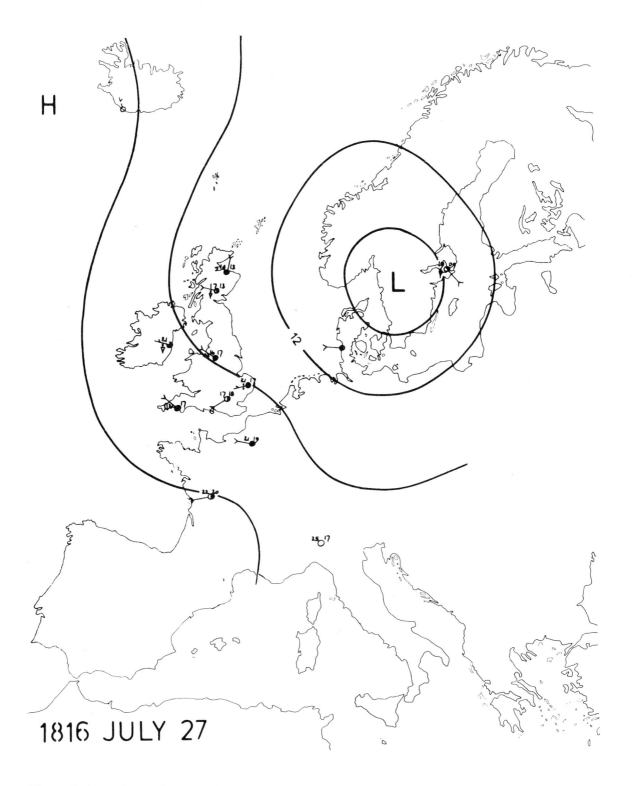

Figure 5: Synoptic weather map for 27 July 1816 illustrating the Lamb British Isles North Westerly weather type.

More recently, in the reconstructions of monthly pressure patterns for Europe back to 1780 based on principal components regression techniques, Jones, Wigley and Briffa presented a map of pressure anomalies for July 1816 that shows an unusually large negative area, in excess of seven millibars, over the British Isles and southern North Sea (Jones et al. 1987).

All these works strongly indicate that some very pronounced regional anomalies occurred in the circulation over Europe in July 1816. Can we discover more? Yes, because an investigation of weather patterns on a *daily* basis can reveal aspects of atmospheric behaviour that are not possible to detect from studies made on monthly or longer time scales.

Although, as previously stated, the two major observation networks of the 1780s were disbanded in the following decade, a number of the original stations continued in operation, while others were newly established during the early part of the nineteenth century. Using a nucleus of such data (readily on hand in the Climatic Research Unit), a run of daily weather maps for July 1816 was specially prepared for this book.

The charts have been analyzed and classified with Professor Lamb according to his system of British Isles weather types (Lamb 1972). This scheme aims to represent the main types of circulation patterns prevalent over the British Isles, namely: Westerly (W), North Westerly (NW), Northerly (N), Easterly (E), Southerly (S), Anticyclonic (A) and Cyclonic (C). Since the British Isles are centrally placed in the mid-latitude westerly wind belt, as well as being located in one of the sectors around the northern hemisphere most frequently affected by blocking of this flow, variations in the circulation over the more extensive North Atlantic-European region are also well registered by this classification. The classification for July 1816 is given in Table 1.

A statistical analysis of the classification (Table 2) shows that the circulation over the British Isles during July 1816 was strongly dominated by Cyclonic weather types (three times more frequent than usual). Of the other patterns, Northwesterly types were also more prevalent than usual (over twice as frequent); Southerly types about average; Westerly types, however, were about one third of the normal frequency, while Anticyclonic, Northerly and Easterly types were totally absent.

Typical examples of the two predominant weather types, Cyclonic and Northwesterly, are shown in Figures 4 and 5.

In Table 3 the frequencies of the Lamb British Isles weather types in July 1816 are compared with those for 1868-1967, 1781-85 and 1785.

This shows that frequency values for July 1816 are nearer to the averages for 1781-85 (a period in the Little Ice Age) than those of the standard period, 1868-1967. In particular, the circulation of 1816 closely parallels that of 1785 with its notable increases in Cyclonic and North Westerly types and corresponding decreases in Anticyclonic and Westerly types.

The Lamb British Isles weather types are also used to determine the *PSCM* indices of: progression, meridionality and cyclonicity, which provide a ready means of indicating the general character of the circulation over the region for a period of a month or more (Murray and Lewis 1966).

Table 1: Lamb British Isles Weather Types for July 1816.[1]

1	C	11	CNW	21	C
2	C	12	NW	22	SW
3	C	13	W	23	CS
4	C	14	C	24	C
5	C	15	C	25	C
6	C	16	C	26	W
7	C	17	C	27	NW
8	C	18	C	28	NW
9	C	19	C	29	CNW
10	C	20	S	30	NW
				31	U

[1] C Cyclonic, CNW Cyclonic North Westerly, NW North Westerly, S Southerly, SW South Westerly, CS Cyclonic Southerly, W Westerly, U Unclassified.

Table 2: Lamb British Isles Weather Types. Monthly Frequencies for July 1816 with Long-Period Mean Percentage Values Given in Brackets for Comparison.[1]

		Days	%	%
W	1½ 1	2½	8	(26)
NW	½ 111½ 1	5	16	(7)
N	—	0	0	(7)
E	—	0	0	(4)
S	1 ½ ½	2	6	(5)
A	—	0	0	(24)
C	1111111111½ 1111111½ 11½	20½	66	(22)
U	1	1	3	(5)

[1] W Westerly, NW North Westerly, N Northerly, E Easterly, S Southerly, A Anticyclonic, C Cyclonic, U Unclassified.

In July 1816: P=-12 or -3; S=-3 or -1; C=+41 or +42 and M= 11 or 13[1]
That is: $P_{12}\ S_{34}\ C_5\ M_3$

This shows that the circulation over the British Isles in July 1816 was characterized by blocked or quasi-stationary cyclonic weather systems. The C index value of +41 or +42 is far greater than the maximum value of +30 (1936) in the official long-period record from 1861. Interestingly this record was also broken in the 1780s when the C index in July 1785 was +33.

In July 1785: P=-3; S=-14; C=+33; and M=14
That is: $P_2\ S_1\ C_5\ M_3$

---

[1] The slightly differing results are due to dealing with a run of charts from a single isolated month, resulting in a certain lack of synoptic continuity at the beginning and end of the series.

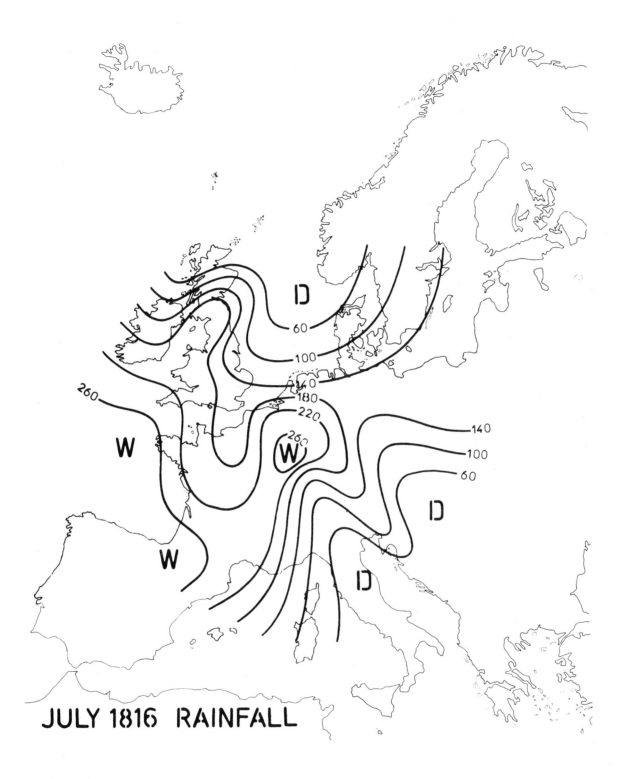

Figure 6: Rainfall anomalies (%) for July 1816.

Table 3: Lamb British Isles Weather Types. July Frequencies for 1816 and 1785; Period Average Frequencies for 1868-1967 and 1781-85.[1]

	Number of Days						
	W	NW	N	E	S	A	C
1816	2.5	5.0	0.0	0.0	2.0	0.0	20.5
1868-1967	8.1	2.2	2.2	1.2	1.5	7.4	6.8
1781-85	6.9	3.9	0.8	0.7	2.7	6.4	8.2
1785	1.0	6.5	2.0	0.0	0.0	1.5	18.0

[1] W Westerly, NW North Westerly, N Northerly, E Easterly, S Southerly, A Anticyclonic, C Cyclonic.

Thus there is a striking similarity (blocked and very cyclonic) in the *PSCM* "signatures" of 1816 and 1785.

As rainfall over England and Wales has been found to be closely correlated with the C-index, it is not surprising that very heavy falls of rain occurred over the region in July 1816 (Figure 6). The map, however, shows that it was not uniformly wet over the British Isles or continental Europe. For instance, while rainfall over southwestern Ireland, southern Wales, southwestern England, most of France, parts of Belgium, Holland and western Germany exceeded 200% of normal, northwestern Scotland, Orkney, Shetland, Denmark, Norway and Italy were drier than usual. Contemporary accounts confirm this contrasting pattern of wet and dry regions:

Europe
> Melancholy accounts have been received from all parts of the Continent of the unusual wetness of the season; property in consequence swept away by inundation, and irretrievable injuries done to the vine yards and corn crops. In several provinces of Holland, the rich grass lands are all under water, and scarcity and high prices are naturally apprehended and dreaded. In France, the interior of the country has suffered greatly from the floods and heavy rains.
>
> "The Norfolk Chronicle", 20 July 1816

Ireland
With depressions centred over or near Ireland for most of the month, the weather over the country was very unsettled and wet. Apparently, all parts had more rain than usual, with the extreme southwest probably having more than twice the normal amount (Figure 6).

> The summer and autumn were excessively wet and cloudy. ... the sun was in general obscured by clouds during the months of July, August and September.

> Great thunderstorms occurred during the month of July, accompanied with hail of an unusually large size. These storms were general throughout the country.

July - wet, great storms, and inundations in England and Scotland, as well as throughout this country ... The month was, without, perhaps, the exception of a single day, a continuity of showers of hail or rain, and at the same time very cold.

Snow remained on some of the hills in Scotland until the middle of July, during which month great thunderstorms occurred in England.

In consequence of the incessant rain, there is a great blight in the wheat crop, particulary in Wicklow and Tipperary: the rain was so severe that scarcely any corn was left standing. For many years so untoward a season had not been experienced, not one week of fine weather since May. Eight weeks of rain in succession. Hay and corn crops in a deplorable state. The grains of corn in many places are covered with a reddish powder like rust, which has proved very destructive to the crop, especially in the counties of Kilkenny and Antrim.[1] The wheat crop was especially injured. Great floods occurred in the Boyne.

The fields of corn presented a lamentable appearance, in many places being quite black. Before the crop was reaped, re-vegetation had commenced, and green shoots were perceived on the fields.

The harvest of grain was uncommonly late both in this country and in England; corn remained uncut during the latter parts of October and November, and much of it was altogether lost. The cold of this season proved highly injurious to the crop of potatoes also. These, which constitute the principal or only food of the poor in most parts of the country, were small and wet, and probably more defective in nutriment than the grain.

The potato crop both in England and Scotland was defective.

"The Census of Ireland," 1851

Denmark

This month for the most part good weather. Quite warm 21° and frequent rain, although this did not do any harm. On the other hand in Germany and Switzerland terrible damage occurred with rivers flooding. This was caused by persistent rain ... whole tracts of land were under water. The hay harvest was also ruined in England.

"A Jutland Weather Diary" (Ribe)

---

[1] This may have been the result of volcanic aerosol particles being washed out of the atmosphere by the rain which, in turn, might have been intensified by the increase in condensation nuclei. Editor's note: Perhaps the possibility of fungal rust should be considered also.

Western Russia and the Baltic Sea Coast
> The city of St. Petersburg [Leningrad] has for a month past suffered by drought and prayers for rain have been offered up at Riga and Dantzig while Germany is devastated by inundations and the churches of Paris are filled with suppliants praying the Almighty for dry weather.
>
> "Records of the Seasons"

Conclusion

One of the main objectives of this book has been to determine to what extent the Tambora eruption in 1815 affected world climate. Already we know that some mid-latitude regions of the northern hemisphere, such as eastern North America and western Europe, were much cooler than normal in the following year, 1816. There is an interesting parallel in the 1780s when it is estimated that annual mean temperatures in mid-latitudes fell by 1.3°C after eruptions in Iceland and Japan in 1783. However, there appear to be two major points of difference: the timing and length of cooling. By all accounts it appears that, unlike 1815-16, the cooling signal in the mid-1780s was strongest not in the year immediately following the eruption but in 1785, two years after the event. Nevertheless I have shown that there were some notable similarities in the circulation patterns of the two cold years, 1816 and 1785. Furthermore, the marked increase in cyclonicity over the British Isles in July 1816 is in accordance with Lamb's (1977) finding that there is a tendency for the subpolar low-pressure zone (the "Icelandic Low") to be displaced southwards over the British Isles during the first July after a great eruption, resulting typically in cold wet summers over the region. Another area of cyclonic activity near Newfoundland gave similar weather conditions over eastern North America. However, the volcanic signal apparently soon died away, with temperatures recovering to above normal values by 1818. On the other hand, after high-latitude eruptions (e.g., those of 1783), pressure and related temperature anomalies in mid-latitudes appear to persist longer - the circulation patterns determined for July in the cold year of 1785 confirm this trend.

Acknowledgements

Drs. A.E.J. Ogilvie and P.D. Jones kindly helped in processing various historical weather data from the archives of the Climatic Research Unit. Observations from Dublin and France were kindly supplied by Dr. J.G. Tyrrell (University College, Cork) and Dr. D. Hubert (Observatoire de Meudon), respectively.

References

Baker, T.H. 1883. *Records of the Seasons, Prices of Agricultural Produce and Phenomena Observed in the British Isles*. Simpkin, Marshall and Co., London.

Dublin. 1856. *The Census of Ireland for the Year 1851*. H.M.S.O., London.

Jones, P.D., T.M.L. Wigley and K.R. Briffa. 1987. Monthly mean pressure reconstructions for Europe (back to 1780) and North America (to 1858). *DOE Technical Report* No. 37, United States Department of Energy, Carbon Dioxide Research Division, Washington, D.C.

Kington, J. 1988. *The Weather of the 1780s Over Europe*. Cambridge University Press, Cambridge.

Lamb, H.H. 1972. British Isles weather types and a register of the daily sequence of circulation patterns, 1861-1971. *Geophysical Memoirs* No. 116. H.M.S.O., London.

_____. 1977. *Climate: Present, Past and Future*, Volume 2, *Climatic History and the Future*. Methuen, London.

Ludlam, F.H. 1966. *The Cyclone Problem: A History of Models of the Cyclonic Storm*. Imperial College of Science and Technology, London.

Murray, R. and R.P.W. Lewis. 1966. Some aspects of the synoptic climatology of the British Isles as measured by simple indices. *Meteorological Magazine* 95:193-203.

Von Rudloff, H. 1967. *Die Schwankungen und Pendelungen des Klimas in Europa seit dem Beginn der regelmässigen Istrumenten-Beobachtungen (1670)*. Vieweg, Braunschweig.

The Climate of Europe during the 1810s with Special Reference to 1816

K.R. Briffa[1] and P.D. Jones[1]

Abstract

The long climatic records available for Europe are used to place the seasonal temperature, precipitation and sea-level pressure anomaly maps for 1816 into their longer-term context. The prevailing climate of the decade of the 1810s (1810-19) is also described with reference to modern climatic normals. The 1810s were probably one of the coldest decades recorded over Europe since comparable records began about 1750. It was only the weather during the spring and, more particularly, the summer of 1816 that was highly anomalous with respect to both recent normals and those for the 1810s.

Tree-ring-based reconstructions of temperature for a 'summer' (April-September) season are available in the form of anomaly maps back to 1750. They indicate that the summer of 1816 was the coldest since 1750 in Britain, that it was the second coldest (after 1814) in central Europe and that in Scandinavia conditions were near normal.

Introduction

Many studies have considered the weather extremes that occurred during the summer of 1816, the so-called "year without a summer" (Landsberg and Albert 1974; Stommel and Stommel 1979). Studies have tended to concentrate on the particular season itself, rather than considering the weather and climate of the rest of 1816 and the decade of the 1810s.

In this article we propose to make use of the long records of temperature, precipitation and mean sea-level pressure (MSLP) available for most of Europe. We will describe seasonal anomaly maps for 1816 with respect to twentieth century reference periods and in relation to those of the 1810s (defined here as 1810-19). We also compare the climate of the 1810s to recent reference periods.

Finally, previously published maps of mean April-September temperature reconstructed from a network of maximum-latewood-density tree-ring chronologies in Europe are reproduced for each of the years 1810-19.

Data

Instrumental recording of air temperature and precipitation totals extends back in Europe to the late seventeenth century. Most of the pre-twentieth century data have been assembled in computer compatible form in data archives. Here we use the compilation of air temperature and precipitation data produced by Bradley *et al.* (1985). This archive contains temperature data for 46 stations in Europe with series that extend over most of the years of the 1810s (Table 1; Figure 1). Of these 46 stations, 12 do not have comparable data through to and encompassing the twentieth century. We can still use these more restricted data, however, to compare the average temperature of 1816 to that of the 1810s.

---

[1] Climatic Research Unit, University of East Anglia, Norwich NR4 7TJ, U.K.

Table 1: Names and Locations of Stations with Temperature Data for the 1810s Continuous to the Present Day.

		Lat.(°N)	Long.
1.	Trondheim	64.3	10.5E
2.	Stockholm	59.4	18.1E
3.	Torneo	66.4	23.8E
4.	Woro	63.2	22.0E
5.	Gordon Castle	57.6	3.1W
6.	Edinburgh	55.9	3.2W
7.	Manchester	53.4	2.3W
8.	Greenwich	51.5	0
9.	Copenhagen	55.7	12.6E
10.	De Bilt	52.1	5.2E
11.	Basel	47.6	7.6E
12.	Geneva	46.2	6.2E
13.	Montdidier	49.7	2.6E
14.	Chalons	48.9	4.4E
15.	Paris	48.8	2.5E
16.	Strasbourg[1]	48.6	7.6E
17.	Nice	43.7	7.2E
18.	Berlin	52.5	13.4E
19.	Karlsruhe[1]	49.0	8.4E
20.	Stuttgart[1]	48.8	9.2E
21.	Regensberg[1]	49.0	12.1E
22.	Augsburg[1]	48.4	10.4E
23.	Munchen[1]	48.1	11.7E
24.	Hohenpeissenberg[1]	47.8	11.0E
25.	Kremuenster[1]	48.1	14.1E
26.	Wien Hohe Warte	48.2	16.4E
27.	Innsbruck	47.3	11.4E
28.	Klagenfurt	46.7	14.3E
29.	Prague	50.1	14.3E
30.	Leobschutz	50.2	17.8E
31.	Gdansk	54.4	18.6E
32.	Warsaw	52.2	21.0E
33.	Wroclaw	51.1	17.0E
34.	Budapest	47.5	19.0E
35.	Udine	46.0	13.1E
36.	Turin	45.2	7.7E
37.	Milan	45.4	9.2E
38.	Padua	45.4	12.0E
39.	Bologna	44.5	11.5E
40.	Rome	41.7	12.5E
41.	Palermo	38.1	13.4E
42.	Arkhangel	64.6	40.6E
43.	Leningrad	60.0	30.3E
44.	Vilnjus	54.6	25.3E
45.	Kazan	55.8	49.1E
46.	Kiev	50.5	30.5E

[1] Not labelled on Figure 1.

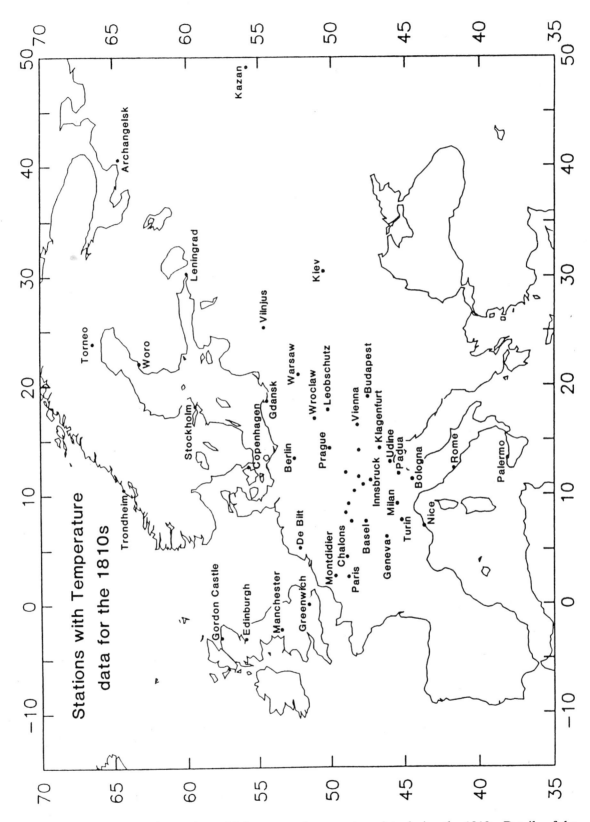

Figure 1: Locations of the 47 sites with instrumental temperature data during the 1810s. Details of the sites are given in Table 1 (from Bradley *et al.* 1985).

Although the Bradley *et al.* (1985) compilation contains details of the sources of the data and the methods, where known, by which the observations were made, it does not consider the long-term homogeneity of the individual station data sets. The homogeneity of the station temperature data used here has, however, been assessed by Jones *et al.* (1985).

For precipitation, the Bradley *et al.* (1985) compilation contains data for 29 sites covering the 1810s. Assessment of the homogeneity of the precipitation data is a considerably more difficult task than for air temperature. Although the stations used here have not been assessed for homogeneity, the data for 27 of the sites are among the 180 or so homogeneous European precipitation records assembled by Tabony (1980, 1981). Only the data for Warsaw and Prague are not in this set. The locations of the 29 precipitation sites we have used are shown (Figure 2; Table 2).

Some of these early temperature and precipitation series have been used in conjunction with early station pressure records by Jones *et al.* (1987) to reconstruct gridded monthly-mean mean sea-level pressure values (MSLP) over Europe extending back to 1780. Jones *et al.* used a principal components regression technique that involves fitting equations expressing MSLP at individual grid points in terms of pressure, temperature and precipitation series at all stations in the predictor network. The fitting was carried out over a 75-year (1900-74) 'calibration' period, and the reliability of the gridded reconstructions was assessed by comparing the estimated data with actual observations over an independent 'verification' period, 1873-99. Jones *et al.* (1987) showed that over Europe, between 65-40°N and 10°W-30°E, the reconstructions are of high quality with 80% or more of the variance of the observed pressure data being explained in each of the separate monthly reconstructions.

From this bank of reconstructed pressures we have extracted the data for individual months and averaged them to produce maps of MSLP anomalies for the four standard seasons of the year 1816.

Anomaly Maps

Figure 3 shows seasonal temperature anomaly maps for 1816 with respect to the reference period 1951-70. Winter in this and subsequent figures is taken to be December 1815 to February 1816. All four seasons are shown to have been generally cooler in 1816 compared with the reference period. Warmer conditions were experienced only over northern Mediterranean coasts and European parts of the Soviet Union, and then only in spring, summer and autumn. The most anomalously cool regions were Scandinavia and northern British Isles (during winter, spring and autumn) and central Europe (in summer).

Figure 4 shows seasonal precipitation anomaly maps for 1816 (expressed as percentages of the 1921-60 reference period). Most regions of western Europe were drier than normal except for summer. Below normal precipitation is evident over central and southern Europe in winter and to some extent in spring and autumn. During summer the only relatively dry areas were southern Italy and northwestern Scotland.

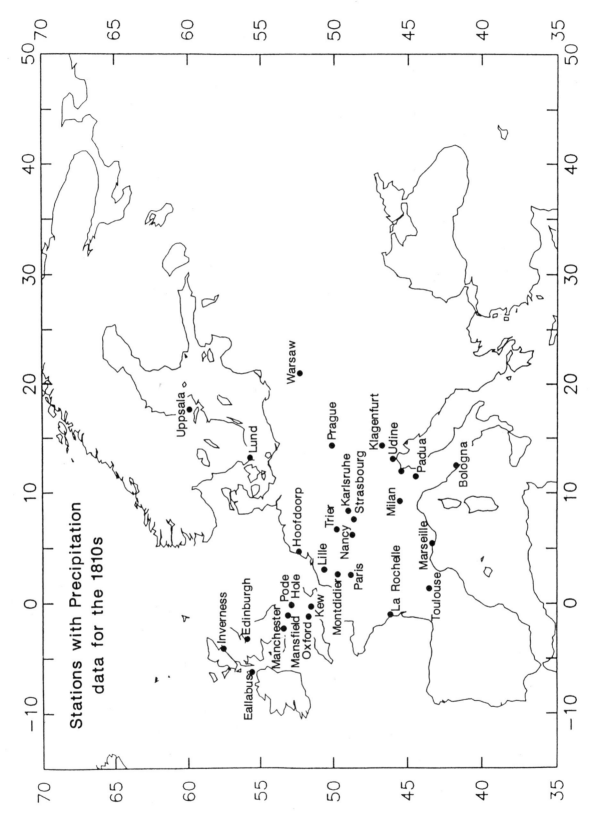

Figure 2: Location of the 29 sites with precipitation-gauge data during the 1810s. Details of the sites are given in Table 2 (from Bradley et al. 1985).

Table 2: Names and Locations of Stations with Precipitation Data for the 1810s Continuous to the Present Day.

	Lat.(°N)	Long.
1. Uppsala	59.9	17.6E
2. Lund	55.7	13.2E
3. Inverness	57.5	4.2W
4. Eallabus	55.6	6.2W
5. Edinburgh	55.5	3.2W
6. Manchester	53.4	2.3W
7. Mansfield	53.1	1.1W
8. Podehole	52.8	0.1W
9. Kew	51.5	0.3W
10. Oxford	51.7	1.2W
11. Hoofdoorp	52.3	4.7E
12. Lille	50.6	3.1E
13. Montdidier	49.7	2.6E
14. Paris	48.8	2.5E
15. Nancy	48.7	6.2E
16. Strasbourg	48.6	7.6E
17. La Rochelle	46.1	1.1W
18. Toulouse	43.6	1.4E
19. Marseille	43.3	5.4E
20. Trier	49.8	6.7E
21. Karlsruhe	49.0	8.4E
22. Klagenfurt	46.7	14.3E
23. Prague	50.1	14.3E
24. Warsaw	52.2	21.0E
25. Udine	46.0	13.1E
26. Milan	45.4	9.2E
27. Padua	45.4	12.0E
28. Bologna	44.5	11.5E
29. Rome	41.7	12.5E

In Figures 5-8 we show similar seasonal anomaly maps for temperature and precipitation, placing the 1810s in the context of modern reference periods, and 1816 in relation to the 1810s. The 1810s (Figure 5) were colder than the recent reference period during winter and autumn but were somewhat milder during spring and summer, particularly over eastern and southern Europe. The relative coolness of the 1810s with respect to 1951-70 in winter and spring means that 1816 was less anomalous when viewed against this decade as a whole (Figure 7). For precipitation, the 1810s were generally drier than the 1921-60 reference period (Figure 6), other than over the British Isles and Scandinavia in spring and Italy during summer and autumn.

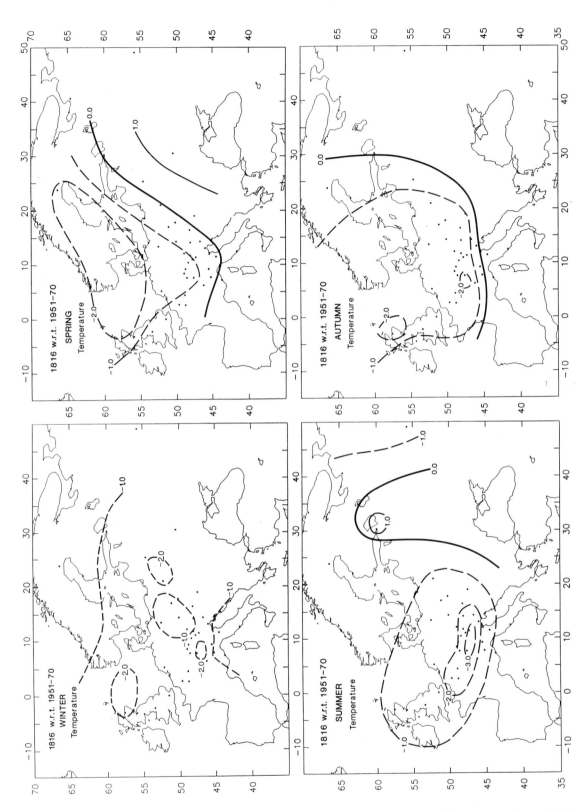

Figure 3: Seasonal temperature anomaly maps in degrees Celsius for 1816 with respect to the 1951-70 reference period. Winter is the average for December 1815 to February 1816. Spring (March-May), summer (June-August) and autumn (September-November).

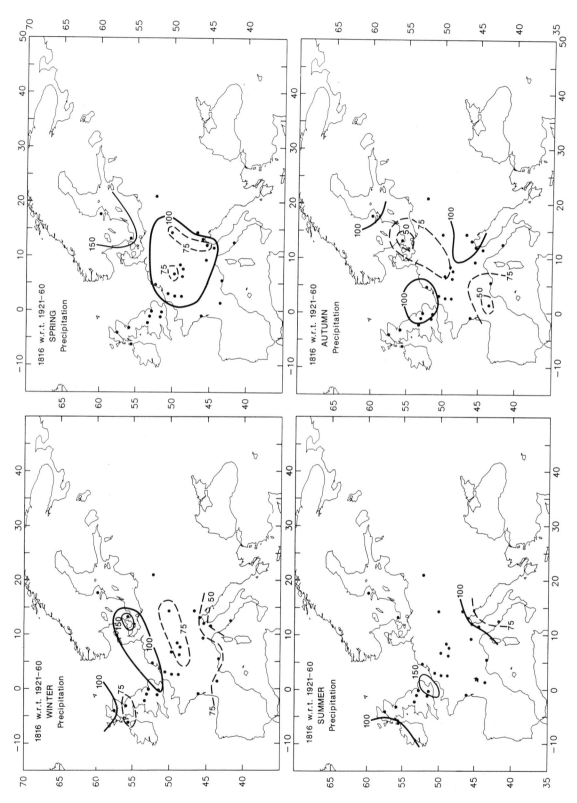

Figure 4: Seasonal precipitation departures for 1816: values expressed as percentages of the 1921-60 reference period mean.

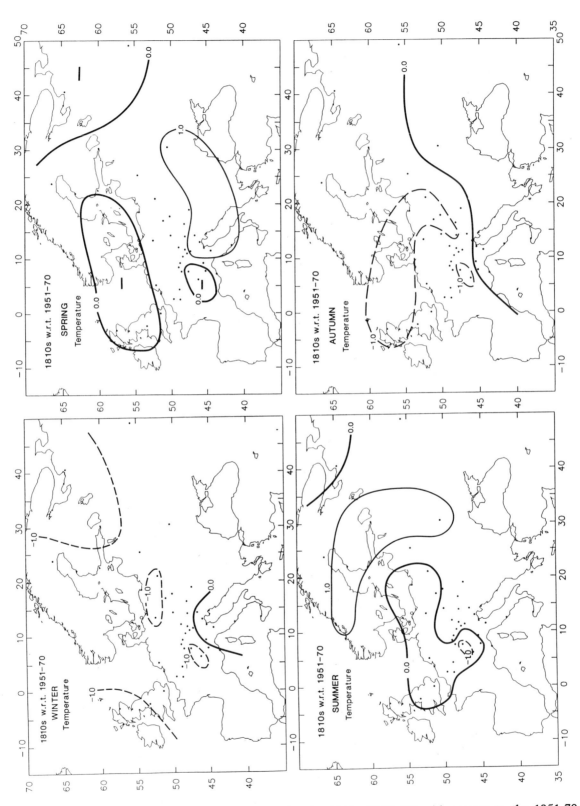

Figure 5: Seasonal temperature anomaly maps for the 1810s (1810-19) with respect to the 1951-70 reference period.

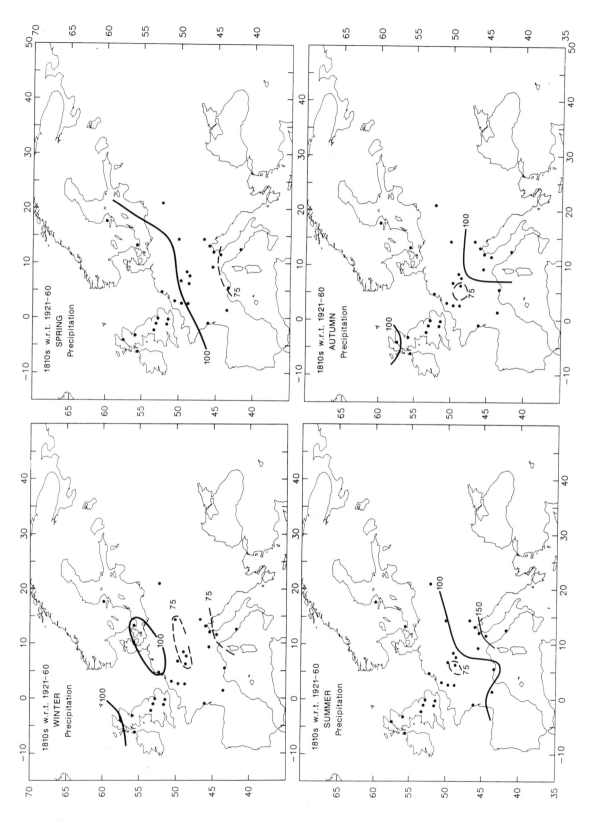

Figure 6: Seasonal precipitation departures for the 1810s: values expressed as percentages of the 1921-60 reference period mean.

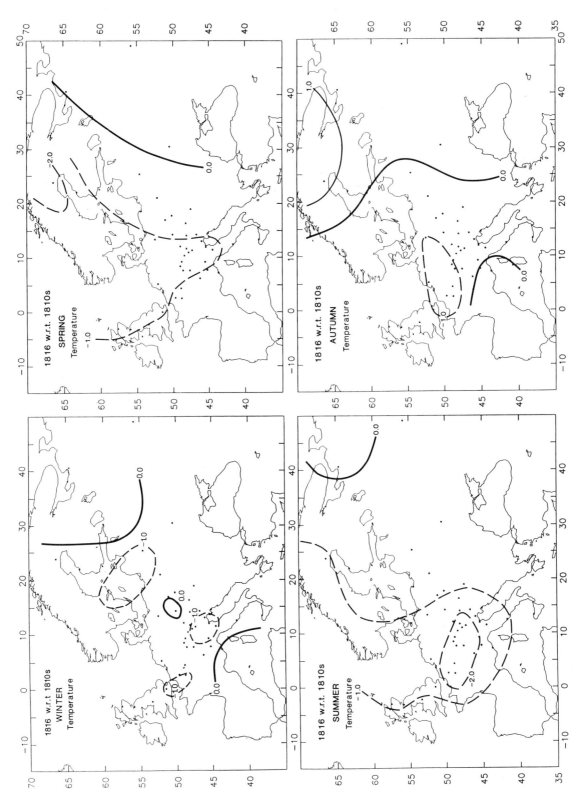

Figure 7: Seasonal temperature anomaly maps for 1816 with respect to the average for the 1810s (1810-19).

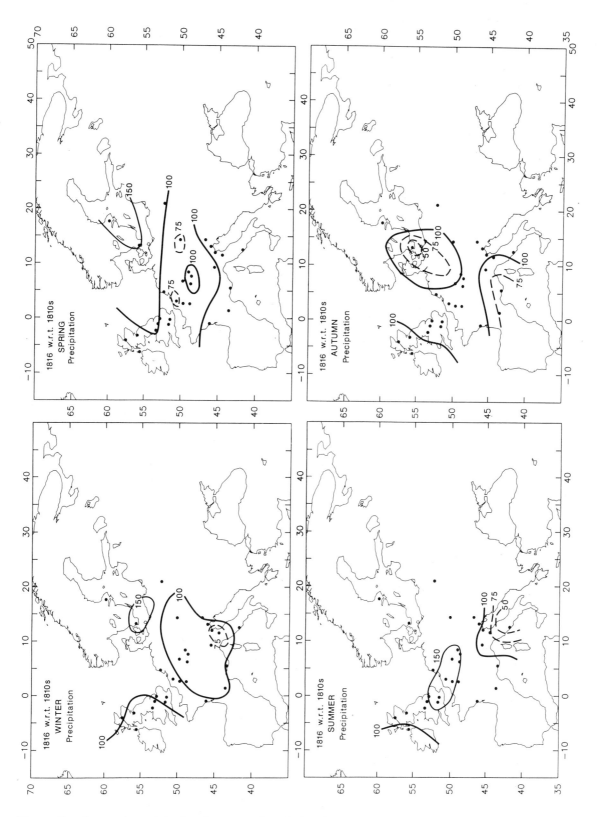

Figure 8: Seasonal precipitation departures for 1816: values expressed as percentages for the mean value for the 1810s.

Comparison of the 1816 Seasonal Temperature, Precipitation and MSLP Anomaly Maps

Winter (Figure 9)
During the winter of 1815-16 all of Europe was affected by anomalously low pressure with respect to the 1941-70 reference period. Both southern and western Europe were affected by greater advection from eastern Europe, which would tend to bring drier and cooler conditions to these regions. Enhanced northerly circulation over western Europe resulted in below normal precipitation in all areas except north-facing coasts.

Spring (Figure 10)
Europe is again shown to be almost entirely under the influence of lower-than-normal pressure. The negative pressure anomaly in this season is centred over northern France, implying that both Britain and Scandinavia experienced increased easterly and northeasterly weather, from Finland and the Gulf of Bothnia, leading to cold temperatures. Drier conditions over continental Europe may have resulted from the stagnation of a number of depressions over this region. Wetter than normal weather over northern Europe was associated with a greater degree of air flow over adjacent seas.

Summer (Figure 11)
Virtually the whole of Europe was affected by anomalously low pressure centred over northern Germany and Denmark. Milder conditions prevailed over European parts of the Soviet Union because of the influence of increased southerly flow across these areas. Britain and the rest of western Europe were affected by anomalous northerly and northwesterly airflow bringing cooler temperatures. The coldest conditions of the summer occurred in northern Alpine regions. Over Scandinavia, in contrast to winter and spring, conditions were near to the recent normal. Precipitation was considerably greater over northern France and southern England.

The summer (June-August) of 1816 was the coldest recorded in the Central England temperature series (Manley 1974; updated in Jones 1987). Temperatures were 2.2°C colder than the 1931-60 average. The Manley series extends back to 1659 (though with slightly lower reliability before 1721).

Autumn (Figure 12)
Again most of Europe was under the influence of anomalously low pressure, although less intense than in the other seasons. The centre of the anomaly was located over Poland, whereas pressure was near normal over Ireland and Scotland. Enhanced northerly and northeasterly air circulation over Scandinavia, Britain and central Europe, particularly north of the Alps would have led to cooler than normal conditions in all of these regions. Southern Europe and European parts of the Soviet Union were milder as a result of anomalous westerly and southerly airflow, respectively.

Precipitation anomalies are consistent with the circulation patterns. Reduced westerly airflow would give drier conditions in northern Germany and southern France. Enhanced north and northeasterly flow is consistent with above average precipitation over northern coastal Sweden and some southern North Sea coasts.

Inferences from Tree-Ring Parameters

Beside the long instrumental-based climatic records available for Europe back to the early 1800s, detailed year-by-year maps of 'summer' (April-September) temperature across Europe have also

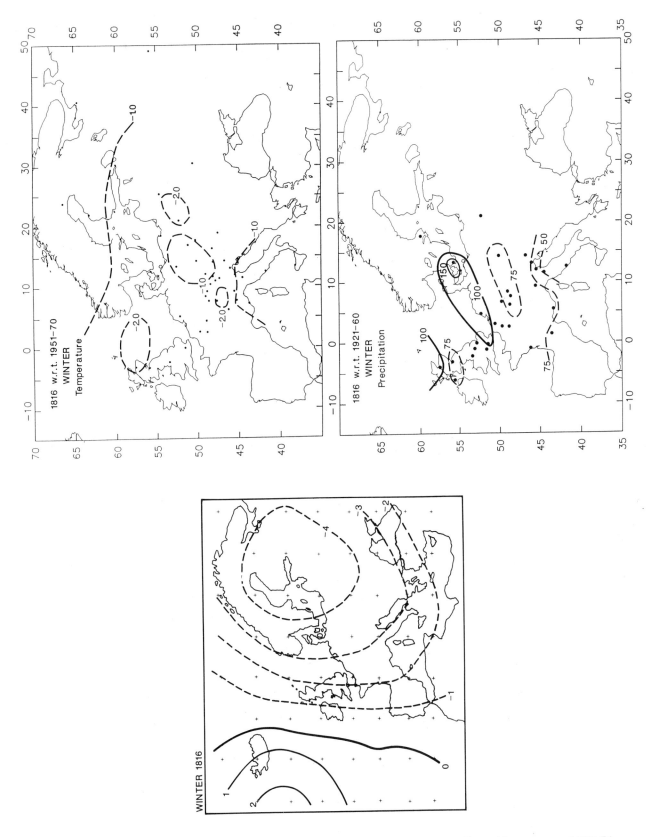

Figure 9: Climate anomaly maps: Winter. Mean sea level pressure anomalies with respect to 1900-74, air temperature anomalies with respect to 1951-70, precipitation as percentages of the 1921-60 reference period.

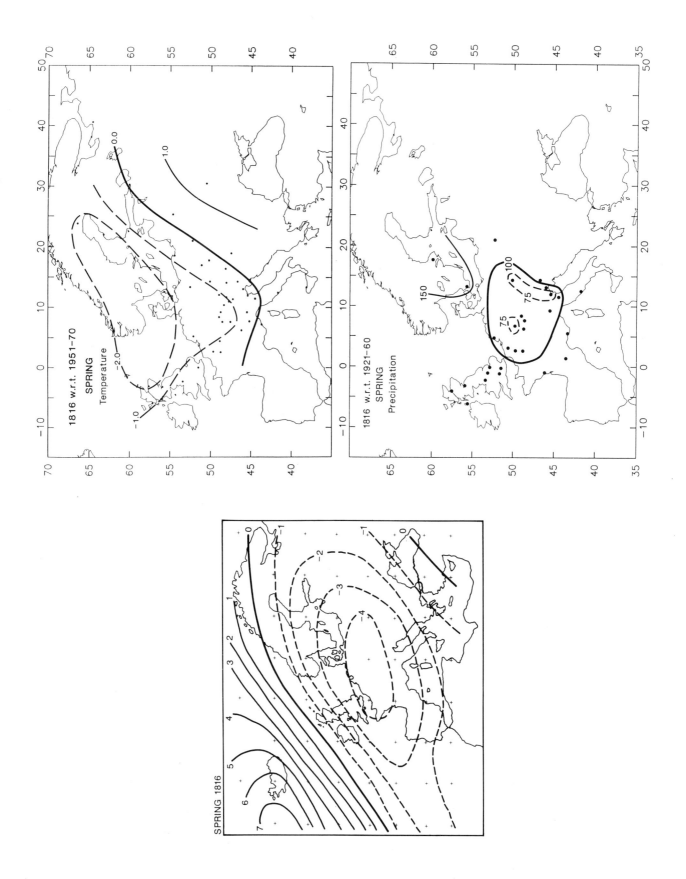

Figure 10: Climate anomaly maps: spring.

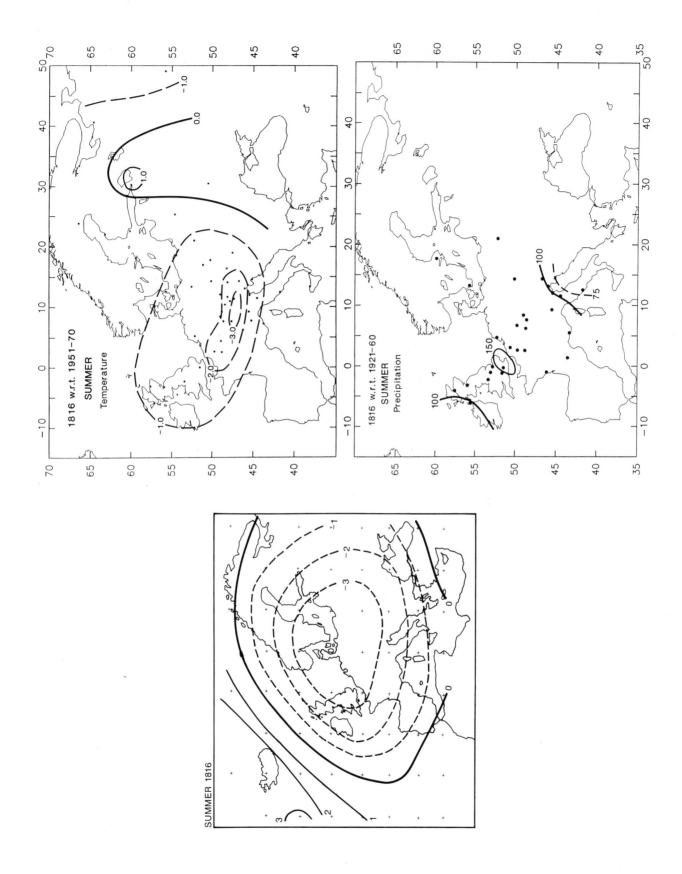

Figure 11: Climate anomaly maps: summer.

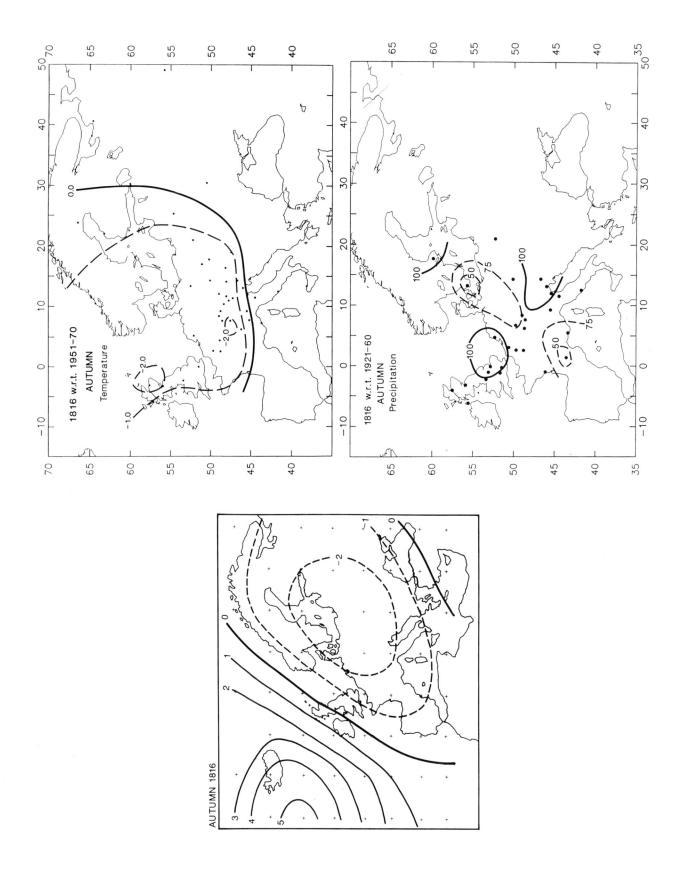

Figure 12: Climate anomaly maps: autumn.

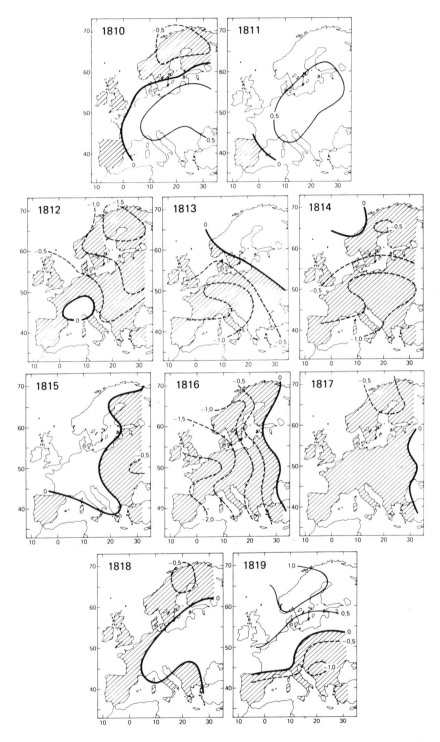

Figure 13: Reconstructions of 'summer' (April to September) temperature based on maximum latewood density chronologies from trees at 37 high-altitude and high-latitude sites across Europe. Further details of the reconstructions are in Briffa *et al.* (1988) and Schweingruber *et al.* (1989).

been produced from a network of tree-ring chronologies (Briffa *et al.* 1988; Schweingruber *et al.* 1989). These maximum latewood density chronologies, 37 in all from trees at high altitude or high latitude, provide reliable reconstructions of summer half-year temperature for the area between 45 to 70°N and 0 to 30°E (i.e. most of western Europe north of the Mediterranean) back to 1750. There are instrumental data for only a few stations (in London and central Europe) back as far as 1750.

The dendroclimatic reconstructions show that the summer of 1816 was cold, particularly in the United Kingdom and central Europe. They also show that summer in Scandinavia was not unusual. These results agree with instrumental temperature data. In the United Kingdom, the relative cold of the summer of 1816 was unmatched in any other year from 1750 to the present, and in central Europe there has been only one colder year - 1814.

All of the reconstructed maps, plotted as anomalies in degrees celsius from the 1951-70 reference period mean, have been published by Schweingruber *et al.* (1989). In Figure 13 we reproduce the maps for each of the 'summers' from 1810 to 1819. It is clear from these maps that cool conditions prevailed over almost all of Europe during the summers of 1812, 1814, 1816, 1817 and, to a lesser but still significant extent, 1813. The magnitudes of the negative temperature departures were however clearly greatest in 1816, and the most extreme of these (temperature departures below -2°C) were experienced in southern Britain and northern France.

Conclusions

The climate and weather of 1816 was indeed unusual over Europe, particularly during the summer. The evidence indicates generally cool conditions over the whole of western Europe during all four seasons of this year. The cold was particularly marked in summer when it was also wetter everywhere in Europe except in the eastern Mediterranean. Temperature and precipitation maps for the four seasons of 1816 are consistent with the reconstructed circulation maps. When considered in the context of the decade of the 1810s however, only the weather during the summer can be classed as extreme. The 1810s were an exceptionally cool/dry decade compared with modern reference periods.

Tree-ring reconstructions also show the 'summer' (April to September) of 1816 to have been cool, most particularly over Britain. These reconstructions show that summers from 1810 to 1819 over much of western Europe were generally cool. However, as with the instrumental data, the tree-ring reconstructions indicate that summers as cool as that of 1816 also occurred in each of the years from 1812 to 1814, i.e. before the eruption of Tambora.

Acknowledgements

The authors are grateful to Dr. F.H. Schweingruber for Figure 13.

References

Bradley, R.S., P.M. Kelly, P.D. Jones, C.M. Goodess and H.F. Diaz. 1985. A climatic data bank for northern hemisphere land areas, 1851 to 1980. *United States Department of Energy Technical Report* TR017, United States Department of Energy, Carbon Dioxide Research Division. Washington, D.C. 335 pp.

Briffa, K.R., P.D. Jones, T.M.L. Wigley, J.R. Pilcher and M.G.L. Baillie. 1986. Climate reconstruction from tree rings: Part 2, Spatial reconstruction of summer mean sea-level pressure patterns over Great Britain. *Journal of Climatology* 6:1-15.

Briffa, K.R., P.D. Jones and F.H. Schweingruber. 1988. Summer temperature patterns over Europe: a reconstruction from 1750 A.D. based on maximum latewood density indices of conifers. *Quaternary Research* 30:36-52.

Jones, D.E. 1987. Daily Central England temperature: recently constructed series. *Weather* 42:130-133.

Jones, P.D., S.C.B. Raper, B.D. Santer, B.S.G. Cherry, C.M. Goodess, R.S. Bradley, H.F. Diaz, P.M. Kelly and T.M.L. Wigley. 1985. A grid point surface air temperature data set for the northern hemisphere, 1851-1984. *United States Department of Energy Technical Report* TR022. United States Department of Energy, Carbon Dioxide Research Division. Washington, D.C. 251 pp.

Jones, P.D., T.M.L. Wigley and K.R. Briffa. 1987. Monthly mean pressure reconstructions for Europe (back to 1780) and North America (to 1858). *United States Department of Energy Technical Report* TR037. United States Department of Energy, Carbon Dioxide Research Division. Washington, D.C. 99 pp.

Landsberg, H.E. and J.M. Albert. 1974. The summer of 1816 and volcanism. *Weatherwise* 27:63-66.

Manley, G. 1974. Central England temperatures: monthly means 1659-1973. *Quarterly Journal of the Royal Meteorological Society* 100:389-405.

Schweingruber, F.H., K.R. Briffa and P.D. Jones. 1989. (in preparation).

Stommel, H. and E. Stommel. 1979. The year without a summer. *Scientific American* 240:134-140.

Tabony, R.C. 1980. A set of homogeneous European rainfall series *Met 0.13 Branch Memorandum No.* 104. United Kingdom Meteorological Office, Bracknell.

_____. 1981. A principal component and spectral analysis of European rainfall. *Journal of Climatology* 1:283-291.

The 1810s in the Baltic Region, 1816 in Particular: Air Temperatures, Grain Supply and Mortality

J. Neumann[1]

Abstract

The mean acidity of the ice core from Crête, central Greenland, for the layer dating to 1816, one year after Tambora's eruption, has been found by Hammer et al. (1980) to be nearly three times greater than that of the layer dating to 1884, one year after Krakatau's eruption. Despite the aforementioned fact, air-temperature data of the Baltic meteorological stations that took observations both in the 1810s and the 1880s (Copenhagen, Gothenburg, Stockholm, Trondheim and Uppsala), do not show that the coldness of 1816 relative to 1814 was any greater than that of 1884 relative to 1882. Moreover, the year 1812 was much colder than 1816 when the two are compared with 1814 at all Baltic stations, although no known important eruption took place shortly before 1812. It seems plausible that the plumes reaching the Baltic Region following the two eruptions were too 'thin' to have produced any appreciable effect on air temperatures.

An examination of data on grain harvests in Denmark, Finland, Norway and Sweden does not indicate that either in 1816 or 1817 there was any noteworthy crop failure. In contrast, the year 1812 (a cold year) was marked by shortfall of the harvest, in consequence of which in 1813 there was a partial famine in Norway, partly because of war conditions (blockade by the British Navy) it was hard to get supplies from abroad.

Mortality data are also available for the above four countries. Mortality was relatively high in 1812 and/or 1813, but not in 1816-17.

No harvest or mortality data are available for Russia. Lists of famines in Russia show none in 1816. In 1817 there was a price rise in a limited area of the Empire.

All-in-all, the Baltic Region did not suffer from Tambora's eruption unlike the lower mid-latitudes of western and central Europe. It is suggested that the Baltic Region, as well as southern European Russia, were spared as they were crossed by air masses whose stratosphere had become depopulated of small volcanic particles, while the troposphere became cleansed of particles through washout by rain previously.

Introduction

Did the particle cloud originating and developing (Appendix 2, No. 1) from Tambora's eruption in 1815 reach the relatively high latitudes of the Baltic Region? Was the year 1816 without a summer in the region? Was there, perhaps a famine in either 1816 or 1817, or both, as in many areas of western, southwestern and central Europe and North America?

---

[1] Emeritus, Department of Atmospheric Sciences, The Hebrew University, Jerusalem, Israel. In 1986-90 visiting with the Department of Meteorology, University of Copenhagen, Denmark.

Hammer *et al.* (1980, Figure 1) have shown that the acidity of the ice layer dating to 1816 at the Crête site in central Greenland (71°N) amounts to 5μ equiv. H^+ per kg ice. For comparison, the acidity of the layer dating to 1884, the year following Krakatau's eruption, amounts only to 2.5 (here and below, same units as above). If we subtract from these figures the background activity of 1.2 ± 0.1, then the contribution of 1816 works out nearly three times as great as that of 1884. Observations for 1816 are consistent with the assumption that aerosols of the Tambora eruption did reach a high northern latitude, at least on the western side of the North Atlantic.

The acidity accretion of 3.8 ± 0.1 μ equiv. H^+ per kg ice in 1816 really is a small mass for a full year's deposition. Only the sensitive physical method applied by Hammer *et al.* (1980) allowed its measurement to such a high degree of precision. We do not know how to turn this deposition into figures representing an atmospheric plume of particles. Even if it turned out that the plume reaching Greenland extended over an air layer of appreciable thickness and particle concentration, there is no guarantee that the plume reaching the Baltic Region was similarly 'powerful' for scattering back to space a significant fraction of direct solar radiation.

In order to examine the possible effects of Tambora's eruption on the Baltic Region, we shall consider the available air-temperature data, as well as figures on grain supply, export and import, and mortality in the 1810s.

Air Temperatures

In view of the fact that monthly means of temperature tend to fluctuate about 'normals' and 'normals' based on a series of years fluctuate themselves, both in magnitude and sign, I propose to look at temperature differences between the two years flanking 1815, and compare these with the parallel differences between 1884 and 1882. As too little is known about the exposure of thermometers in the early nineteenth century and the effects of the particular exposure, I assume that treating differences of adjacent years will minimize the effects of exposure differences from present-day standards.

Since Tambora ejected a much greater mass of material than Krakatau in 1883 (Appendix 2, No. 2; Sigurdsson and Carey, this volume) and, since Tambora more severely effected parts of Europe and North America (Post 1977), I expect that the air-temperature reduction in 1816 should have been greater than in 1884 - unless other meteorological processes overshadowed the effects of volcanic particles, if any.

Air-temperature measurements in the Baltic Region covering both the 1810s and 1880s are available for Copenhagen, Gothenburg, Stockholm and Uppsala, but I will also consider data for Trondheim. Baltic stations that have data for the 1810s only are St. Petersburg and Vöyri (63°08'N, 22°14'E, near Finland's west coast). Temperatures for April through September for Stockholm and St. Petersburg are shown in Table 1; those of other stations will be found in Appendix 1. Although the year 1816 was called a 'year without a summer', data on the spring months are included because of their importance for the growth of young plants. The tables cover several years of the 1810s, on the one hand, and the years 1882 and 1884, on the other - where available. Basel's data are included for comparison's sake.

Table 1: Mean Monthly Temperatures (°C), April through September, of Stockholm, St. Petersburg and Basel (the Latter for Comparison), and Differences of 1816's Temperatures from Those of Some Previous Years as Well as Differences of 1884 from 1882.[1]

Year(s)	A	M	J	J	A	S	Mean of year
Stockholm							
1809	-0.2	11.3	14.4	17.6	18.3	13.0	4.8
1812	-1.2	6.5	13.9	14.5	16.6	9.1	4.1
1814	5.3	7.1	12.2	19.2	16.7	10.6	4.4
1816	2.7	6.0	15.1	18.5	14.7	11.8	4.9
1817	2.5	10.7	14.2	17.3	15.2	12.6	5.7
1882	3.9	9.8	14.3	17.0	17.4	13.2	6.7
1884	2.7	8.1	12.6	16.8	14.6	14.1	5.9
(1816)-(1814)	-2.6	-1.1	+2.9	-0.7	-2.0	+1.2	+0.5
(1884)-(1882)	-1.2	-1.7	-1.7	-0.2	-2.8	+0.9	-1.2
(1816)-(1809)	+2.9	-5.3	+0.7	+0.9	-3.6	-1.2	+0.1
(1816)-(1812)	+3.9	-0.5	+1.2	+4.0	-1.9	+2.7	+0.8
St. Petersburg							
1809	-1.7	7.4	14.8	16.4	16.0	11.3	1.2
1812	1.3	6.8	13.7	17.9	19.6	7.7	2.5
1814	1.6	6.0	15.3	20.7	17.0	10.3	2.6
1816	3.6	7.5	15.1	19.1	14.7	12.4	3.5
1817	2.0	11.6	14.0	19.1	17.5	8.1	3.5
(1816)-(1814)	+2.0	+1.5	-0.2	-1.6	-2.3	-2.1	+0.9
(1816)-(1809)	+5.3	+0.1	+0.3	+2.7	-1.3	+1.1	+2.3
(1816)-(1812)	+2.3	+0.7	+1.4	+1.2	-4.9	+4.7	+1.0
Basel							
(1816-1814)	-2.4	+0.1	-1.4	-3.4	-1.4	+0.9	-0.7
(1884)-1882)	-1.7	+0.6	-1.7	+3.1	+2.4	+1.6	+0.3

[1] Stockholm - Hamberg (1906, pp. 13-14). Also *Historisk Statistik för Sverige*, (1959, Volume II, pp. 3-4).
St. Petersburg - Wahlén (1881, p. 16).
Basel - Bider *et al.* (1959, pp. 407-409).
A = April, M = May, etc.

Differences for the means for the April-September seasons for all stations are shown in Table 2. All the differences (1816)-(1814) are negative. So are the differences (1884)-(1882), except for Basel. In Trondheim's case the 'Tambora difference' is more negative than the 'Krakatau difference'; in the case of Stockholm and Uppsala, the Krakatau difference is more negative; in Gothenburg's case, the two are about equal. Hence, the aforementioned sets of differences do not indicate a marked Tambora effect at all the stations. Moreover, as can be seen (Table 2), at all Baltic stations the spring-summer season of 1812 was colder than that of 1816. Stothers (1984,

Figure 4) shows, too, that the stations situated in northwestern and northern Europe along the 5°C annual isotherm were colder in 1808, 1809 and 1812 than 1816. Evidently the years 1807-11 were not characterized by sizable volcanic eruptions: Lamb's (1972, Table 10.3) Dust Veil Index for the aforementioned four years is much lower than for Tambora, so there is little reason to attribute the coldness of 1809 and 1812 to volcanic particles. Because the first two or three decades of the nineteenth century produced several cold years or winters, it is doubtful that the minor coldness of 1816 relative to 1814 in the Baltic Region was a sequel of Tambora.

Table 2: Mean Seasonal (April through September) Differences of Air Temperatures (°C).

Station	(1816) - (1814)	(1884) - (1882)	(1816) - (1809)	(1816) - (1812)
Trondheim	-0.73	-0.48	-0.63	+0.68
Gothenburg	-0.37	-0.38	-1.08	+0.75
Copenhagen	-0.42	-0.20	-1.17	+0.30
Uppsala	-1.0 (?)	-1.3 (?)	___	___
Stockholm	-0.38	-1.12	-0.93	+1.57
Vöyri	-0.30	___	small	+1.92
St. Petersburg	-0.45	___	+1.37	+0.9
Basel	-1.27	+0.72	___	___

In consequence of the appreciable fluctuations of temperature from one year to the next, also in periods when no important volcanic eruptions occur, I consider comparison of temperatures does not yield convincing results regarding possible volcanic effects (Appendix 2, No. 3) - at least not in the period covered here. Harvest data can supply more trustworthy inferences.

Grain Supply, Mortality - Some General Remarks

The only country of the Baltic Region for which both harvest estimates and grain import-export data are to be had for the 1810s is Sweden. Finland comes next, since data are available on the difference between exports and imports as well as 'qualitative' estimates of the degree of success of the harvests.

In pre-industrial societies of Europe death rates rose to a maximum either in the year of a serious crop failure or in the year following; in other cases, the rise was, or may have been, indirectly due to a poor diet, (e.g., the consumption of tree bark, roots, etc.) which promoted endemic diseases. The consequences of a crop failure were especially grave when a dearth-stricken country could not obtain supplies from abroad (e.g., in the case of wars like that of Norway in 1812-13: see below).

Sweden - Grain Supply

Figures on grain harvests in Sweden in the 1810s are published in *Historisk Statistik för Sverige* (1959, Volume II, p. 46, Table E16). Data on import and export of grain are listed in Tables 1 and 2 of the monograph *Spannmålshandel och Spannmålspolitik i Sverige 1719-1830* (Grain Trade and Grain Policy in Sweden 1719-1830) by the Swedish economist K. Åmark (1915, pp. 354-355 and 356-357). The import-export quantities of flour are so small that they were neglected here.

Export-import data were collected through the Customs Authority and appear to be fairly reliable. Because the compilers-editors of the *Historisk Statistik* (Volume II, 1959, p. 17°) warn that the harvest estimates are not very reliable, I base my comments on the import-export data. These data are independent of the harvest-estimate figures. All the aforementioned data are listed in Table 3.

Table 3: Estimates of Grain Harvest and Data on Excess of Import Over Export of Grain for Sweden in the 1810s in Thousands of Metric Tons.[1,2]

A	B	C	D
1811	512	24	536
1812	473	42	515
1813	530	96	626
1814	568	62	630
1815	581	21	602
1816	527	22	549
1817	545	19	564
1818	473	54	527
1819	532	40	572
1820	654	1	655

[1] A - year, B - grain harvest, C - excess of import over export of grain, D - B+C.

[2] Grain harvest sources: *Historisk Statistik för Sverige* (1959, Volume II, Table E16, p. 46); Import-export: Åmark (1915, Tables 1 and 2, pp. 354-355 and 356-357).

The excess of imports over exports was very high in 1813 (96,000 metric tons). In fact, a reference to Åmark's Table 1 indicates that not since 1785 (winter 1783-84 was rather cold, a fact mentioned in Appendix 2, No. 3) were imports as high as in 1813. Åmark (1915, pp. 106-108) repeatedly states that 1812 was a year of crop failure ('missvaxt') (Appendix 2, No. 4). This failure necessitated increased imports the following year. No mention is made of a dearth in 1816 and 1817. Table 3 shows, indeed, that in the two post-Tambora years, imports were but one-fifth to one-quarter of the imports of 1813.

As has been stated, the spring and summer of 1812 were cold in the Baltic countries. At Stockholm the mean temperature of April, May and July was 6.6°C, whereas the corresponding mean of 1816 was 9.1° and that of 1817, 10.2°. Thus the growing seasons of the two post-Tambora years were warm. Similarly, Gothenburg's temperature data (*Historisk Statistik*, 1959, Volume II, p. 6) indicate that in 1816 and 1817 spring and summer were warmer than those of 1812, though the differences were smaller than in Stockholm's case.

The economic historian Sommarin (1917) published a monograph *The Economic Development of the Scanian Agriculture* (1917; in Swedish). As the title indicates, the emphasis of the work is on Scania, Sweden's southern region. Sommarin classifies the harvests of Malmöhus and Kristianstad, the two southernmost counties, into seven groups: 0 to 1 = crop failure, ..., 4 = average, ..., 6 = abundant, rich. In a Table (pp. 208-209) he puts the harvests of both 1816 and 1817 into group "6", that is, abundant. In contrast, the harvest of 1811 is estimated by him as "3.4" and that of 1812 as "5.1"; the three years 1813-15 were considered "a little above average". I will discuss 1811 later under Denmark.

A commentary on the meteorological background of the below-average harvest in southern Sweden in 1811 is offered by rainfall data for Lund, the university city in Malmöhus. Precipitation records at Lund began in 1748 at the Astronomical Observatory of the University. The rainguage was placed on the Observatory's roof, 20 m above ground, during the period 1775-1867 (Tidblom 1875-76, p. 65). This is an objectionable exposure, but the results of the measurements are still of value in comparing one year with another. Cumulative rainfall for April through July (the monthly figures are found in Tidblom 1875-76, pp. 66-68), the main growing season, follow:

Year	Cumulative Rainfall (mm) at Lund, April through July
1811	98
1816	247
1817	158

Clearly, the rainfalls of 1816 and 1817 were much more abundant than that of 1811.

Sweden - Mortality
Volume I of the *Historisk Statistik för Sverige 1720-1967* (1969) includes population data for Sweden. On p. 93 the death rates for 1811-20 are given (see Table 4, with similar data for the other Baltic countries, excepting Russia).

In the 1810s, 1816 had the lowest mortality, whereas 1812 had the highest rate (Table 4). One of the reasons for the low mortality in 1816 is that the largest population-increase rate of the decade occurred from 1815 to 1816 (incidentally, 1815 was a good crop year). In any case, mortality was not high in the two post-Tambora years. That the crop failure of 1812 did not cause famine or high mortality was largely due to the fact that Sweden was able to import large quantities of grain (partly from Swedish Pomerania and from Danzig), being a kind of trading centre for Polish grain that was conveniently shipped on the Vistula. Swedish Pomerania was adjacent to the southern coast of the Baltic. Sweden's ability to import large quantities of grain in 1813 was mainly due to her siding with the British-Russian coalition against Napoleon.

Norway - Grain Supply
No crop-production or grain import-export data are available for Norway for the 1810s. However, Norway suffered a grave crop failure in 1812 (e.g., Steen 1933, p. 339; Drake 1969, p. 60), and, unlike in Sweden, 1813 was a year of famine. The British blockade of Norway

Table 4: Mortality in the Baltic Region, 1811-20, in ‰ of Population.[1]

Year	Denmark	Finland	Norway	Sweden
1811	24.4	30.8	25.5	28.8
1812	27.0	23.9	21.3	30.3
1813	22.8	27.3	29.5	27.4
1814	24.7	26.6	22.6	25.1
1815	21.6	26.0	29.8	23.6
1816	20.7	23.4	19.4	22.7
1817	19.0	24.1	17.7	24.3
1818	18.9	25.1	19.1	24.4
1819	19.5	27.3	19.7	27.4
1820	20.9	25.3	18.9	24.5

[1] Sources: Denmark: Andersen (1973, p. 300); Finland: Turpeinen (1979, p. 108); Norway: Drake (1969, p. 194); Sweden: *Historisk Statistik för Sverige* (1969, p. 93).

virtually cut off the country from foreign sources, including Denmark which used to supply about 25% of Norway's grain needs (Derry 1965, p. 488). Neither 1816 nor 1817 were years of dearth.

The historian Steen (1933, p. 339) writes that 1812 was the blackest year that the poor of Norway had to experience. Spring was late. In many places, the grain could not be sown before June. Another historian, Mykland (1978, p. 239) states that on 9-10 August there was a strong nightfrost, in consequence of which the crop was totally destroyed in the northern part of the country. Returning to Steen's account, in some parts of Norway in September snow covered the unripe grain.

The famine was so severe that Prince Frederick of Hessen, Norway's Vice-Governor, wrote King Frederick VI, in Copenhagen recommending that the British be asked for an armistice (Mykland 1978, p. 228) "not to have to succumb to that most frightful scourge, famine". The King did not accept the recommendation. From May to autumn 1813, bread riots broke out in many areas of Norway (Lindval 1952, p. 129).

A reference to the temperature data of Trondheim (Appendix 1) shows how much lower the temperatures were in 1812 compared with those of 1811 and with some of 1816.

Norway - Mortality
Drake (1979, p. 291) publishes a diagram showing the mortality rates in dioceses of Norway and in the country as a whole. The diagram shows that mortality reached a peak in 1813 in the Akershus and Trondheim dioceses (Akershus includes Oslo), with another peak in 1809 - also a cold year. The fact that the curve for the country exhibits a peak in the same years as Akershus and Trondheim dioceses is due to the fact that those dioceses comprised 67% of Norway's population (Drake 1979, p. 290). As to the peak for Akershus, Drake (1969, p. 71) states: "A number of major peaks in the death rate [in Norway] ... occurred in years when the grain harvest

failed over large parts of the country and it is noticeable in each that the Akershus diocese, where the diet was based more on grain than elsewhere, suffered particularly badly."

Drake's diagram shows that there was no rise in mortality in either 1816 or 1817. A reference to Trondheim's temperature data (Appendix 1) indicates that April, May and July 1816 and 1817 were warm relative to 1812 - just as at Stockholm.

Denmark - Grain Supply

Denmark was a grain-producing and exporting country during this period. According to information received from Professor Claus Bjørn, historian of agriculture at the University of Copenhagen, no harvest figures and no complete sets of data are available on exports to various countries in the 1810s. The only data available relate to the export of grain to England from 1816 on. The Danish economist Falbe-Hansen (1889, p. 16) publishes the following grain-export figures for 1816-25:

Year	Quarters
1816	14,900
1817	149,000
1818	342,000
1819	123,000
1820	147,000

These quantities are small, but indicate that there was no shortfall of harvest either in 1816 or in 1817: the export of 1817 was made possible by the harvest of 1816.

Professor Bjørn has drawn my attention to an interesting volume of 1811-38 records of a tenant-farmer Søren Pedersen. Pedersen lived at Havrebjerg in southwestern Sjælland. He was not only a farmer but also the Executive Officer of his parish. As an Executive Officer, he kept accounts and records of events, at the end of each year recording his thoughts on the character of the year's weather, crops, prices, etc. In his reflections on 1816 (Pederson 1983, pp. 214-215) he writes:

> "This year was again a lovely fertile year, not only on account of the quantity of fodder but also because the kernels were big. This was important for our country as this commodity was very much coveted in foreign lands. The price was high so that the monetary situation of the country has improved and one assumed that the bad times of war [the Napoleonic Wars] are over. The rate of exchange was nearly par and one thought that the good old times have returned and this not only because of our fortunate grain trade with other countries where there has been a general crop failure. Denmark has been selling this year for many millions of rigsdaler [old Danish monetary unit] grain to other nations."

Probably his reference to crop failure in other countries and the demand for Danish grain reflect "the year without a summer" in western and central Europe.

In his contemplations on 1817 (Pedersen 1983, pp. 223-224) we read the following: "This year was, I suppose, not as happy as the previous year, neither with respect to fertility nor the prices obtained for our produce. But, it was not a year of famine, not a barren year, rather an average year."

In contrast, in his review of 1811, he states (Pedersen 1983, pp. 168-170): "The year 1811 ... a hard and barren year. There was a shortage of both grain and hay because of the too dry summer. This led to very high prices, even of commodities of poor quality ... The public authorities conducted enquiries how much each peasant had reserves so that they can distribute the food, if necessary."

We see from Pedersen's reflections that 1816 and 1817 were good to average crop years, whereas 1811 was a year of drought (Appendix 2, No. 4) and crop failure, at least in a part of Denmark. A reference to Copenhagen's air-temperature data in Appendix 1 shows that 1811 was, indeed, a warm year. As to rainfall, this was measured in Copenhagen in the 1810s on the top of a 36-m high tower (The Round Tower) in the city centre. However, the exposure of the gauge was unsatisfactory. Some reliance can be placed in the precipitation measurements at Lund, southern Sweden, which, as has been pointed out previously is but a short distance from Sjælland across the Danish Sound. We have mentioned also the Lund rainfall data for spring-summer 1811 showing that rainfall was low - as stated by Pedersen for southwestern Sjæland.

That Pedersen's statement concerning a poor harvest in 1811 applies not only to his neighbourhood, also follows from a statement in minutes of a meeting of "Sjælland's Provisioning Committee" on 31 August 1811. Dr. Helle Linde (personal communication 1988) of the Copenhagen City Archives has kindly brought to my attention a passage in the minutes which translates as follows: "It is known that the [price of] grain fell substantially during the first summer months when the crops looked promising. But it has turned out soon that the harvest of this year will not be better and that it will be even less satisfactory than the harvest of the year-before-last."

The listed prices of grain and other commodities in the Danish newspaper *Danske Statstidende* for 1809 and 1811 indicate a clear rise compared with spring 1811 and those of 1809 (Appendix 2, No. 5).

Denmark - Mortality
Figures for Denmark (Andersen 1973, p. 300), and parallel data for Finland, Norway and Sweden show that mortality in Denmark was relatively low in both 1816 and 1817, whereas it was high in 1812 (Table 4).

Finland - Grain Supply
Data are available on the excess of export over import of grain as from 1812, collected by the Customs Authority. Additionally, 'qualitative' harvest estimates are recorded in reports of the Provincial Governors to the 'Senate' (the semi-autonomous Government of Finland under Tsarist rule).

Data on excess of exports over imports have been published by the Finnish agricultural historian Soininen (1974, Table 24, pp. 190-191), and the Provincial Governors' reports have been studied and summarized by Johanson (1924, see especially pp. 82-85). The export-import figures for 1812-20 are listed (Table 5a,b). The large excesses of imports in 1813 and 1819 suggest poor harvests in 1812 and in 1818. These inferences are in close agreement with estimates of the Provincial Governors who state that 1812 was a year of severe crop failure, whereas 1818 was a year of crop failure. 1815 was likely to be a good year and 1816 and 1817 did not excel, but the Governors do not classify them as years of crop failure. Table 5b gives the amounts of grain exports to Sweden from Finland, showing that during 1816-18 exports were appreciable, whereas in 1819 (after the crop failure of 1818) exports to Sweden were reduced.

According to Johanson (1924, pp. 82-83), the spring of 1812 was very cold - as were all the Baltic countries then (see section on Air Temperatures above, and temperature data in Appendix 1). He adds that the summer of 1812 was cool and rainy, leading to a severe shortfall of the harvest. The same author states that the harvest of 1816 was normal, except in the Viipuri (Viborg) and Vaasa (Vasa) counties. In 1817 the spring and summer were rainy, but the crops did not suffer.

Finland - Mortality

In Table 4 are listed Finnish mortality data as published by Turpeinen (1979, Table 1, p. 108). Mortality was relatively high in 1811, and, after a major drop in 1812, it rose in 1813. Another rise, compared with the flanking years took place in 1819. Presumably, the peaks are related to crop failures of the previous years. On the other hand, we note that mortality was low in 1816, though a slight rise is seen in 1817 and 1818.

Table 5a: Excess of Exports over Imports of Bread Cereals and Oats, in Thousands of Hectolitres, from Finland, and Classification of Grain Harvests by Provincial Governors.[1]

Year	Bread cereals	Oats	Total	Harvest
1811	—	—	—	crop failure
1812	49.5	10.9	60.4	severe crop failure
1813	-179.6	10.5	-169.1	—
1814	-7.3	-5.0	-12.3	abundant crop
1815	13.2	-6.5	6.7	—
1816	98.5	0.1	98.6	—
1817	-66.5	-7.2	-73.7	—
1818	-141.5	-12.5	-154.0	crop failure
1819	-274.7	-31.4	-306.1	—
1820	-57.5	-9.1	-66.6	—

[1] Sources: Export-Import - Soininen (1974, Table 24, pp. 190-191). Harvest estimates - Johanson (1924, pp. 82-85).

Russia

No figures on grain production, nor on grain export-import are available, but there is information on years of famine and, in some cases, on years when the crop failed. References to famines in Russia can be found in a bibliography prepared by D.R. and V. Kazmer (1977) titled *Russian Economic History, a Guide to Information Sources*. The most comprehensive English paper on famines in Russia is that by Kahan (1968). Other articles in English are by Dando (1981) and a brief review paper by Robbins (1979) in *The Modern Encyclopedia of Russian and Soviet History*. All three list dates and brief remarks on the degree of severity of the famines.

Table 5b: Export of Bread Cereals, Oats and Peas from Finland to Sweden Alone, in Metric Tons.[1]

Year	Export
1815	16,300
1816	73,000
1817	53,000
1818	62,000
1819	16,000
1820	13,000

[1] Source: Johanson (1924, p. 165).

The most detailed list (Kahan 1968, pp. 367-375) includes a good discussion. As to the early nineteenth century, Kahan says that grain yield was low in 1812 near Moscow and in Siberia, and that in 1817 there was a drought in the non-black soil region, causing a rise in prices. Compared with the specifications given of other dearth years, 1817 must have been a light case. In any case, no dearth or famine is reported for 1816. Neither Dando's nor Robbins' lists mention dearth in 1817. Sorokin (1975, p. 179) writes that in 1812, 2½ million rubles were appropriated by the Government for the purchase and distribution of bread to the starving population of Moscow province, and that in 1813 the provinces of Kaluga and Smolensk were granted 6 million rubles for similar purposes. Such shortages were, in all probability, a sequel of the cold year 1812. No mention is made of dearth in 1816 (Appendix 2, No. 6).

In addition to the above works in English, several Russian encyclopedias published in Tsarist times toward the end of the nineteenth century furnish rather long articles (unlike the *Great Soviet Encyclopedia*) on famines in Russia and Europe. None mentions famine in Russia in 1816 and 1817 though one remarks on famine in Germany in 1817.

Marshall (1833 p. 95) quotes figures on grain exports in 1791-1825 from northwestern and northern Russian ports (Libau = Liepāja, Riga, St. Petersburg and Archangel). While quantities were not large, there was a manifold increase from 1816 to 1817. As it was the custom of Governments to prohibit exports in years of dearth, there could not have been serious shortages in the Empire. On the other hand, the column for Riga shows that in 1818 and 1819 no exports were allowed.

Grain Prices

Prices are not always good indicators of harvests. Governments, including local ones, used to control prices, subsidize sales to the poor and, in some cases released stocks earmarked for military purposes to the general public. For example, in Norway in 1813, quantities of grain were taken by the Regent from army stocks. Nevertheless, I shall make an exception and quote Abel (1974, pp. 318-319) who points out that from 1815 to 1817 the rise in the price of rye in Sweden (rye being a staple food of the population) amounted to only 20-30%. At Danzig, the principal grain-trade centre of the Baltic Region, which was selling to countries near and far, the rise was 18%. These figures are to be compared with the following figures for price rise in some of the Tambora-stricken western European countries (e.g., England 150%, France 185%).

Why the Baltic Region Was Not Reached by Large Numbers of Volcanic Particles from Tambora's Eruption

Why was the Baltic Region not affected by Tambora's eruption of 1815? It is puzzling, indeed, that while western and central Europe south of 50-55° were gravely hit, the Baltic Region and European Russia, including its south, were not. Because the acidity accretion of $3.8 \pm 0.1\ \mu$ equiv. H^+ per kg ice in the 1816 layer of central Greenland (71°) really is relatively small, perhaps aerosol plumes of Tambora crossing the Baltic Region were too 'thin' to produce a noticeable impact on air temperatures and agriculture.

The following comments are based in part on computed data and in part on speculation. Speculation is inevitable for 1816 since we have no other meteorological data than surface air temperatures at an admittedly small number of stations.

Direct Solar Radiation (DSR)

Table 6 lists the amounts of DSR reaching a horizontal surface at the top of the atmosphere in the summer and in the winter half-years respectively (summer half-year: 21 March to 23 September). The figures were computed by List (1958, p. 418) on the assumption that the solar constant equals 1.94 cal cm^{-2} min^{-1}.

I will compare the figures for Latitude 60° with those for Latitude 45°. Oslo, Stockholm, Helsinki and St. Petersburg all are situated near the former.

In the summer half-year the DSR reaching a horizontal surface at the top of the atmosphere at Latitude 60° is about 10% less than Latitude 45°. However, when account is taken of the depletion of DSR along the longer atmospheric path to the stratosphere at 60°, the difference between the two latitudes is bound to be greater. This means that in the atmosphere of the higher latitudes any **volcanic particles have less DSR to scatter back to space**. If we assume that the two latitudes compared have, hypothetically, the same size distributions of particles, and that they have the same physical-chemical characteristics of the particles, then **the reduction of air temperature in the higher latitudes will be less than in the lower mid-latitudes**.

In the winter half-year the difference in DSR at the top of the atmosphere is 47%, and the difference at stratospheric levels is **much** greater. We can expect, consequently, that in the winter half year volcanic particles in the higher latitudes will scatter back to space very little DSR and, concomitantly, the diminution of air temperature will be very small. Even in the lower mid-latitudes, where the year 1816 was 'without a summer', the winter shows minor effects.

Volcanic Particles

Kondratyev *et al.* (1983) provide a useful review of our knowledge of the nature and characteristics of volcanic particles. In the case of powerful eruptions both particulate and gaseous matter are injected into the stratosphere which the atmospheric circulation can carry great distances. (We are not concerned here with the 'close fall-out' of heavy ash.) The gaseous matter is predominantly SO_2 which undergoes with time (~transport by winds) photo-oxidation

Table 6: Direct Solar Radiation at the Top of the Atmosphere (in cal cm^{-2}) in Summer and Winter Half-Years on the Assumption that the Solar Constant Equals 1.94 cal cm^{-2} min-1. (Summer half-year: 21 March to 23 September[1]).

Latitude	Summer half-year	Winter half-year
70°	134,540	13,040
60°	144,610	32,610
50°	156,030	56,980
45°	160,790	69,360
40°	164,620	81,510
30°	169,220	104,570

[1] Source: List (1958, p. 418).

converting it into sulphuric acid. The formation of sulphuric acid involves the heterogeneous nucleation on small mineral particles and homogeneous nucleation, the latter if a high degree of supersaturation of H$_2$SO$_4$ vapour prevails (e.g., Hofmann and Rosen, 1983b, p. 327). It is probably correct to say that high supersaturation conditions can occur in the stratosphere rather than in the troposphere. In the conversion process mineral particles coated with sulphuric acid and sulphuric acid droplets are produced (Appendix 2, No. 7). In addition to the process of formation itself, the fact is of special importance to our considerations that sulphuric acid particles/droplets tend to grow with time. As to the sulphuric-acid coating growth, I have quoted (Appendix 2, No. 1) Mossop's communication concerning the volcanic particles from Agung's eruption; see also the papers of Hofmann and Rosen (1982, 1983a and b) concerning particles from other eruptions.

As the coating grows and the particles become larger, their terminal velocity of fall increases. However, the process of enhancement of terminal velocity is counteracted by the increasing air density. The terminal velocity is given by Stokes' equation:

$$V = \frac{2}{g} \frac{a^2 g}{\nu} \left(\frac{\rho'}{\rho} - 1 \right), \tag{1}$$

provided that the Reynolds number, Re,

$$\frac{2aV}{\nu} = \frac{4}{g} \frac{a^3 g}{\nu^2} \left(\frac{\rho'}{\rho} - 1 \right) \tag{2}$$

is appreciably less than 1, a condition that is amply satisfied by our particles. In the above equations a is the radius of the supposedly spherical particles or droplets and ρ is the density of the particles/droplets. The other symbols are standard.

In the troposphere the terminal velocity of the 'large' particles that have fallen out from the stratosphere will decrease (or, continue decreasing) as they carry on falling, unless the growth of the particles outweighs the effect of increasing air density. However, the chances of growth of the sulphuric-acid coating, or that of the sulphuric acid droplets, are less favourable in the troposphere than in the stratosphere since most of the SO_2 of large eruptions is injected into the stratosphere. Assuming, for purposes of illustration, no growth of coating of the particles in the troposphere, and no coagulation and or washout by rain, a particle of radius of 1 μm and a density $\rho = 2$ g cm^{-3} will have in the International Standard Atmosphere at 10 km (T = -50°C, $\rho = 0.413 \times 10^{-3}$ g cm^{-3}, $\nu = 0.092$ cm^2 s^{-1}) $V = 0.117$ cm s^{-1}, while at 5 km (T = -17.5°C, $\rho = 0.736 \times 10^{-3}$ g cm^{-3}, $\nu = 0.105$ cm^2 s^{-1}) $V = 0.057$, that is, half the terminal velocity at 10 km. Thus, in a climatic zone where there is little precipitation (and, if other assumptions hold), the concentration of volcanic particles will tend to increase with time in the troposphere - at least for a while.

On their way to Europe (and North America) the Tambora particles had to cross the subtropical high-pressure belt of the northern hemisphere. In many areas of that belt there is very little precipitation, especially in summer. Once moved into the middle latitudes, the tropospheric air that became enriched by volcanic particles in the high-pressure belt, is liable to lose many of the particles by washout, a process that is orders-of-magnitude more efficient than the process of 'dry fall-out'. Additionally, the stratospheric air reaching the mid- and higher latitudes from the high-pressure belt was likely to have become depopulated of small particles as these grew and fell into the troposphere.

If our speculations are essentially correct, then the Baltic Region should have been crossed by air poor in sulphuric-acid particles and these relatively few particles could not scatter back to space more than a very small fraction of the DSR.

Abel (1974, pp. 318-319) points out that Tambora effects decreased from south to north and from west to east in Europe. How did it happen that southern European Russia was spared? After all, the latitudes of southern Russia are roughly the same as those of western and central Europe that bore the brunt of Tambora's eruption. I suggest that a consideration of the direction of tropospheric winds resolves the enigma, assuming that the directions were much the same as in recent decades. The tables in the *Handbook of Geophysics and Space Environments* (1965, pp. 4-48) show that in summer, in the Longitudes of western and central Europe in Latitudes 40° and 50°, in the higher troposphere, the winds flow from west to west-southwest. These winds would 'carry' to southern Russia air largely depleted of volcanic particles by precipitation in the more western areas of the continent. Precipitation data for a few stations in western Europe show that the summer of 1816 was definitely not dry (Table 7).

The fact that the low mid-latitudes of western and central Europe suffered greatly but not the Baltic Region, despite the relatively short distance between them (from 1,000 to 1,500 km **and under**; in comparison Europe is about 10,000 km from Tambora), suggests that a drastic factor cleaned the air of the low mid-latitudes of many of its volcanic particles. Washout by rain is such a factor.

Absorption of DSR by Sulphuric-Acid Particles
It is worth looking at the possibility that the apparent failure of **supposedly** numerous volcanic particles to reduce temperatures in the Baltic Region was due to absorption of the DSR by the sulphuric-acid particles, and that this hypothetical absorption left little DSR to scatter back to space. In this context my (Neumann 1973, pp. 96-97) laboratory measurements of the absorption spectra of water solutions of sulphuric acid at three high concentrations, viz. 49, 73 and 98% (Appendix 2, No. 8), as a function of wave length of radiation are worth considering. The

Table 7: Precipitation (mm) in 1815, 1816 and 1817 at Some Stations in the Lower Mid-Latitudes of Europe.[1]

	A	M	J	J	A	S
Milan						
1815	76	113	85	181	116	10
1816	84	76	109	74	79	54
1817	6	77	48	105	84	54
Paris						
1815	30	29	79	32	15	32
1816	13	38	54	97	51	63
1817	1	65	102	59	50	62
Kew						
1815	68	58	48	45	45	30
1816	53	55	60	108	63	55
1817	3	115	35	108	68	23

[1] Sources: Milan, Brunt (1925, p. 277); Paris, Garnier (1974, p. 51); Kew (London), Wales-Smith (1971, p. 359). A = April, M = May, etc.

results, obtained by advanced instrumentation, show that sulphuric acid solutions have very low (nearly zero) absorption for light at wave lengths between 0.3 and 1.5 μm. Since about 90% of the DSR resides in that wave-length range, clearly there could have been no important absorption by sulphuric acid particles of DSR.

The result that concentrated water solutions of sulphuric acid have virtually no absorption in the range of DSR, is confirmed by an independent statement of Deirmendjian (1973, p. 293, item (iii)). Referring to the physical and chemical characteristics of volcanic particles, he writes: " ... the particles may have been composed of nonabsorbing (or very weakly absorbing) dielectric material."

Conclusion

I suggest that data quoted in this paper, especially those on grain supply and mortality, indicate conclusively that Tambora's eruption of 1815 had slight effect in the Baltic Region. Additional support for this conclusion is rendered by the fact that Post's monograph *The Last Great Subsistence Crisis in the Western World* (1977) does not quote data indicative of any significant effect on the region (Appendix 2, No. 9).

A plausible explanation for the observation that the Baltic Region showed no perceptible effects of Tambora's eruption is that the air reaching the region was cleansed of volcanic particles by precipitation in western and central Europe. I think that much the same principle applies to southern European Russia which, like the Baltic Region, was not harmed by the eruption.

Acknowledgements

I thank the following scientists for their ready assistance: Professor Claus Bjørn, Institute of History, University of Copenhagen, for information on literature on agriculture in Denmark in the 1810s; Drs. H.B. Clausen, D.A. Fisher and C.U. Hammer, Department of Glaciology, Institute of Geophysics, University of Copenhagen, for assistance with literature and comments; Dr. C.R. Harington, Coordinator "Climate in 1816" Meeting, Paleobiology Division, Canadian Museum of Nature, Ottawa, Canada, for his invitation to prepare this paper; Lars Landberg, graduate student, Department of Meteorology, Institute of Geophysics, University of Copenhagen, for translation from the Danish; Dr. Helle Linde, Copenhagen's City Archive, for information and some passages from documents relating to 1811; Dr. Jarl Lindgrén, Population Research Institute, Helsinki, for assistance with literature; Susanne Lindgrén, Adviser, Secretariat, Nordic Council of Ministers, Copenhagen, for past and present discussions on the subject of 1816 in the Baltic Region and for literature references; Torben Pedersen, Department of Meteorology, Institute of Geophysics, University of Copenhagen, for translation from the Danish; Dr. Arvo M. Soininen, formerly Lecturer in Agricultural History, University of Helsinki, for comments and for copies from the literature; Dr. Oiva Turpeinen, historical demographer, Institute of History, University of Helsinki, for comments and literature; Dr. Cynthia Wilson, Gillingham, Kent, England, for her readiness to present the substance of this paper at the "Climate in 1816" Meeting, Ottawa, Canada, June 1988.

The following libraries (and their librarians) were helpful with literature: **Copenhagen:** The Danish Veterinary and Agricultural Library; The Royal Veterinary and Agricultural University; Library of Denmark's Statistics; Royal Library; The University of Copenhagen's (a) Library of the Institute of Economic Research, (b) Library of Humanities, Fiolstræde and Njalsgade divisions, (c) Library of Slavic Philology; **Göttingen:** University of Gottingen's Library; **Helsinki:** University of Helsinki's Library/Slavica.

References

Abel, W. 1974. *Massenarmut und Hungerkrisen im vorindustriellen Europa. Versuch einer Synopsis*. Paul Paray, Hamburg and Berlin.

Bider, M., M. Schüepp and H. von Rudloff. 1959. Die Reduktion der 200 jährigen Basler Temperaturreihe. *Archiv für Meteorologie Geophysik und Bioklimatologie* B, 9:360-411.

Brunt, D. 1925. Periodicities in European weather. *Philosophical Transactions of the Royal Society of London*, Series A, 225:247-302.

Deirmendjian, D. 1973. On volcanic and other particulate turbidity anomalies. *In: Advances in Geophysics*, Volume 16. Academic Press, New York and London. pp. 267-296.

Dettwiller, J. 1981. Les températures annuelles à Paris durant les 300 années. *La Météor* No. 25:103-109.

Dyer, A.J. 1971. Anisotropic diffusion coefficients and the global spread of volcanic dust; reply. *Journal of Geophysical Research* 76(3):757. (Reply to comments by B.R. Olemesha).

Garnier, M. 1974. *Longues séries de mesures de précipitations en France*, Zone 1 (Nord, Région parisienne et Centre). *Mémoires de Météorologie Nationale*, No. 53, Fasc. 1, Paris.

Hammer, C.U., H.B. Clausen and W. Dansgaard. 1980. Greenland ice sheet evidence of post-glacial volcanism and its climatic impact. *Nature* 288:230-235.

Handbook of Geophysics and Space Environments. 1965. S.L. Valley (ed.). McGraw-Hill, New York.

Hofmann, J. and J.M. Rosen. 1982. Balloon-borne observations of stratospheric aerosol and condensation nuclei during the year following the Mt. St. Helens eruption. *Journal of Geophysical Research* 87:11,039-11,061.

_____. 1983a. Stratospheric sulfuric acid fraction and mass estimate for the volcanic eruption of El Chichón. *Geophysical Research Letters* 10:313-316.

_____. 1983b. Sulfuric acid droplet formation in the stratosphere after the 1982 eruption of El Chichón. *Science* 222:325-327.

Kondratyev, K.Ya., R.D. Bojkov and B.W. Boville. (Editors). 1983. *Volcanoes and Climate*. Report of a Meeting at the Institute for Lake Research, Academy of Sciences, Leningrad, USSR. World Climate Project 54 (WMO/TD No. 166). ICSU-WMO.

Langway, C.C., Jr., H.B. Clausen and C.U. Hammer. 1988. An interhemispheric volcanic time marker in ice cores from Greenland and Antarctica. *Annals of Glaciology* 10:1-7.

Lamb, H.H. 1972. *Climate: Present, Past and Future*, Volume I. Methuen & Co., London.

List, R.J. 1958. *Smithsonian Meteorological Tables*. Sixth Revised Edition. *Smithsonian Miscellaneous Collections*, Publication 4014. Smithsonian Institution, Washington.

Manley, G. 1974. Central England temperatures. Monthly means 1659-1973. *Quarterly Journal of the Royal Meteorological Society* 100:389-405.

Marshall, J. 1833. *A Statistical Display of the Finances, Navigation and Commerce of the United Kingdom of Great Britain and Ireland, etc.* Haddon, London.

Mossop, S.C. 1963. Stratospheric particles at 20 km. *Nature* 199(3):325-326.

_____. 1964. Volcanic dust collected at an altitude of 20 km. *Nature* 203:824-827.

_____. 1965. Stratospheric particles at 20 km altitude. *Geochimicha et Cosmochimica Acta* 29:201-207.

Neumann, J. 1973. Radiation absorption by droplets of sulfuric acid water solutions and by ammonium sulfate particles. *Journal of Atmospheric Sciences* 30:95-100.

_____. 1974. The sizes of stratospheric volcanic particles over southeast Australia after Mt. Agung's eruption in 1963. *Quarterly Journal of the Royal Meteorological Society* 100:384-388.

Post, J.D. 1977. *The Last Great Subsistence Crisis of the Western World*. Johns Hopkins University Press. Baltimore and London.

Sorokin, P. A. 1975. *Hunger as a Factor in Human Affairs*. University of Florida Press, Gainesville.

Stothers, R.B. 1985. The great Tambora eruption in 1815 and its aftermath. *Science* 224:1191-1198.

Wales-Smith, B.G. 1971. Monthly and annual totals of rainfall representative of Kew, Surrey, from 1697 to 1970. *Meteorological Magazine* 100:345-360.

World Weather Records. 1944. *Smithsonian Miscellaneous Collections* 79. Smithsonian Institution, Washington.

Denmark

Andersen, O. 1973. Dødelighedsforholdene i Danmark 1735-1839 (Mortality rates in Denmark 1735-1839). *Nationaløkonomisk Tidsskrift* 2:277-305.

Falbe-Hansen, V. 1889. *Stavnsbaands-Løsningen og Landreformerne Set Fra Nationaløkonomiens Standpunkt* (Abolition of Land-Boundness and Land Reforms from the Point of View of National Economy). G.E.C. Gad., Copenhagen.

Pedersen, S. 1983. *En Faestebondes Liv. Erindringer og Optegnelser af Gårdfaester og Sognefoged Soren Pedersen* (Life of a Tenant-Farmer. Reminiscences and Records of Estate-Leaser and Parish Executive Officer Søren Pedersen). Manuscript record edited by Karen Schousboe. Landhistorisk Selskab (Danish Agricultural History Society), Copenhagen.

Finland

Johanson, V.F. 1924. *Finlands Agrarpolitiska Historia*, Volume I. *Från 1600-Talet till Ar 1870* (Finland's Agropolitical History, Volume I. From the 1600s to year 1870). Lantbruksvetenskapliga Samfundets i Finland Meddelanden No. 13 (Society of Scientific Agriculture No. 13). Helsingfors.

Soininen, A.M. 1974. *Vanha Maataloutemme. Maatalous ja Maatalousväestö Suomessa Perinnäisen Maatalouden Loppukaudella 1720-Luvulta 1870-Luvulle* (Old Traditional Agriculture in Finland in the Eighteenth and Nineteenth Centuries). Historical Researches, Publications of the Finnish Historical Society No. 96. Helsinki. (Published also in Journal of the Scientific Agricultural Society of Finland 46, Supplement.)

Turpeinen, O. 1979. Fertility and mortality in Finland since 1750. *Population Studies* 33:101-114.

Wild, H. 1881. See References under **Russia**.

Norway

Birkeland, B.J. 1949. Old Meteorological Observations at Trondheim. Atmospheric Pressure and Temperature During 185 Years. *Geofysiske Publikasjoner* 15, No. 4. Oslo.

Derry, T.K. 1965. Scandinavia In: *The New Cambridge Modern History*, Volume IX. *War and Peace in an Age of Upheaval, 1793-1830*. C.W. Crawley (ed.). Cambridge University Press, Cambridge. pp. 480-490.

Drake, M. 1969. *Population and Society in Norway 1735-1865*. Cambridge University Press. Cambridge.

_____. 1979. Norway, In: *European Demography and Economic Growth*. W.R. Lee. (ed.). Croom Helm, London. pp. 284-318.

Linvald, A. 1952. *Kong Christian VIII, Norges Statholder 1813-1814* (King Christian VIII, Norway's Regent 1813-1814). Gyldendal, Copenhagen.

Mykland, K. 1978. *Norges Historie*, Volume 9. *Kampen om Norge 1784-1814* (History of Norway, Volume 9. War about Norway). J.W. Cappelen, Oslo.

Steen, S. 1933. *Det Norske Folks Liv og Historie Gjennem Tidene*, Volume VII (Life and History of the Norwegian People through the Ages, Volume VII). H. Aschehoug & Co. (W. Nygaard), Oslo.

Russia

Dando, W.A. 1981. Man-made famines: some geographical insights from an exploratory study of a millennium of Russian famine. *In: Famine: Its Causes, Effects and Management*. J.R.K. Robson. (ed.). Gordon and Breach Science Publishers. New York, London, Paris. pp. 139-154.

Kahan, A. 1968. Natural calamities and their effect upon food supply in Russia (An introduction to a catalogue), *In: Jahrbücher für Geschichte Osteuropas*, Volume 16. Munchen. Ost-Europa Institut. Harrasowitz. Wiesbaden. pp. 353-377.

Kazmer, D.R. and V. Kazmer. 1977. *Russian Economic History, A Guide to Information Sources*. Gale Research Company, Detroit.

Robbins, R.G., Jr. 1979. Famine in Russia. *In: The Modern Encyclopedia of Russian and Soviet History*, Volume II. J.L. Wieczynski (ed.). Academic International Press, New York. pp. 45-51.

Wahlén, E. 1881. *Der Jährliche Gang der Temperatur in St. Petersburg nach 188-jährigen Tagesmitteln*. Repertorium für Meteorologie, Volume VIII, Number 7. Imperial Academy of Sciences, St. Petersburg.

Wild, H. 1881. *Die Temperatur-Verhältnisse des Russischen Reiches*. Supplementband zum Repertorium für Meteorologie. Imperial Academy of Sciences, St. Petersburg.

Sweden

Hamberg, H.E. 1906. Moyennes Mensuelles et Annuelles de la Température à l'Observatoire de Stockholm. *Kungliga Svenska Vetenskapsakademiens* 40, No. 1.

Historisk Statistik för Sverige. 1969. Del. I. *Befolkning* (Population) 1720-1967. Second Edition. Central Statistical Bureau, Stockholm.

Historisk Statistik för Sverige. 1959. Del. II. *Vaderlek, Lantmäteri, Jordburk, Skogsbruk, Fiske T.O.M. År 1955* (Climate, Land Surveying, Agriculture, Forestry, Fisheries to Year 1955). Central Statistical Bureau, Stockholm.

Åmark, K. 1915. *Spannmålshandel och Spannmålspolitik i Sverige 1719-1830* (Grain Trade and Grain Policy in Sweden 1719-1830). I. Marcus, Stockholm.

Sommarin, E. 1917. *Det Skånska Jordbrukets Ekonomiska Utveckling 1801-1914.* Volume I. (Economic Development of the Scanian Agriculture 1801-1914). Skrifter Utgivna av de Skånska Hushållningsållskapen (Publications of Scanian Economic Society), Lund.

Tidblom, A.V. 1875-76. *Einige Resultate aus den Meteorologischen Beobachtungun, Angestellt auf der Sternwarte zu Lund in den Jahren 1741-1870. Acta Universitatis Lundensis* 12.

Appendix 1

Air-temperature data (°C) for Copenhagen, Gothbenburg, Kiev, Trondheim, Uppsala, Vöyri and Warsaw for some of the years of the 1810s and for 1882 and 1884 - where available. Although Kiev and Warsaw are not part of the Baltic Region, their data are cited to reinforce the observation that the effects of the Tambora eruption seem to have diminished northward and eastward in Europe. This observation, made previously by Abel (1974, p. 319), appears to be supported by the temperature data. For example at Kiev, 1814 was about as warm as 1812 relative to 1816; at Warsaw 1816 was substantially warmer than 1812. Vöyri (western Finland) was considerably colder in 1812 than in 1816. Hence, it is uncertain that any coldness of 1816 relative to 1814 was due to Tambora. In the tables, A = April, M = May, etc.

Year(s)	A	M	J	J	A	S	Mean of year
Copenhagen							
1811	4.8	13.6	17.5	19.0	17.0	13.8	8.8
1812	2.4	9.4	14.8	14.7	16.6	12.1	6.6
1814	6.0	7.8	13.7	17.8	16.2	12.8	6.4
1816	5.0	8.1	13.7	17.0	15.0	13.0	6.8
1817	4.3	11.0	14.8	15.7	15.7	14.9	7.9
(1816)-(1814)	-1.0	+0.3	0.0	-0.8	-1.2	+0.2	+0.4
(1816)-(1811)	+0.2	-5.5	-3.8	-2.0	-2.0	-0.8	-2.0
(1816)-(1812)	+2.6	-1.3	-1.1	+2.3	-1.6	+0.9	+0.2
1882	6.9	11.0	14.6	17.5	16.2	14.3	8.4
1884	4.8	10.9	14.4	17.5	16.7	14.9	8.2
(1816)-(1814)	-1.0	+0.3	0.0	-0.8	-1.2	+0.2	+0.4
(1884)-(1882)	-2.1	-0.1	-0.2	0.0	+0.5	+0.6	-0.2
(1816)-(1811)	+0.2	-5.5	-3.8	-2.0	-2.0	-0.8	-2.0
(1816)-(1812)	+2.6	-1.3	-1.1	+2.3	-1.6	+0.9	+0.2

Year(s)	A	M	J	J	A	S	Mean of year
Gothenburg							
1811	4.6	13.1	16.6	19.8	17.1	13.2	8.2
1812	2.1	10.2	14.7	15.0	17.3	11.1	6.3
1814	6.2	9.3	14.4	18.2	16.9	12.5	6.0
1816	5.3	8.5	14.8	18.4	15.9	12.4	6.5
1817	4.3	11.1	15.4	16.5	15.6	15.3	7.5
1882	6.1	11.4	15.3	17.6	16.8	14.0	8.2
1884	5.6	9.9	14.5	17.8	16.6	14.5	8.2
(1816)-(1814)	-0.9	-0.8	+0.4	+0.2	-1.0	-0.1	+0.5
(1884)-(1882)	-0.5	-1.5	-0.8	+0.2	-0.2	+0.5	0.0
(1816)-(1811)	+0.7	-4.6	-1.8	-1.4	-1.2	-0.8	-1.7
(1816)-(1812)	+3.2	-1.7	+0.1	+3.4	-1.4	+1.3	+0.2
Kiev[1]							
1812	4.9	12.7	18.2	21.2	19.8	13.6	?
1814	8.6	11.9	17.3	21.6	19.8	13.4	7.4
1816	7.2	13.6	19.7	18.5	17.9	15.7	7.7
(1816)-(1814)	-1.4	+1.7	+2.4	-3.1	-1.9	+2.3	+0.3
(1816)-(1812)	+2.3	+0.9	+1.5	-2.7	-1.9	+2.1	?
Trondheim							
1811	2.8	10.1	13.1	16.2	13.5	10.1	5.5
1812	0.1	6.4	11.1	11.9	13.8	7.8	3.5
1814	5.3	6.1	10.4	15.0	13.4	9.4	3.8
1816	2.8	6.3	11.4	14.1	11.9	8.7	3.8
1817	2.7	7.9	11.4	13.2	11.9	10.4	4.6
1882	3.0	8.6	13.2	15.6	15.1	12.0	5.5
1884	4.2	7.0	11.4	14.4	15.5	12.1	6.0
(1816)-(1814)	-2.5	+0.2	+1.0	-0.9	-1.5	-0.7	0.0
(1884)-(1882)	+1.2	-1.6	-1.8	-1.2	+0.4	+0.1	+0.5
(1816)-(1812)	+2.7	-0.1	+0.3	+2.2	-1.9	+0.9	+0.3
Uppsala							
1812		-1.1	6.3	13.3	13.6	15.0	8.9.0
1814		5.2	6.8	12.7	18.8	16.1	10.9.4
1816		3.1(?)	5.7(?)	13.9	17.0	13.8	11.0.4
1817		3.2	10.8(?)	13.1	16.8	14.5	11.9.3
1882		3.1	9.7	13.8	16.7	16.6	11.8.8
1884		2.2	7.5	11.8	16.1	13.6	12.9.9
(1816)-(1814)	-2.1(?)	-1.1(?)	+1.2	-1.8	-2.3	+0.1	0.0
(1884)-(1882)	-0.9	-2.2	-2.0	-0.6	-3.0	+1.1	-0.9
(1816)-(1812)	+4.2(?)	-0.6(?)	+0.6	+3.4	-1.2	+2.1	+0.4

[1] No data have been published for Kiev for the 1880s; no data are available for all the months of 1812 and, consequently, the annual mean for that year cannot be given.

Year(s)	A	M	J	J	A	S	Mean of year
Vöyri[1]							
1809	-1.3	8.7	15.3	17.7	17.4	10.1	2.4
1811	-0.3	8.2	17.6	19.0	15.4	9.1	3.4
1812	-2.3	6.8	12.5	15.6	16.7	7.4	2.1
1814	3.2	6.9	14.6	19.7	16.2	9.4	2.8
1816	1.1	8.9	14.4	19.1	14.2	10.5	3.2
1817	0.9	10.5	13.0	18.3	14.8	10.5	3.0
(1816)-(1814)	-2.1	+2.0	-0.2	-0.6	-2.0	+1.1	+0.4
(1816)-(1809)	+2.4	+0.2	-0.9	+1.4	-3.2	+0.4	+0.8
(1816)-(1811)	+1.4	+0.7	-3.2	+0.1	-1.2	+1.4	-0.2
(1816)-(1812)	+3.4	+2.1	+1.9	+3.5	-2.5	+3.1	+1.1
Warsaw[2]							
1812	3.5	12.4	17.5	18.5	17.7	11.3	5.7
1814	8.7	9.9	15.4	20.1	17.7	10.6	6.0
1816	6.4	12.2	16.8	16.9	16.3	12.6	6.3
1817	2.4	13.4	17.0	17.5	18.0	12.6	7.1
(1816)-(1814)	-2.3	+2.3	+1.4	-3.2	-1.4	+2.0	+0.3
(1816)-(1812)	+2.9	-0.2	-0.7	-1.6	-1.4	+1.3	+0.6

[1] Data for Vöyri are available only for the years 1800-24.

[2] No data for Warsaw for the 1880s.

Appendix 2

Notes

1. By the term 'developing' I mean to the process whereby, e.g., the SO_2 and SO_3 gases emanating from a volcano 'catalyze' to sulphuric acid in the course of their transport and diffusion. Mossop (1963, 1964, 1965) sampled the stratospheric particles after Mount Agung's (Bali) 1963 eruption and found that they were mostly of a 'dual' structure: a mineral core around which accreted a water-soluble coating. Mossop (1964) states that the most likely composition of the coating was sulphuric acid. Mossop (personal communication 1973) indicated that the process of coating growth was such that a year after the eruption the volume of coating was about 10 times that of the core (see Neumann 1974, second footnote to p. 385). It is reasonable to assume that Tambora's particles developed similarly. In addition to SO_4^{2-} anions, other anions, such as NO_3^- and Cl^- were involved.

2. Langway et al. (1988, p. 7) estimate that the quantity of stratospheric H_2SO_4 from Tambora was about an order-of-magnitude greater than that from Krakatau's eruption. Kondratyev et al. (1983) estimate that the stratospheric H_2SO_4 originating from Tambora's 1815 eruption

amounted to 150-200 in units of 10^6 tons, while Krakatau's 1883 eruption yielded 50-55, in the same units. Thus, on these estimates, the mass of H_2SO_4 connected with Tambora was something like three to four times greater than that associated with Krakatau. Kondratyev *et al.* (1983, Table 1) do not mention the authors of the foregoing estimates so it is not possible to weigh their arguments. A strong point in favour of the estimates of Langway *et al.* (1988) is that their acidity determinations were made at more than one place (South Pole; Crête and Dye 3 in Greenland), and that they used data on well-documented, atmospheric nuclear explosions by the United States and the Soviet Union. For example, the type of activities, amounts released ('source strength'), decay characteristics of radioactive particles, place and time of explosion, etc. are known. Moreover, in the documented cases information is available on deposition of the total β-activity. Possible shortcomings of the method used by Langway *et al.* (1988) may be that in the eruptions of Tambora and Krakatau, the nuclear explosions occurred at different times and places: the global transport and dispersal processes may have varied from one case to the other.

3. Apparently Trondheim, Norway is exceptional. Birkeland (1949, Table 1, pp. 18-21) shows that the lowest annual temperature of the whole series 1761-1946 (3°C) was reached in 1784. That low temperature may have been caused by the eruption of the Icelandic volcano Laki in June 1783 (see Wood, this volume). Kondratyev *et al.* (1983, Table 1) provide an estimate that, in the aftermath of the eruption, some 10^8 tons of stratospheric H_2SO_4 was produced. Though that mass was 33 to 50% less than the yield of Tambora in 1815, it was the **largest yearly signal** of ice-core acidity in the whole series spanning 553 to 1972 measured at Crête, Greenland (Hammer *et al.* 1980, Figure 1). No doubt, the size of the signal was due, at least in part, to the proximity of Crête and Laki (650 km). Also in the central England record (Manley 1974, pp. 393-398), one of the lowest annual temperatures occurred in 1784, and it is worth noting that central England is about the same distance from Iceland as Oslo. Similarly, the series of annual temperatures 1680-1980 for Paris (Dettwiller 1981, p. 107) show that one of the lowest occurred in 1784. However, we cannot be sure that the low temperatures of central England and Paris in 1784 were connected with Laki's eruption. On the other hand, the cold winter of 1783-84 at Stockholm (Hamberg 1906, p. 12), where WNW to NW winds are frequent, may have been connected with Laki's eruption.

4. Crop failure caused by drought is infrequent in the Nordic countries. More commonly crop failure is due to cold springs or springs and summers; or a superfluity of rain in summer.

5. The *Danske Statstidende* was the forerunner of the present Copenhagen daily *Berlingske Tidende*.

6. A comparison of data of the meteorological stations of the Russian Empire operating in 1812 (Kiev, Reval (Tallinn), Riga, St. Petersburg (Leningrad), Vilnius, Vöyri and Warsaw) indicates that, although the year on the whole was rather cold, October was somewhat milder than the long-term average. About mid-October Napoleon's Grande Armée began its retreat from Moscow. November was relatively cold and December was particularly cold. Much of the heavily decimated army reached Vilnius in December amidst a cold wave, with air temperatures around -20°C. The army had crossed the Niemen River in June on its way to Moscow (almost on the day of the month when the Germans crossed the same river into the Soviet Union in 1941). When the Grande Armée entered Russia, officers and men were well-dressed and well-fed; when they reached Vilnius in December on their way back, the remainder were in tatters and suffering from exposure and famine.

7. It is of interest to note that in 1888, long before the volcanic particles could be sampled and long before various remote-sensing techniques became available, the Krakatau Committee of the Royal Society of London suggested that the volcanic 'dust' consisted of 'condensed gaseous products of the eruption (other than water) such as sulphurous and hydrochloric acid'. (Deirmendjian 1973, p. 274). Dyer (1971, p. 757) cites a personal communication from Mossop to the effect that after Agung's eruption in 1963, high-flying aircraft were found to be thickly coated with a water-soluble material, most likely sulphuric acid.

8. Hofmann and Rosen (1983a, p. 315) state that volcanic particles they observed over Texas after El Chichón's eruption (Mexico, spring 1982) were primarily in two layers. At 25 km, the aerosol was composed of an 80% H_2SO_4 solution, while the lower aerosol layer at 18 km was composed of a 60-65% H_2SO_4 solution.

9. The Danish tenant-farmer and Parish Executive Officer Søren Pedersen (see the section Denmark - Grain Supply) wrote that 1816 was a productive ['fertile'] year and that high prices were obtained for grain "as this commodity was very much coveted in foreign lands". In other words, the harvest failure of the countries affected by Tambora's eruption profited the Baltic lands which were not afflicted and which were able to export surplus grain. A passage from a report on grain production and export in another country facing the Baltic Sea *(Möglinsche Annalen der Landwirtschaft*, Berlin, 1818, Volume II, pp. 540-552) is worth quoting. The report (p. 546) 'Landwirtschaftliche Bericht aus Meklenburg-Schwerin vom ersten Julius 1818' translates as follows: "After so many dreary years the last two years [1816 and 1817] were very favourable for Mecklenburg. Everything cheers up and gets a bright appearance and looks forward with confidence to the future. The import ban of England, which was so harmful initially to our country, has now turned to our advantage. It drove the grain prices high in England, and, consequently, our prices ... The rise in prosperity can also be seen in the surplus of capital at low interest rates."

The Years without a Summer in Switzerland: 1628 and 1816

Christian Pfister[1]

Years without Summer - 1628 and 1816[2]

These two summers embody extreme cases of a type that occurred quite frequently within the Little Ice Age. In 1816, the thermometer remained well below the average of many years in all three summer months: with 13.4°[C] that June just reached today's average temperature for May, and that July and August with 14.9° and 14.8°, respectively, reached the average of a very cold June now. Without exception, the other months were extremely cold as well; moreover, April-May and July-August were extremely wet: in Bern there was precipitation on 52 days in the summer months. Lakes Neuenburg, Murten, and Biel [Neuenburgersee, Murtensee, Bielersee] formed a contiguous body of water throughout the summer. In all summer months snow reached down to altitudes around 1000 m; on 6 June [1816] the snowcover extended from the Rigi down to Weggis. In the Rhine Valley at Bünden [Bündner Rheintal] snow persisted from 8 to 10 June down to the 1050 m level; every two weeks it snowed farther down. In the night from 30 to 31 August Salis heard avalanches thundering down from the Calanda massif. Cattle could not be let out onto the higher alpine meadows (Pfister 1976; p. 84 seq.).

At the earliest locations in the central plain [Mittelland] the rye harvest could not start until early August, the cherries in the Rhine Valley at Bünden [Bündner Rheintal] ripened only by the end of that month, and the vines had barely finished flowering by that time in some places. At the end of September the potato fields were as green as in July. In the higher locations potatoes, wheat and oats could not mature any more and ended up covered by snowfalls starting in late October. The green grapes had to be picked out of the snow around St. Martin's Day [31 October]. In the Siggental in Aargau they were dragged into a fountain trough and worked over with a flail before dumping them into a wine-press. From a comparison of proxy data and weather reports (Table 1/19) we can assume that the summer of 1628 can hardly have been any warmer.

---

[1] Historisches Institut, Universität Bern, Engehaldenstrasse 4, CH-3012 Bern, Switzerland.
[2] Excerpts from *Klimageschichte der Schweiz 1525-1860* (Band I:140-141. 1984. Verlag Paul Haupt, Bern and Stuttgart) translated and included here with the author's permission.

Table 1/19: Comparison of Proxy Data and Weather Reports in the Summers of 1628 and 1816 (Deviations from the Mean of Many Years).

	1628	1816
Start of vine flowering	11 July (+33 days)	8 July (+29 days)
Driving cattle to alpine pasture	8-10 July (Adelboden)[1]	6 July (Grindelwald)[2]
Beginning of rye harvest	31 July (+17 days)	13 August (+32 days)
Beginning of grape harvest	29 October (after frost!) (+22 days)	9 November (+32 days)
Must yields (deviation from trend)	-77%	-99%
Wine quality	very poor	very poor
Latewood density (index values)	846 (-154)	809 (-191)
Snowfall events at levels below approximately 2000 m (June-August)	"Covered the mountain, i.e., the Engstligenalp (1964 m) within seven weeks (i.e., to the end of August) 23 times"[1]	10-15[3]

Sources (unless quoted otherwise in the Climhist documentation):

[1] Bärtschi, 1916, p. 3.
[2] Strasser, 1890, p. 190.
[3] Observations in the Calanda Region (Pfister, 1976, p. 32).

Reference

Pfister, C. 1976. Die Schwankungen des Unteren Grindelwaldgletschers im Vergleich mit historischen Witterungsbeobachtungen und-messungen. *In:* Messerli, B. *et al.*: Die Schwankungen des Unteren Grindelwaldgletschers seit dem Mittelalter. Ein interdisziplinärer Beitrag zur Klimageschichte. *Zeitschrift für Gletscherkunde* 11(1):74-90.

Climatic Conditions of 1815 and 1816 from Tree-Ring Analysis in the Tatra Mountains

Zdzisław Bednarz[1] and Janina Trepińska[1]

Abstract

Dendroclimatological analyses carried out on the stone pine (*Pinus cembra* L.) in the Tatra Mountains indicate that the first half of the nineteenth century was abnormally cold and rainy. In this period - the last episode of the Little Ice Age, the year 1816 (quoted by climatologists as "the year without a summer") claims special attention. Tree-ring widths of the stone pine are abnormally narrow due to low air temperature and high rainfall during the summer of 1816. Severe social and economic consequences (e.g., the food deficit and rising prices) of the climatic anomalies of 1816 in the Polish Carpathian Mountains were reported by Polish newspapers. The climatic anomalies of 1815 and 1816 are probably associated with the world's greatest recorded eruption, that of the volcano Tambora on the island of Sumbawa, Indonesia during April 1815. Similar climatic anomalies (cold and rainy weather) were observed in 1912 and 1913 after eruption of the volcano Katmai in Alaska in June 1912. Also, in this case, the exceptional climatic character of these two years was associated with abnormally narrow rings in European high mountain trees and a rapid drop in maximum density of the wood.

Introduction

Many dendroclimatological investigations from different parts of the world indicate that tree-ring widths can be used as source of information about past climatic changes (Fritts 1976; Eckstein and Aniol 1981; Schweingruber 1983). The method is particularly sensitive to extreme climatic conditions that exert a decisive effect on tree growth. The alpine timberline is especially informative to dendroclimatologists, because that is where the main climatic factor limiting tree growth is low air temperature during a short growing season (Eckstein and Aniol 1981; Schweingruber 1983; Bednarz 1984).

Among trees of the upper forest border in the Tatra Mountains, stone pine (*Pinus cembra* L.) deserves special attention. Analysis of the relationship between tree-ring widths of the stone pine from a few sites in the Tatras with average monthly air temperature (1911 to present), showed a close relationship between tree-ring widths and June-July temperatures. Heavy precipitation in those months also limits tree growth (Bednarz 1984). On the basis of these relationships, June to July temperatures in the Tatra Mountains were reconstructed for 1741-1911. The reconstructed June and July mean air temperatures show a few cold periods (Figure 1). The greatest and longest cold period, justly called the last episode of "Little Ice Age", occurred in the Tatras and in the Alps during the first half of nineteenth century. The coldness is documented by advances of Alpine glaciers then (LeRoy Ladurie 1967; Messerli *et al.* 1978; Bircher 1982) (Figure 1).

---

[1] Department of Forest Botany and Nature Protection, Agricultural University, 31 425 Kraków, Al. 29-Listopada 46, Poland.

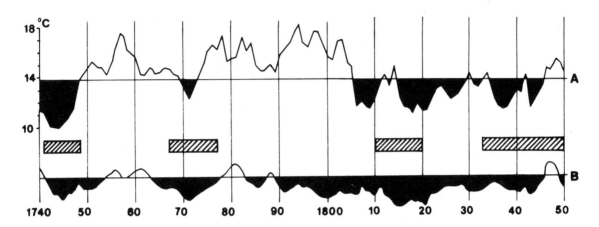

Figure 1: Reconstructed mean air temperatures for June-July from 1740 to 1850. A- Tatra Mountains, B- Alps (summer data from Eckstein and Aniol 1981). Known glacial advances are indicated by diagonal hatching.

Climatic Conditions in 1815 and 1816 in the Tatra Mountains

Toward the close of the Little Ice Age, 1816 (sometimes called "the year without summer") deserves special attention (Stommel and Stommel 1979). Severe social and economic consequences of the climatic anomalies of the years 1815 and 1816, in the Polish Carpathian Mountains were reported in Polish newspapers. Apart from news about food deficits and rising prices, remarks on weather conditions can also be found. For example, *Gazeta Lwowska* (The Lwów Newspaper) for 1816 reports:

> "In the information on this year's harvest received from Biala on October 13, 1816, we read that for over 30 years the inhabitants have not remembered such bad weather conditions. From autumn 1815 rains did not stop almost to the present day, except for a few weeks of better weather from the end of August to the middle of September" (Sadowski 1980).

Also in western Europe (Germany, France and neighbouring countries) the weather was abnormally bad. After the long and frosty winter of 1815-16, an exceptionally rainy summer began. From May 1816 rains did not stop until August, damaging the crops. Thus, food was imported to western Europe from Russia and North America (Borisenkov and Pasuckij 1988). In contrast to the Tatras and Alps, in lower areas of Poland (Cracow, Warsaw) the weather in 1815 and 1816 was normal or almost normal (Sadowski 1980; Trepińska 1988). In the period of instrumental recording which began in the Tatra Mountains in 1911, a similar drastic drop in air temperatures is noted for 1912-26.

Volcanos and Climate

The years 1912 and 1913 are of particular interest. Owing to a deep growth depression observed in all investigated Tatra stone pines, 1912 and 1913 are particularly useful as marker years for dendrochronologists (Bednarz 1975). Like 1816, 1912 and 1913 were characterized by exceedingly low air temperatures and heavy rainfall in summer. The exceptional character of these two years was stressed by Schweingruber (Schweingruber *et al.* 1979), who observed that a rapid drop in the maximum density of wood in trees from high mountain sites throughout Europe coincided with the June 1912 eruption of Katmai in Alaska. The influence of this eruption on growth processes of mountain trees is probably due to the fact that from the end of June until August 1912 the air transparency dropped to 0.56, the normal being 0.77 (Kalitin 1938). Such a phenomenon was observed for two years. The decrease of air transparency resulted in a more than 20% reduction in direct solar radiation (Griggs 1928; Lamb 1970; Budyko 1971). Many climatologists (Lamb 1970; Budyko 1971; Stommel and Stommel 1979; Stothers 1984; Kelly and Sear 1984) associate the weather anomalies of the first half of the nineteenth century with a period of increased volcanic activity.

Among severe volcanic eruptions, particular attention should be paid to that of Tambora in April 1815 - "the largest and deadliest volcanic eruption in recorded history" (Stothers 1984).

References

Bednarz, Z. 1975. Geographical range of similarities of annual growth curves of stone pine (*Pinus cembra* L.) in Europe. *In: Bioecological Fundamentals of Dendrochronology: Symposium Materials.* T.T. Bitvinskas (ed.). XII International Botanical Congress, Leningrad, July 1975. pp. 75-83.

_____. 1984. The comparison of dendroclimatological reconstructions of summer temperatures from the Alps and Tatra Mountains from 1741-1965. *Dendrochronologia* 2:63-72.

Borisenkov, E.P. and V.M. Pasuckij. 1988. *Tysiaczletniaja letopis nieobyczajnych jawlenij prirody.* Myśl, Moskva. 522 pp.

Bircher, W. 1982. Zur Gletscher - und Klimageschichte der Saastales; Glazialmorphologische und dendroklimatologische Untersuchungen. *Physische Geographie* 9:1-233.

Budyko, M.I. 1971. *Climate and Life.* Leningrad. 470 pp. (In Russian. Also published in English by Academic Press, New York, 1974).

Eckstein, D. and R.W. Aniol. 1981. Dendroclimatological reconstruction of summer temperatures for an alpine region. *Mitteilungen der forstlichen Bundesversuchsanstalt* 142:391-398.

Fritts, H.C. 1976. *Tree Rings and Climate.* Academic Press, London. 567 pp.

Griggs, R.F. 1928. *Das Tal der zehntausend Dampfe.* Brockhaus, Leipzig. 334 pp.

Kalitin, N.N. 1938. *Aktinometriia.* Gidrometeoizdat, Leningrad-Moskva. 324 pp.

Kelly, P.M. and C.B. Sear. 1984. Climatic impact of explosive volcanic eruptions. *Nature* 311:740-743.

Lamb, H.H. 1970. Volcanic dust in the atmosphere, with a chronology and assessment of its meteorological significance. *Philosophical Transactions of the Royal Society of London* A266:425-533.

LeRoy Ladurie, E. 1967. Histoire du climat depuis l'an mil. Flammarion, Paris. 381 pp.

Messerli, B., P. Messerli, C. Pfister and H.T. Zumbuhl. 1978. Fluctuations of climate and glaciers in the Bernese Oberland, Switzerland, and their geoecological significance, 1600 to 1975. *Arctic and Alpine Research* 10:247-260.

Sadowski, M. 1980. Czy rzeczywiście "rok bez lata". *Problemy* 5(410):33-36.

Schweingruber, F.H. 1983. *Der Jahrring: Standort, Methodik, Zeit und Klima in der Dendrochronologia*. Verlag Paul Haupt, Bern und Stuttgart. 234 pp.

Schweingruber, F.H., O.U. Braker and E. Schar. 1979. Dendroclimatic studies on conifers from central Europe and Great Britain. *Boreas* 8:427-452.

Stommel, H. and E. Stommel. 1979. The year without a summer. *Scientific American* 240:134-140

Stothers, R.B. 1984. The great Tambora eruption in 1815 and its aftermath. *Science* 224:1191-1198.

Trepińska, J. 1988. Many years run of air pressure and temperature in Cracow against the background of their variability in Europe. *Uniwersytetu Jagiellonskiego. (Roztrawy habilitacyjne; Series nr. 140)* Kraków. pp. 140-169.

Major Volcanic Eruptions in the Nineteenth and Twentieth Centuries and Temperatures in Central Europe

Vladimir Brůžek[1]

Abstract

This paper deals with temperatures in Prague (1771-1987) as related to the major volcanic eruptions (11) in the nineteenth and twentieth centuries, particularly that of Tambora in April 1815. The circulation, which is the primary cause of temperature changes, was observed with the help of the synoptic periods of the German Hess-Brezowski classification, known as the Grosswetterlagen, for 1881 to 1987. Altogether 24 follow-up months after 11 eruptions were studied, i.e., 264 months. However, positive temperature deviations (in Klementinum) predominate over negative ones 154 to 110. Therefore it cannot be said that volcanic eruptions are followed by cooling in central Europe within the next two years.

Observing the circulation with the help of Grosswetterlagen, we can see in the absolute majority of cases, that the drop in temperature in individual months was caused by advection of cooler air masses into central Europe. Because of the geographical extent of the Grosswetterlagen classification that covers an area larger than Europe, we may conclude that western Europe also had cooler weather for the same reasons.

Therefore I assume that the cooler summer of 1816 in western as well as central Europe following the Tambora eruption on Sumbawa in April 1815 was caused by cold air mass advection from the north, especially to western Europe where cooling was more marked than in our area. The general circulation plays the primary role.

Introduction

Atmospheric pollution by volcanic dust that might cause temperature decreases in certain regions on our planet is a problem, and the focus of considerable attention. It is assumed that enormous pollution might result in fundamental climatic changes, like the ones due to the explosions of huge meteorites in the Earth's geological past.

This paper deals with temperatures in Prague as related to major volcanic eruptions in the nineteenth and twentieth centuries, particularly that of Tambora on Sumbawa in April 1815. The circulation, which undoubtedly is the primary cause of temperature changes, was observed with the help of the synoptic periods of the German Hess-Brezowski classification, known as the Grosswetterlagen, for 1881 to 1987. Furthermore, a series of monthly temperature means measured at the Prague Klementinum station between 1771 and 1987 were used.

Processing Method

When studying the problem, those eruptions were selected during which at least 0.5 km^3 of material was ejected. In the resulting table (Table 1), the month of the eruption was denoted as

---

[1] Czech Hydrometeorological Institute, Na Šabatce 17, 143 06, Prague 4, Komořany.

0, and for months -1 to +24, deviations from the monthly temperature means for each of the 11 selected events were recorded. For each month, average deviations were determined for volcanoes to the west as well as to the east of central Europe and, finally, their mean.

Table 1A: Data on Major Volcanic Eruptions in the Nineteenth and Twentieth Centuries.

Volcano	Locality	Date of Eruption	Material Ejected (km^3)
1. Tambora	Sumbawa	April 1815	(80+)
2. Coseguina	Nicaragua	20 Jan. 1835	25
3. Mount Shasta	California	December 1860	2
4. Krakatau	Indonesia	26 Aug. 1883	17 (approx.)
5. Santa Maria	Guatemala	November 1902	10
6. Katmai	Alaska	2 June 1912	15
7. Anizapu	Chile	summer 1932	20
8. Bezimyanny	Kamchatka	20 March 1956	2
9. Vesuvius	Italy	28 Oct. 1979	0.5
10. St. Helen's	Washington	18 May 1980	15
11. El Chichón	Mexico	28 March 1983	0.6

For months showing a negative temperature deviation of at least -1°C, I determined the number of days on which the synoptic situation occurred that brought cooler weather, or below-average temperatures in that month. Of the overall number of 29 Grosswetterlagen synoptic periods we formed, e.g., for the summer months, a group of Wc, Ws and Ww situations together with all (nine) northern and northwestern ones, including troughs and lows in central Europe. These represent a cooler circulation. Similarly, for winter months, we observed the occurrence of eastern, northeastern and northern situations accompanied by below-average temperatures, etc. The data were tabulated, together with the name of the volcano, date of eruption and quantity of ejected material, as well as temperature deviations for individual months (as measured at Prague Klementinum), and the average monthly deviations for eruptions to the west and east of central Europe and their mean.

Discussion

Among the 11 episodes reviewed (except for the eruption of Laki on Iceland in 1873 with 12.5 km^3 of ejected material, where the month of the event was not known to the author), the Tambora episode was followed by a series of eight months with negative deviations - from May 1816 to December 1816. The relatively largest deviations were found for June (-1.2°C), July (-1.2°C), August (-1.7°C) and September (-1°C). In the remaining months, the deviations ranged from -0.3 to -0.9°C. However, in summer months, these negative deviations do not mean anything out of the ordinary. The absolute extremes of the Klementinum series are -4.7°C in 1923 for June, -4.6°C in 1771 for July, -3.2°C in 1833 for August and -5.0°C in 1912 for September. The

Table 1B: Monthly Temperature Deviations (°C) at Prague Klementinum as Related to Major Volcanic Eruptions (Table 1A).

Volcano[2]	Months[1]									
	-1	0	+1	+2	+3	+4	+5	+6	+7	+8
1.	2.7	1.1	1.2	0.8	-2.0	-1.1	-1.2	0.4	-0.7	-2.0
2.	1.5	1.9	2.2	0.3	-0.3	0.8	0.8	1.9	0.5	1.1
3.	-2.9	-1.6	-4.0	2.1	1.6	-2.3	-2.6	1.4	0.2	1.1
4.	-1.0	-1.2	-0.1	0.1	0.6	0.7	4.3	2.1	1.3	-2.5
5.	-1.8	-3.4	-4.6	0.6	3.9	2.8	-2.6	0.1	-1.1	-1.2
6.	-0.9	0.0	0.3	-2.9	-5.0	-2.5	-1.8	2.4	0.1	0.6
7.	-1.8	0.9	1.3	2.5	0.6	0.7	-0.2	-1.8	0.6	1.3
8.	-9.8	-1.2	-1.6	-0.1	-2.1	0.0	-1.5	0.6	0.2	-1.7
9.	-0.4	-1.1	1.1	4.8	-1.6	2.9	0.7	-1.9	-2.2	-0.4
10	-1.9	-2.2	-0.4	-2.6	-0.1	-0.7	-0.1	0.0	1.0	-0.1
11.	0.0	2.4	-1.0	1.2	1.5	1.8	1.1	2.7	1.6	2.1
Pw[3]	-0.9	-0.6	-0.9	0.5	0.0	0.4	-0.6	0.9	0.0	0.5
Pe[3]	-2.5	-0.1	0.2	0.8	-0.7	0.1	0.4	0.3	0.4	-1.2
P[3]	-1.5	-0.4	-0.5	0.6	-0.3	0.3	-0.3	0.7	0.1	-0.2

	+9	+10	+11	+12	+13	+14	+15	+16
1.	2.5	-0.6	0.2	0.5	-0.7	-1.2	-1.2	-1.7
2.	-0.4	-3.7	-1.6	-0.9	0.2	3.9	-0.2	-2.9
3.	-0.5	0.3	1.1	-1.1	-1.1	-0.6	2.4	2.1
4.	0.1	-3.1	0.3	-0.9	0.4	-1.1	-1.8	2.3
5.	-0.9	-0.7	0.5	1.5	-0.7	0.5	1.7	0.6
6.	3.2	0.2	-0.6	-1.2	-3.1	-2.5	-1.2	-0.7
7.	-1.4	-0.8	-1.5	1.1	0.3	-0.3	0.7	-0.5
8.	1.7	1.5	3.8	3.0	0.4	-2.4	2.0	0.9
9.	-2.6	-0.1	-0.7	-0.1	0.0	1.0	-0.1	0.9
10.	0.9	4.8	0.1	1.1	0.9	-1.0	0.2	0.5
11.	2.9	6.1	-1.2	2.4	2.4	0.6	1.0	3.7
Pw[3]	0.4	1.0	-0.3	0.2	-0.2	1.0	0.5	0.6
Pe[3]	0.7	-0.8	0.7	0.9	0.1	-1.3	-0.1	0.3
P[3]	0.5	0.4	0.0	0.5	-0.1	0.2	0.3	0.5

	+17	+18	+19	+20	+21	+22	+23	+24
1.	-1.0	-0.7	-0.9	-0.3	3.1	4.2	0.5	-4.2
2.	0.5	0.3	0.1	-0.4	1.9	-0.6	1.8	1.2
3.	1.7	-1.0	-0.4	-1.0	0.2	1.7	0.7	-0.2
4.	-1.3	1.7	0.0	2.7	-1.4	2.2	0.2	-2.0
5.	1.1	-1.0	-0.6	1.4	0.4	-1.5	-0.4	-0.2
6.	2.7	2.5	-2.3	-0.1	1.7	1.7	-1.3	-1.3
7.	-4.0	1.7	1.3	2.4	2.7	1.5	1.2	1.4
8.	-1.6	-1.7	0.0	1.9	0.4	0.6	2.7	-3.4
9.	4.8	0.1	1.1	0.9	-1.0	0.2	0.5	0.5
10.	0.5	2.0	-1.4	-2.1	0.0	2.4	-1.0	1.2
11.	1.2	0.5	1.5	-0.1	0.5	3.0	0.5	0.2
Pw[3]	1.8	0.5	-0.3	-0.2	0.5	1.0	0.1	0.2
Pe[3]	-2.0	0.3	0.1	1.7	1.2	2.1	1.2	-2.1

[1] Months are given before and after (-1 to +24) the eruption of Tambora in April 1815 (0 in this scheme).

[2] Volcano numbers are keyed to names and other data in Table 1A.

[3] Pw and Pe denote the arithmetic mean of temperature deviations for cases of seven volcanoes situated to the west and four volcanoes situated to the east of central Europe, respectively. P denotes the arithmetic mean of temperature deviations for all 11 volcanoes studied.

summer of 1816, with its average temperature deviation of -1.37°C at Prague Klementinum, was not extremely cold, unlike western Europe. The deviations found for the months under study do not nearly approach the above extremes. Regarding the other cases, there does not appear any uninterrupted, longer series of negative deviations, except for Vesuvius in 1979, following the eruption of which seven months (April to October 1980) exhibited a negative deviation. However, only three months showed a deviation from -1.9 to -2.6°C, while the others deviated by only -0.1 to 0.7°C. So, the figures for 1816 are not extraordinary.

Altogether 24 follow-up months after 11 eruptions were studied (i.e., 264 months). However, positive temperature deviations predominate over the negative ones by 154 to 110. Therefore it cannot be said that volcanic eruptions are followed by cooling in central Europe within the next two years.

Nor do the average monthly temperature deviations, or their means, show any cooling down in the cases of volcanic eruptions both to the west and east of central Europe; rather the opposite. In the western case, 18 deviations were positive and six negative, whereas in the eastern case 17 were positive and seven negative. In all 11 cases, 17 positive and seven negative deviations were recorded within two years after the month of the eruption.

Let us observe the circulation with the help of the Grosswetterlagen synoptic periods (since 1881), in the interval of -1 to +24 months surrounding the month in which the eruption took place, including that month. In the following, months showing a deviation of at least -1°C or more will be considered. In the eight cases under review, negative deviation ranging from -1 to -3.4°C occurred five times in the month of eruption. In the preceding month, a negative deviation from -1 to -9.8°C also occurred five times. Counting the number of days with cooler synoptic periods in a given month, we may conclude that in the eruption months there were 24 to 27 days with cooler circulation; the figure is 18 to 28 for the month prior to the eruption. Among the 192 months which followed the eruptions, in 48 instances the temperature deviation was at least -1°C. Of these, only seven months included a mere seven to 12 days with cooler circulation. In the remaining 41 cases, the number of such days was 15 to 30.

In the absolute majority of cases, the drop in temperature in individual months was caused by advection of cooler air masses into central Europe. Because of the geographical extent of the Grosswetterlagen classification that covers an area larger than Europe, we may conclude also that the western European continent had cooler weather for the same reasons.

Probably the cooler summer of 1816 in western as well as central Europe following the Tambora eruption in April 1815 was caused by atmospheric circulation that brought cold air masses from the north and northwest, especially to western Europe where cooling was substantially greater than in our area. Where climatic change is concerned, atmospheric pollution by volcanic dust perhaps plays a secondary role, the general circulation of the atmosphere being of prime importance.

Conclusions

This work has been inspired by the questions connected with the cold summer of 1816 in western Europe following the 1815 Tambora eruption. This topic was the subject of an international conference "The Year without a Summer? Climate in 1816" held in Canada in June 1988.

On the basis of the Klementinum temperature series and the Hesse-Brezowski classification of synoptic periods, I conclude that the summer cooling of 1816 in central and western Europe was caused primarily by atmospheric circulation - volcanic dust being a factor of secondary importance.

Acknowledgement

I acknowledge the effort of Dr. Zdeněk Kukal (Central Geological Institute of Prague) for collecting and making available the required data on volcanoes.

Asia

Climate over India during the First Quarter of the Nineteenth Century

G.B. Pant[1], B. Parthasarathy[1] and N.A. Sontakke[1]

Abstract

A detailed examination of climatic conditions over the Indian subcontinent for 1801-25 has been made using instrumental data, historical records and proxy records such as food-grain prices.

The long summer monsoon (June to September) rainfall records of the east coast station, Madras, indicate four deficient years of 1813, 1815, 1820 and 1824, with one excess year 1818. However, the west coast station, Bombay, reported one deficient year, 1824 and two excess years 1817 and 1822 respectively. The rainfall series of these stations are significantly correlated with All-India rainfall, and further, these data are the earliest available instrumental records for the nineteenth century. Therefore, an attempt is made to assess the rainfall conditions over All-India during the period.

Documentary reports indicate that frequently India had faced famine conditions due to failure of monsoon rains in large parts of the country during: 1799-1804; 1811-14; and 1819-25.

The rainfall series is also examined in relation to the occurrence of El Niño phenomena, which are now established to be significantly related to the southwest monsoon. After examining all relevant available records, we infer that 1816 was a normal monsoon rainfall year, without any report of droughts or floods over the country.

Introduction

The primary focus of this paper is examination of the climatic variability over the Indian region for 1801-25, which includes the year without summer, 1816. The information is extracted from the available data, such as: (1) instrumental records; (2) historical evidence of climate-related phenomena like famines, floods, droughts, etc.; (3) other indicators like food-grain production or prices; and (4) information on El-Niño phenomena in the South Pacific.

The search for instrumental records from old manuscripts for elements like pressure, temperature and rainfall indicated that pressure and temperature are not recorded during 1801-25. However a few isolated rainfall records for Calcutta in 1784, and continuous records for Madras in 1813 and Bombay in 1817 are available. An examination of several relevant reports [e.g., those of the Indian Famine Commission (1880), Fact-finding Committee (1973); and details of famines reported by Majumdar (1963), Masefield (1963), and Walford (1878)] suggests that much information about climatic conditions over the Indian region is available for the period mentioned[2].

A rare document on Delhi wheat prices during 1763-1836 (an indirect climatic indicator for that region) was also examined. Based on these data, a broad inference about climatic change, particularly the occurrence of major droughts, if any, can be ascertained.

---

[1] Indian Institute of Tropical Meteorology, Post Box No. 913, Pune-411008, India.
[2] Editor's note: additional information in the Workshop section, this volume pp. 538-541.

Period of Instrumental Record

The first astronomical observatory in India was established in Madras in 1792, and the earliest meteorological observations occurred in September 1793 (India Meteorological Department 1976). Systematic and continuous monthly rainfall records for Madras (13°N, 80°E) are available from 1813. The second reliable, continuous monthly rainfall record for Bombay (19°N, 73°E) is available from 1817 (Eliot 1902).

During the monsoon months (June-September) over India, rainfall accounts for about 70 to 90% of the total annual amount for India. Extreme monsoon rains have catastrophic effects on the Indian economy. Therefore, the monsoon rainfall series for Madras and Bombay during 1813-1988 and 1817-1988, respectively, are discussed in detail. The summer monsoon rainfall series for All-India, with a fixed 306 well distributed rain-gauge stations over all parts of India except the small hilly region, has been constructed from 1871 to present by Mooley and Parthasarathy (1984) and Parthasarathy et al. (1990). This series is extended back to 1813 by interrelating (with the help of all the available stations' monsoon-rainfall series) with the All-India series for 1871-1984, and by applying appropriate methodology accounting for the variance due to the small number of stations (Sontakke 1990). The All-India rainfall series for 1813-25 prepared on the basis of a smaller number of stations is also discussed.

Madras Monsoon Rainfall

Madras is the east coast station representing the southeastern peninsula of India. The correlation coefficient between monsoon rainfall series of Madras and All-India (1871-1984) is 0.39, which is significant at the 0.1% level. Therefore, a detailed investigation of this oldest monsoon rainfall series has been carried out. Figure 1 shows the standardized (1871-1970) summer monsoon rainfall series of Madras for the period 1813-1988. The mean ($\bar{R}$) monsoon rainfall of Madras is 38.6 cm, standard deviation (S) is 12.2 cm, and coefficient of variation (CV) is 31.6%.

Extreme events of deficient and excess rainfall are identified with the criteria of $\geq (\bar{R}+S)$ values as excess and $\leq (\bar{R}-S)$ as deficient. Figure 1 shows that 1818 is the excess rainfall year and 1813, 1815, and 1824 are the deficient rainfall years. The year 1816 is an above-normal rainfall year.

Bombay Monsoon Rainfall

Bombay is the west coast station representing the western region of India. The correlation coefficient between Bombay and the All-India monsoon rainfall series (1871-1984) is 0.46, which is significant at the 0.1% level. Figure 1 shows the standardized (1871-1970) summer monsoon rainfall series for Bombay for the period 1817-1988. The mean ($\bar{R}$) monsoon rainfall for Bombay is 179.6 cm, standard deviation (S) is 48.1 cm, and coefficient of variation is 26.8%.

The excess and deficient rainfall years are identified using the same criteria as for Madras. The two years 1817 and 1822 are excess rainfall years, and 1824 is the deficient rainfall year.

All-India Monsoon Rainfall

Figure 1 shows the All-India rainfall series for 1871-1988 and 1813-25 (a part of the reconstructed rainfall series 1813-1988) in its standardized form. The mean of this series is 85.3 cm, standard deviation is 8.2 cm, and coefficient of variation is 9.6%.

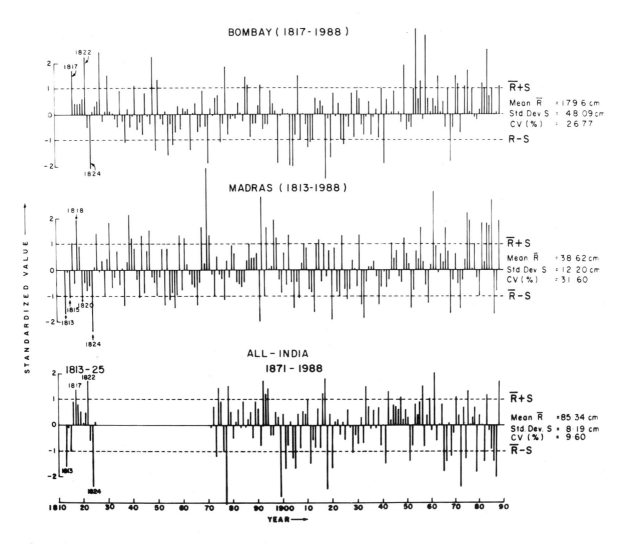

Figure 1: Summer monsoon rainfall series for Bombay (1817-1988), Madras (1813-1988) and All-India (1813-1825 and 1871-1988) expressed as standardized values.

We note that the three years 1813, 1815 and 1824 are deficient rainfall years, whereas 1817 and 1822 are excess rainfall years. 1816 has the standardized value of the order of 0.92 which shows that rainfall is much above the mean value, however it is not in the flood category as defined earlier.

Climate from Historical Evidence

A thorough search has been made of all available records concerning the abnormally deficient or excess rainfall years over various regions of India during 1801-25. Figure 2 shows details of different meteorological regions of India. Below are the consolidated details of historical information for 1801-25 for these regions. In most cases the meteorological phenomena (conditions) relevant to climatic conditions such as: late onset of monsoon rains; deficient/excess rainfall; frost; hailstorms; cyclones; and thunderstorms are reported yearly. For 1816, none of the reports has yielded information about abnormal climatic events, which suggests that this was a normal year.

1802: It has been reported that the core region of monsoon rainfall failure lies over Gujarat and Maharashtra states, with extension to the Rajasthan and Karnataka side.

1803: During March-May, crops are reported to have been spoiled over Uttar Pradesh due to hailstorms. The monsoon rains were isolated from June to the middle of August: because of this, hot summer-type winds are reported mainly from the northern regions of India. However, some crops were saved due to adequate rainfall in September.

1806: Drought conditions reportedly occurred over Tamilnadu State and adjoining coastal Andhra Pradesh.

1811: Reportedly, the monsoon rainfall over Tamilnadu was not favourable for crops.

1812: Complete failure of monsoon rains over Gujarat State was reported. To some extent drought conditions were reported over Tamilnadu, Andhra Pradesh and Madhya Pradesh State.

1813: Reportedly famine due to drought was experienced in Rajasthan, south Haryana, Gujarat, Saurashtra and Kutch, Tamilnadu and Andhra Pradesh.

1819: It has been reported that the monsoon onset was late over eastern Uttar Pradesh, and that state had heavy rains afterward. A heat wave was noticed during May/June due to lack of rains.

1820: Reportedly the state of Rajasthan and parts of Uttar Pradesh were affected by frost during early spring.

1822: Destruction of crops and a heavy death toll resulting from a severe cyclonic storm followed by strong tidal waves were reported from Bengal State.

1823: Southwest and northeast monsoon rains were reportedly below normal, or greatly deficient, over Tamilnadu and adjoining areas of Andhra Pradesh.

1824: The southwest monsoon rainfall was reported to be very much below average (i.e., about one third of the normal) over Tamilnadu and adjacent Andhra Pradesh. Widespread considerable scarcity was noted due to failure of rains over Maharashtra and adjacent areas of Karnataka. Delhi and neighbourhood districts in Haryana also experienced severe drought. Thus the drought in 1824 was widespread and severe in India.

1825: The crops in Uttar Pradesh State were badly affected due to failure of monsoon and winter rainfall. Heavy thunderstorms were reported from Madhya Pradesh State.

Other Indications of Rainfall Anomalies

Many kinds of environmental evidence (e.g., food-grain production or prices, and climatic events from other regions of the globe known to be monsoon-related like the El Niño phenomenon) can yield information for a region. Details of these are examined and discussed below.

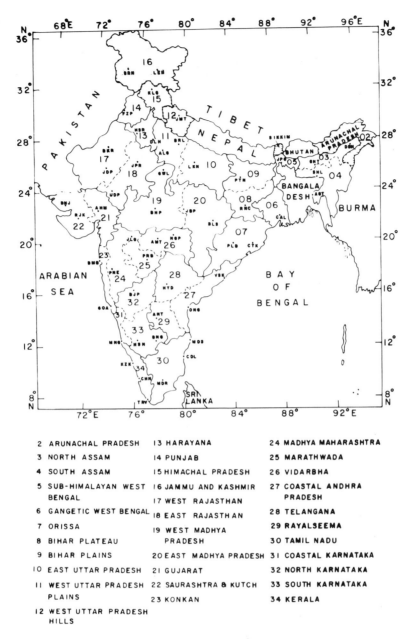

Figure 2: Details of Meteorological Subdivisions of India.

Delhi Wheat Prices

One of the indicators of reduced seasonal rainfall is the loss of food-grain production, and subsequent rise in prices. In recent decades, the anomalies in food-grain production and prices, over and above the technological trend, are found to be highly related to seasonal rainfall (Parthasarathy *et al.* 1988). Although the data on prices during the periods under consideration from one sector of India cannot be taken as representative of the whole country due to political barriers and lack of efficient transport, major droughts of widespread nature are generally reflected in these series. One such rare document concerns Delhi wheat prices for 1763 to 1835 (Roy 1972). Figure 3 shows the prices in Rupees per 1,000 seers (the mode of measurement used in those days, 1 seer = 2.0571 lb). As seen from this curve, the period under consideration shows rather stable wheat prices within a radius of about 1,000 km around Delhi.

Some inferences can be drawn from Figure 3. Wheat prices were very high during the years that had created a crisis (or famine conditions probably due to failure of rains) for the buyers: 1763; 1773; 1782-83; 1792-93; 1803-04; 1826; 1834; and 1836. The price of food-grain for the Delhi region in 1816 does not reflect stress on crops to the extent of being a drought year or an excess year.

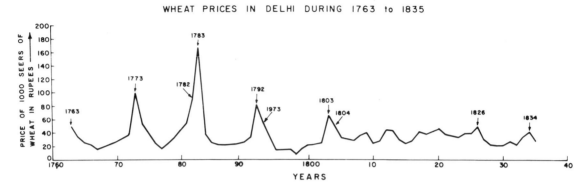

Figure 3: Wheat prices at Delhi during 1763-1835.

El-Niño Phenomena

El-Niño, the warming of the Pacific Ocean mixed layer from the coast of Peru to the international date line is a well-known phenomenon to meteorologists and oceanographers. It affects the weather over many parts of the tropical regions. Further, many workers have shown that it is strongly related to Indian monsoon rainfall. Parthasarathy and Sontakke (1988) noted that for the period 1871-1980 it is highly correlated with monsoon rainfall, and there is an average drop of 11% in monsoon rainfall during El Niño years. The evidence of El Niño occurrences over the past four and a half centuries compiled by Quinn et al. (1987) is used in this study. It is seen from their work that 1814 and 1828 are strong and very strong El Niño years respectively, and 1806-07, 1812, 1817, 1819, 1821 and 1824 are moderate El Niño years. There is no indication that 1815 or 1816 are El Niño years, therefore, we infer that the All-India monsoon rainfall was normal during 1816.

Conclusions

An examination of the relevant records for climatic conditions over India at the beginning of the nineteenth century, indicates that:

1. 1824 had deficient monsoon rainfall over large parts of India.

2. During the first quarter of the nineteenth century, large-scale monsoon failures were of comparatively moderate nature.

3. Analyses of Delhi wheat prices and the El Niño event do not support the existence of abnormal climatic conditions over India during 1816.

Acknowledgements

We are grateful to Mr. D.R. Sikka, Director, Indian Institute of Tropical Meteorology for the facilities to carry out this study. The rainfall data used have been obtained through sources in the India Meteorological Department. Thanks are due to Mrs. J.V. Revadekar for typing the manuscript.

References

Eliot, J. 1902. Monthly and annual rainfall of 457 stations in India to the end of 1900. *India Meteorological Department Memoirs* 14:1-709.

Fact-finding Committee for Survey of Scarcity Areas, Maharashtra State. 1973. Volume I. Government of Maharashtra, Bombay. 310 pp.

Indian Famine Commission. 1880. *Famine Relief, Part I. Measures of Protection and Prevention.* Blue Book, London. (C-2591 and C-2735).

Majumdar, R.C. 1963. The History and Culture of the Indian People. Part-I. Second Edition. Bhartiya Vidya Bhavan. pp. 835-836.

Mani, A. 1975. History of rainfall measurement. Volume III. *Proceedings of a Symposium of the Second World Congress on 'Water Resources' held at New Delhi, December 1975.* Central Board of Irrigation and Power, New Delhi. pp. 393-402.

Masefield, G.B. 1963. *Famine: Its Preventation and Relief.* A Three Crown Book, Oxford University Press. 159 pp.

Mooley, D.A. and B. Parthasarathy. 1983. Fluctuations in All-India summer monsoon rainfall during 1871-1978. *Climatic Change* 6:287-301.

Parthasarathy, B., A.A. Munot and D.R. Kothawale. 1988. Regression model for estimation of Indian foodgrain production from summer monsoon rainfall. *Agricultural and Forest Meteorology* 42:167-182.

Parthasarathy, B. and N.A. Sontakke. 1988. El Niño/SST of Puerto Chicama and Indian summer monsoon rainfall: statistical relationships. *Geofisica Internacional* 27(1):37-59.

Parthasarathy, B., N.A. Sontakke, A.A. Munot and D.R. Kothawale. 1990. Vagaries of Indian monsoon rainfall and its relationships with regional/global circulations. *Mausam* 41(2):301-308.

Quinn, W.H., V.T. Neal, and S.E. Antunez de Mayolo. 1987. El Niño occurrences over the past four and a half centuries. *Journal of Geophysical Research* 92(C13):14,449-14,461.

Roy, S. 1972. A rare document on Delhi wheat prices 1763-1835. *The Indian Economic and Social History Review* 9(1):91-99.

Sontakke, N.A. 1990. Indian summer monsoon rainfall variability during the longest instrumental period 1813-1988. M.Sc. thesis (submitted), University of Poona. 145 pp.

Walford, C. 1878. The famines of the world: past and present. *Journal of the Statistical Society* 41, Part 3:433-526.

Evidence for Anomalous Cold Weather in China 1815-1817

Pei-Yuan Zhang[1], Wei-Chyung Wang[2] and Sultan Hameed[3]

Abstract

The extraordinarily cold summer of 1816 recorded in New England, Canada and western Europe has been attributed to the atmospheric effects of the Mount Tambora eruption in April 1815. We have carried out a comprehensive search of historical Chinese manuscripts to obtain information on weather conditions in China in the years following the eruption. The types of documents examined include: official histories; gazettes; memoranda to the emperor; and personal diaries. Episodes of anomalously cold and stormy weather were noted from winter 1815-16 to summer 1817 in 14 provinces (Heilongjiang, Jilin, Liaoning, Neimongol, Hebei, Shandong, Henan, Jiangsu, Zhejiang, Fujian, Anhui, Jiangxi and Taiwan) and three cities (Beijing, Tianjin and Shanghai). Crop production in several provinces was significantly below normal in 1816 and 1817. The results suggest that the anomalous weather patterns, attributed to the effects of Tambora, lasted until the summer of 1817 and occurred over widespread regions of China.

Introduction

It is of interest to establish the geographical extent of the climatic anomaly associated with the Tambora eruption. The nature of the perturbation of atmospheric circulation invoked to understand the observed effects would be quite different if these were confined to a limited region as opposed to the case of a global event. Also, the duration of the atmospheric effects following the eruption may be an indication of, among other things, the amount of sulphur dioxide released by the volcano.

In the literature on the effects associated with Tambora a viewpoint emerges that these effects were confined to the two sides of the Atlantic Ocean, and were observed only in the summer of 1816.

Under the sponsorship of the United States Department of Energy an extensive program has been undertaken at the Institute of Geography, Academia Sinica to search historical manuscripts for information on past climatic changes. In this research, documentary evidence has been found showing that unusually cold conditions prevailed in China following the Tambora eruption (see also Huang, this volume; Thompson and Mosley-Thompson, this volume). Observations of contemporary writers indicate that anomalously cold and stormy weather was experienced in widespread regions of China until at least the summer of 1817. The abnormal conditions were noted not only in the temperate zones but extended to southernmost China. These perturbations of weather were accompanied by shortfalls in food production. Reports of abnormal conditions in China are presented later, followed by a discussion of available information from North America supporting the results from China that the climatic impact of Tambora lasted until the summer of 1817, being experienced in tropical as well as temperate regions.

---

[1] Institute of Geography, Academia Sinica, Beijing, People's Republic of China.
[2] Atmospheric and Environmental Research, Inc., 840 Memorial Drive, Cambridge, Massachusetts 02139, U.S.A.
[3] Institute for Atmospheric Sciences, State University of New York, Stony Brook, New York 11794, U.S.A.

Historical Writings in China

It is well known that many historical documents are available in China describing details of life and environment as observed by the writers of various periods. Several different types of manuscripts exist. One commonly-found category, that may be called local gazette, originates from the widely followed practice of local and provincial governments in China of compiling official histories. A prominent citizen or a government official was commissioned for the purpose. These accounts of local histories were compiled usually at intervals of several decades and provide descriptions of geography, hydrology, agriculture and climatic conditions, as well as social and political conditions.

Official histories were also compiled by the central government in Beijing where tradition dictated that each new dynasty of emperors publish an official record of the previous dynasty. A large government bureau was maintained for the purpose. This office archived documents and books relating to events from the previous era.

Various branches of government communicated with the emperors in Beijing by writing memoranda. A large number of the memoranda have survived. Their contents indicate the keen interest in climatic data taken by the rulers of China throughout recorded history. We may assume that this interest was derived mostly from the realization that weather fluctuations affected food production and economic welfare. The central government had several independent channels through which such information was received. One was the hierarchy of the civilian administration that has existed in China for most of the last 2,000 years. Designated officials in each locality recorded information on agriculture and weather, especially on moisture, and sent it to higher levels of government. Similar information was gathered by the military authorities and communicated to Beijing independently. In addition, the emperors frequently sent "ombudsmen" to the provinces, and these travelling officials sent back reports describing local conditions.

Another rich source of information on past climates is found in personal diaries. A large number of diaries kept by prominent citizens have survived in China, and most contain weather observations. Some diaries list phenological data such as the dates in spring of first blooming of certain flowers and the dates in autumn of freezing of lakes. An analysis of the reliability of climatic information obtained from the historical writings is presented by Wang *et al.* (in preparation).

Reports of Anomalous Weather during 1815-17

An example of an historical document referring to the summer of 1816 is shown in Figure 1. It is a memorandum written by the Gexia, (an official in the central government of Tibet in Lhasa) to the headman of Doilungdeqen, a town 20 km west of Lhasa. It refers to three days of snowfall in June 1816. A translation of the memorandum follows:

Ao-ka (memorandum)
Reply to Zongdui Weiga from Gexia

Your report has been received, in which you mentioned: "In June last year [converted from the Tibetan lunar calendar], it snowed for three days and nights in this region, so that we had no autumn harvest. Additionally, a large number of houses here collapsed, and in the north many houses were damaged and are still in danger of collapse. We can hardly feel at ease. Someone was put in charge of this matter, but to no effect. The houses south of Doilungdeqen are also in danger of collapse, a few of them have collapsed...".

If both you, Zongdui, and the landlords had paid more attention to maintenance every year, the houses would not have collapsed like this. It is really caused by your paying no attention in normal times. So much damage has been done because of your negligence and making excuses that you deserve disciplinary action. But considering that precipitation has been extremely high this year, you might escape punishment if you pay immediate attention to repairs.

Fire Cow Year (1817) [illegible] month

Figure 1: A Tibetan memorandum that cites three days of snowfall in June 1816.

Other reports of unusual weather observations during 1815 to 1817 found in the historical documents are as follows. The provinces referred to are indicated in Figure 2.

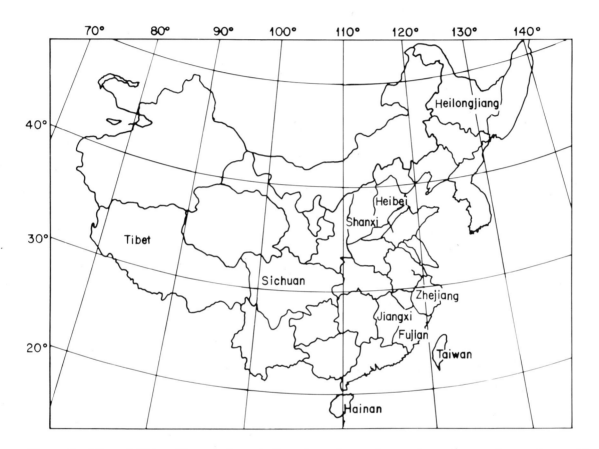

Figure 2: Map of China. The provinces indicated are those where written reports of anomalous cold conditions during 1815-17 were found.

Zhejiang Province
Jiaxing Area
1817: It was reported from the Jiaxing area (Wuxing and Deqing) that there were very few sunny days from January 1817 to November 1817. Rice spoiled on a large scale and food prices soared.

Jiangxi Province
Jiujiang Area
1816: Several feet of snow covered the ground around Jiujiang, Hukao, and Duchang after the heavy snowfall in February. Torrential rain fell in Hukao and Duchang in the summer. The rain continued for one month in Duchang.

1817: Pengze - Lightning killed seven or eight people within 10 days in May. Heavy rain, fierce wind and hail hit the area. Many trees were uprooted. Cold wind blew from the north in late June; it was very cold the night of the 29th. Snow covered the top of the mountain Douhou the next morning.

Hukao - Wind uprooted big trees in late April. It was very cold in late June.

Xingzi - it was very cold in early July because of the heavy rain with cold wind. Some people wore fur coats and stayed by the fireside.

Pingchan Area (Anyuan)
1816: It was usual to have very light snow in the winter and spring, but heavy snowfall continued for three days (8-10 February) in 1816.

Yichun Area (Wenzai)
1816: Heavy snow fell in February 1816. Water buffalo killed by the cold.

For comparison we note that instrumental records of 1951 to 1980 from Jiangxi province showed the maximum snow depth in Jiujiang to be 25 cm (3 and 4 February 1969) and 21 cm in Hukao and Pengze on the same days.

Fujian Province
Jianyang Area (Guangze)
1816: Heavy snow fell on 7 February 1816. Ice formed. There was serious flooding in April 1816.

Ningde Area
1815: Three feet of snow fell in Fuan in January 1815 [note that this is a pre-Tambora event].

1816: Snow fell on 4 January 1816; more than 30 cm of snow covered the ground.

In the modern records at the Ningde Meteorological Station, the maximum snow depth of 1 cm was observed on 28 January 1971 and 6 cm on 14 December 1975.

Taiwan

Hsinchu and Miaoli
1815:   One-inch-thick ice formed on the ground in December 1815.

Changhua
1816:   Ice was observed in December 1816.

Heilongjiang Province
Harbin Area
1817:   Hulan - Drought in summer and early frost in autumn resulted in poor harvest.

Qiqihar - Widespread damage caused by frost in July.

In the modern records (1951-80) at Qiqihar the average number of frost days is two in May and zero in June, July and August. During this period, the average date of first frost is 21 September; the earliest observation was on 9 September, and the latest on 3 October.

Hainan Island
Wanning
The winter of 1815-16 was dry and very cold. More than half of the trees died.

Ding'an
Cold and rain lasted throughout the winter of 1815-16. Frost killed rice. Most of the vegetation was destroyed. The Climate Atlas of China (published by the Central Meteorological Agency, 1985) shows no frost days in the plain of Hainan Island.

Hebei Province
Shijiazhuang Area
1817:   There was very little rainfall from summer through autumn. Agricultural production was damaged by drought, hail and frost. The harvest was scanty in 111 counties.

Shanxi Province
Datong Area
1817:   Early frost in 1817 caused widespread damage to 27 counties around Datong, Huairen area. Refugees appeared.

Sichuan Province
The normal winter of Sichuan Province is not very cold. Rain is more common than snow in the winter, but there was snow in a dozen or so counties in November 1815. Xuyong, Luzhou, Maozhou, Guan Xian, Jintang, Wenjiang, Leshan, Emei, Santai, Shehong, Yongning, and Nanchuan had two additional snowfalls. It was very unusual. Only 10-20% of Sichuan did not get any snow; but frequent rain was reported in these areas.

Damage to Agricultural Production

Historical manuscripts in China contain detailed information on the state of the harvest in each county. Quantitative estimates of the harvests for each season were made by local officials and reported usually in units of tenths of normal expectations. The reports from the period 1815-17 indicate severe damage to crops in widespread areas in autumn 1816, summer 1817 and autumn 1817. Reported shortfalls for the autumn harvests of 1816 and 1817 in the various provinces are shown in Table 1. The results are given in terms of the number of counties in each province reporting harvests of less than 60% or less than 50% of contemporary normal values.

The geographical distribution of the damage to the autumn 1816 crop is shown in Figure 3 and for the summer 1817 crop in Figure 4. The degree of damage is given in terms of four categories: light (1-5 counties reporting serious damage), medium (5-20 counties), heavy (21-30 counties) and severe (more than 30 counties).

Table 1: Autumn Harvest Conditions.[1]

	<60%		<50%	
	1816	1817	1816	1817
Shanxi	6	1F	2	4
Shaanxi	5	1	---	---
Gansu	16	30	6	12
Jiangsu	---	---	1F	---
Anhui	2	---	5	7
Hubei	11	4	1	1
Hunan	2	---	---	---
Zhejiang	1F	---	1	---
Hebei	---	P	---	---
Henen	---	P	---	---
Guangxi	---	1	---	---
Fujian	4	1	---	1
Guangdong	18	19	10	5

[1] The number refers to county, e.g., 6 means six counties; 1F refers to Fu (a district) and P refers to the whole Province.

Evidence from North America

The evidence presented in the two previous sections suggests that the perturbation of climate following the eruption of Tambora was global. Reports of abnormally cold conditions in China ranged throughout its latitudinal extent, from Heilongjiang province in the north to Hainan Island in the south. Moreover, reports of unusually cold conditions start in winter 1815-16 (ice in Taiwan and snowfall in Sichuan province) and persist until summer 1817.

Most of the literature on the climatic effects of Tambora has focused on the cold summer of 1816 noted in New England (e.g., Baron, this volume) and in western Europe (e.g., Kington, this volume; Briffa and Jones, this volume; Guiot, this volume), leading to the impression that it was a trans-North Atlantic event that lasted one season. However, considerable historical evidence suggests that the anomalously cold conditions extended to regions beyond New England in North America, and beyond the summer of 1816. Some of this evidence is presented below.

LaMarche and Hirschboeck (1984) have shown that extraordinary cooling produced by major volcanic atmospheric dust veils are registered as frost damage in tree rings. Examination of accurately dated tree-ring sequences in bristlecone pines from seven localities in the Great Basin and Rocky Mountains showed frost damage in both of these regions in 1817.

David Thomas (1819) travelling through the mid-western United States during the summer of 1816 noted the "unexampled coldness of the summer" in Ohio. Soon after arriving in Pittsburgh in the first week of June he wrote in his diary "On the day of our arrival in this city, we had several thundershowers from the west. The weather then became clear, and for three days we had brisk gales from the north-west, of unusual severity for summer. The surface of the river was rolled into foam, and each night was attended by considerable frost. Indeed it still continues."

Frost was experienced in South Carolina in mid-May 1816. Subnormal cold weather was also noted in August with a succession of cold fronts. This resulted in shortfalls in agricultural production in parts of the State (Haynsworth 1818; *Niles Weekly Register*, 21 September 1816). Crop failures in Alabama in 1816 caused enough hardship to force many settlers to leave the territory (Breckenridge 1898-99).

Crop failures attributed to unusually cold weather in 1816 were noted in Cuba (Chladni 1819; *North American Review* 1816) and the British West Indies (*The Times* 1817).

Considerable evidence exists that unusually cold weather was experienced in North America after the summer of 1816.

A figure (Stommel and Stommel 1983, p. 23) shows mean June temperatures at New Haven, Connecticut for the period 1790-1860. Stommel and Stommel have pointed out that the low of slightly more than 60°F (15.6°C) marked the unusual cold of the summer of 1816. It is also worth noting in the figure, however, that temperature in June 1817 was only half a degree warmer than in June 1816, indicating the persistence of unusually cold conditions in the region for at least a year after the summer of 1816. The intervening winter of 1816-17 produced record-breaking cold weather in New England lasting from mid-January to early March (Ludlum 1966). The extreme cold weather began with an anticyclone which arrived in the region on 12 January. A low temperature of -34°F (-36.7°C) was registered at Hallowell, Maine on 15 January and the condition was widespread enough to be called "probably the greatest cold ever registered in the United States" up to that time. The second major cold outbreak of the season occurred on 26 January and was experienced in Maine, Vermont, New Hampshire and Connecticut. February 1817 was one of the coldest months on record in all parts of New England. As shown in Figure 5, at New Haven it was the coldest February since the records began in 1780 and registered a temperature anomaly of -8.6°F (-22.5°C) that was not exceeded until February 1836 following the Coseguina eruption in Nicaragua in 1835. Similar record cold departures were observed at New Bedford and Williamstown, Massachusetts.

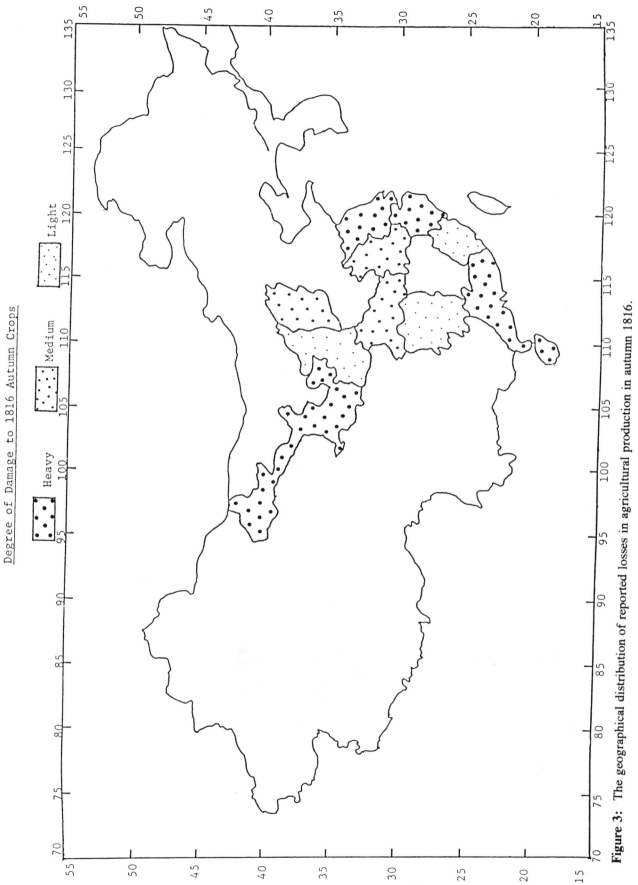

Figure 3: The geographical distribution of reported losses in agricultural production in autumn 1816.

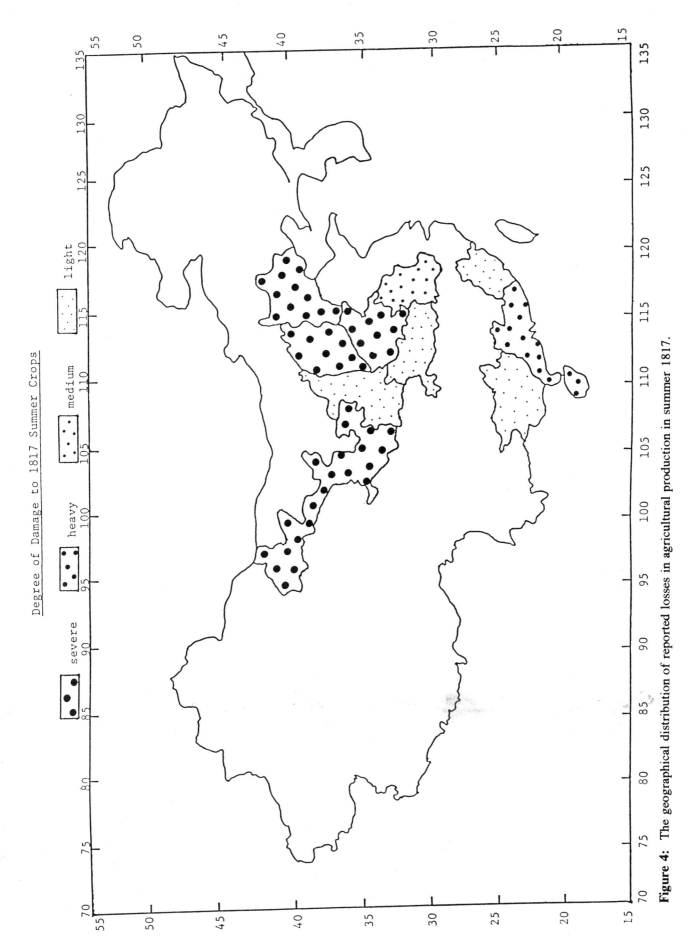

Figure 4: The geographical distribution of reported losses in agricultural production in summer 1817.

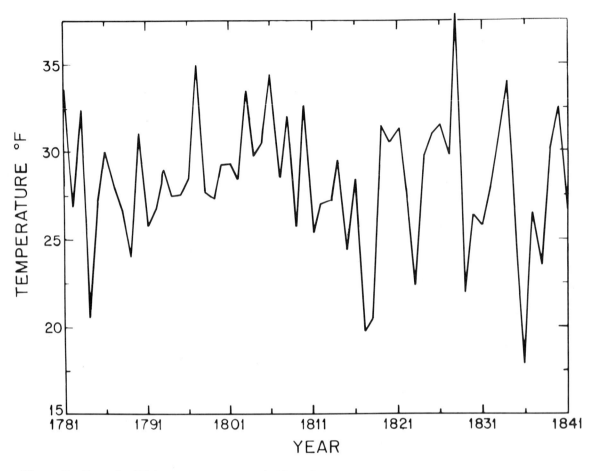

Figure 5: Record of February temperatures in New Haven, Connecticut. Abnormally low temperatures in 1816 and 1817 are noticeable.

Acknowledgements

This work was performed under the auspices of the CO_2 Research Division, Office of Basic Energy Sciences, United States Department of Energy.

References

Breckenridge, R. (1898-99). Diary 1816, *Transactions of the Alabama Historical Society* II:149.

Chladni, E.F.F. 1819. Ueber die Urfachen des nasskalten sommer von 1816, und zum Theil auch 1817. *Annalen der Physik, (Leipzig).* LXII:133-136.

Haynsworth, J. 1818. Account of the winter epidemic of 1815, 16 and 17, as it appeared in Salem and Claremont counties. *New York Medical Repository, New Series*, IV:8.

LaMarche, V.C. and K.K. Hirschboeck. 1984. Frost rings in trees as records of major volcanic eruptions. *Nature* 307:121-126.

Ludlum, D.M. 1966. *Early American Winters 1604-1820*. American Meteorological Society, Boston, Massachusetts.

Niles Weekly Register. 21 September 1816. p. 64.

North American Review. July 1816. p. 285.

Stommel, H. and E. Stommel. 1983. *Volcano Weather*. Seven Seas Press, Newport, Rhode Island. 177 pp.

The Times (London). 10 March 1817. p. 2; 23 June 1817, p. 3; 18 July 1817, p. 2.

Thomas, D. 1819. *Travels Through the Western Country in the Summer of 1816*. Auburn, New York. pp. 55-109.

Wang, W.C., P. Zhang and S. Hameed. (in preparation). Extraction of climate information from Chinese historical writings.

Was There a Colder Summer in China in 1816?

Huang Jiayou[1]

Abstract

Was there a colder summer in China in 1816? The data describing anomalous weather in this year have been collected from local annals compiled at 20 stations within the country. They show that severe floods occurred in most parts of China. Dryness/wetness data for 1811-20 have been analyzed using principal component analysis. The dominant mode shows that the lowest value (corresponding to the flood grade of wetness) is in 1816. A study of the atmospheric circulation during this period showed that the Subtropical High over the western Pacific reached its weakest point during the last few centuries. There is sufficient evidence to indicate that there was a colder summer in China in 1816.

Introduction

The year without a summer (1816) has been discussed by workers in various fields (e.g., climatologists, tree-ring experts, historians, geographers, glaciologists, biologists and volcanologists). It is important to gain the clearest picture possible of climate and weather sequences in different regions around the world in 1816 or the period about that year. What happened in China?

Was there a cold summer in 1816: was it colder than in other years? There are few instrumental records in China, as in other regions of the world. Fortunately, among the enormous Chinese historical writings there are abundant climatic descriptions of great value for studying past climate. Rainfall was described more frequently than temperature or other similar phenomena associated with it in these writings. Summer data are especially well treated. According to my analysis of the records, the correlation between temperature and precipitation is negative in summer. The study indicates that when flooding occurred in summer, climate in the region would be cold. Therefore, floods or abundant precipitation generally indicate a cold summer in that year. The basis of my discussion is drawn from ancient Chinese writings and data on dryness/wetness obtained therein.

Data

The data used in this paper consist of two parts. First are the Chinese historical documents for 1816. Second are data on dryness/wetness that were published by the Central Meteorological Institute and other agencies (1984). Dryness/wetness for each year covering the 510 years between 1470 and 1979 are classified into five grades: 1 - very wet and 5 - very dry, both of which occur 5% of the time; 2 - wet and 4 - dry, both of which occur 30% of the time; and 3 - normal occurring with a frequency of about 30%. Climatic phenomena in grade 1 are defined by continued strong rainfall with flooding over a large area. Frequent local rainfall of great strength occurs in grade 2. Data concerning wet grades were representative of summer rainfall, mainly because very strong rainfall usually occurred in summer. I use data selected from 20

---

[1] Department of Geophysics, Peking University, Beijing 100871, The People's Republic of China.

stations in China located in the lower basins of the Yangtze and Yellow rivers (Table 1). Information on dryness/wetness grade at the stations is listed in Table 2.

Table 1: Stations and Their Locations.

Number	Name	Location
1	Taiyuan	(37.8°N, 112.5°E)
2	Beijing	(39.9°N, 116.4°E)
3	Tianjin	(39.1°N, 117.2°E)
4	Baoding	(23.4°N, 111.3°E)
5	Changzhou	(38.3°N, 116.8°E)
6	Handan	(36.6°N, 114.4°E)
7	Anyang	(36.0°N, 114.3°E)
8	Luoyang	(34.6°N, 112.4°E)
9	Zhengzhou	(34.7°N. 113.6°E)
10	Jinan	(36.6°N, 117.0°E)
11	Linyi	(35.0°N, 118.3°E)
12	Heze	(35.2°N, 115.4°E)
13	Xuzhou	(34.2°N, 117.1°E)
14	Yangzhou	(32.4°N, 119.4°E)
15	Bengbu	(32.9°N, 117.1°E)
16	Jiujiang	(29.7°N, 117.3°E)
17	Nanchang	(28.6°N, 115.9°E)
18	Yichang	(30.6°N, 111.2°E)
19	Jiangling	(30.3°N, 112.1°E)
20	Yeuyang	(29.3°N, 113.1°E)

Was There Anomalous Weather in 1816?

Many anomalous weather phenomena occurred over China in 1816. According to descriptions of Chinese historical writings, at the beginning of the year heavy snow occurred in the Yangtze River Basin (e.g., Jiujiang). From early spring to early summer the basin underwent severe flooding spreading to the Hai River Basin. Many descriptions in the historical documents concern heavy rainfall and severe flooding at Xuzhou and Nanchang. Reports for mid-summer mention that people in the lower Yellow River Basin suffered from severe floods. For example: "Heavy rain occurred in June (traditional Chinese calendar) and fierce flood in the area near Luoyang"; "Flood was in Zhang River. A lot of farmland around it had been covered by water in Anyang"; "A long spell of wet weather from January to June (traditional Chinese calendar) and disaster due to the flood occurred in Heze"; "Heavy rain in June, the area about 2800 square [Chinese] miles covered by water 4 - 5 Chi (Chinese measure of depth, about 1.3 - 1.6 m) near Baoding. There were a long series of rainy days during the summer in the Baoding-Tainjing area. The people of 53 surrounding counties suffered from the flood."

Table 2: Grades of Dryness/Wetness at Stations during the Period 1811-20.

Station Numbers	Wetness/Dryness Grades									
	1811	1812	1813	1814	1815	1816	1817	1818	1819	1820
1	5	4	4	5	2	2	5	3	1	2
2	4	5	4	3	3	2	5	3	2	3
3	3	4	4	3	3	2	5	3	2	3
4	4	4	4	4	3	1	5	4	3	4
5	2	3	4	3	2	2	2	2	2	2
6	4	4	5	3	3	2	4	2	2	2
7	4	5	5	4	2	1	3	2	2	3
8	3	3	5	4	3	1	3	2	4	3
9	4	4	5	3	2	2	3	3	2	2
10	4	4	5	4	2	2	2	1	2	3
11	5	2	5	3	2	1	2	2	1	2
12	5	5	5	4	2	1	2	2	2	2
13	4	5	4	3	2	1	3	3	2	2
14	2	2	2	5	2	2	2	2	1	1
15	4	2	2	4	3	1	2	4	2	2
16	4	2	2	4	3	2	2	2	2	5
17	4	2	2	2	3	2	3	3	3	5
18	4	3	2	3	3	2	3	3	3	3
19	4	3	4	5	3	2	3	3	3	3
20	4	3	5	4	3	1	3	3	4	5

During 1816, eight and 12 stations in the selected series are classified as very wet and wet, respectively (Table 2). So I infer that the lower basins of Yangtze and Yellow rivers were very cold that summer.

Was Climate Abnormal during the 1810s?

Were cold summers common during 1811-20? To answer this question, the series of drought and flood grades for the 20 stations mentioned above were analyzed. The frequencies of occurrence of the five dryness/wetness grades are shown in Table 3.

Clearly the frequency occurrences of grades 1 or 2 in 1816 are higher than in other years analyzed. This implies that the degree of wetness in 1816 is outstanding during the 1810s.

Furthermore, the two-dimensional data of the dryness/wetness grade series at 20 stations in the 1810s can be treated by principal component analysis (cf. Richman 1986). This operation is based on the correlation matrix produced by the normalized series of dryness/wetness grade. The total variance equals the number of the stations. Table 4 shows the eigenvalues of the matrix and the explained variances (EV) in the first principal components (PCs).

Table 3: Frequency Occurrence of the Five Dryness/Wetness Grades at the Stations (1811-20).

Year	Dryness/Wetness Grades				
	1	2	3	4	5
1811	0	2	2	13	3
1812	0	5	5	7	3
1813	0	5	0	7	8
1814	0	1	8	8	3
1815	0	10	10	0	0
1816	8	12	0	0	0
1817	0	7	8	1	4
1818	1	7	9	2	0
1819	3	11	4	2	0
1820	1	8	7	1	3

Table 4: Eigenvalues and Explained Variances of Principal Components.[1]

	PC1	PC2	PC3	PC4	PC5
Eigenvalue	9.943	3.274	2.287	1.861	1.106
EV	0.497	0.164	0.114	0.093	0.051

[1] For basic data see preceding tables.

According to North et al. (1982), the significant PCs can be determined. They showed that the standard error of an eigenvalue is roughly $\lambda (2/n)^{1/2}$, where n is the sample size. Thus, unless the difference between two neighboring eigenvalues is greater than $\delta\lambda$, the two associated PCs cannot be properly resolved. The significance test shows that none of them is significant except the first principal component. Therefore, PC1 is the main mode and representative of the characteristic of variation in dryness and wetness. Table 5 gives the values of the first principal component for each year of the decade 1811-20.

On examining Table 5, the value for 1816 is the lowest for the years analyzed. Because the corresponding eigenvectors are more positive, it implies that the extreme grade of wetness during the decade occurred in 1816. Obviously, the chance of the occurrence of severe flooding was greater then than in other years of the decade.

Atmospheric Circulation

According to Huang et al. (1985), the occurrence of wetness or dryness in China is frequently associated with the atmospheric circulation - especially that of eastern Asia. In summer the relationship between the grade of dryness/wetness and the Subtropical High is close. Correlations between the western-limit index (the most westerly longitude reached by the 588 DM contour)

Table 5: First Principal Component of the Dryness/Wetness Grade (1811-20).

Year	PC1
1811	0.314
1812	0.257
1813	0.499
1814	0.270
1815	-0.223
1816	-0.562
1817	0.073
1818	-0.153
1819	-0.315
1820	-0.160

and wetness is positive over most of China. When the Subtropical High retreats toward the east, floods frequently occur near the border. When the Subtropical High weakens, floods often occur in the Yellow River Basin.

In fact, the western-limit index in the 1810s was the lowest of any decade during the period 1771-1820, and the strength index (represented by the areally-weighted sum of the number of grid points in the region enclosed by the 588 DM contour) was even at its lowest value for the longer period of 1471-1980. Clearly, lower grades of dryness/wetness occurred in the 1810s. Therefore, I suggest that the severe floods of 1816 were due to the relative weakness of the atmospheric circulation at the time.

Conclusions

The atmospheric circulation in eastern Asia in the 1810s was weaker than other decades during the last few centuries. The western margin of the Subtropical High retreated eastward and its strength diminished during that decade. Considering the background of atmospheric circulation, it is not surprising that floods occurred frequently in eastern China in the 1810s. Particularly severe floods occurred in the lower Yangtze and Yellow River basins in 1816. The pronounced rainfall and a long series of raindays was associated with colder summer weather in 1816. Therefore, I conclude that there was a cold summer in 1816.

References

Central Meteorological Institute, Peking University, Nanking University, etc., 1984. *Yearly Charts of Dryness/Wetness in China for the Last 500-Year Period.* Atlas Press, Beijing. 382 pp.

Huang Jiayou and Wang Shaowu. 1985. Investigations on variations of the Subtropical High in the Western Pacific during historic times. *Climatic Change* 7:427-440.

North, G.R., T.L. Bell and R.F. Cahalan. 1982. Sampling errors in the estimation of empirical orthogonal functions. *Monthly Weather Review* 110:699-706.

Richman, M.B. 1986. Review article, rotation of principal components. *Journal of Climatology* 6:293-355.

The Reconstructed Position of the Polar Frontal Zone around Japan in the Summer of 1816

Yasufumi Tsukamura[1]

Abstract

In the summer of 1816 the islands of Japan did not experience cool weather. There were no recorded crop losses anywhere in Japan. I attempt to reconstruct the polar frontal zone in 1816 in and around Japan - the western part of the Pacific Polar Fronts. The summer weather situations in Japan are mainly controlled by seasonal migration of the polar frontal zone. Daily positions of polar fronts are reconstructed by selecting a suitable map from recent years, which was most similar in weather distribution to the individual weather maps for 1816. The July-August polar frontal zones in 1816 were estimated to have been located near the position in the hot summer months - from northern Japan to northern China. The northward shift of polar frontal zones occurred somewhat earlier than in recent years. Therefore, the mid-summer season in Japan in 1816 came earlier and persisted for longer than for a modern comparative period.

Introduction

The year 1816 is well known as "the year without a summer" in Europe and North America, but in Japan it was not so unusual. Therefore, no Japanese climatologists have focused on the weather situation of that year.

Mikami and Tsukamura (this volume) showed the weather situation in the summer of 1816 around Japan: they could not present any evidence for unusual cool weather from historical documents.

I attempt to reconstruct the position of the polar frontal zone affecting the weather situation of Japan. This polar frontal zone is referred to as the Pacific Polar Frontal Zone. Geographical deviation and seasonal changes in the position of this zone associated with climatic change are discussed.

Data and Method

The daily positions of polar fronts in 1816 are reconstructed as follows:

1. Daily weather distribution maps in Japan for historical years (1815-40) were prepared from old diaries. Maps for recent years (1975-84) were taken from meteorological observation data.

2. Any recent weather maps which were similar to those in the historical years were selected.

3. In order to select the recent map that was most similar to the corresponding map in the historical years, the following points were considered: (1) the similarity of the day-to-day change in weather distribution pattern between the historical and present maps; (2) whether

---

[1] Department of Geography, Tokyo Metropolitan University, Fukazawa 2-1-1, Setagaya-ku, Tokyo 158, Japan.

the descriptions of wind direction and temperature on the historical weather maps present good agreement with those in the selected recent synoptic maps.

The best analogue thus selected was regarded as the map that had the same synoptic situation as the corresponding day in the historical years. Therefore, the position of the polar fronts which were taken from the synoptic maps at 9:00 a.m. on selected days were used as the position on historical days (Figure 1).

Reconstructed Position of the Polar Frontal Zone around Japan in Each Month, from June to September

A polar frontal zone is defined as an area where polar fronts are most frequently located. Since the polar frontal zone around Japan changes its position seasonally, monthly geographical deviations are shown from June to September (Figures 2-5). The positions of polar frontal zones in hot summer months and cool summer months for recent years are also shown in the same figures to compare with the position in 1816.

June (Figure 2)
The polar frontal zone in a cool summer month (e.g., June 1983) was located along the latitude of 30°N. In a hot summer month (e.g., June 1979) there were two polar frontal zones in parallel along the Eurasian continent. The polar frontal zone in June 1816 was estimated to have shifted widely in the cool summer month and in the hot summer month.

July (Figure 3)
The polar frontal zone in a cool summer month (e.g., July 1980) was located along the southern coast of Japan from central China. In a hot summer month (e.g., July 1978) the polar frontal zone was located from northern Japan to central China through the Sea of Japan. The polar frontal zone in July 1816 is estimated to have been located near the position of the hot summer month.

August (Figure 4)
The polar frontal zone in a cool summer month (e.g., August 1980) was located along the southern coast of Japan from central China. The polar frontal zone in a hot summer month (e.g., August 1978) was located from northern Japan to northern China through the Sea of Japan. The polar frontal zone in August 1816 is estimated to have been located near the position of the hot summer month.

September (Figure 5)
The polar frontal zone in a cool summer month (e.g., September 1981) was located in southern Japan and did not reach to the Eurasian continent. In a hot summer month (e.g., September 1975) there were two polar frontal zones. One was located from northern Japan to central China through the Sea of Japan, and the other in southern Japan. In 1816 there were two polar frontal zones (as in the hot summer month) but the polar frontal zone in 1816 that was estimated to have been located in southern Japan was more frequent than in the hot summer month.

As mentioned above, the polar frontal zones in July and August 1816 were estimated to have been located near the position in the hot summer months. The polar frontal zones in July and August 1816 were located from northern Japan to northern China through the Sea of Japan.

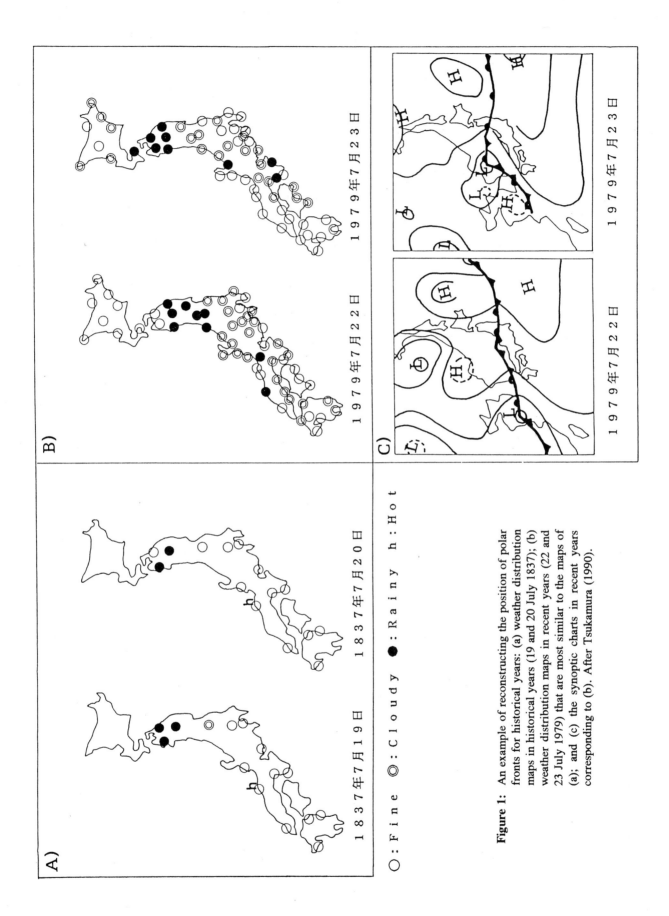

○:Fine ◎:Cloudy ●:Rainy h:Hot

Figure 1: An example of reconstructing the position of polar fronts for historical years: (a) weather distribution maps in historical years (19 and 20 July 1837); (b) weather distribution maps in recent years (22 and 23 July 1979) that are most similar to the maps of (a); and (c) the synoptic charts in recent years corresponding to (b). After Tsukamura (1990).

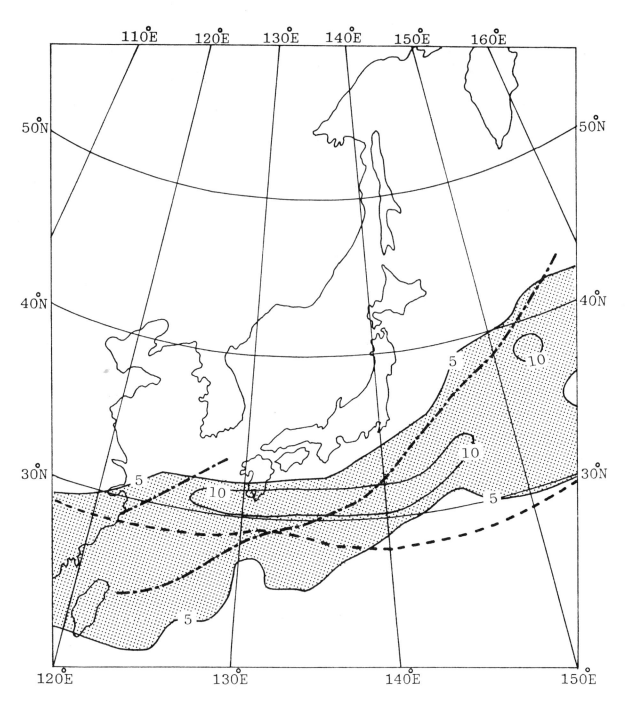

Figure 2: Frequency of daily polar fronts in June 1816. Areas where polar fronts most frequently appeared are defined as the polar frontal zone. The lines of dashes and dots indicate the position of the polar frontal zone for a hot summer month (June 1979), and the dashed line indicates its position for a cool summer month (June 1980).

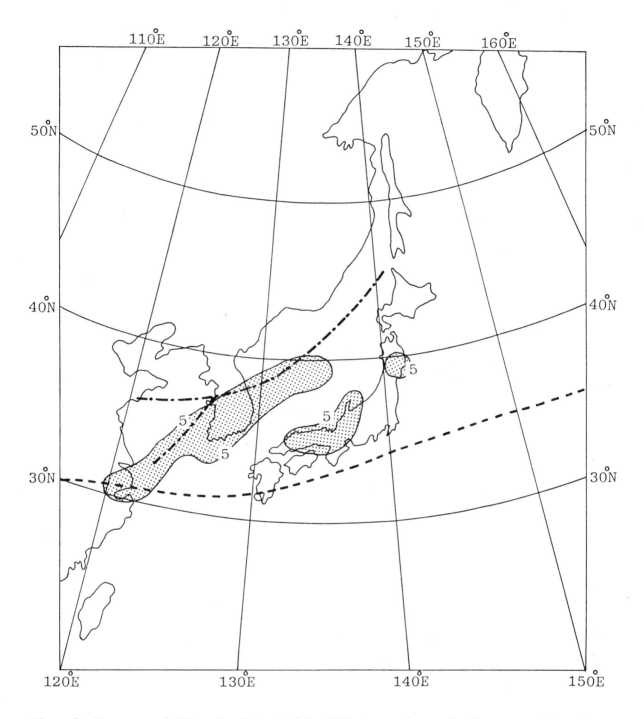

Figure 3: Frequency of daily polar fronts in July 1816. Areas where polar fronts most frequently appeared are defined as the polar frontal zone. The lines of dashes and dots indicate the position of the polar frontal zone for a hot summer month (July 1978), and the dashed line indicates its position for a cool summer month (July 1980).

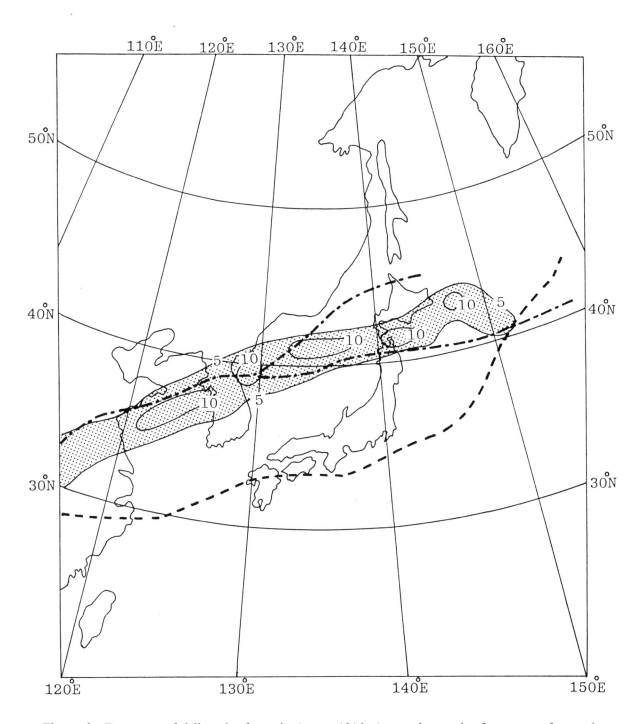

Figure 4: Frequency of daily polar fronts in August 1816. Areas where polar fronts most frequently appeared are defined as the polar frontal zone. The lines of dashes and dots indicate the position of the polar frontal zone for a hot summer month (August 1978), and the dashed line indicates its position for a cool summer month (August 1980).

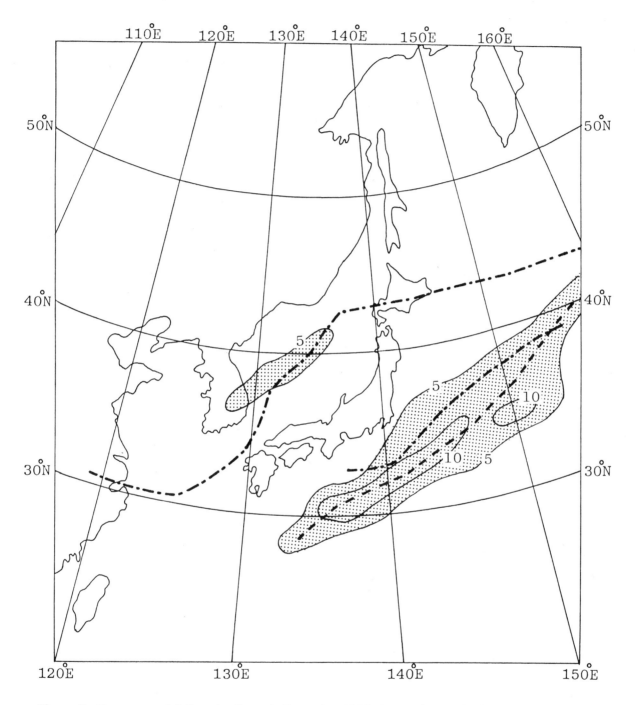

Figure 5: Frequency of daily polar fronts in September 1816. Areas where polar fronts most frequently appeared are defined as the polar frontal zone. The lines of dashes and dots indicate the position of the polar frontal zone for a hot summer month (September 1975), and the dashed line indicates its position for a cool summer month (September 1981).

Seasonal Changes in the Position of the Polar Frontal Zone

The polar frontal zone in Japan generally moves northward in accordance with the seasons from March to July. At the end of August it moves southward. To show seasonal movement of the position of the polar frontal zone, the four months were divided into twelve 10-day periods. Then the daily polar front frequencies were counted by 10-day period and by latitude for 1816 (Figure 6). The most frequent position is the polar frontal zone. The frequency of polar fronts for a recent 10-year period (1970-79) is also shown for comparison (Figure 6).

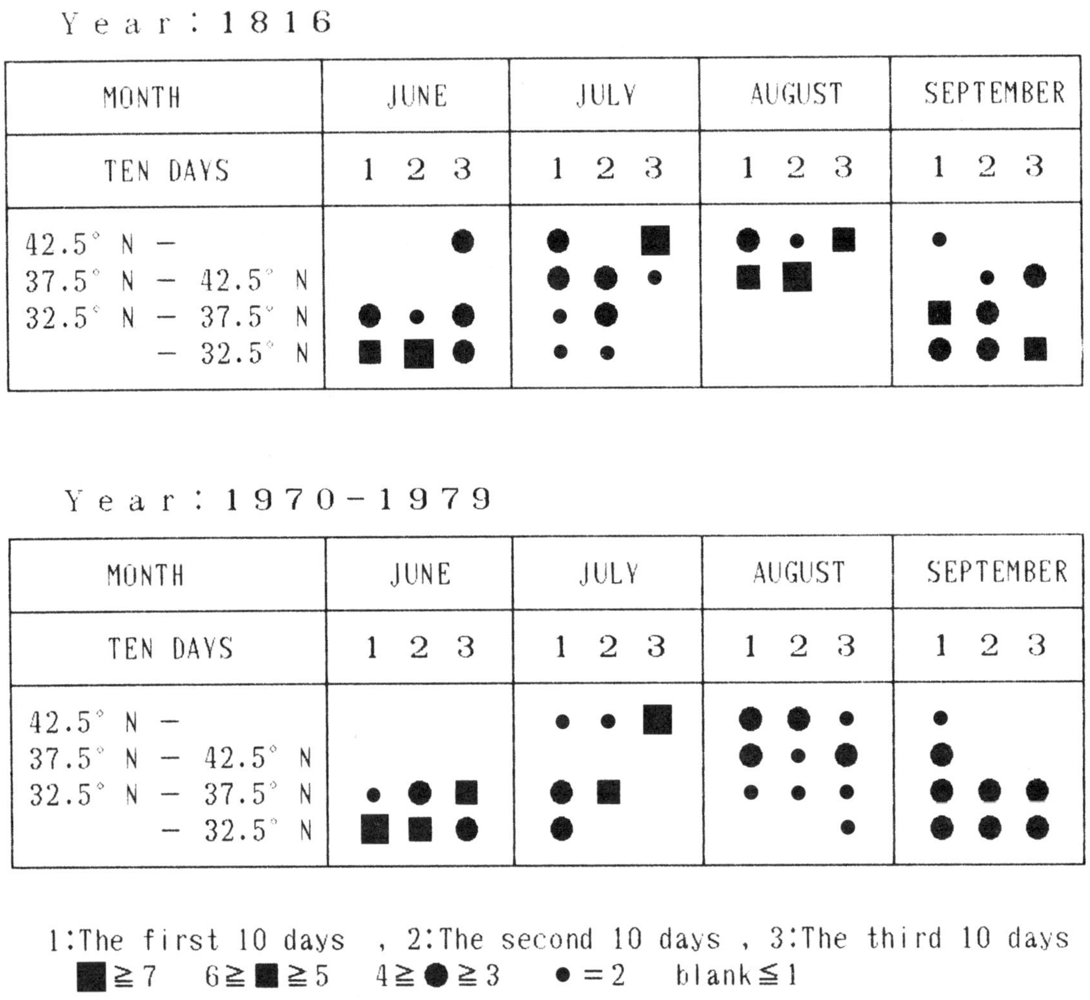

Year: 1816

MONTH	JUNE			JULY			AUGUST			SEPTEMBER		
TEN DAYS	1	2	3	1	2	3	1	2	3	1	2	3
42.5° N —					●		●	■		●	•	■	•		
37.5° N — 42.5° N				●	●	•					•	■	■		•	●
32.5° N — 37.5° N	●	•	●	•	●						■	●	
— 32.5° N	■	■	●	•	•					●	●	■

Year: 1970-1979

MONTH	JUNE			JULY			AUGUST			SEPTEMBER		
TEN DAYS	1	2	3	1	2	3	1	2	3	1	2	3
42.5° N —				•	•	■	●	●	•	•		
37.5° N — 42.5° N							●	•	●	●		
32.5° N — 37.5° N	•	●	■	●	■		•	•	•	●	●	●
— 32.5° N	■	■	●	●					•	●	●	●

1:The first 10 days , 2:The second 10 days , 3:The third 10 days
■ ≧ 7   6 ≧ ■ ≧ 5   4 ≧ ● ≧ 3   • = 2   blank ≦ 1

Figure 6: The daily polar-front frequencies counted by 10-day periods and by latitude.

Northward shift of the polar frontal zone in 1816 was somewhat earlier than in recent years. Therefore the beginning of the mid-summer season in 1816 was estimated to have been earlier than in the 1970-79 period.

Discussion and Conclusion

The polar frontal zone over East Asia in 1816, and the weather situation associated with it, are at the heart of this paper.

The polar frontal zones in July and August 1816 were estimated to have been located near the positions in hot summer months of recent years, that is from northern Japan to northern China across the Sea of Japan. Consequently, the wetness/dryness distribution map over East Asia, which is referred to by Mikami and Tsukamura (this volume), shows the wetness pattern in northern Japan and in northern China. Furthermore, because the estimated polar frontal zone in 1816 moved northward earlier than in 1970-79, and the rainy season in Beijing was long (Feng 1980), evidently, the polar frontal zone stagnated from northern Japan to northern China. In other words, the mid-summer season in Japan in 1816 came earlier and persisted longer than for a modern comparative period.

Acknowledgement

The author is grateful to Professor Ikuo Maejima and Associate Professor Takehiko Mikami of Tokyo Metropolitan University for their advice and suggestions. Some weather records were quoted from Maejima *et al.* (1983) and Mizukoshi (1986). Others were derived from the "Historical Weather Database Project" (represented by Dr. Minoru Yoshimura of Yamanashi University).

References

Feng, L. 1980. The rainy season and its variation in Beijing during the last 255 years. *Acta Meteorologica Sinica* 38(4):341-350.

Maejima, I. 1967. Natural season and weather singularities in Japan. *Geographical Reports, Tokyo Metropolitan University* 2:77-103.

Maejima, I., M. Nogami, S. Oka and Y. Tagami. 1983. Historical weather records at Hirosaki, northern Japan, from 1661 to 1868. *Geographical Reports, Tokyo Metropolitan University* 18:113-152.

Mizukoshi, M. 1986. *Historical Weather Records in Kinki-Tokai Districts, Central Japan, June and July (1781-1870)*. Disaster Prevention Research Institute, Kyoto University. 84 pp.

Tsukamura, Y. 1990. Southward deviation of the polar frontal zone in East Asia in the summers of the early 19th century. *Geographical Review of Japan* 63:4-9.

Wilson, C. 1985. Daily weather maps for Canada, summers 1816 to 1818 - a pilot study. *In: Critical Periods in the Quaternary Climatic History of Northern North America. Climatic Change in Canada 5*. C.R. Harington (ed.). *Syllogeus* 55:191-218.

The Climate of Japan in 1816 as Compared with an Extremely Cool Summer Climate in 1783

T. Mikami[1] and Y. Tsukamura[1]

Abstract

Climates of Japan in the late eighteenth and early nineteenth century are analyzed based on the daily weather records in old diaries and the record of freezing dates of Lake Suwa. The results indicate that the summer climate of Japan in 1816 was rather hot and dry due to the northward expansion of the subtropical anticyclone and that the winter climate in 1815-16 was severe which suggests volcanic effects. Unlike this situation, it was extremely cool and moist in the summer of 1783, which caused heavy damage to rice crops and led to the great historic famine. It is also a year renowned for major volcanic eruptions in Japan and Iceland. Thus, 1783 is properly regarded as "the year without a summer" in Japan.

Introduction

The year 1816 is well known for its exceedingly cold summer climate in North America and Europe (e.g., Milham 1924). The effects of the Tambora eruption in 1815 have also been discussed (e.g., Stothers 1984). However, there are no reliable data to confirm a cold summer in 1816 in Japan and East Asia.

In this paper, we attempt to clarify the climate in 1816 quantitatively based on historical documents. We also compare the climate in 1816 with those periods in the late eighteenth and early nineteenth century when severe famines due to unusually cold weather occurred frequently.

Compilation of Historical Weather Documents

Weather Records in Old Diaries

The data we used for reconstructing historical climates were, for the most part, daily weather records in old official diaries that had been kept during the Edo Era when Japan was under the control of feudal clans (Mikami 1988). Figure 1 shows a typical example of weather records in the diary of the feudal clan Isahaya at Nagasaki, southwestern Japan. The weather description usually follows the date as shown on the right hand in Figure 1 where we can see that the weather on 13 and 14 August 1783 (Gregorian calendar) was fine with rain in the afternoon and rainy, respectively.

Weather records in old diaries are available in various parts of Japan (Figure 2). They include not only official records such as those of feudal clans, shrines and temples but also some private records (e.g., personal diaries). The total number of records we obtained has reached 20, including one for Seoul, South Korea. In order to discuss regional weather situations, we divided the whole country into five districts from A to E.

---

[1] Department of Geography, Tokyo Metropolitan University, Fukazawa 2-1-1, Setagaya-ku, Tokyo 158, Japan.

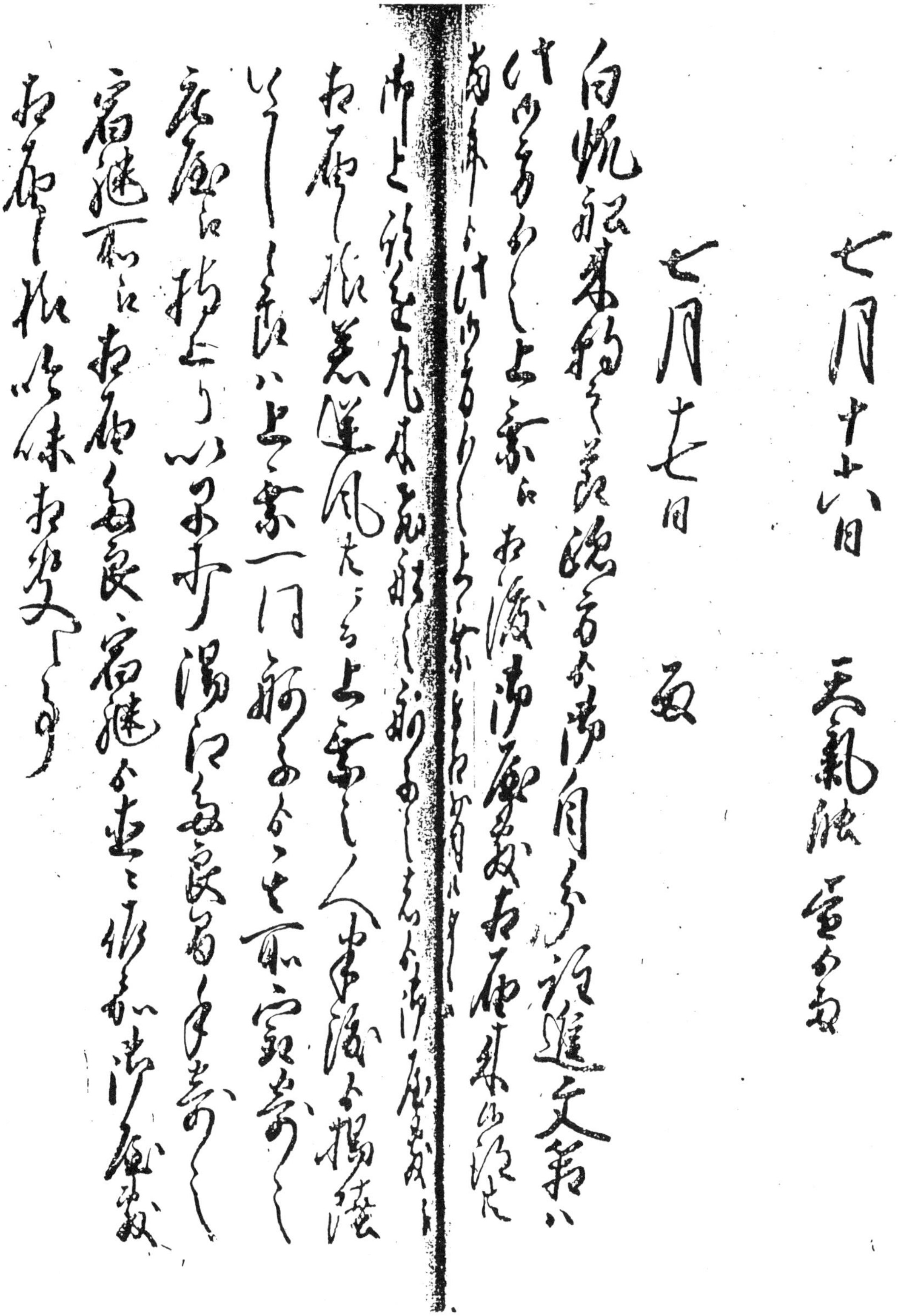

Figure 1: An example of weather descriptions in old diaries.

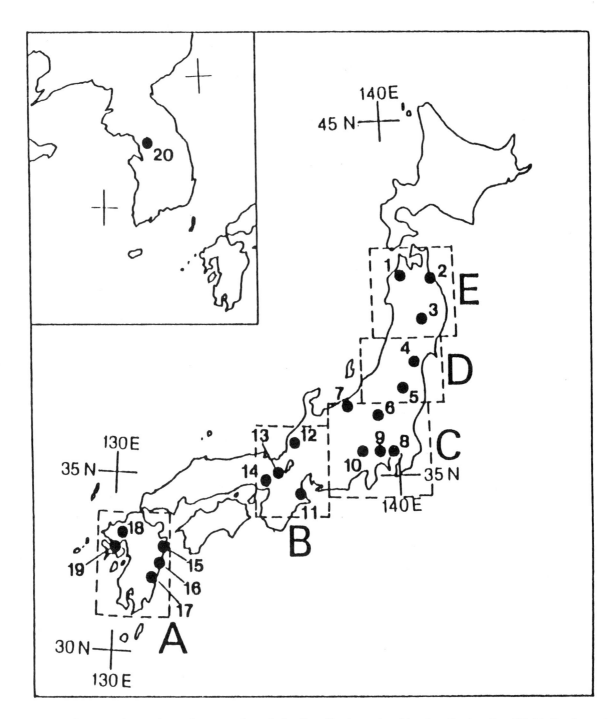

Figure 2: Locations of weather records and the five districts. A - Kyushu district; B - Kinki district; C - Kanto-Chubu district; D - Southern Tohoku district; E - Northern Tohoku district. 1 - Hirosaki; 2 - Hachinohe; 3 - Morioka; 4 - Sendai; 5 - Koriyama; 6 - Nikko; 7 - Takada; 8 - Tokyo; 9 - Hachioji; 10 - Kofu; 11 - Ise; 12 - Sabae; 13 - Kyoto; 14 - Ikeda; 15 - Usuki; 16 - Nobeoka; 17 - Sadowara; 18 - Saga; 19 - Isahaya; 20 - Seoul.

Weather-Pattern Calendar

Among the weather records in old diaries, data concerning precipitation are probably the most important factor in reconstructing historical climates. The way of describing precipitation varies from diary to diary; for example: "rainy all day", "cloudy with occasional rain", "heavy rain", "showery", "snowy", "little rain in the afternoon", etc. If we try to discuss spatial weather patterns or to connect them with recent instrumental data, these qualitative data should be standardized by converting them into quantitative terms. Therefore, we attempt to define the daily-weather pattern by a combination of numerical codes. We distinguish rainy areas from areas without rain by denoting 1 for rain and 0 for no rain in each district from A (left) to E (right) as shown in Figure 3.

Thus, we have completed the weather-pattern calendar in June, July, August and September for 50 years from 1771 to 1820. Table 1 shows an example of a weather-pattern calendar for 1783, which is characterized by frequent rainy days in July and August. We also derived a table which indicates the number of days with rain in July and August for each district including Seoul (denoted by S) from the weather-pattern calendars (Table 2). The year-to-year variation in the number of days with rain in August for District A (southwestern Japan) is delineated with a bar graph (Figure 4).

Wet/Dry-Pattern Maps

In order to clarify the spatial pattern of the degree of rainy-weather situations, the number of days with rain in each district was grouped into five grades according to the probability of occurrence frequencies: grade 1 (very wet), grade 2 (wet), grade 3 (normal), grade 4 (dry) and grade 5 (very dry), where the probabilities are 0.1, 0.2, 0.4, 0.2 and 0.1 respectively. By plotting the grade number for each district, we obtain a map showing wet/dry patterns (Figures 5, 6).

In China, wetness/dryness charts for the last 500 years were compiled using various kinds of historical documents (Central Meteorological Institute 1981). Wang and Zhao (1981) examined the characteristics of spatial and temporal variations of droughts and floods in China for 1470-1979 based on the above-mentioned charts. Therefore, it is possible to draw a composite map of wet/dry patterns over East Asia by combining the Chinese chart with our map.

Climate in 1783 and 1816

Climatic Variability during 1771-1820

As indicated in Figure 4, the number of days with rain in August for southwestern Japan shows large year-to-year variabilities. The years 1774, 1783, 1787, 1788 and 1814 were very wet (grade 1) with more than 12 rainy days. In particular, it was extremely wet in 1783 with 16 rainy days. On the other hand, the years 1781, 1800, 1808 and 1809 were very dry with less than two rainy days that could be categorized as grade 5. The number of days with rain in August 1816 was only three, which corresponded to grade 4 (dry).

Extremely Cool Summer Climate in 1783

The year 1783 is well known for the severest famine in Japan's history. Mikami (1983, 1987) made an attempt to clarify the 1781-90 climate of Japan based on ancient weather records in official diaries.

The results indicated that climate in 1783 was characterized by exceptionally cool and rainy weather in summer, which have also been clarified by the weather-pattern calendar (Table 1). As indicated in Figure 5, extreme wetness enveloped the whole of Japan, and it was also wet in

Seoul, Korea. In July, it was wet in central and southwestern Japan, but very dry in the Korean Peninsula.

The extraordinary wetness in August 1783 is estimated to have been caused by the stagnation of polar frontal zones along the Pacific coast of Japan due to the southward shift of subtropical anticyclones (North Pacific Anticyclones or NPA).

Hot and Dry Summer in 1816
In Japan, little attention has been paid to the summer weather in 1816, because we find no special climatic-hazard records for this year. Therefore, we tried to reconstruct the climate in 1816 quantitatively on the basis of weather records. The method for reconstruction is the same as that in 1783. Figure 6 shows the wet/dry patterns in the summer of 1816, where dry areas are clearly denoted for most of Japan in August, although wet areas extend from northern Japan to the Korean Peninsula. In July, however, it is nearly normal except for Korea.

From these results, we conclude that the summer weather of Japan in 1816 was rather hot and dry, especially in August. The wet area extending from northern Japan to the Korean Peninsula would correspond to the polar frontal zone. In other words, the dry area which expanded south of the wet area is considered to have been introduced by the northward expansion of subtropical anticyclones (NPA). The location of polar fronts for summer 1816 is discussed by Tsukamura (this volume).

Cold and Snowy Winter in 1815-16
Long-term records of the freezing dates of Lake Suwa in central Japan have been used as an efficient indicator of winter severity in Japan (Lamb 1977; Tanaka *et al.* 1982). Since winter temperatures at the city of Suwa and the dates of complete freezing of Lake Suwa (days after 1 December) have positive correlations, it is possible to estimate winter-temperature series at Suwa in the historical period (Tanaka *et al.* 1982). Figure 7 shows complete freezing dates of Lake Suwa during the winters 1800-01 and 1870-71. In the 1815-16 winter, the complete freezing date of Lake Suwa was 31 December, which was far earlier than averages in this period. Therefore, the climate in winter 1815-16 was cold in central Japan, although there are some uncertainties for the data reliability during this period (Tanaka and Yoshino 1982).

Among the weather records from old diaries, snowfall descriptions are effective for estimating winter severity. Figure 8 indicates year-to-year variability of the number of snowy days and the snowfall ratios at Yokohama. In winter 1815-16, approximately eight months after the Tambora eruption, both the number of snowy days and the snowfall ratio were much higher than average.

These results affirm that climate during the winter of 1815-16 was also very severe as indicated in Figure 8. However, it would not be appropriate at the moment to attribute the cold and snowy weather in 1815-16 to the effect of the Tambora eruption in 1815.

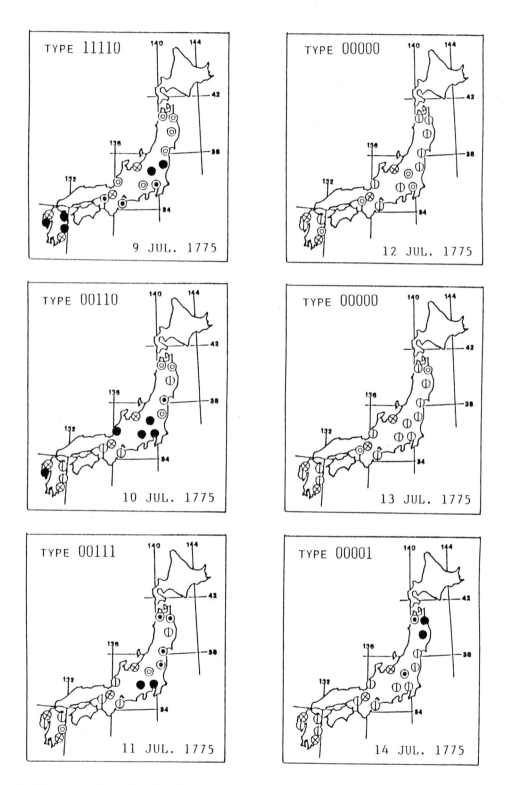

Figure 3: Examples of weather-distribution maps and their types by the combination of numerical codes.
①: fine; ◉: cloudy; ⊙: showery; ●: rainy; ⊗: data missing.

Table 1: An Example of a Weather-Pattern Calendar (June-September 1783).

DAY	JUN.	JUL.	AUG.	SEP.
1	00000	11000	00000	00011
2	00001	00110	00000	00000
3	00000	00001	00000	00000
4	00000	10000	00000	01110
5	00000	11000	00100	11110
6	00000	11000	11000	11111
7	11000	00000	11111	11111
8	01000	11111	10110	01111
9	00000	11001	11111	01101
10	00001	11100	11111	00101
11	11111	11000	11111	00000
12	11111	00100	00000	00000
13	00001	00000	10011	00110
14	00011	00110	10000	00111
15	00001	01100	11100	00100
16	00000	01100	01001	00110
17	11000	11111	11111	00111
18	11111	01111	01001	00000
19	00111	10000	00001	00000
20	00100	00000	10000	00000
21	00001	00000	10000	00000
22	00000	00000	00001	00000
23	00000	00000	00011	00000
24	11100	00000	00100	00000
25	00111	00000	00000	00000
26	00000	11000	11000	00000
27	00100	11100	10011	01000
28	00111	01111	11111	01111
29	00000	00011	11111	00110
30	10000	01101	00000	00000
31		00001	00100	

Table 2: Number of Days with Rain in August for Each District (Seoul and A-E) from 1771-1820.

YEAR	S[1]	A	B	C	D	E	Mean
1771	4	9	10	12	4	4	7.8
1772	10	3	2	8	10	9	6.4
1773	7	4	9	9	11	11	8.8
1774	12	12	8	11	13	7	10.2
1775	0	6	2	5	6	4	4.6
1776	8	9	6	10	8	12	9
1777	12	8	1	3	6	13	6.2
1778	9	5	3	8	10	6	6.4
1779	10	8	13	6	7	17	10.2
1780	7	6	2	0	0	8	3.2
1781	11	1	3	8	7	8	5.4
1782	9	7	8	5	2	4	5.2
1783	13	16	13	13	11	15	13.6
1784	10	10	5	1	0	2	3.6
1785	6	10	5	4	7	6	6.4
1786	6	10	8	11	9	6	8.8
1787	15	14	15	7	5	7	9.6
1788	6	13	12	8	3	4	8
1789	12	6	7	4	6	5	5.6
1790	10	5	2	0	3	4	2.8
1791	15	5	6	4	2	8	5
1792	7	7	3	6	2	5	4.6
1793	4	8	9	11	5	9	8.4
1794	4	4	7	10	3	2	5.2
1795	10	9	9	8	5	11	8.4
1796	9	7	1	7	0	1	3.2
1797	9	4	4	3	2	3	3.2
1798	4	8	12	6	2	6	6.8
1799	13	3	3	2	0	0	1.6
1800	14	2	7	2	1	5	3.4
1801	7	5	7	7	5	4	5.6
1802	11	9	8	3	5	7	6.4
1803	5	8	12	8	4	8	8
1804	6	8	13	11	5	5	8.4
1805	12	10	8	3	2	4	5.4
1806	12	5	9	1	1	8	4.8
1807	13	3	8	6	4	6	5.4
1808	6	1	7	11	3	7	5.8
1809	6	1	6	5	1	3	3.2
1810	10	8	9	3	1	8	5.8
1811	8	6	9	13	4	7	7.8
1812	5	7	17	20	13	9	13.2
1813	9	10	11	6	4	9	8
1814	13	12	10	8	4	10	8.8
1815	7	9	10	12	3	7	8.2
1816	12	3	4	3	1	9	4
1817	13	9	2	4	1	3	3.8
1818	11	8	4	2	0	3	3.4
1819	9	3	3	2	2	2	2.4
1820	6	4	12	4	0	5	5
Mean	8.94	6.96	7.28	6.48	4.26	6.52	6.3

[1] Seoul district.

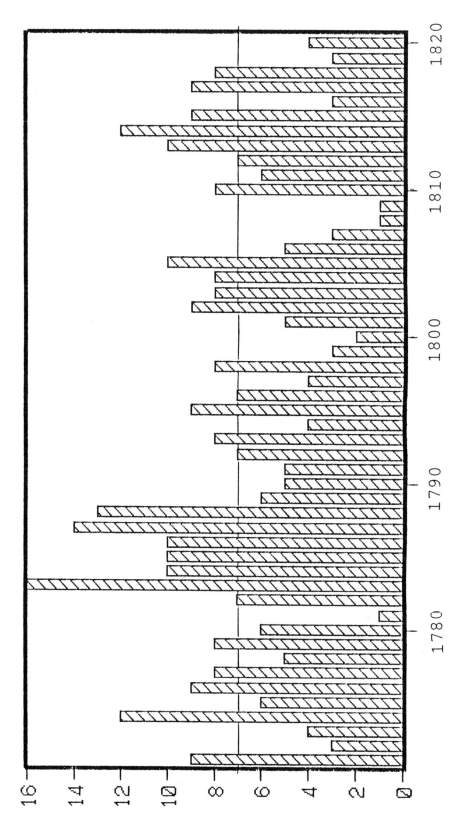

Figure 4: Variations in the number of days with rain in August for District A (southwestern Japan).

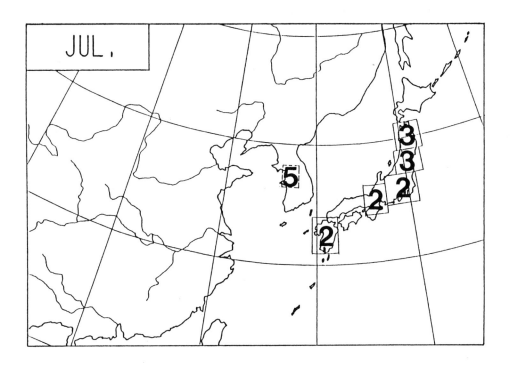
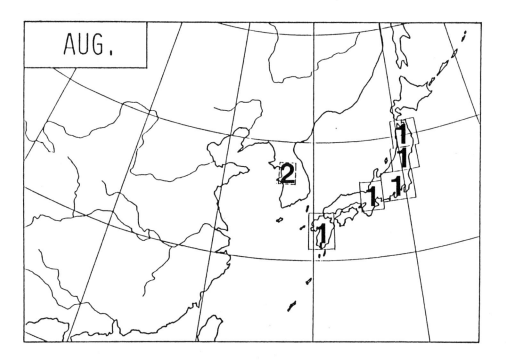

Figure 5: The wet/dry-pattern map for 1783. 1: very wet; 2: wet; 3: normal; 4: dry; 5: very dry.

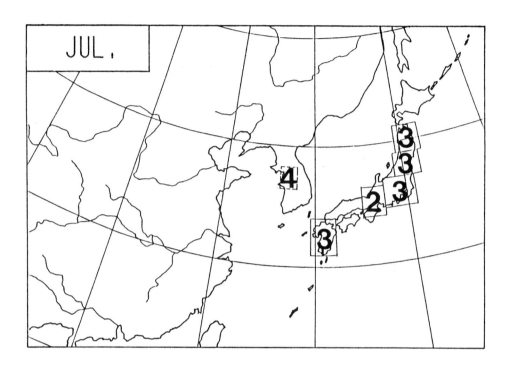
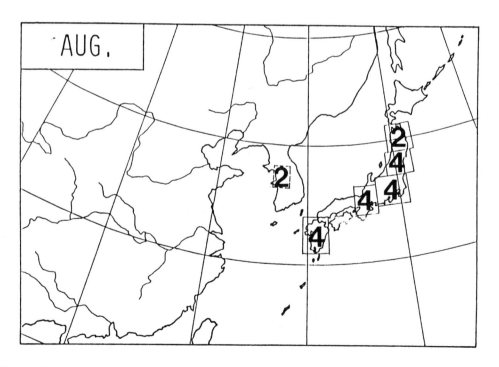

Figure 6: The wet/dry-pattern map for 1816. 1: very wet; 2: wet; 3: normal; 4: dry; 5: very dry.

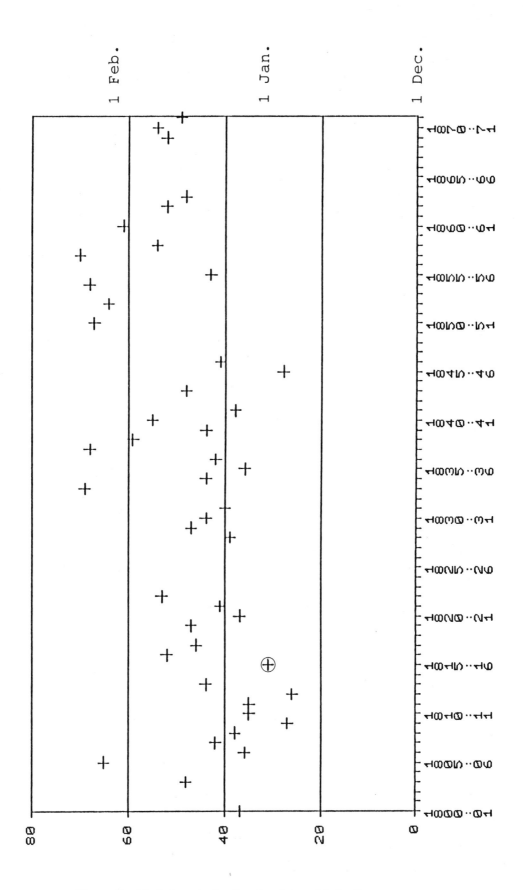

Figure 7: Variations in the freezing dates of Lake Suwa.

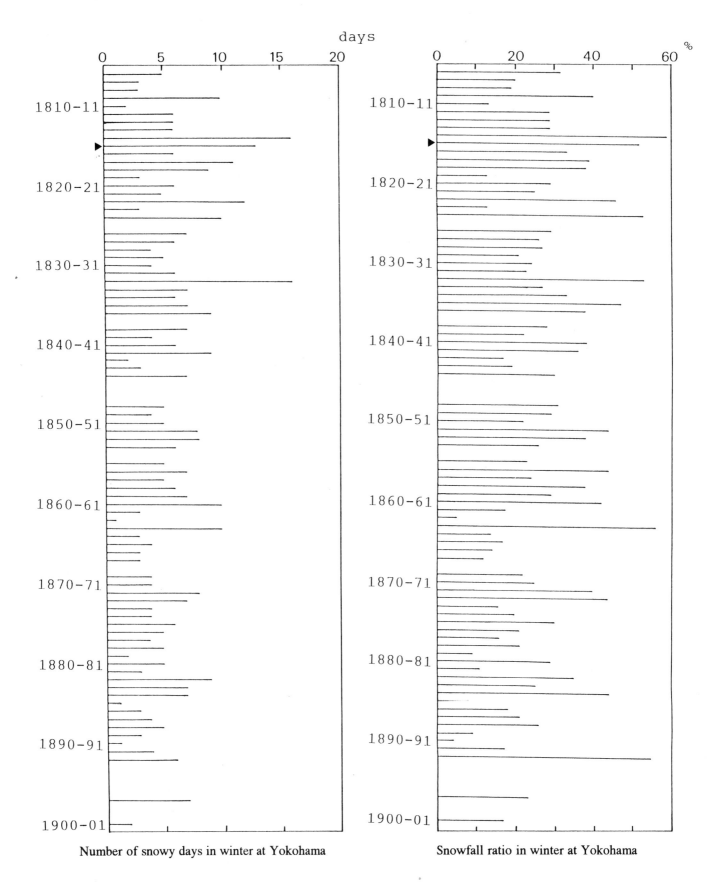

Figure 8: Variations in the number of snowy days and snowfall ratios at Yokohama.

Conclusions

The 1816 summer in Japan was not sufficiently cold to be called "the year without a summer". It was rather hot and dry in August with a few rainy days throughout Japan except for northern areas. On the other hand, a very cold summer occurred during 1816 in Europe and North America, although there may have been some regions where conditions were normal or rather warm then. The amount of data available to reconstruct past climates in Japan and East Asia is far smaller than in Europe. Therefore, we are apt to overestimate climatic conditions for Europe and North America, and to underestimate those for East Asia and other parts of the world.

In Japan, 1783 is famous as the historic famine year. Many historical documents refer to the dreadful conditions of the severe famine resulting from the poor rice harvest. Apparently, exceptionally cold and rainy summer weather occurred in 1783, while it was rather hot and dry in Europe. Volcanic eruptions that might have exacerbated this climatic anomaly were remarkable in Iceland and Japan during that year (Wood, this volume).

Historical weather records of Japan indicate that it was cold and snowy in the winter of 1815-16. Probably the Tambora volcanic signal appeared earlier in East Asia than in Europe and North America. However, we need more detailed regional information on past climates to discuss the effect of volcanic activities on global climate.

References

Arakawa, H. 1954. On five centuries of freezing dates of Lake Suwa (36°N, 138°E) in central Japan. *Chigakuzassi (Journal of Geography)* 63:193-200. (In Japanese with English abstract).

Central Meteorological Institute of China. 1981. *Yearly Charts of Dryness/Wetness in China for the Last 500-Year Period*. Atlas Press, Beijing. 332 pp.

Lamb, H.H. 1977. *Climate: Present, Past and Future* (Volume 2). Methuen, London. 835 pp.

Mikami, T. 1983. Classification of natural seasons in Japan for summer half years 1781-90 based on the seasonal march of weather. *Chigakuzassi (Journal of Geography)* 92:105-115. (In Japanese with English abstract).

_____. 1987. Climate of Japan during 1781-90 in comparison with that of China. *In: The Climate of China and Global Climate*. D. Ye, C. Fu, J. Chao and M. Yoshino (eds.). China Ocean Press, Beijing. pp. 63-75.

_____. 1988. Climatic reconstruction in historical times based on weather records. *Geographical Review of Japan* (Series B) 61:14-22.

Milham, W.I. 1924. The year 1816 -- the causes of abnormalities. *Monthly Weather Review* 52:563-570.

Stothers, R.B. 1984. The great Tambora eruption in 1815 and its aftermath. *Science* 224:1191-1198.

Tanaka, M. and M.M. Yoshino. 1982. Re-examination of the climatic change in central Japan based on freezing dates of Lake Suwa. *Weather* 37:252-259.

Wang Shao-Wu and Zhao Zong-Ci. 1981. Droughts and floods in China, 1470-1979. *In: Climate and History: Studies in Past Climates and Their Impact on Man.* T.M.L. Wigley, M.J. Ingram and G. Farmer (eds.). Cambridge University Press, London. 530 pp.

Southern Hemisphere

Evidence for Changes in Climate and Environment in 1816 as Recorded in Ice Cores from the Quelccaya Ice Cap, Peru, the Dunde Ice Cap, China and Siple Station, Antarctica

Lonnie G. Thompson[1] and Ellen Mosley-Thompson[1]

Abstract

The climate and environmental records preserved in three ice cores from three quite different sites (the tropical Quelccaya Ice Cap, Peru, the subtropical Dunde Ice Cap, China, and the Antarctic Ice Sheet, Siple Station) are examined for the period 1808-21. Emphasis is placed upon identifying potential changes in climate following the eruption of Tambora in 1815. Ice cores provide a multifaceted record of past variations in the Earth's climate and environment.

Historical documentation is scarce or absent for these three sites where the dust concentrations, electrical conductivity, net accumulation, oxygen isotopic abundances and seasonal ranges for 1808-21 are compared with the modern levels. For Siple Station Cl^- and SO^{2-}_4 levels are included as polar precipitation most closely represents atmospheric background levels.

The climatic and environmental variations of the early 1800s are discussed in the perspective of variations over the last 500 years available from each of these three ice-core sites. These records indicate that while there is an apparent global signal produced by the 1815 eruption of Tambora as recorded in the ice cores, the climatic response at each site is quite complex. The Dunde and Quelccaya sites reveal a cooling trend underway before the eruption; although the eruption may have strengthened the cooling trend that culminated in the coldest isotopic year (1819 at all three sites) of the decade 1810-20. Moreover, the year 1819-20 is the isotopically coldest year in the entire 1500-year record from the two Quelccaya ice cores. On the other hand, the mean isotopic temperatures for the 1810-20 decade appear to have been warmer at Siple Station in comparison to the modern "norm".

Dust records indicate no significant signal in insoluble particle concentrations above the high levels that characterized much of the Little Ice Age. A significant peak in conductivity for the Dunde Ice Cap, China records and in sulphate, excess sulphate and conductivity for the Siple Station, Antarctica records occur in 1817.

Introduction

Reliable meteorological observations for climate reconstruction prior to 1850 are scarce or absent for most of the globe. Ice caps and ice sheets provide an archive of atmospheric history which pre-dates human observations. It is recognized that much of the important climatic history of the Earth never reaches the polar regions and thus, would be lost if the poles were the only sites from which ice cores were recovered. Ice-core records from three sites: the Quelccaya Ice Cap, Peru (13°56'S; 70°50'W: 5670 m a.s.l.), the Dunde Ice Cap, China (38°06'M; 96°24'E: 5325 m a.s.l.) and Siple Station, Antarctica (75°55'S; 84°15'W: 1054 m a.s.l., Figure 1), are examined on an annual basis from 1808 to 1820 to ascertain the potential impact of the 1815 eruption of Tambora on climate and the environment.

---

[1] Byrd Polar Research Center, Ohio State University, Columbus, Ohio 43210, U.S.A.

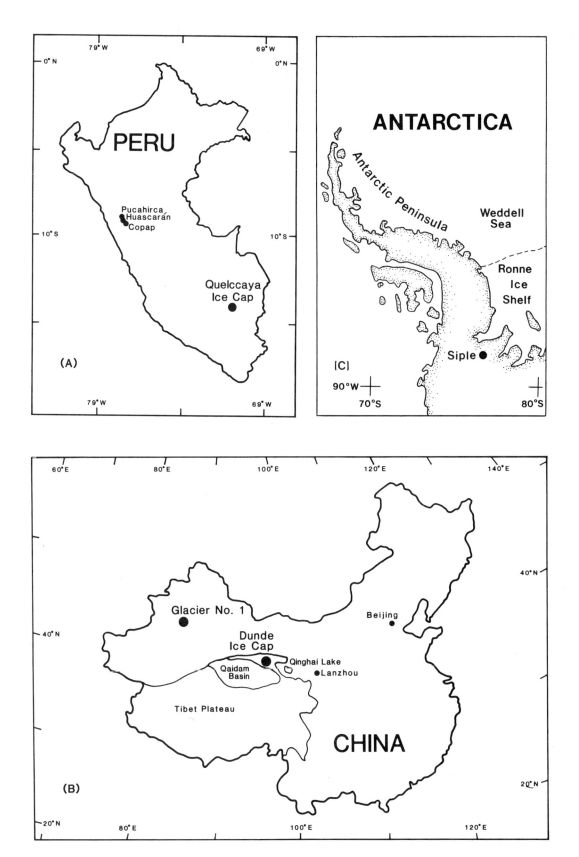

Figure 1: Ice core location maps: (A) Quelccaya Ice Cap, Peru (B) Dunde Ice Cap, China and (C) Siple Station, Antarctica.

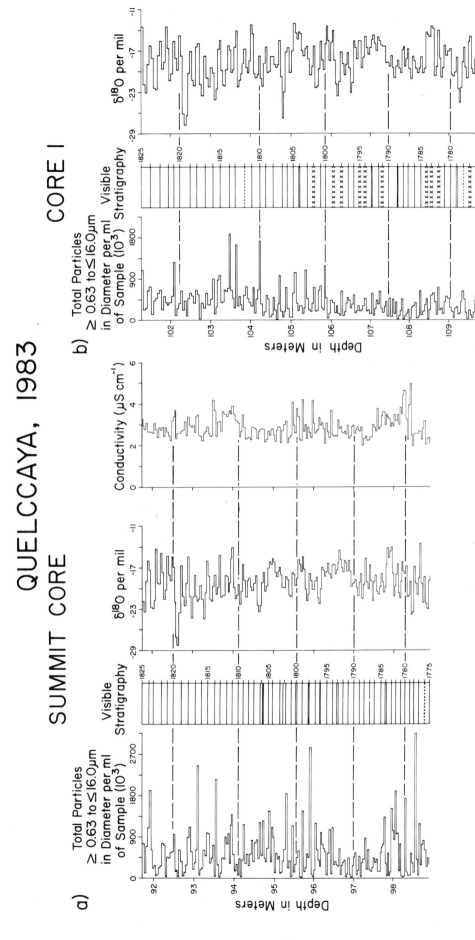

Figure 2: The stratigraphic parameters used to date ice cores are illustrated for the period 1775-1825 in the Quelccaya summit core (A) and core 1 (B). Annual signals are recorded in microparticle concentrations (particles from 0.63 to 16.0 μm in diameter per ml sample), oxygen isotopic ratios, electrical conductivity (summit core only), and visible stratigraphy. For stratigraphy, a single solid line represents a normal dry-season dust layer; a single dashed line and a double dashed line represent very light and light dust layers, respectively. Series of Xs

For each site, identical samples were analyzed for microparticle concentrations and size distributions, electrical conductivity and oxygen isotopic abundances. The microparticle, conductivity and chemistry measurements were made under class 100 clean room conditions at Ohio State University, and $\delta^{18}O$ analyses were conducted at the University of Copenhagen and at the University of Washington.

Dating of the ice cores was accomplished using several stratigraphic features that exhibit seasonal variability. For the Dunde and Quelccaya cores, stratigraphic dating was possible using visible, annual dust layers in conjunction with annual variations in microparticle concentrations, conductivity, chemistry and oxygen isotopes. Using Quelccaya as an example, Figure 2 illustrates how both the Quelccaya and Dunde cores were dated. Here the annual record over a 50-year period from 1775 to 1825 for two cores, the summit core and core 1, is illustrated. Thus, the dating is confirmed through a series of cross-checks and allows evaluation of the reproducibility of the ice-core records (Thompson et al. 1986). The initial dating of the Siple core is based only on the seasonality of $\delta^{18}O$ (Mosley-Thompson et al. submitted). Sulphate and nitrate concentrations also exhibit an excellent seasonal signal (Dai et al. submitted), and the final time scale will be produced using cross-checks among these three constituents ($\delta^{18}O$ SO^{2-}_4, NO_3^-).

The Little Ice Age recorded in the northern hemisphere earlier in this millennium was characterized by colder temperatures and expanded glaciers. Little Ice Age dates, determined from historical and proxy climatic records, vary depending on location and observed parameters (Thompson et al. 1986; Grove 1988). However, here we used the term Little Ice Age for the period 1500-1880.

The eruption of Tambora (8°S, 118°E) on 10 and 11 April 1815 was the largest and most deadly eruption in recorded history (Stothers 1984). Moreover, it is believed to have been the greatest ash eruption in the last 10,000 years. It is significant that this eruption can be seen in records presented here for both Greenland (Hammer et al. 1978) and Siple Station, Antarctica. A potential temperature response to this eruption is seen in oxygen isotope records at such widely dispersed ice-core sites as the Andes of Peru and the Qinghai-Tibetan Plateau of China. These records, when viewed in terms of the well-documented period of unusual temperatures in parts of North America and Europe, including "the year without a summer, 1816" (Milham 1924; Hoyt 1958; Stommel and Stommel 1983; and Stothers 1984), strongly suggest a global impact. The instrumental, historical, and proxy data (including the ice-core data presented here) also indicate that although global in scale, the specific climatic response at any given site was quite complicated. This was documented in the northern hemisphere instrumental records (Deirmendjian 1973; Landsberg and Albert 1974). Thus, the first step to understanding the impact on global climate of a volcanic eruption the magnitude of Tambora requires the compilation of existing data sets and the acquisition of new data sets from regions where such information is absent.

In Figures 3-6 and Tables 1 (A-C), the period 1808-21 is compared to modern "norms" for the Quelccaya Ice Cap, Peru, the Dunde Ice Cap, China, and the Antarctic Ice Sheet at Siple Station. These figures and tables allow comparisons of: (1) large, small and total particles and electrical conductivity indicating variations in concentration of insoluble and soluble dust, respectively; (2) relative changes in precipitation as recorded by average net accumulation in metres of ice equivalent; and (3) proxy temperatures as recorded by oxygen isotopes. These data are presented both as annual means and annual ranges between extreme values. In Figure 5 and Table 1 (C), Cl^-, SO^{2-}_4 and excess SO^{2-}_4 values are presented for 1808-21 for Siple Station, Antarctica. Remote from local sources, Antarctica provides an ideal site for recording major volcanic eruptions.

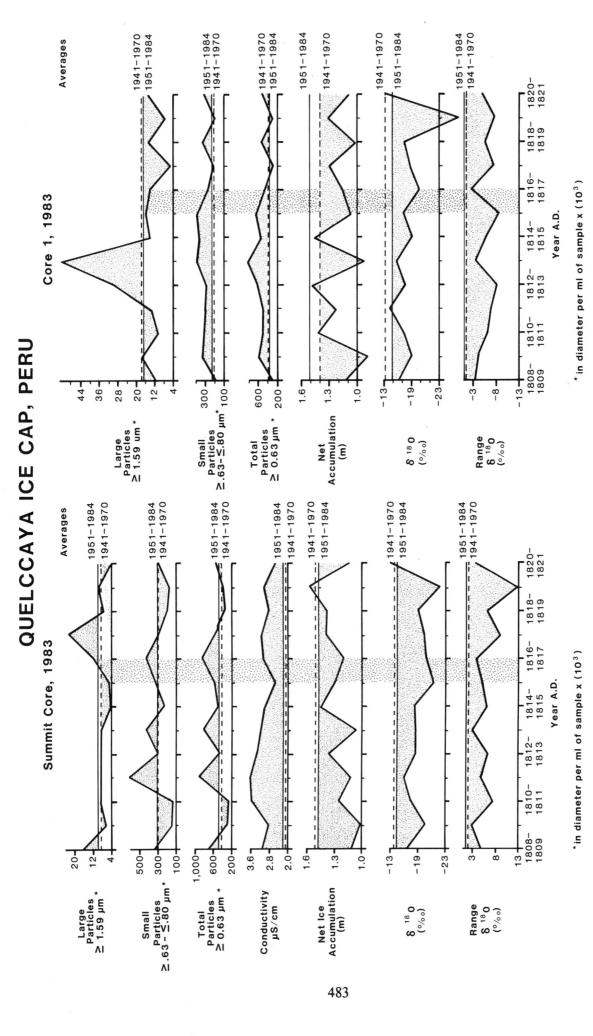

Figure 3: Quelccaya Ice Cap, Peru ice-core record of summit core and core 1 for the period 1808-21 of microparticle concentrations, conductivity, net ice accumulation, $\delta^{18}O$ and range of $\delta^{18}O$. Solid line indicates the 1951-84 mean and the dashed line indicates the 1941-70 mean.

Table 1 (A): Quelccaya Ice Cap, Peru.[1]

Summit Core	$\delta^{18}O‰$	.63-.80 μ	≥.63 μm	≥1.59 μm	Cond. (μS/cm)	Accumulation Ave-ice (m)	Range $\delta^{18}O‰$
1941-1970 31 years	-17.4	300734	475840	8308	2.145	1.516	1.71
1951-1984 33 years	-17.6	305948	489378	9441	2.098	1.486	1.54
1808-1821	-19.9	316271	551253	9278	2.900	1.259	5.70
Δ	-2.5	+15537	+75413	+163	+0.755	-0.257	+3.99

Core 1	$\delta^{18}O‰$	.63-.80 μm	≥.63 μm	≥1.59 μm	Accumulation Ave-ice (m)	Range $\delta^{18}O‰$
1941-1970 31 years	-17.2	215641	382217	17401	1.398	1.31
1951-1984 33 years	-17.9	230380	357874	17243	1.513	1.27
1808-1821	-19.7	295731	516310	16778	1.166	5.86
Δ	-2.5	+80090	+134073	-623	-0.232	+4.55

[1] Tables 1 (A), (B) and (C) represent modern averages for 1941-70 and 1951-84, averages for 1808-21 and the differences between 1941-70 and 1808-21 period (Δ) of $\delta^{18}O$, number of small, medium and large diameter particles per ml of sample, conductivity, average accumulation in metres of ice equivalent and the range of $\delta^{18}O$ for the three sites mentioned. In addition, for the Siple Station, Antarctica site, modern averages for of 1965-85, averages for 1808-21, and the differences between the modern average and the 1808-21 period (Δ) of Cl^-, SO^{2-}_4 and excess SO^{2-}_4 are presented.

The Quelccaya Ice Cap is situated immediately east of the high dry Altiplano (Thompson *et al.* 1984) whereas the Dunde Ice Cap has the Gobi Desert to the north and west and the dry Qaidam Basin to the southwest. Thus the dust records at both sites are dominated by local sources (Thompson *et al.* 1988a). Figure 3 and Table 1 (A) compare the 1808-21 period with two modern "norms", 1951-84 and 1941-70 for two cores from the Quelccaya Ice Cap, Peru. The year 1815-16 is shaded. In Peru 1810-20 is characterized by high dust concentrations, more negative oxygen isotopic ratios, and reduced accumulation. Note that these are characteristic of most of the Little Ice Age on this Ice Cap (Thompson *et al.* 1986; Thompson *et al.* 1988c; Thompson and Mosley-Thompson, in press). Thus, this decade is believed to have been colder, drier and dustier than present at this tropical site. Although there is no obvious indication of the Tambora eruption in the dust record, this is not unexpected in view of the predominance of the Altiplano as a major local source of dust. However, the oxygen isotope records show a cooling that reaches a minimum in 1819-20. This minimum is the lowest in the entire 1500 years preserved in the two ice cores.

Figure 4 and Table 1 (B) compare the 1808-21 period with the modern "norms" 1951-84 and 1941-70 for the Dunde Ice Cap, China. The year 1815-16 is shaded. 1810-20 is characterized in central China with lower (more negative) oxygen isotope values, low accumulation and higher dust concentrations. Again, these features are also characteristic of much of the Little Ice Age in this part of the world, and are consistent with records from the Quelccaya Ice Cap. Thus, the ice-core record suggests that conditions were colder, drier and dustier than present. However, the period following the eruption of Tambora shows: (1) a decrease in oxygen isotopes, with the most negative values occurring in 1819-20; and (2) an increase in conductivity (soluble dust) in 1817-18.

For Siple Station, Antarctica, Figure 5 and Table 1 (C) compare the characteristics of 1808-21 period with the modern "norms" 1941-70, except for anion concentrations for which the period 1965-85 is used. The decade 1810-20 is characterized by less insoluble dust, and more soluble dust, generally less accumulation, higher (less negative) oxygen isotope values, generally lower chloride values and higher sulphate and excess sulphate levels than found in the modern record. Thus, the only common features of the Little Ice Age with those in the China and Quelccaya ice cores are the higher electrical conductivity values and lower accumulation. The isotopic values may reflect warmer temperatures near the Siple Station, Antarctica during this period. The $\delta^{18}O$ is also subject to other interpretations possibly related to changes in sea-ice extent and hence, changes in distance from the moisture source. Following the eruption of Tambora in 1815-16, there is a major increase in conductivity, sulphate and excess sulphate between 1817 and 1819 that can probably be attributed to the eruption of Tambora. The mean oxygen isotope values actually rise (less negative) after the eruption with the lowest (most negative) values of the decade occurring in 1819-20, consistent with the Dunde and Quelccaya ice-core results.

Figure 6 illustrates the interannual variability in sulphate and oxygen isotopes for the period 1810-27. These plots of actual sample measurement illustrate the distinct seasonal record in both the sulphate (shown for two cores, A and B), and oxygen isotopic ratios used to date these cores. Evidently the Siple Station, Antarctica record has the clearest signal of the Tambora eruption, which is recorded in the sulphate signature and is reproducible in two cores. The sulphate deposited during 1817-19, essentially two years after the eruption, is consistent with the time lag in transport of stratospheric radioactivity to Antarctica (Lambert *et al.* 1977). However, the oxygen isotope record shows no apparent response to this eruption. The only meteorological parameter that potentially shows a change following the eruption is a rather sharp decrease in accumulation in 1817, which may or may not be related to the eruption of Tambora.

Discussion

Ice sheets and ice caps are now recognized as libraries for proxy climatic and environmental histories that can be used to extend historical records. Attempts to describe global climatic changes have been hampered by the lack of historical records from the southern hemisphere and the tropics. The Dunde, Quelccaya and Siple Station ice-core records allow, for the first time,

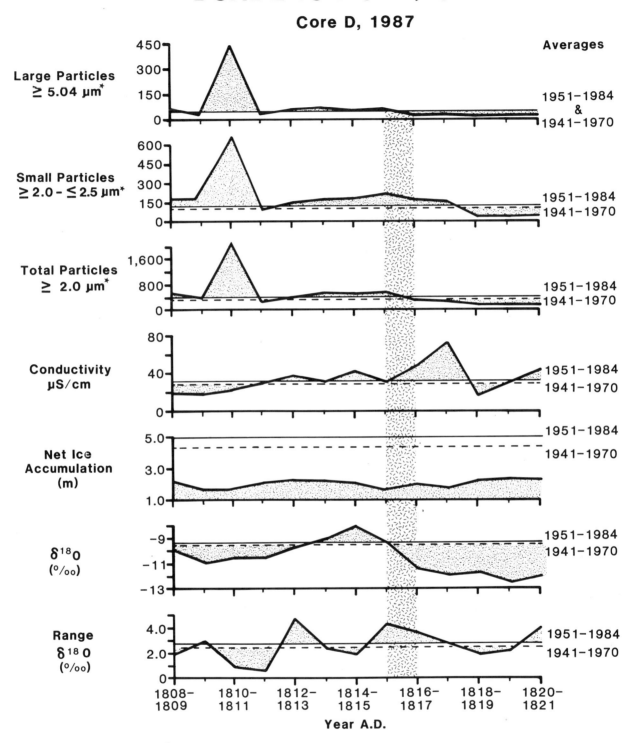

Figure 4: Dunde Ice Cap, China ice-core record for the period 1808-21 of annual variations in microparticle concentrations, conductivity, net ice accumulation, $\delta^{18}O$ and range of $\delta^{18}O$. Solid line indicates the 1951-84 mean and the dashed line indicates the 1941-70 mean.

Table 1 (B): Dunde Ice Cap, China.[1]

Core 1	$\delta^{18}O$‰	2.0-2.52 μm	≥2.0 μm	≥5.04 μm	Cond. (μS/cm)	Accumulation Ave-ice (m)	Range $\delta^{18}O$‰
1941-1970 31 years	-9.6	94106	245031	35052	26.946	0.43	2.25
1951-1984 33 years	-9.5	106728	275643	35820	28.220	0.50	2.56
1808-1821	-10.8	+152769	+424538	+66154	+32.300	0.20	2.51
Δ	-1.2	+58663	+179507	+31102	+5.54	-0.23	-0.26

[1] For explanation see Table 1 (A) footnote.

annual reconstruction of events at all three sites. These records permit a preliminary assessment of change and environmental variations for the early 1800s in general, and the year 1816, in particular.

In the subtropical and tropical sites of Dunde and Quelccaya, respectively, the period 1808-21 was colder, drier and dustier than modern levels. These conditions are deduced from a 14% decrease in oxygen isotope values at both the Dunde and Quelccaya sites, an average increase of 60% for insoluble particles on the Dunde Ice Cap, an average increase of 20% for insoluble particles on the Quelccaya Ice Cap, an average increase of 35% in soluble dust for both Dunde and Quelccaya, and an accumulation decrease of 50% for Dunde and 17% for Quelccaya when compared to the modern period 1941-84. For both Quelccaya cores the 1910-20 decade is characterized by the most negative $\delta^{18}O$ (coldest temperatures). The lowest mean decadal values are -20.14 in the summit core and -20.01‰ in core 1, which are 0.63 and 0.67‰ lower than the next coldest decade (1750-60). Additionally, the lowest $\delta^{18}O$ values measured in both cores occur during the southern hemisphere wet season of 1819-20. The lowest values are -28.25‰ (summit core) and -27.90‰ (Core 1) which are 2.65‰ and 1.02‰ lower than the next lowest value in their respective cores. The Dunde oxygen isotope values are low for the decade, with the low value of -12.80‰ occurring in 1819-20. However, in the China record, many years during the Little Ice Age have lower values, with 1779 and 1852 exhibiting significantly lower values of -14‰.

Particle analyses for the Quelccaya and Dunde ice cores for the decade 1810-20 show no significant change in particle concentrations above the high levels that characterize much of the Little Ice Age. Probably abundant local dust sources mask any potential change due to distance from particle sources such as the eruption of Tambora. The liquid conductivity profile for the Dunde site shows a prominent peak for the decade in the year 1817-18.

The $\delta^{18}O$ records from Siple Station, Antarctica, have an average oxygen isotope value for the 1810-20 decade which is 0.50‰ higher than the modern "norm" (1941-70). Moreover, these

Table 1 (C): Siple Station, Antarctica.

	$\delta^{18}O$‰	.63-.80 μm	≥0.63 μm	≥1.59 μm	Cond. (μS/cm)	Accumulation H₂0 eq. (m)	Range $\delta^{18}O$‰
1941-1970 30 years	-29.94	1347	3089	257	2.16	0.540	-6.43
1951-1984 33 years	-29.85	NA	NA	NA	2.16	0.556	-7.21
1808-1821	-29.44	777	1905	176	2.33	0.468	-6.55
Δ	+0.50	-570	-1184	-81	+0.17	-0.07	1.12

	Cl^-	SO_4^{2-}	Excess SO_4^{2-}
1965-1985 20 years	2.46	0.77	1.37
1808-1821	1.77	1.35	1.17
Δ	-0.69	+0.58	-0.20

[1] For explanation see Table 1 (A) footnote.

records show a major electrical conductivity and SO_4^{2-} peak associated with the years 1817-19. Thus, while the Siple Station record contains the strongest physical evidence of the Tambora eruption, there is little evidence of a coding response in the oxygen isotopes, and in fact, the $\delta^{18}O$ values actually *rise* (warming) after the eruption. On the other hand, note that 1819 ($\delta^{18}O$ -13.5‰) contains the lowest annual value for the decade which is consistent temporally with the Quelccaya and Dunde ice-core records.

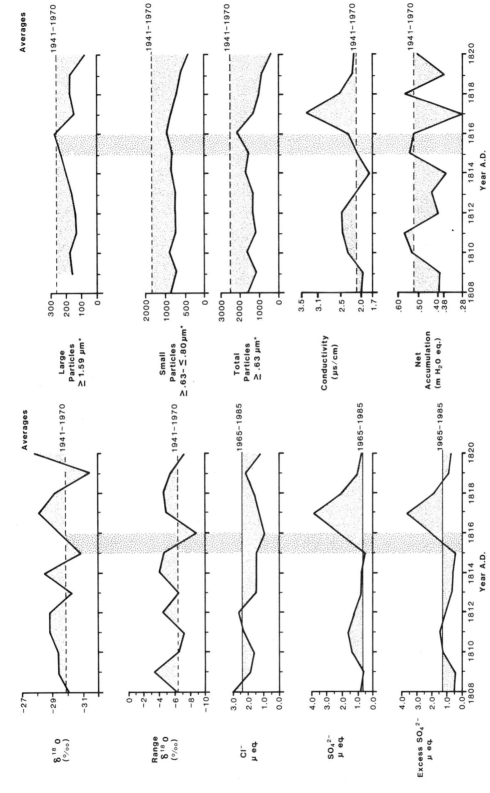

Figure 5: Siple Station, Antarctica, ice-core record of $\delta^{18}O$, range of $\delta^{18}O$, Cl^-, SO_4^{2-} and excess SO_4^{2-} concentrations, electrical conductivity and net accumulation for the period 1808-20. The solid line is the 1965-85 mean for anion concentrations and the dashed line is the 1941-70 mean for all other analyses.

*in diameter per ml of sample x (10^3)

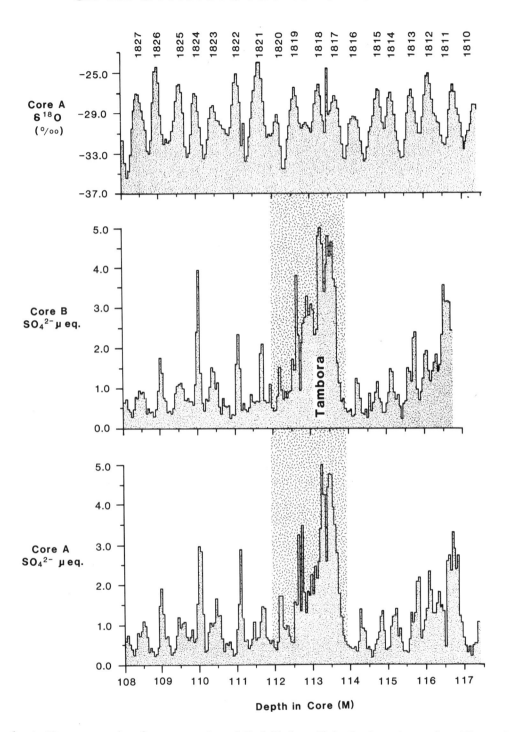

Figure 6: A 10-meter section from cores A and B drilled at Siple Station, Antarctica. These show individual sample measurements of SO_4^{2-}, concentration and demonstrate the reproducibility (and hence the reliability) of the chemistry record. Individual analyses of $\delta^{18}O$ are also given for core A. The annual variation in $\delta^{18}O$ and SO_4^{2-} data are readily apparent and illustrate the methods by which the cores were dated. Tambora stands out in both cores A and B as a period of enhanced concentrations of SO_4^{2-} for 1817-19.

Acknowledgements

Research was supported by the United States National Science Foundation, Division of Atmospheric Sciences ATM 82-13601A02 and ATM 85-19794A03, Division of Polar Programs, DPP 8410328 and the National Geographic Society, Grant No. 3323-86. The $\delta^{18}O$ measurements were conducted at the University of Washington under NSF Grant DPP-8400574 and by Professor W. Dansgaard, University of Copenhagen. We thank the many people who have assisted in the laboratory and field seasons associated with the individual projects.

References

Dai, J., E. Mosley-Thompson, L.G. Thompson and J.K. Arbogast. 1989. Chloride, sulfate and nitrate in snow at Siple Station, Antarctica, 1965-85. (Submitted to *Journal of Glaciology*).

Dermendjian D. 1973. On volcanic and other particular turbidity anomalies. *Advances in Geophysics* 16:267-296.

Grove, J.M. 1988. *The Little Ice Age*. Methuen, London and New York. 498 pp.

Hammer, C.U., H.B. Clausen, W. Dansgaard, N. Gundestrup, S.J. Johnsen and N. Reeh. 1978. Dating of Greenland ice cores by flow models, isotopes, volcanic debris and continental dust. *Journal of Glaciology* 20(82):3-26.

Hoyt, J.B. 1958. The cold summer of 1816. *Annals of the Association of American Geographers* 48(2):118-131.

Lambert, G., B. Ardouin, J. Sanak, C. Lorius and M. Pourchet. 1977. Accumulation of snow and radioactive debris in Antarctica: a possible refined radiochronology beyond reference levels. *International Association of Scientific Hydrology Publication (IAHS* Publication) 118:146-158.

Landsberg, H.E. and J.M. Albert. 1974. The summer of 1816 and volcanism. *Weatherwise* 27:63-66.

Milham, W.I. 1924. The year 1816 - the causes of abnormalities. *Monthly Weather Review* 52(12):563-570.

Mosley-Thompson, E., L.G. Thompson, J.F. Paskievitch and P.M. Grootes. 1989. Glaciological studies at Siple Station (Antarctica) and potential ice core paleoclimatic record. (Submitted to *Journal of Glaciology*.)

Stothers, R.B. 1984. The great Tambora eruption in 1815 and its aftermath. *Science* 224(4654): 1191-1198.

Stommel, H. and E. Stommel. 1983. *Volcano Weather, The Story of 1816, The Year Without a Summer*. Seven Seas Press, Newport, Rhode Island. 177 pp.

Thompson, L.G., E. Mosley-Thompson, P.M. Grootes, M. Pourchet and S. Hastenrath. 1984. Tropical glaciers: potential for ice core paleoclimatic reconstructions. *Journal of Geophysical Research* 89(D3):4638-4646.

Thompson, L.G., E. Mosley-Thompson, W. Dansgaard and P.M. Grootes. 1986. The "Little Ice Age" as recorded in the stratigraphy of the tropical Quelccaya Ice Cap. *Science* 234:361-364.

Thompson, L.G., X. Wu, E. Mosley-Thompson and Z. Xie. 1988a. Climatic ice core records from the Dunde ice cap, *Annals of Glaciology* 10:178-182.

Thompson, L.G., E. Mosley-Thompson, X. Wu, and Z. Xie. 1988b. Wisconsin/Würm glacial stage ice in the subtropical Dunde Ice Cap, China. *GeoJournal* 17(4):517-523.

Thompson, L.G., M. Davis, E. Mosley-Thompson and K. Liu. 1988c. Pre-Incan agricultural activity recorded in dust layers in two tropical ice cores. *Nature* 336:763-765.

Thompson, L.G. and E. Mosley-Thompson. (in press). One-half century of tropical climatic variability recorded in the stratigraphy of the Quelccaya Ice Cap, Peru. *In: Aspects of Climate Variability in the Pacific and Western Americas*. D.H. Peterson (ed.). Geophysical Monograph Series, American Geophysical Union, Washington, D.C.

Changes in Southern South American Tree-Ring Chronologies following Major Volcanic Eruptions between 1750 and 1970

Ricardo Villalba[1] and Jose A. Boninsegna[1]

Abstract

The effect of the volcanic eruptions of Asama and Laki in 1783, Tambora in 1815, Coseguina in 1835, Armagura 1846, Krakatau in 1883, Tarawera in 1886, Santa María, Soufriere and Pelée in 1902, Nilahue in 1955, Agung in 1963 and an unknown eruption occurring around 1808, is examined on temperature-sensitive tree-ring chronologies of southern South America. These records include two chronologies from the subtropical area (27°S), 12 chronologies from the temperate middle latitude region (37° to 41°S), and 21 chronologies from the subantarctic region (54° to 55°S).

A uniform tree-ring width decrease in all the studied chronologies is indicated following the Agung eruption in 1963. The Santa María, Soufriere and Pelée eruptions in 1902 provoked a significant decrease in tree-ring variations of subtropical and middle latitudes. On the other hand, the Tambora and Krakatau volcanic episodes affected more clearly variations in the subantarctic chronologies. No apparent changes in any of the tree-ring records follow the Coseguina, the Armagura and the Tarawera eruptions. Based on an application of Student's t-test to tree-ring indices, on 18 occasions (out of a possible 46) the tree-ring index average for the three-year period after the eruption is significantly (at the 10% level) lower than the tree-ring index average for the three-year period before the eruption, but only in four cases are the tree-ring indices significantly higher after the eruption. An increase in tree-ring indices after volcanic eruptions is significantly less probable than a decrease.

Introduction

Major explosive eruptions inject large amounts of ash and sulphur aerosols into the stratosphere and upper troposphere. In addition to causing rare atmospheric optical phenomena, such events can have pronounced effects on climate that may persist for several years (Lamb 1970, 1972). It is generally agreed that a tropospheric cooling would be expected following volcanic eruptions sufficiently powerful to inject large amounts of gases and particles into the stratosphere. However, attempts to define such cooling on the basis of long-term temperature records have not been completely convincing (Taylor *et al.* 1980; Angell and Korshover 1985).

Coolings associated with volcanic eruptions should be reflected in the annual-ring patterns of temperature sensitive tree-ring records. To check this hypothesis, more than 50 chronologies from South America were chosen as a basis for analyzing the impact of 10 volcanic episodes between 1780 and 1970: the six major volcanic eruptions with the largest dust-veil indices according to Lamb (1970), and four additional eruptions clearly recorded in acidity profiles from several Antarctic locations (Legrand and Delmas 1987).

---

[1] Laboratorio de Dendrocronología, CRICYT Mendoza, C.C. 330-5500, Mendoza, Argentina. The present address of the first author is: Department of Geography, University of Colorado, Campus Box 260, Boulder, Colorado 80309, U.S.A.

Tree-Ring Chronology Database

An important set of tree-ring chronologies from southern South America, between 32 and 43°S, has been developed by LaMarche *et al.* (1979 a, b) employing *Araucaria araucana* and *Austrocedrus chilensis*. Additional chronologies have been produced by the Laboratorio de Dendrocronología of Argentina. In subtropical northwestern Argentina (23 to 27°S), 12 chronologies have been developed using *Juglans australis* and *Cedrela* sp. (Villalba *et al.* 1985, 1987). In the Patagonian Andes, several chronologies have been built employing *Fitzroya cupressoides* (Boninsegna and Holmes 1985; Villalba unpublished), and *Nothofagus* species (Boninsegna *et al.* unpublished). Also, new *Araucaria araucana* chronologies have recently been developed. From this set, 34 temperature-sensitive chronologies were selected for analyzing the effect of volcanic eruptions on tree growth. Table 1 shows 31 of the chronologies selected, the species employed and some characteristics of sample sites. The geographical location of these selected chronologies is shown in Figure 1.

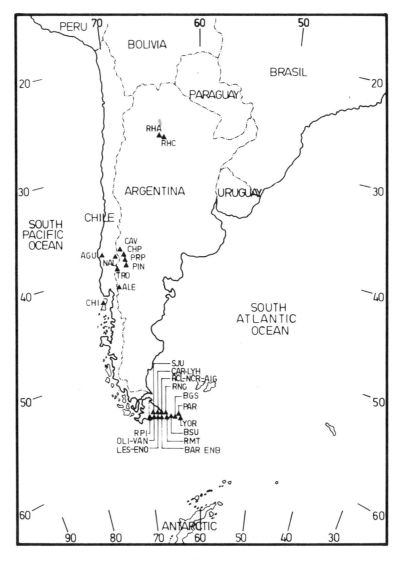

Figure 1: Sites of selected tree-ring chronologies used in this study. Names of sites and other information are listed, by map code, in Table 1.

In order to better evaluate the impact of volcanic eruptions on tree growth, the chronologies with a common climatic signal were joined by area in a mean regional chronology. The chronology with the strongest response to climate was employed in other cases.

Despite the similar response to climate of the two temperature-sensitive chronologies in subtropical Argentina, the *Juglans australis* chronology at Río Horqueta was chosen as a basis for analyzing volcanic episodes due to the stronger relation to summer temperature variations (Villalba *et al.* 1987). At middle latitudes (36 to 38°S) the *Araucaria araucana* chronologies were grouped into a mean *Araucaria* chronology. The Río Alerce chronology was selected as representative of the *Fitzroya cupressoides* tree-ring variations at 41°S (Villalba unpublished). Also the *Nothofagus* chronologies of Tierra del Fuego (54 to 55°S) were joined in a mean regional chronology.

The correlation and response functions of the subtropical *Juglans* (27°S), temperate mean *Araucaria* (37°S), temperate *Fitzroya* (41°S), and subantarctic mean *Nothofagus* (54°S) chronologies are shown in Figure 2. Each response and correlation function (Fritts 1976; Blasing *et al.* 1984) includes 16 weights for average regional monthly temperature and 16 weights for total regional monthly precipitation, from January of the year prior to the season of growth, through the April concurrent with the growth.

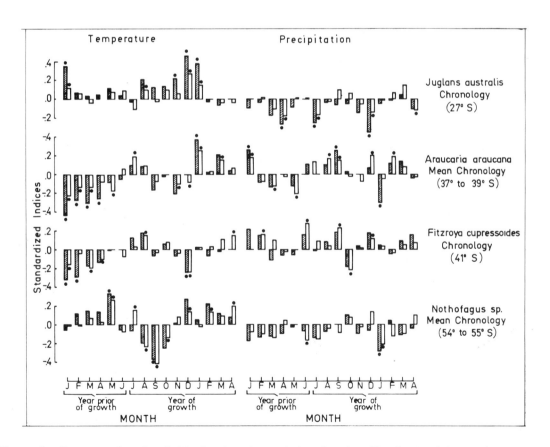

Figure 2: Response function (white bars) and correlation function (line bars) of the *Juglans australis* (27°S), the mean *Araucaria araucana* (37° to 39°S), the *Fitzroya cupressoides* (41°S) and the mean *Nothofagus* sp. chronologies (53° to 54°S). Black dots denote response-function elements or correlation coefficients significant at 95% (one dot) or at 99% (two dots) confidence level.

Table 1: Site and Chronology Characteristics.

Site			Geographical Loc.			Chronology					
Code	Name	Genus	Lat. S	Long. W	Alt. (m)	Tree N	Radii N	Aut. Cor.[1]	SD.[2]	Mean Sen.[3]	Time Span	Source[4]
RHA	Rio Horqueta	*Juglans*	27°10'	65°53'	1850	22	40	0.623	0.346	0.26	1783-1981	Mendoza
RHC	Rio Horqueta	*Cedrela*	27°10'	65°52'	1650	18	24	0.713	0.542	0.29	1729-1982	Mendoza
AGU	P. del Aguila	*Araucaria*	37°50'	73°02'	1300	14	42	0.015	0.160	0.12	1242-1975	Arizona
NAL	Nalcas	*Araucaria*	38°20'	71°29'	1420	8	30	0.648	0.238	0.14	1375-1975	Arizona
CAV	Caviahue	*Araucaria*	37°52'	71°01'	1540	14	29	0.649	0.194	0.13	1444-1974	Arizona
CHP	Chenque Pehuen	*Araucaria*	38°06'	70°51'	1650	19	48	0.559	0.191	0.14	1246-1974	Arizona
PRP	Pino Hachado	*Araucaria*	38°38'	70°45'	1400	12	25	0.704	0.226	0.14	1459-1974	Arizona
PIN	Primeros Pinos	*Araucaria*	38°53'	70°37'	1620	11	38	0.653	0.237	0.16	1140-1974	Arizona
TRO	Lago Tromen	*Araucaria*	39°36'	71°23'	1250	13	22	0.512	0.181	0.14	1385-1983	Mendoza
ALE	Rio Alerce	*Fitzroya*	41°10'	71°56'	1100	23	48	0.716	0.242	0.15	864-1985	Mendoza
CHI	Tichihue, Chiloe	*Fitzroya*	42°30'	73°50'	750	21	33	0.862	0.405	0.16	1386-1987	Mendoza
SJU	Ea. San Justo	*Nothofagus*	54°03'	68°34'	250	38	65	0.322	0.217	0.20	1723-1985	Lamont
CAR	Estancia Carmen	*Nothofagus*	54°26'	67°55'	250	28	54	0.371	0.290	0.28	1726-1986	Mendoza
LYH	Lago Yehuin	*Nothofagus*	54°28'	67°43'	200	16	25	0.468	0.285	0.25	1731-1986	Mendoza
RCL	Rio Claro	*Nothofagus*	54°30'	67°48'	150	18	32	0.393	0.279	0.27	1739-1986	Mendoza
MCR	Ea. Maria Crist.	*Nothofagus*	54°30'	67°05'	300	19	30	0.355	0.248	0.25	1743-1986	Mendoza
AIG	A. Isla Grande	*Nothofagus*	54°31'	67°25'	200	21	35	0.621	0.338	0.26	1639-1986	Mendoza
RMG	Rio Malengena	*Nothofagus*	54°31'	66°10'	15	24	42	0.440	0.301	0.28	1705-1986	Mendoza
LES	Lago Escondido	*Nothofagus*	54°39'	67°52'	700	33	72	0.797	0.359	0.33	1575-1984	Lamont
EMO	Est. Microondas	*Nothofagus*	54°41'	67°50'	700	15	20	0.531	0.214	0.16	1664-1984	Mendoza
VAN	Valle Andorra	*Nothofagus*	54°47'	68°11'	250	22	33	0.538	0.196	0.14	1593-1984	Mendoza
OLI	Monte Olivia	*Nothofagus*	54°45'	68°05'	350	10	16	0.576	0.191	0.17	1767-1908	Mendoza
RPI	Rio Pipo	*Nothofagus*	54°47'	68°28'	50	28	33	0.459	0.194	0.15	1657-1984	Lamont
BAR	Barlovento	*Nothofagus*	54°55'	68°10'	40	10	12	0.569	0.300	0.14	1852-1985	Mendoza
EHB	Ea. Harberton	*Nothofagus*	54°55'	67°20'	30	17	23	0.413	0.376	0.35	1700-1985	Mendoza
RMT	Rio Moat 1	*Nothofagus*	54°54'	66°55'	50	13	24	0.726	0.247	0.14	1528-1986	Mendoza
RMT	Rio Moat 2	*Nothofagus*	54°54'	66°55'	50	20	30	0.623	0.219	0.14	1504-1985	Lamont
YOR	Bahia York	*Nothofagus*	54°50'	64°20'	60	30	30	0.448	0.232	0.19	1647-1986	Mendoza
PAR	Puerto Parry	*Nothofagus*	54°50'	64°22'	20	14	25	0.545	0.258	0.19	1726-1986	Mendoza
BCS	Bahia Crossley	*Nothofagus*	54°42'	64°40'	20	13	18	0.390	0.263	0.26	1715-1986	Mendoza
BSU	B. Buen Suceso	*Nothofagus*	54°50'	65°12'	35	9	15	0.510	0.212	0.16	1751-1986	Mendoza

[1] Aut. Cor.: first order autocorrelation.

[2] SD.: Standard Deviation.

[3] Mean Sen.: Mean Sensitivity.

[4] Sources: Mendoza (Laboratorio de Dendrocronologia, CRICYT-Mendoza, Argentina); Arizona (Laboratory of Tree-Ring Research, University of Arizona); Lamont (Tree-Ring Laboratory, Lamont-Doherty Geological Observatory, Palisades, New York).

In the subtropical chronology, response and correlation functions show that above-average temperature in late spring and summer of the current year favour tree growth. On the other hand, both response and correlation functions, indicate that *Araucaria araucana* and *Fitzroya cupressoides* growth is largely controlled by thermic conditions in the summer of the previous year. Also, the dendroecological analysis (Kienast and Schweingruber 1986) shows a strong inverse relationship between the radial growth of both species and temperature in the previous growth season. In Tierra del Fuego, temperature of late winter and early spring is negatively linked to growth of *Nothofagus* species (Boninsegna et al. unpublished).

Tree-Ring Indices Variations from 10 Years before to 10 Years after the Volcanic Episodes

Taking into account the work of Angell and Korshover (1985), mainly based on Lamb (1970), the six major volcanic episodes with the largest dust-veil indices between 1780 and 1970: Asama-Laki, Tambora, Coseguina, Krakatau, Santa María-Soufriere-Pelée and Agung, were considered in this study. Besides, due to the relative proximity of South America to Antarctica, four additional volcanic eruptions were selected: Armagura, Tarawera, Nilahue and an another large-scale event not recorded historically. They have been clearly determined in the acidity profile at Dome C, Antarctica (74°S, 125°E) by Legrand and Delmas (1987). Table 2 shows the latitude and longitude, month and year of the 10 volcanic eruptions considered in this study.

Table 2: Volcanic Episodes Considered in This Study.

Volcano	Latitude	Longitude	Date	DVI[1]	VEI[2]
Asama, Japan	36°N	138°E	August 1783	300	4
Laki, Iceland	64°N	17°W	June-December 1783	700	4
Unknown eruption	—	—	Around 1808	—	-
Tambora, Indonesia	8°S	118°E	April 1815	3000	7
Coseguina, Nicaragua	13°N	88°W	January 1835	4000	5
Armagura, Nicaragua	18°S	174°W	October 1846	1000	3?
Krakatau, Indonesia	6°S	106°E	August 1883	1000	6
Tarawera, New Zealand	38°S	176°E	July 1886	—	5
Santa María, Guatemala	14°N	92°W	October 1902	600	6
Soufriere, St. Vincent	13°N	61°W	May 1902	300	4?
Pelée, Martinique	15°N	61°W	May 1902	100	4
Nilahue, Patagonia	40°S	70°W	October 1955	—	4
Agung, Indonesia	8°S	116°E	March 1963	800	3

[1] DVI: dust veil index from Lamb (1972).

[2] VEI: volcanic explosivity index from Simkin et al. (1981).

The variation in tree-ring indices from 10 years before to 10 years after the aforementioned eruptions is analyzed below. According to Angel and Korshover (1985) the 21-year period allows us to obtain a reasonable picture of interannual variability during the period of the eruptions. The

Araucaria and *Fitzroya* chronologies were lagged by one year in agreement with the response and correlation function results. Except for the *Juglans australis* chronology, all the tree-ring records considered are inversely related to temperature variations, therefore they have been **inverted** in order to clarify the figures. According to this arrangement every decrease in tree-ring index is associated with a decrease in temperature.

Finally, standardized regional chronologies were averaged to produce composite tree-ring index signals. The superposition clarifies the volcanic signal by averaging out features not common to each chronology. To avoid biasing the results toward subtropical chronology where the variability is greater, the data were standardized before averaging.

Asama-Laki in 1783

Figure 3 shows the tree-ring indices before and after the Asama-Laki eruptions. In the *Araucaria* mean chronology only the index following the eruptions was low, while in the *Fitzroya* chronology the seven years following the eruption were the lowest of the 21-year period. At subantarctic latitudes, a small increase in the first tree-ring index after the eruptions is followed by a gradual decrease during the five following years.

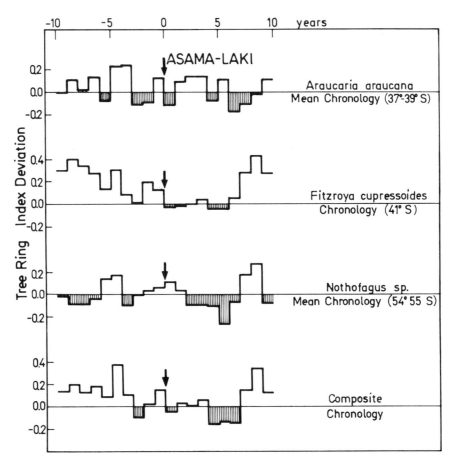

Figure 3: Tree-ring index deviations from 10 years before to 10 years after the 1783 eruptions of Asama and Laki, in the mean *Araucaria araucana* (37° to 39°S), the *Fitzroya cupressoides* (41°S) and the mean *Nothofagus* (54° to 55°S) chronologies. Vertical arrows indicate the year of the eruptions.

Unknown and Tambora Eruptions in 1808 and 1815 Respectively

Figure 4 shows the variation in tree-ring indices 10 years before and after the unknown eruption around 1808, and 10 years before and after the Tambora eruption. The tree-ring index of *Araucaria* (37° to 39°S) and *Fitzroya* (41°S) chronologies after the unknown volcanic episode were the lowest of the 27-year period. In subtropical latitudes, the lowest index of the comparison period occurred two years after the proposed eruption. On the other hand, in the subantarctic chronology, the tree-ring index after the unknown eruption was only a little lower than it was in the previous year.

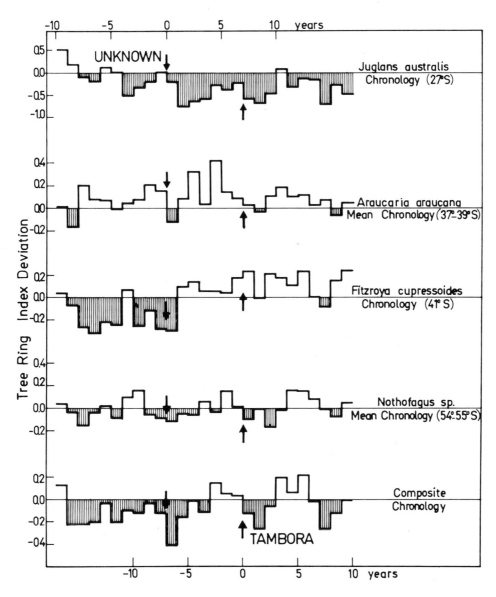

Figure 4: Tree-ring index deviations from 10 years before the 1808 large-scale volcanic event registered at Dome C, Antarctica, to 10 years after the 1815 eruption of Tambora, in the *Juglans australis* (27°S), the mean *Araucaria araucana* (37° to 39°S), the *Fitzroya cupressoides* (41°S) and the mean *Nothofagus* (54° to 55°S) chronologies. Vertical arrows indicate the years of the eruptions.

Except for the *Fitzroya* chronology, in all other chronologies the three indices after the Tambora eruption were lower than the three indices before it. The effect on tree-ring decrease is larger in subantarctic latitudes, where the third index following the eruption is the lowest of the 20-year period. Even though a gradual tree-ring decrease in the composite chronology is observed after the eruption, evidence for a volcanically induced tree-growth decrease is not so compelling.

Coseguina and Armagura Eruptions in 1835 and 1846 Respectively

Figure 5 shows the variations in tree-ring indices 10 years before and after the Coseguina and Armagura eruptions. In subtropical latitudes, no significant decrease in ring width is indicated after the Coseguina eruption. At temperate latitudes, although a decrease in tree-ring indices is observed in the *Fitzroya* chronology, an increase is indicated in the *Araucaria* mean chronology. Finally, in the *Nothofagus* mean chronology at Tierra del Fuego a decrease in tree-ring indices follows the Coseguina eruption.

No clear evidence for a volcanic induced tree-growth decrease is observed at any latitude at the time of the Armagura eruption.

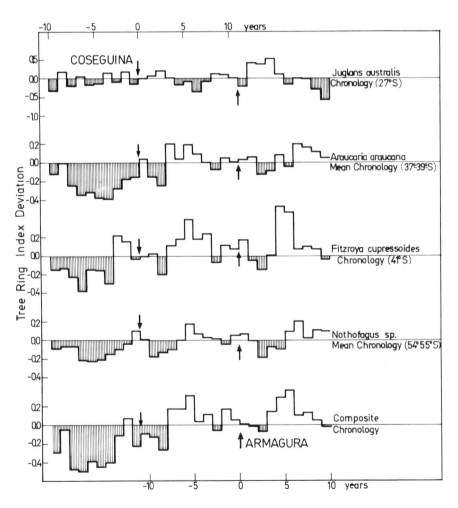

Figure 5: Tree-ring index deviations from 10 years before the 1835 eruption of Coseguina to 10 years after the 1846 eruption of Armagura, in the *Juglans australis* (27°S), the mean *Araucaria araucana* (37° to 39°S), the *Fitzroya cupressoides* (41°S) and the mean *Nothofagus* (54° to 55°S) chronologies. Vertical arrows indicate the years of the eruptions.

Krakatau and Tarawera Eruptions in 1883 and 1886 Respectively

Figure 6 shows the variation in tree-ring indices 10 years before and after Krakatau and Tarawera eruptions. As with Coseguina, there is little evidence of tree-growth decrease in subtropical and *Araucaria* forest latitudes following the Krakatau eruption, but a gradual decrease is indicated in *Fitzroya* and subantarctic chronologies.

The evidence for a volcanically-induced decrease immediately following the Tarawera eruption is relatively poor in all regions considered. A little decrease is observed in subtropical latitudes.

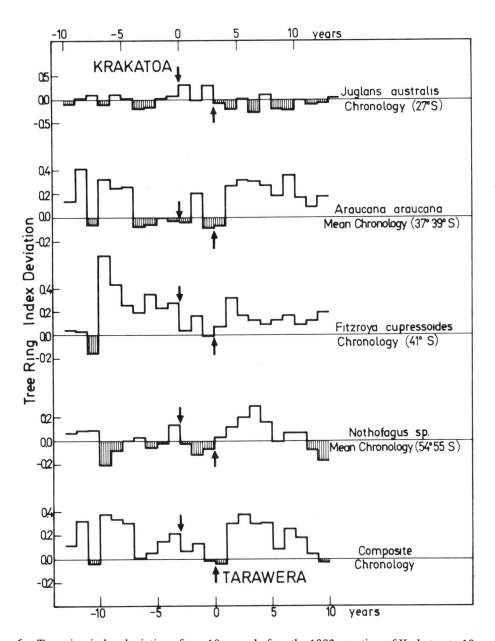

Figure 6: Tree-ring index deviations from 10 years before the 1883 eruption of Krakatau to 10 years after the 1886 eruption of Tarawera, in the *Juglans australis* (27°S), the mean *Araucaria araucana* (37° to 39°S), the *Fitzroya cupressoides* (41°S) and the mean *Nothofagus* (54° to 55°S) chronologies. Vertical arrows indicate the years of the eruptions.

Santa María, Soufriere and Pelée Eruptions in 1902

Figure 7 shows the variation in tree-ring indices before and after the Santa María- Soufriere-Pelée eruptions. A pronounced decrease in tree-ring indices is observed in subtropical latitudes following the eruptions, the years after the eruptions being the lowest indices since the 1810-20 period. In temperate latitudes there was a long period of low tree-ring growth prior to the eruption. However, both in *Araucaria* and *Fitzroya* chronologies the tree-ring index after the eruptions was extremely low. Particularly in the *Araucaria* mean chronology, the year after these volcanic episodes is the lowest of the 1780-1970 period. While a gradual decrease is indicated after the eruptions in subantarctic latitudes, the same appears more significant in the composite mean chronology.

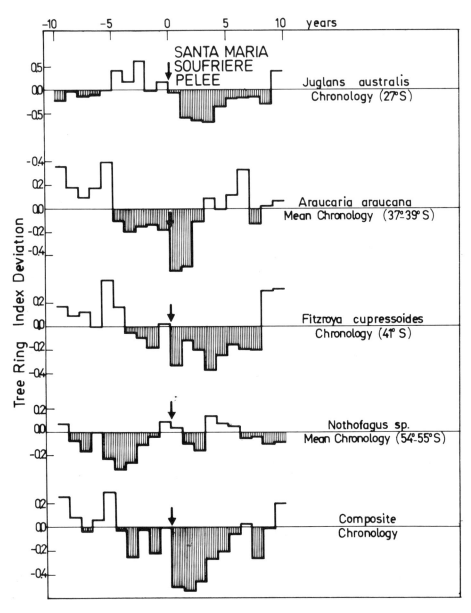

Figure 7: Tree-ring index deviations from 10 years before to 10 years after the 1902 eruptions of Santa María-Soufriere-Pelée, in the *Juglans australis* (27°S), the mean *Araucaria araucana* (37° to 39°S), the *Fitzroya cupressoides* (41°S) and the mean *Nothofagus* (54° to 55°S) chronologies. Vertical arrows indicate the year of the eruptions.

Nilahue and Agung Eruptions in 1955 and 1963 Respectively

Figure 8 shows the variations in tree-ring indices 10 years before and after the Nilahue and Agung eruptions.

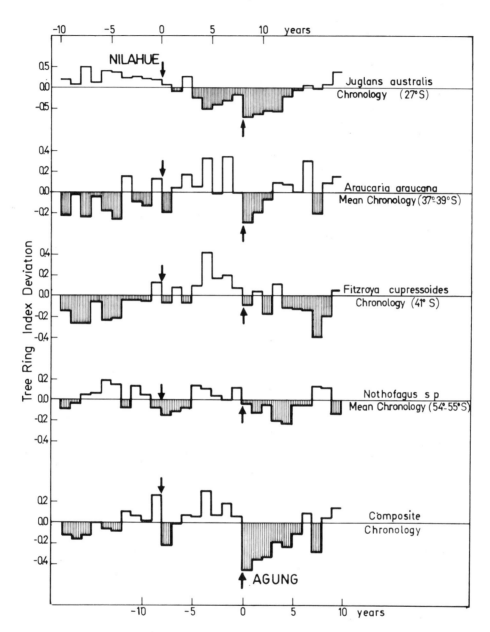

Figure 8: Tree-ring index deviations from 10 years before the 1955 eruption of Nilahue to 10 years after the 1963 eruption of Agung, in the *Juglans australis* (27°S), the mean *Araucaria araucana* (37° to 39°S), the *Fitzroya cupressoides* (41°S) and the mean *Nothofagus* (54° to 55°S) chronologies. Vertical arrows indicate the years of the eruptions.

Although a decrease in the first tree-ring index following the Nilahue eruption is indicated in subtropical, temperate and subantarctic latitudes, the evidence for a volcanically-induced tree-ring decrease is not quite so compelling. The decrease is more pronounced at temperate [*Araucaria* (37° to 38°S) and *Fitzroya* (41°S)] latitudes, probably due to the proximity of Nilahue volcano (40°S).

A larger-than-expected decrease, considering the magnitude of the eruption, is observed following the Agung volcanic episode in the whole region studied. In both *Juglans* and *Araucaria* mean chronologies, the year after the eruption is the lowest of the 20-year period. In general, the tree-ring decreases following the Agung eruption are the most impressive of the 10 volcanic episodes considered in this study (Table 3).

Estimated Significance of Changes in Tree-Ring Indices Related to Volcanic Eruptions

In order to determine the significance of differences between average tree-ring index deviations for the three-year periods immediately before and after volcanic episodes, the Student's t-test was used (Dunn and Clark 1974, p. 52). The three-year periods were chosen in agreement with the average stratospheric lifetime of volcanic aerosol (Mass and Schneider 1977; Rampino and Self 1982; Kelly and Sear 1984). Advantages and disadvantages of Student's t-test applied to changes in temperature before and after volcanic eruptions are presented by Angell and Korshover (1985).

Table 2 shows the results obtained by this statistical method for each of the four regional chronologies considered in this study. On 18 occasions (out of a possible 46) the tree-ring indices are observed to be significantly (at the 10% level) lower after the eruption than before the eruption. Only in four cases were tree-ring indices significantly higher after the eruption than before it. Similar analysis applied to the composite mean chronology yield three cases of significant decrease at the 5% level (Tambora, Santa María-Soufriere-Pelée and Agung), out of a possible 10.

Discussion

Volcanic Eruptions during the Pre-Instrumental Period

Considering the high latitudes of the Asama (36°N) and particularly the Laki eruption (64°N), a volcanic influence on temperature decrease in middle and high latitudes of South America is doubtful. Although a decrease in tree-ring indices appears in the *Fitzroya* (41°S) and *Nothofagus* (54°S) chronologies following the eruption, the volcanically-induced effect is uncertain. Also, a decrease in tree-rings induced by the unknown eruption around 1808 is not so compelling. However, some effect is indicated mainly in *Juglans* and *Araucaria* chronologies. Legrand and Delmas (1987) observed a variable deposition of the unknown-eruption debris over Antarctica, the deposition being less evident in West Antarctica (South America-South Atlantic sector).

A statistically-significant decrease in tree-ring indices follow the Tambora eruption is indicated in subtropical, temperate (38°S) and subantarctic chronologies, as well as in the composite chronology (Table 2). However considering the magnitude of the Tambora eruption, the observed volcanic effect on tree-rings is weaker than expected.

Angell and Korshover (1985) considered that the cooling influence of volcanic eruptions on hemispheric temperatures might be moderate if a tropical warming, associated with an El Niño event, occurred simultaneously or shortly after the volcanic episode. According to Aceituno (1987), in summers during the mature phase of an El Niño event (negative Southern Oscillation phase) anomalously warm water in the Eastern Pacific is consistent with positive air temperature departures in subtropical and central Argentina (22 to 30°S). During strong El Niño events, like the one in 1982-83, positive temperature departures were observed as far as 45°S (Aceituno 1987, p. 91).

	Asama	Unknown	Unknown	Tambora	Coseg.	Coseg.	Arma.	Kraka.	Taraw.	S. María-Sou.-Pel.		Nilahue	Agung
	1783-4	1807-8	1808-9	1815-6	1834-5	1835-6	1846-7	1883-4	1886-7	1901-2	1902-3	1955-6	1963-4
Juglans	—	-0.34*	-0.53***	-0.27**	0.08	—	-0.10	0.22*	-0.29**	-0.69**	—	0.16	0.33***
Araucaria	0.05	-0.05	-0.06	-0.14*	0.13*	0.08	-0.04	0.04	0.09	-0.15*	—	0.02	-0.22**
Fitzroya	-0.13**	-0.07	0.19	0.05	-0.03	-0.22*	-0.05	-0.22**	0.13	-0.13	—	-0.04	-0.22**
Nothofagus	-0.01	-0.08	0.07*	-0.13*	-0.12*	—	-0.11*	-0.09	0.26***	0.15*	-0.05	-0.15*	-0.13**
Composite Chronology	-0.03	-0.09	—	-0.23**	-0.00	—	-0.09	-0.08	0.17	-0.43***	—	-0.14	-0.50***

[1] A negative value means that tree-ring indices were lower after the eruption. Based on Student's t-test, differences significant at the 10%, 5% and 1% levels are indicated by one, two and three asterisks, respectively. For the unknown eruption two possible intervals were evaluated (1807-08 and 1808-09). Taking into account the response function of the *Nothofagus* mean chronology, two intervals were considered for the Santa María-Soufriere-Pelée eruptions (1901-02 and 1902-03).

Table 3: Tree-Ring Index Difference between the Three-Year Interval before and after Volcanic Episodes.[1]

Searching for a possible El Niño event in times of the Tambora eruption, Lough and Fritts (1985) suggest that the Southern Oscillation low-index in 1815-16 might have been of comparable magnitude to the 1982-83 event. Also, Quinn *et al.* (1978), based on a conjunction of meteorological and environmental data, have found evidence of a strong El Niño in 1814 and a moderate one in 1817. Therefore, any atmospheric cooling resulting from the eruption of Tambora may have been moderated by El Niño in 1814-17. In the same way, Angell and Korshover (1985) suggest that the induced cooling of the Krakatau eruption was smoothed by the atmospheric warming related to the strong and prolonged El Niño in 1884. Further, they proposed that the decrease in sea-surface temperature (SST) in the eastern equatorial Pacific (0 - 10°S) shortly after the Santa María-Soufriere-Pelée and Agung volcanic episodes, perhaps allowed the cooling induced by these eruptions to become more evident.

In the northern hemisphere, the temperature decrease following the Coseguina eruptions is one of the most impressive (Lamb 1972; Angell and Korshover 1985). However, no clear evidence of a decrease in tree-ring indices is observed after the Coseguina eruption in southern South America. A possible, but debatable, explanation for this difference could be a highly unbalanced distribution of Coseguina debris between the two hemispheres. No historical El Niño event seems to have been associated with the Coseguina eruption.

The decrease in tree-ring indices possibly related to the Armagura eruption was clearest in subantarctic latitudes. A moderate El Niño event is suggested by Quinn *et al.* (1978) for the year of the Armagura volcanic episode. Finally, no clear effect in tree-ring indices is indicated following the Tarawera eruption. Kelly and Sear (1984) found a notable northern-hemisphere increase in temperature after the Tarawera eruption.

Volcanic Eruptions during the Instrumental Meteorological Period
The Santa María-Soufriere-Pelée eruptions were not followed by an obvious temperature decrease in North America and Europe (Angell and Korshover 1985), but in southern South America they produced the second most impressive tree-ring decrease. The volcanically-induced effect is more remarkable at subtropical and temperate latitudes. An unbalanced distribution of volcanic debris between the two hemispheres probably caused this.

Legrand and Delmas (1987) assigned the Nilahue eruption the weakest of the nine peaks in the acidity profile at Dome C, Antarctica. High-latitude eruptions (>40°) usually disperse aerosols over only a portion of the hemisphere, and have much less global impact than near-equatorial eruptions (Rampino and Self 1982). A local, rather than a hemispheric cooling, would be associated with the Nilahue eruption.

The Agung eruption of 1963 produced the most consistent and well-defined tree-ring decrease in all four latitudinal regions considered in this study. Analyses of magmatic volatiles indicate that the Agung eruption was proportionately richer in SO_2, and Cl than either Tambora or Krakatau (Rampino and Self 1982). The sulphur amount is considered to be a more critical measure of climatic significance. Besides, Legrand and Delmas (1987) suggest an approximate two-thirds and one-third distribution of Agung debris between the southern and northern hemispheres, respectively. Between 30° and 60°S, a temperature decrease of 0.64°C from 1962 to 1964 was indicated by Rampino and Self (1982) in connection with a very large optical depth change in the southern hemisphere - 10 times bigger than in the northern hemisphere. All these facts are consistent with the observed tree-ring decrease.

A reasonable surface coverage of meteorological information in South America is available since 1900. In order to verify the volcanically-induced effect on tree growth of the Santa María-Soufriere-Pelée, Nilahue and Agung eruptions, three regional temperature records were obtained for subtropical, temperate and subantarctic regions of southern South America. Monthly temperature records at Jujuy (24.5°S), Salta (24.8°S), Rivadavia (24.3°S), Tucumán (26.8°S), Villa Nogués (26.9°S) and Tafí del Valle (26.9°) stations were grouped into a subtropical temperature record. Likewise, Chos Malal (37.4°S), Cipolleti (39.0°S), Valdivia (39.9°S), Bariloche (41.2°S), Maquinchao (41.3°S) and Esquel (42.2°S) stations were joined into a temperate record, and Lago Argentino (50.3°S), Rio Gallegos (51.6°S), Punta Arenas (53.3°S) and Ushuaia (54.9°S) were grouped into a subantarctic record.

Table 4 presents the temperature differences three years before and after the Santa María-Soufriere-Pelée, Nilahue and Agung eruptions, and their significance for the subtropical, temperate and subantarctic temperature records. According to the response and correlation functions, in the subtropical and temperate latitudes the summer (November to April) temperature difference was considered. On the other hand, in the subantarctic region the winter (May to October) temperature difference was evaluated. A decrease in temperature is observed following all the eruptions, statistically significant for Agung in all latitudes, for Nilahue in subtropical and subantarctic latitudes, and for Santa María-Soufriere-Pelée eruptions in subtropical latitudes.

These results agree with the previous tree-ring analysis, showing the consistent decrease of tree-growth in all southern South American latitudes after the Agung eruption, the more local after-effect of the Nilahue volcanic episodes, and the larger impact of the Santa María-Soufriere-Pelée event in subtropical latitudes.

Table 4: Seasonal Temperature Differences (°C) between the Three-Year Intervals Immediately before and after Volcanic Eruptions Occurring in the Subtropical (24° to 27°S), Middle (37° to 42°S), and High (50° to 55°S) Latitudes of South America.[1]

	Santa María	Nilahue	Agung
Subtropic (summer)	-1.39*	-0.96***	-0.76***
Mid-latitude (summer)	-0.41	-0.31	-0.61*
High-latitude (winter)	-0.57	-0.01	-0.67*

[1] Based on Student's t-test, differences at the 10%, 5% and 1% levels are indicated by one, two and three asterisks, respectively.

Conclusions

The main findings from this study of tree-ring changes in southern South America following the major volcanic episodes between 1750 and 1970 are:

1. The magnitude, the duration and the geographical extension of the volcanically-induced tree-ring decreases in southern South American chronologies are related to the type of eruption (concentration of sulphur gases), as well as to the hemispheric distribution of debris.

2. Based on statistical tests, an increase in tree-ring indices after volcanic eruptions is significantly less probable than a decrease. We also find that a large volcanic episode is not always related to a uniform decrease in temperature in all regions.

3. It seems that tropical and extratropical warming in South America associated with El Niño events tends to mask the cooling due to volcanic eruptions. El Niño events in 1814 and 1817 (Quinn *et al.* 1978) bracketing the Tambora eruption, as well as the strong El Niño event in 1884 following the Krakatau eruption, may have moderated the aftereffects of these eruptions.

Acknowledgements

This research was supported by CONICET-Argentina Grant number 2309/87-003. We are grateful to M.A. Soler for editing the English version.

References

Aceituno, P. 1987. On the interannual variability of South American climate and the Southern Oscillation. Ph.D. dissertation, University of Wisconsin, Madison. 128 pp.

Angell, J.K. and J. Korshover. 1985. Surface temperature changes following the six major volcanic episodes between 1780 and 1980. *Journal of Climate and Applied Meteorology* 24:937-951.

Blasing, T.J., A.S. Solomon and D.N. Duvick. 1984. Response functions revisited. *Tree-Ring Bulletin* 44:1-17.

Boninsegna, J.A. and R.L. Holmes. 1985. *Fitzroya cupressoides* yields 1534-year long South American chronology. *Tree-Ring Bulletin* 45:37-42.

Boninsegna, J.A., J.M. Keegan, G.C. Jacoby, R. D'Arrigo and R.L. Holmes. (in press). Dendrochronological studies in Tierra del Fuego, Argentina. (Submitted for publication in *Quaternary of South America and Antarctic Peninsula*.)

Dunn, O.J. and V.A. Clark. 1974. *Applied Statistics: Analysis of Variance and Regression*. John Wiley & Sons, New York. 387 pp.

Fritts, H.C. 1976. *Tree Rings and Climate*. Academic Press, London. 567 pp.

Kelly, P.M. and C.B. Sear. 1984. Climatic impact of explosive volcanic eruptions. *Nature* 311:740-743.

Kienast, F. and F.H. Schweingruber. 1986. Dendroecological studies in the Front Range, Colorado, U.S.A. *Arctic and Alpine Research* 18:277-288.

LaMarche, V.C., R.L. Holmes, P.W. Dunwiddie and L.G. Drew. 1979a. *Tree-Ring Chronologies of the Southern Hemisphere*. Volume 1: *Argentina*. Chronology Series V. University of Arizona, Tucson.

_____. 1979b. *Tree-Ring Chronologies of the Southern Hemisphere.* Volume 2: *Chile.* Chronology Series V. University of Arizona, Tucson.

Lamb, H.H. 1970. Volcanic dust in the atmosphere: with a chronology and assessment of its meteorological significance. *Philosophical Transactions of the Royal Society of London* A266:425-533.

_____. 1972. *Climate: Present, Past and Future.* Volume 1: *Fundamentals and Climate Now.* Methuen, London. 613 pp.

Legrand, M. and P.J. Delmas. 1987. A 220-year continuous record of volcanic H_2SO_4 in the Antarctic Ice Sheet. *Nature* 327:671-676.

Lough, J.M. and H.C. Fritts. 1985. The Southern Oscillation and tree rings: 1600-1961. *Journal of Climate and Applied Meteorology* 24:952-965.

Mass, C. and S.H. Schneider. 1977. Statistical evidence on the influence of sunspots and volcanic dust on long-term temperature records. *Journal of Atmospheric Sciences* 34:1995-2004.

Quinn, W.H., D.O. Zopf, K.S. Short and R.T. Kou Yang. 1978. Historical trends and statistics of the Southern Oscillation, El Niño, and Indonesia droughts. *Fishery Bulletin* 76:663-678.

Rampino, M.R. and S. Self. 1982. Historic eruptions of Tambora (1815), Krakatau (1883), and Agung (1963), their stratospheric aerosols and climatic impact. *Quaternary Research* 18:127-143.

Simkin, T., L. Siebert, L. McClelland, D. Bridge, C. Newhall and J.H. Latter. 1981. *Volcanoes of the World. A Regional Directory, Gazetteer, and Chronology of Volcanism during the Last 10,000 Years.* Hutchinson Ross Publishing Co., Stroudsburg, Pennsylvania. 240 pp.

Stothers, R.B. 1984. The great Tambora eruption in 1815 and its aftermath. *Science* 224:1194-1198.

Taylor, B.L., T. Gal-Chen and S.H. Schneider. 1980. Volcanic eruptions and long-term temperature records: An empirical search for cause and effect. *Quarterly Journal of the Royal Meteorological Society* 106:175-199.

Villalba, R. (in press). Climate fluctuations in northern Patagonia during the last 1000 years as inferred from tree-ring records. (Submitted to *Quaternary Research*).

Villalba, R., J.A. Boninsegna and D.R. Cobos. (in preparation). A tree-ring reconstruction of summer temperature between A.D. 1500 and 1974 in western Argentina. Third International Conference on Southern Hemisphere Meteorology and Oceanography. Buenos Aires, Argentina. November, 1989.

Villalba, R., J.A. Boninsegna and R.L. Holmes. 1985. *Cedrela angustifolia* and *Juglans australis*: two tropical species useful in dendrochronology. *Tree Ring Bulletin* 45:25-35.

Villalba, R., J.A. Boninsegna and A. Ripalta. 1987. Climate, tree-growth and site conditions in subtropical northwest Argentina. *Canadian Journal of Forest Research* 17:1527-1539.

Tree-Ring Chronologies from Endemic Australian and New Zealand Conifers 1800-30

Jonathan Palmer[1] and John Ogden[1]

Abstract

Tree-ring chronologies derived from endemic conifers in Australia and New Zealand covering the period 1800-30 are presented and discussed in the light of the few climatic reconstructions based upon them. The majority of New Zealand chronologies show narrow rings in 1817, the third summer after the eruption of Mount Tambora. The latter part of the 1810-20 decade, and the 1820s, appear to have been characterized by several cooler-than-average summers. However seen in the context of the full chronologies, the patterns for this period are not exceptional, and it is concluded that additional comparative studies of ring-width patterns following other major eruptions are required before safe conclusions can be drawn.

Introduction

Dendroclimatology is the study of past climate as reflected in the annual growth rings of trees. Different tree species have different climatic responses and different lags. Various statistical techniques are available for establishing which climatic parameters have the strongest influence on ring widths, and for reconstructing past climates (Fritts 1976). Our simple premise in this paper is that extremely narrow rings common to several chronologies from one species over most of its distribution are likely to reflect some aspect of the macroclimate. Narrow rings may reflect cold cloudy summers for some species (e.g., those at timberline), or warm dry periods for others (e.g., lowland mesic species). However, if such patterns occur in several species they suggest that exceptional conditions characterized the years in question, and imply that poor growing conditions were widespread.

We present plots of standardized tree ring widths (indices) for the period 1800-30 from published and unpublished chronologies from endemic conifers in New Zealand and Tasmania. (LaMarche *et al.* 1979a, b; Ogden 1982; Ahmed and Ogden 1985; Norton and Ogden 1987). The plots are averages from different numbers of chronologies for different species.

For each species all chronologies covering the relevant time period have been included. A few conifers with < 3 chronologies available have been excluded. New Zealand chronologies derived from *Nothofagus* species are discussed elsewhere (Norton, this volume).

The Tambora eruption occurred in April 1815 (Lamb 1970). In the southern hemisphere the tree-growth period (spring-summer) overlaps two calendar years. By convention a growth ring is designated by the year in which growth commenced. Thus "1815" refers to growth occurring over the period October 1815 to May 1816 (approximately), "1816" to the same period in 1816-17, etc.

---

[1] Department of Botany, Auckland University, Auckland, New Zealand.

Results

1. The geographical coverage of chronologies is good for the cool and warm temperate zones (New Zealand, North and South Islands, Tasmania) but there are no suitable chronologies from mainland Australia or the tropical regions (Figures 1, 2).

2. The main features of the chronologies in the period around 1815 are summarized in Table 1. Average plots are presented in Figure 2.

3. The majority of New Zealand tree-ring chronologies show below-average ring widths in 1817. This represents the summer of 1817-18; the third summer following the eruption. There was a slight recovery in 1818, further depression in 1819, and sometimes for a few years longer.

4. In the case of *Agathis australis* in northern New Zealand, narrow rings such as 1817-18 may reflect a cold prior winter and/or wet overcast conditions in spring (September-October 1917). (Response function analyses; Ogden and Ahmed 1988).

5. The Tasmanian chronologies show reduced growth in 1814 - i.e., before the April 1815 eruption of Tambora. *Phyllocladus aspleniifolius* shows reduced growth also in the summer 1816-17. Response functions for this species suggest a cold winter in 1816 and/or a warm summer in 1815-16. Dendroclimatic temperature reconstructions for Tasmania indicate below-average summer temperatures for the 1816-20 period (LaMarche and Pittock 1982).

6. Summer temperature reconstructions derived from the New Zealand *Phyllocladus* chronologies (Palmer, unpublished results) agree closely with those derived from *Nothofagus* by Norton *et al.* (1989). Norton (this volume) tentatively concludes that the period 1810-25 was cooler than average. Cool summers occurred in 1818, 1820 and 1822. The coolest year in the 30-year period under discussion was 1829. 1814 was a warm summer.

Table 1: Summers with Poor Growth in Tasmanian and New Zealand Tree-Ring Chronologies.[1]

Species	n		Summers with Poor Growth
Agathis australis	9	NZ	1817-18 but poorer 1819-20
Libocedrus bidwillii	8	NZ	1816-17/17-18; declining until 1823
Phyllocladus trichomanoides	5	NZ	1819-20. (slight drop 1817-18)
Phyllocladus glaucus	4	NZ	1817-18; 1819-20
Phyllocladus aspleniifolius	10	T	1814-15; 1816-17
Athrotaxis cupressoides	3	T	1814-15; low 1811-21

[1] n = number of chronologies averaged; NZ = New Zealand T = Tasmania.

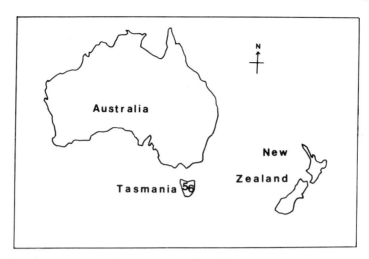

Figure 1a: Map showing the geographical relationships of the region discussed. 5, location of *Athrotaxis cupressoides* chronologies. 6, *Phyllocladus aspleniifolius* chronologies.

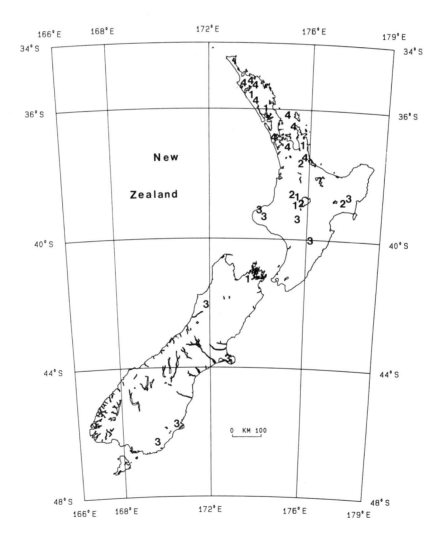

Figure 1b: Map of New Zealand showing main chronology-site locations. 1, *Phyllocladus trichomanoides*. 2, *Phyllocladus glaucus*. 3, *Libocedrus bidwillii*. 4, *Agathis australis*.

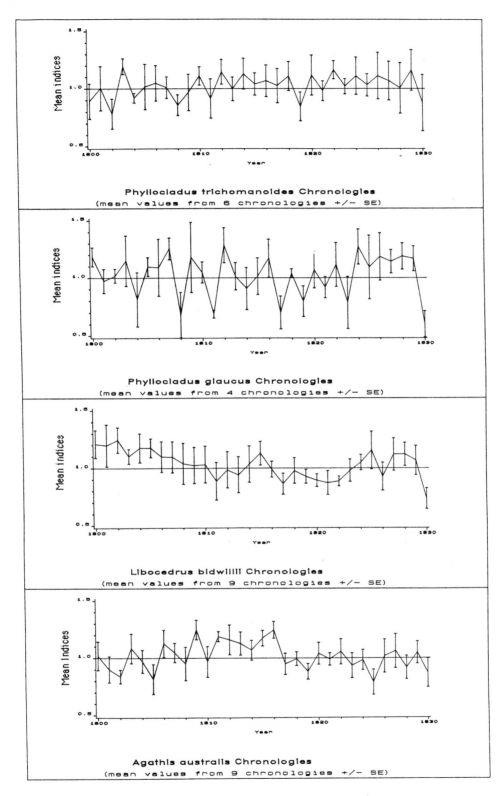

Figure 2: Average standardized chronology plots for six tree species mentioned in text.

Figure 2: Cont'd.

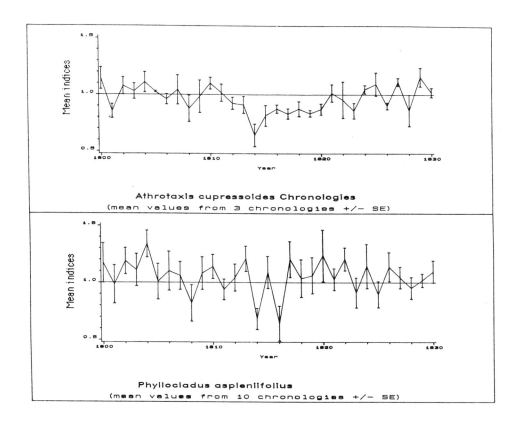

Discussion

The summer of 1817-18 shows marked ring-width reduction in many tree-ring chronologies from New Zealand, but not Tasmania. Temperature reconstructions imply it was a colder-than-average summer. Colder summers occurred in Tasmania also in the latter part of the 1810-20 decade. The 1817 narrow ring represents the third austral growing season after the April 1815 eruption of Mount Tambora. If it is even indirectly a consequence of this eruption, then the general lack of a marked effect in the two previous growing seasons requires explanation. Moreover, the narrow 1814 ring in the Tasmanian chronologies indicates a need for caution in concluding that the austral ring-width patterns for this period reflect the influence of distant eruptions. The results are suggestive, but additional comparative studies of ring-width patterns following other major eruptions are required before safe conclusions can be drawn.

References

Ahmed, M. and J. Ogden. 1985. Modern tree-ring chronologies 3. *Agathis australis* (Salisb) - Kauri. *Tree-Ring Bulletin* 45:11-24.

Fritts, H.C. 1976. *Tree-Rings and Climate*. Academic Press, London. 567 pp.

LaMarche, V.C. Jr., R.L. Holmes, P.W. Dunwiddie and I.G. Drew. 1979a. *Tree-Ring Chronologies of the Southern Hemisphere 4. Australia*. Laboratory of Tree-Ring Research, University of Arizona, Tucson, Arizona.

_____. 1979b. *Tree-Ring Chronologies of the Southern Hemisphere 3. New Zealand*. Laboratory of Tree-Ring Research, University of Arizona, Tucson, Arizona.

LaMarche, V.C. Jr. and A.B. Pittock. 1982. Preliminary temperature reconstructions for Tasmania. *In: Climate from Tree Rings*. M.K. Hughes, P.M. Kelly, J.R. Pilcher and V.C. LaMarche Jr. (eds.). Cambridge University Press, Cambridge. pp. 177-185.

Lamb, H.H. 1970. Volcanic dust in the atmosphere; with a chronology and assessment of its meteorological significance. *Philosophical Transactions of the Royal Society*. A226:425-524.

Norton, D.A., K.R. Briffa and M.J. Salinger. 1989. Reconstruction of New Zealand summer temperatures back to 1730 A.D. using dendroclimatic techniques. *International Journal of Climatology* 9:633-644.

Norton, D.A. and J. Ogden. 1987. Dendrochronology: a review with emphasis on New Zealand applications. *New Zealand Journal of Ecology* 10:77-95.

Ogden, J. 1982. Australasia. *In: Climate from Tree-Rings*. M.K. Hughes, P.M. Kelly, J.R. Pilcher and V.C. LaMarche Jr. (eds.). Cambridge University Press, Cambridge. pp. 90-103.

Ogden, J. and M. Ahmed. 1988. Climate response function analyses of Kauri (*Agathis australis*) tree-ring chronologies in northern New Zealand. *Journal of the Royal Society of New Zealand*. (in press).

New Zealand Temperatures, 1800-30

David A. Norton[1]

Abstract

Instrumental records of New Zealand temperature extend back to 1853; prior to this the only estimates of annual temperature variations come from tree rings. A tree-ring reconstruction of summer temperatures for the decade 1810-20 shows falling temperatures from a maximum of 17.5°C in 1810-11 to 14.3°C in 1817-18. Reconstructed temperature for the 1816-17 summer is 15.5°C, only 0.2°C below the 1800-30 average. 1817-18 is the second coldest reconstructed summer for the period 1750-1850; 1845-46 being the coldest (13.6°C).

Introduction

Few instrumental temperature records in New Zealand extend back before 1900 A.D.; the longest (Dunedin) starts in 1853. Consequently for information on temperature variations prior to the present century we must turn to indirect sources of climatic information, the so-called 'proxy data'. In New Zealand, however, many of the proxy data for palaeoclimatic studies have a rather coarse temporal resolution. Furthermore, the evidence is often contradictory and difficult to interpret in terms of climate (Burrows 1982).

Dendroclimatology, the study of annual variations in the widths of tree-growth rings, has been shown to be a promising tool for reconstructing annual to decadal fluctuations in palaeoclimates (Fritts 1976). In New Zealand, tree rings provide the only estimates of annual temperature variations prior to the instrumental record. In this paper, the development of a tree-ring temperature reconstruction for New Zealand is summarized and particular attention given to temperature variations between 1800 and 1830 A.D.

Tree Rings in New Zealand

Dendrochronological techniques have only been applied with success to New Zealand trees in the last 15 years (Norton and Ogden 1987). Today there are some 70 modern tree-ring chronologies. The main taxa involved are *Nothofagus* (Fagaceae), *Agathis* (Auracariaceae), *Phyllocladus* (Podocarpaceae) and *Libocedrus* (Cupressaceae). These chronologies have been developed from sites throughout New Zealand, and range in altitude from near sea-level to the alpine timberline (~1400m).

Nothofagus Chronologies

Two species of *Nothofagus* (*N. menziesii* and *N. solandri*) have been sampled extensively in South Island, with 30 chronologies, mainly from subalpine sites, developed (Norton 1983a, 1983b). Crossdating between trees and sites has been good, with some narrow growth rings being common to trees up to 400 km apart. However, some differences in growth-ring patterns also occur between the two species. The chronologies have high mean sensitivities and a considerable

---

[1] School of Forestry, University of Canterbury, Private Bag, Christchurch 1, New Zealand.

amount of common variance, which suggests that they offer good potential for palaeoclimatic reconstruction (Table 1).

Table 1: Summary of Subalpine *Nothofagus* Chronology Statistics (from Norton 1983a,b).[1]

Species	N	AC	MS	%Abs	r	Start Year
Nothofagus menziesii	5	0.42	0.31	0.42	0.35	1746
Nothofagus solandri	16	0.50	0.34	1.35	0.45	1634

[1] N, number of chronologies. AC, autocorrelation. MS, mean sensitivity. %Abs, percentage absent rings. r, mean correlation between all pairs of trees in each chronology. Start year, mean year in which chronologies start.

Response Function Analysis

The climatic usefulness of the *Nothofagus* chronologies was assessed using response function analysis (Norton 1984). Principal-components multiple-regression was used to assess the relationship between variations in annual growth ring width and temperature and precipitation variables. The regression equations explained between 37% and 64% of the variance in ring width. The most significant correlations were with temperature during December, February and March; the austral summer.

Temperature Reconstruction

Ten of the subalpine *Nothofagus* chronologies extending back to at least 1730 A.D. were used to develop reconstructions of past temperature (Norton *et al*. 1989). Because the response function analyses showed the strongest growth-climate link to be with December-March temperature, the reconstruction was for these months. A regression model where climate in year i is estimated as a function of tree growth in years i-1, i and i+1 for each chronology, was used. The reconstruction model was developed for 1916-79 and explained 55% of the temperature variance. Verification of this model over an independent period (1853-1915) accounted for 42% of the temperature variance. The result compares well with other studies, and was used to develop a reconstruction of New Zealand summer temperature back to 1730 A.D. Analysis of the spectral properties of the reconstruction indicates that relatively short-period variability (<30 years) is better reconstructed than longer-period variability (>60 years). The reconstruction provides statistically reliable estimates of annual to decadal summer temperature variations.

Temperature Variations During the Period 1800-30 A.D.

This temperature reconstruction provides the opportunity to look more closely at temperatures during 1800-30 A.D., and especially at the decade 1810-20. However, in interpreting these data, it is important to remember that 45% of the temperature variance was **not** explained in calibrating the model and 58% **not** explained in verifying it.

The reconstruction shows temperatures ranging from a high of 17.5°C in the 1810-11 summer to a low of 14.3°C in the 1817-18 summer (Figure 1). A pronounced trend of decreasing summer temperature is evident between these two years. The 1815-16 summer temperature (15.5°C) was only 0.2°C below the 1800-30 average, whereas 1817-18 (14.3°C) was the second coldest for the period 1750-1850 - only 1845-46 was colder (13.6°C).

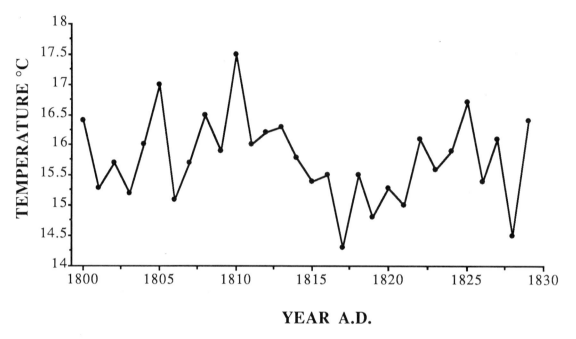

Figure 1: Reconstructed New Zealand summer temperature 1800-30 (from Norton *et al.* 1989).

These data suggest that if a temperature depression did occur in New Zealand as a result of the 1815 Tambora eruption, it was not evident until two to three years later. This result is not inconsistent with analyses of post-eruption temperature depressions (e.g., Self *et al.* 1981) which suggest that hemispheric temperature depressions may occur for periods ranging from one to five years after an eruption and that the maximum temperature depression may lag by up to three years with increasing latitude.

A closer examination of the tree-ring chronologies does however reveal a more complex story, with the two species showing differences in ring-width variations over the period 1810-20 (Figure 2). *N. solandri* ring widths reached a minimum in the 1815-16 summer whereas *N. menziesii* ring widths reached a minimum in the 1817-18 summer (although growth was still depressed in 10/14 of the *N. solandri* chronologies at this time). The reasons for this difference are uncertain. It could be that the two species responded differently to temperature then, although this is somewhat surprising given the similarities in their modern climatic response functions. Alternatively, it may be that the growth depression in 1815-16 was not temperature related. For example, it may have resulted from the occurrence of a mast seed year in *N. solandri*, but not in *N. menziesii*. Pronounced radial-growth depression is likely to occur as a result of this. However, it is not at present possible to answer this question further.

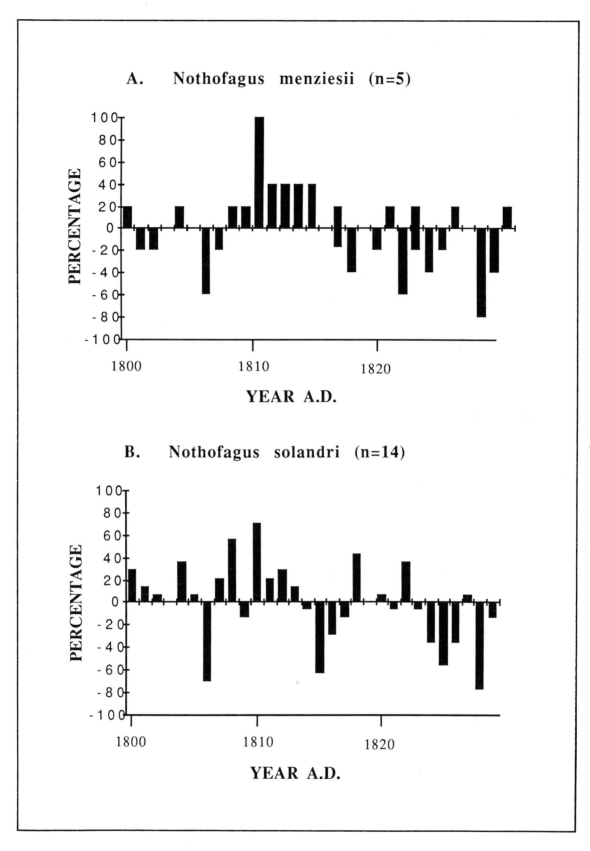

Figure 2: Percentage of chronologies with index values more than or less than one standard deviation from the mean.

Conclusions

Before any definite conclusions can be drawn on the effect of the Tambora eruption on New Zealand temperature, more detailed analyses of New Zealand palaeotemperatures are needed, especially the development of further independent reconstructions. However, it is clear that the period between 1810-25 included one of the cooler episodes between 1730 and the start of instrumental temperature records in 1853, namely 14.3°C in the summer of 1817-18.

References

Burrows, C.J. 1982. On New Zealand climate within the last 1000 years. *New Zealand Journal of Archaeology* 4:157-167.

Fritts, H.C. 1976. *Tree-Rings and Climate*. Academic Press, London.

Norton, D.A. 1983a. Modern New Zealand tree-ring chronologies. I. *Nothofagus solandri*. *Tree-Ring Bulletin* 43:1-17.

_____. 1983b. Modern New Zealand tree-ring chronologies. II. *Nothofagus menziesii*. *Tree-Ring Bulletin* 43:39-49.

_____. 1984. Tree-growth-climate relationships in subalpine *Nothofagus* forests, South Island, New Zealand. *New Zealand Journal of Botany* 22:471-481.

Norton, D.A., K.R. Briffa and M.J. Salinger. 1989. Reconstruction of New Zealand summer temperatures back to 1730 A.D. using dendroclimatic techniques. *International Journal of Climatology* 9:633-644.

Norton, D.A. and J. Ogden. 1987. Dendrochronology: a review with emphasis on New Zealand applications. *New Zealand Journal of Ecology* 10:77-95.

Self, S., M.R. Rampino and J.J. Barbera. 1981. The possible effects of large 19th and 20th century volcanic eruptions on zonal and hemispheric surface temperatures. *Journal of Volcanology and Geothermal Research* 11:41-60.

Summary

Workshop on World Climate in 1816: a Summary and Discussion of Results

C. Wilson[1]

Introduction

The aim of the Workshop was to provide a geographical perspective on global climatic events in 1816. Members of the group organizing the Workshop[2] proposed to do this by mapping direct weather and proxy-weather data from a variety of disciplines together with any other relevant material on the nature of the events, their intensity and timing. Thus, we hoped to be able to see at a glance what was currently known, and where further work was required. We also hoped that the maps might serve to stimulate discussion and speculation on the atmospheric circulation of this period (Harington 1988, p. 5).

In order to assess and compare the different kinds of information, we needed two reference periods: (1) to see the year 1816 in the context of the climate at that time; and (2) to assess the data in terms of modern climate. For the former, we chose 1809-20; the latter 1941-70. A further objective of the Workshop was to provide tabulations of the kinds, resolution and sources of data available for the period 1809-20. However, it has proven impractical to publish this mass of data. Nevertheless, such information is contained in the course of this volume and indicated in this report.

We have acted on several recommendations made during the Workshop:

1. that all data be plotted on a single world map;

2. that some indication should be given as to the origin of the different data points, and how the map evolved;

3. that years other than 1816 be discussed where relevant;

4. that the map should be refined by reference to the original presentations and data sets, as well as pertinent data from references or other reliable sources not previously considered by Workshop participants, with a view to seeing what the combined information tells us at this stage - and in the expectation of second- and third-generation maps.

The map (Figure 1) is essentially for summer 1816, in the northern hemisphere; in the southern hemisphere, where the summer extends across two calendar years, and where data are fewer and often less clear-cut, the season for each data point is indicated. The following text is an annotation to the map. Where reference is made to a paper in this volume, no date is given after the author's name.

---

[1] 90 Holmside, Gillingham, Kent ME7 4BE, United Kingdom. Manuscript received March 1991.
[2] C. Wilson, T. Ball and C.R. Harington.

Summer 1816 in North America/Greenland

For this period there are a large number of weather diaries, many with instrumental temperature records (and more rarely, barometer readings), as well as descriptions in private diaries and official documents of weather and its impact. The Hudson's Bay Company (HBC) records form the largest single archive, providing an early synoptic weather-reporting network centred on the Bay and extending eastward, southward and westward in Canada. A limitation is that coverage for these years is restricted to areas east of the Mississippi in the United States, while in Canada the outbreak of hostilities in June 1816 between the North West and Hudson's Bay companies, over the Athabaska fur trade, curtailed most reporting west of the Red River Valley until at least 1818. Since the Workshop, I have received from Michael Chenoweth[1] a copy of a weather record for the Bering Sea/Kotzebue area (July-September 1816) from the ship's log of von Kotzebue's voyage, published in 1821. Similar data are available for 1817. There are now a rapidly increasing number of tree-ring chronologies, as well as ice-core data from High Arctic Canada and Greenland and geomorphic evidence from Québec.

Western North America

Tree-ring chronologies for 65 arid sites in the **western United States** (Lough) indicate that in 1816 and 1811-20 climate was near the 1901-60 normal. There is no evidence for large-magnitude climatic anomalies. The 1816 summer appears to have been favourable for tree growth in parts of the western United States, possibly associated with moister and cooler conditions. Data also suggest that, after volcanic eruptions, the western United States may be warmer in winter and the east cooler in summer. Working with Douglas-fir chronologies on the **Colorado Plateau** (37°N), Cleaveland finds a pattern of greatly increased ring width, latewood width and density from 1815 to 1817. In 1816, the largest regional-averaged late-season growth occurred since 1487. Above-normal growth occurs when moisture stress is reduced by abnormally low growing-season temperature and/or abundant precipitation. The influence of temperature on tree growth is stronger than precipitation in the late growing season, and Cleaveland suggests that the 1815-17 seasons were prolonged by below-normal temperatures, reducing mid-summer evapotranspiration, together with normal or above-normal precipitation. This climatic effect, which reached a maximum in 1816, probably began shortly after April 1815 and persisted into the 1817 growing season. Nothing of note was found in the chronology at times of other large volcanic eruptions. From a tree-ring index for **northeastern Nevada** (40°N), Smith (1986; 1988) estimated precipitation in 1816 as over 20% greater than the 1932-82 average; this was only exceeded five times in these 50 years, and only once in the derived estimates. As trees in Nevada are known to be more sensitive to dry rather than wet conditions, Smith believes that 1816 was significantly wetter than normal. This implies that troughs were more common than usual over the Great Basin from September 1815 to August 1816.

Farther north in western Canada, tree-ring indices from treeline (temperature-sensitive) sites in and near **Banff and Jasper National parks** (51°-53°N) (Luckman and Colenutt) show no evidence of a summer temperature anomaly associated with Tambora - such as signalled by narrow or light latewood marker rings in northern Québec and Europe. The records do show a sharp decrease in ring width during 1810-20, following a period of relatively high growth

---

[1] M. Chenoweth: Suslo Unit Box 8c, APO New York, New York 09125-5000 (current residence: 8 Gallops Lane, Prestbury, Cheltenham GL52 5SD, United Kingdom) has also provided information for New Orleans, Mississippi and weather extracts for St. Louis, Missouri. An extract from his text concerning the availability of ship's logs is included in the Appendix.

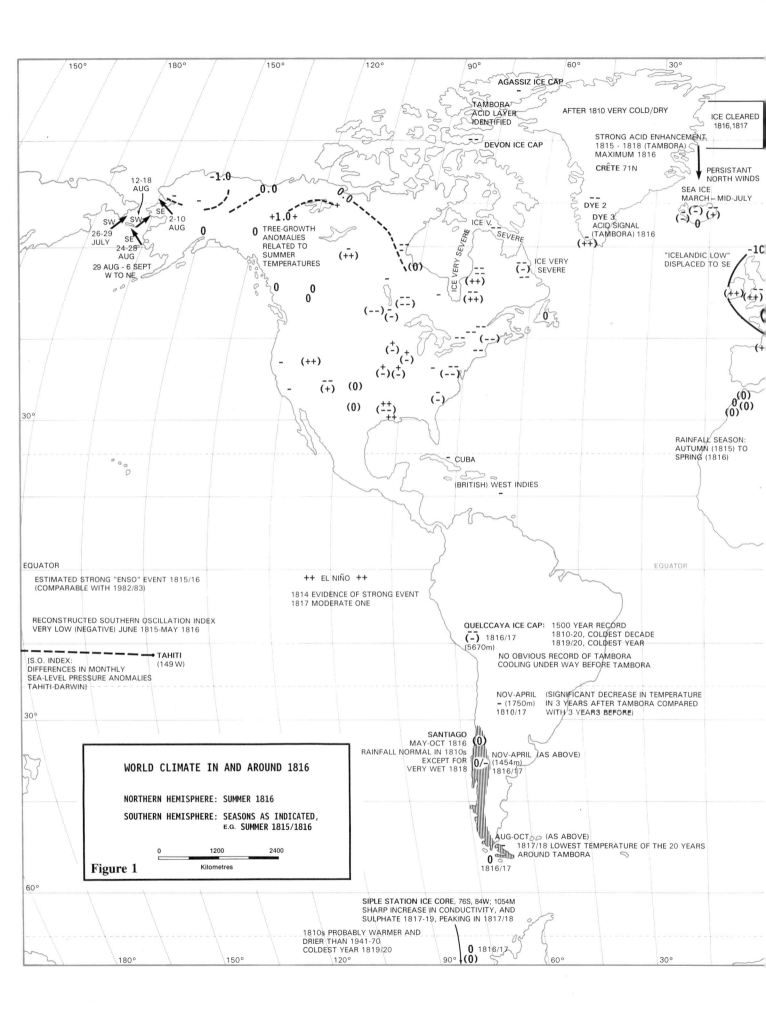

Figure 1. World Climate in and around 1816.
Northern Hemisphere: Summer 1816
Southern Hemisphere: Seasons as indicated, e.g. Summer 1815/1816

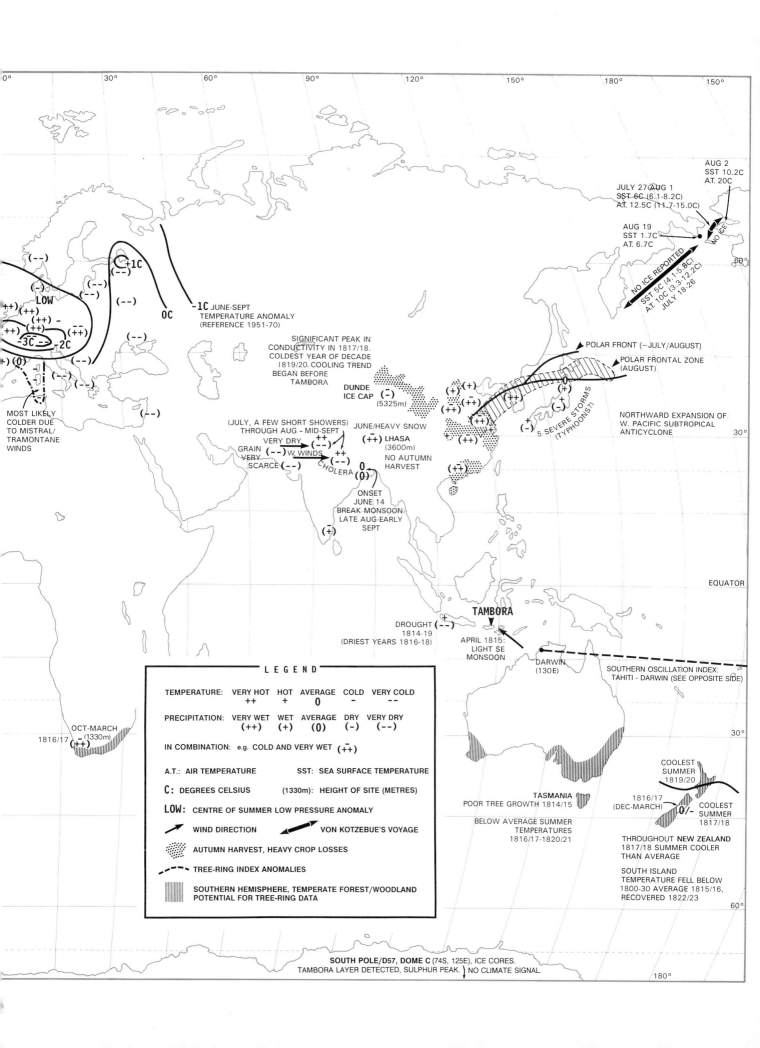

apparently associated with an abrupt deterioration of climate several years before Tambora. If anything, 1816 may have been a little warmer than adjacent years. This view agrees with that of Parker and Geast (1988), who used tree-ring maximum density indices as indicators of August temperature for sites from the Rocky Mountains to Hudson Bay, and the United States border to the northern treeline. They found that, apart from the eastern Hudson Bay coast, the magnitude and extent of the low temperatures in 1816 were not as extreme or extensive as in some other years. Generally, **central and southern Alberta**, and **British Columbia** (Jozsa) appear to have had a near-normal summer in 1816.

White spruce tree-ring width data (24 chronologies) along the northern treeline of the boreal forest (Jacoby and D'Arrigo) offer estimates of summer temperatures for **Alaska and northwestern Canada**, as well as farther eastward. For 1816, a clear pattern emerges indicating that the unusual cold in the east was matched by a warmer season over the north in western Canada, normal temperatures in southeastern Alaska and a colder summer in northern Alaska (Brooks Range). Jacoby and D'Arrigo have mapped the indices in sequence from 1805 to 1824. More severe cold does not reach western Canada and eastern Alaska until 1817. In 1818, the coldest conditions occur in the Northwest Territories. Thus after Tambora, extreme cold affected all of the study area at different times until 1818. Von Kotzebue's (1821) weather entries in the ship's log from 26 July to 14 September 1816 point to a mild summer in **southwestern Alaska and the Bering Sea**. During this time he sailed from St. Lawrence Island, (63°N), in the Bering Sea to Kotzebue Sound, (66°N) (3-14 August), back to St. Lawrence Island (20-30 August), then on to Dutch Harbour in the Aleutians (54°N) (7-14 September). A striking feature of the record is the persistence of SE to SW winds from 26 July, when the barometer was reported high, to 28 August. This period was followed by persistent W to NE winds, predominantly northerly, until at least 6 September. This indicates that the frontal zone lay to the northwest in late July through August, shifting rapidly southeastward at the end of the month. Near the mouth of Kotzebue Sound on 2 August, air temperature was measured at 20°C, and sea temperature at 10.2°C, with light S/SE winds. On 8 August, there was a violent storm from the SE, with the barometer as high as 1016 mb. At St. Lawrence Island, just before the change in regime, there was another violent storm with strong SE winds, but in this case the barometer read 972 mb. No ice was seen in Kotzebue Sound, and none was reported on the entire journey from Kamchatka. Sea-surface temperature ranged from 8°-10°C in the Sound, with corresponding air temperatures between 14.4° and 20°C. No observations were taken when the ship was at anchor. The lowest sea temperature was 1.7°C on 19 August, just north of St. Lawrence Island, with an air temperature of 6.7°C.

Eastern North America and Greenland
Over the eastern half of the continent, evidence from all sources combines to show the effects of the now well-documented outbreaks of Arctic air, in the negative-anomaly pattern from west of Hudson Bay to the Atlantic coast. From our present knowledge, the greatest lowering of temperature was along the **southeastern coast of Hudson/James Bay**. Here, severe ice conditions through the summer, a very high frequency of onshore west-to-northerly winds and arctic airmasses were contributing factors in suppressing the monthly mean summer temperatures to arctic levels, near or beyond the limits for tree growth.

For Great Whale/Fort George and Eastmain (55° - 52°N), weather diaries including instrumental temperature records were kept at the Hudson's Bay Company posts from 1814 to 1821. These readings, which have been calibrated with respect to the modern record and are amply supported by other weather and weather-related information in the Post Journals, indicate a mean temperature anomaly of about -5°C for June to August 1816 at Great Whale (now Kuujjuarapik); the July anomaly approached -6°C (Wilson). In real terms, this effectively reduced the summer temperature to 4.1°C, and the July temperature to 4.7°C from a 1941-70 average of 10.6°C. These low temperatures were associated with a higher frequency of precipitation on this coast, much of it falling as snow even in August, which is unprecedented in the modern record. In 1817, summer temperature was fractionally higher at 4.4°C, although spring and fall were colder than in 1816. Late summer rainfall was heavy at Fort George in 1817. The May-to-October temperatures had begun to decline by the beginning of the 1810 decade from levels close to those of today, then dropped abruptly in 1816 and 1817. The period of maximum cold lasted from October 1815 to March 1818. Values were not to recover entirely until the 1870s. Similar results for this region were arrived at independently by Jacoby (see also, Jacoby et al. 1988) and by Parker et al. (1981) in their tree-ring studies. Filion et al. 1988 identified two light (marker) rings in black spruce northeast of Great Whale during 1816 and 1817, indicating that climatic conditions were inimical to growth (cf. Filion et al. 1986); they also found geomorphic evidence of more humid conditions in landslide activity in the Grande Rivière de la Baleine estuary from 1818. Guiot's findings using a combination of historical records and other proxy data, are also very similar for this area at this time, with the exception of 1817, where no cold signal is found.

The **Québec-Labrador Peninsula** was greatly affected by ice along its western, northern and northeastern coasts through the summer months of 1816. Catchpole's study of sea and river-estuary ice from HBC ships' logs (see also, Faurer) and Post Journals indicates severe ice conditions in **Hudson Strait** in 1816 (ranked 1 most severe, in the 1751-1870 series), in **eastern Hudson Bay** in 1816 and 1817 (ranked 7 and 8 - 1813 was 2) and in **western Hudson Bay** in 1815 (ranked 6). The 1816 season may never have been completely ice-free in **northern James Bay/southeastern Hudson Bay** (Wilson 1985a). In 1817, first-breaking of the river ice in spring was very late at Fort George and Eastmain, Churchill, Albany and Moose, and first partial freezing in autumn was very early at Fort George and Eastmain. For the 1810 decade, river- and sea-ice data show that the cold was first apparent in 1811 and 1812, but the most severe summer cold in that decade occurred in the two years following the Tambora eruption. Catchpole and Hanuta (1989) found the number of concurrences between years with severe ice in Hudson Strait and major eruptions significant at the 99.5% level; in the western part of the Bay the significance was 95%, whereas those years with severe ice in eastern Hudson Bay did not coincide with major eruptions. The coastal ice in the **Labrador Sea** was also unusual in summer 1816, as demonstrated by records and ships' logs of the Moravian Mission on the Labrador coast (Newell). Off northern Labrador/Hudson Strait, it was unlikely that it cleared in the 1816 season. There is evidence of an increased frequency of northerly winds here during summer. In **northeastern Labrador**, spring 1816 had been cold and late at Okkak (57°N), and the whole summer cold and dry. An indication of the magnitude of the cold is also shown by Jacoby and D'Arrigo's tree-ring indices for the area. Such cold, dry summers are usually associated with N to NW winds and lower pressure. There is no information as yet concerning conditions on the **Québec/Labrador plateau**. However, along the eastern coast of Hudson Bay/northern James Bay and in the immediate hinterland, 1816 and 1817 came closer than any others on record to survival of extensive areas of snow and drift through the summer (Wilson 1985a). During the coldest summer on modern record along this coast (1965), negative-temperature anomalies at Great Whale closely paralleled those in the interior at Schefferville, and at Nitchequon, where the median height of the freezing level in July was 600 to 700 m below the decadal average. While there is danger in extrapolating beyond the coastal zone, in July 1816, freezing level could well have been 800 to 900 m lower over the plateau. More extensive areas of snowcover than we know today may have lingered through much of the summer.

In contrast, Newell found no evidence of a cold season in spring or summer 1816 in **southeastern Newfoundland**, either in a weather diary for Trinity (49°N), or from fishing records. In June, there was a higher frequency of SW winds than is normal today. From modern records, these winds are indicators of warmer weather. This suggests normal or above-normal temperatures for the region in 1816. Moravian Mission reports from **southwestern Greenland** (60°N) describe a gloomier season (Newell) - one of almost incessant rain and persistent overcast. Combining the evidence for Labrador, Newfoundland and Greenland, Newell proposes that, in spring and summer, a mean centre of abnormally low pressure (sea level) was located over the Labrador Sea southwest of Greenland, with a trough extending northward into Davis Strait. The main storm tracks crossed southern Labrador into the Labrador Sea, while southeastern Newfoundland lay to the south and was nudged by the northward extension of the Atlantic Subtropical High. Mean sea-level pressure for June 1969, 1971, 1972, 1978 and 1986 offer modern analogues.

Vibe's (1967) work on climatic fluctuations in and around Greenland, based on the varying distribution of a number of animals (trading records) suggests a milder, wetter period in west- and east-central Greenland ending about 1810, followed by a very cold, dry sequence in **northwestern, northern and northeastern Greenland** until about 1860. During the latter period, evidently the ice stagnated in Arctic waters and drift was at a minimum. In Davis Strait, ice carried by the East Greenland current probably didn't advance far to the north. After 1817, most drift ice stayed north of Baffin Bay, which was navigable, and the main whaling took place in parts of the North-West Passage. Dunbar[1] noted during the Meeting (see also Dunbar 1985), with respect to the **Baffin Bay** area, that through the early nineteenth century, whalers had been sailing farther north. In 1817 the North Water was found. The first of a series of British Naval Expeditions set sail for Arctic Canada in 1818. In 1819, ships reached **Melville Island** (110°W). Scoresby (1817) reported to the Royal Society that the ice off northeastern Greenland between 74° and 80°N and from the Greenwich Meridian to beyond 10°W had entirely disappeared in 1816 and 1817, and he could probably have reached the coast of Greenland. He had never before been able to penetrate west of the Greenwich Meridian in this sector.

The Tambora acid layer has been identified in ice cores at Crête (71°N), Dye 2 (67°N) and Dye 3 (65°N) sites on the **Greenland Ice Cap** (Hammer *et al.* 1980), and ice cores from the Agassiz (81°N) and Devon Island ice caps (75°N) in the **Canadian Arctic Islands** Alt *et al.*). At Crête, at strong signal occurred from 1815 to 1818, with a maximum acidity enhancement in 1816. In the Canadian ice cores, the major acid layers are correlated with those in the absolutely-dated Dye 3 core, where the Tambora volcanic signal appears in 1816. From the Agassiz core, temperatures in summer 1816 were near or somewhat below the 1941-70 normal, with a rise to above normal in 1817. Combined Devon cores (five-year averages) show a very cold period from 1810-55; the fall in temperature began about 1800 and reached the lowest value for 200 years in the 1830s. At Dye 2 in Greenland, the 10-year averages similarly indicate that Tambora occurred during a well-established cooling trend. From synoptic studies, Alt *et al.* conclude that the circulation pattern for the summer containing the Tambora signal (1816) is best represented by the 1972 analogue, and that this period of the Little Ice Age may have been dominated by summers like that of 1972. A long, narrow vortex at 500 mb (50kPa) extends from the Siberian side of the central Arctic Ocean across the North Pole deep into Labrador-Ungava and is held tight against Greenland by a strong ridge of high pressure in the Alaska-Beaufort Sea area. The vortex is deeper and narrower than the 1948-78 mean, and is shifted eastward by the ridge over

---

[1] Moira Dunbar, until her retirement a sea-ice expert with the Defence Research Board of Canada.

the Beaufort Sea. This results in a strong, cold northwesterly flow over the Canadian Arctic Islands, with frequent light precipitation and very little melt on the ice caps. The circulation was stronger than usual and distinctly meridional. In the light of the evidence now compiled for the middle and high latitudes of the northern hemisphere in summer 1816, the circumpolar circulation patterns in 1972 (Alt *et al*. Figures 11, 12 and relevant text) provide a useful analogue from a hemispheric point of view. I shall refer to them in the Summary and Conclusions.

From the work of Ball (fur-trade records), Parker and Geast (1988), Jacoby and D'Arrigo (tree-rings) and Guiot (1986) (a combination of both), any relative lowering of summer temperature to the west of Hudson Bay (**Manitoba, Saskatchewan, northeastern Alberta and Keewatin**) was considerably less than to the east of the Bay. A possible exception is Churchill (59°N) on the west Bay coast, where Fayle *et al*., after a detailed study of tree growth in open-forest and forest-tundra sites, conclude that growing-season temperature in 1816 on the more exposed forest-tundra may have been closer to 5°C than the normal 10°C. Generally, they found a decline in growth from about late 1815-16 which persisted for one to three years, with some recovery in late 1817. The growth of the larger trees in the open forest had been declining for over two decades prior to 1816 (Ball found 1800-10 mild at Fort Churchill and York Factory). Guiot's temperature index for summer 1816 at Churchill also suggests marked cooling, although not nearly to the same extent as across the Bay. On the other hand, Jacoby and D'Arrigo's results indicate near-normal conditions. Perhaps this a reminder of the importance of microclimate in the subarctic environment. Certainly it indicates the importance of cross-checking such climatic proxy data where possible.

Southward and southeastward of the Bay, **Ontario, southern Québec and the northeastern United States** lay at times under the influence of unseasonably strong N to NW winds, further chilled by passage over the lingering Bay ice. At least three major cold outbreaks have been chronicled (Baron; Milham 1924; Hoyt 1958). The most famous (5-10 June) brought killing frosts and snow to the St. Lawrence Valley, New England and New York State; nightly frosts were reported in Pittsburgh, and a severe frost in Trenton, New Jersey. During the second (5-9 July), widespread frosts occurred in New England, and below-normal temperatures at least as far south as Philadelphia. During the third cold wave (20-22 August), snow was seen on the mountains in New Hampshire, and there were hard frosts in many localities throughout the region including Trenton and Philadelphia. The effects of these cold episodes on truncating the growing season were felt throughout the region, and were disastrous in those areas climatically marginal for farming (Post 1977; Stommel and Stommel 1983). Other general characteristics of the regional climate in summer 1816 include: (1) a very cold late spring prior to the cold wave in June; (2) mean temperatures of spring and summer months consecutively well-below normal; (3) generally cooler weather with few really warm episodes, and unusually cool nights; (4) a cold early autumn.

Looking briefly across the region: in **northern Ontario, at Osnaburgh** (51°N), Suffling and Fritz describe (from HBC Journals) the extremes of weather from August 1815 to the end of winter 1817. The dry hot summer of 1815 was terminated abruptly on 22 August by a change to cold, wet stormy weather with NW and E winds heralding the early arrival of the geese. After a very cold dry winter 1815-16, spring was two to three weeks late. The gardens were badly injured by the very cold weather and hard frosts of 4-6 and 23-27 June. From July to mid-August 1816 it was cool and rainy with mostly NW winds. On 18 August snow fell - unheard of for this month - and weather continued cold; many geese were flying by mid-September - a month early. By 3 October, the ground was frozen - three weeks early. Winter 1816-17 was again cold, but with heavy snow. Farther south in **Toronto** (44°N), Crowe's extension of the temperature record indicates a clearly-defined drop in July temperature of some 4°C from 1815 to 1816, the coldest July for the 1780-1870 period shown. Temperatures remained low in 1817, rising again in 1818. A period of lower temperatures had begun in 1809. In the cities of **Québec** (47°N) and **Montréal** (46°N), weather diaries and newspaper accounts indicate a very cold late spring in 1816 (Hamilton 1986). The cold wave in early June brought hard frosts, as well as significant snowfall to Québec and a lesser fall to Montréal; the snow seemed to taper off near Kingston. This was a month later than the latest measured snowfall on modern record at both cities. The St. Lawrence Valley may have escaped frosts during the July and August cold spells, but it was the coldest summer of the decade, with all months below average, and with little warm weather throughout. A number of days were close to or below long-period record low temperatures. Summer 1817 was also cold.

Was there really no summer in 1816 in New England? Baron examines the evidence for the **northeastern United States** from eight weather diaries, with instrumental temperature records, and many qualitative data to see whether folklore has exaggerated. With reference to 1790-1839, 1816 was one of a run of years (1812 to 1818) much colder than the reference mean, but not as cold as 1812 in northern New England or 1817 in the southern part. Summer (June-August) mean temperatures at **Brunswick** (44°N) and **New Haven** (41°N) gave similar results. As Milham (1924) suggested, the unique quality of summer 1816 lay in the unbroken sequence of below-normal monthly temperatures from May to September. Baron shows that in each of these months prolonged cold outbreaks had occurred. The growing season in 1816 was by far the shortest of the reference period. Baron also mentions that 1816 had one of the highest percentages of days with fair skies, which in combination with cold fronts also contributed to some low minimum temperatures. In addition, the frequency of westerly "fair-weather bearing" winds was below the reference mean in 1816. The number of days with thunderstorms was about average. Deborah Norris Logan[1], commenting near **Philadelphia** (40°N), mentions the coolness of the weather on 12 days from late June through July (as cool as mid- or late-September); fires were lit, windows kept closed, and on July 17 plants put into the greenhouse for fear of frost. She writes of strong, cold NW winds and frost during 20-22 August. The main features of this cold summer were clearly recorded as far west as **Cincinnati** (39°N) (Horstmeyer, see Appendix), probably near the midwestern limit.

To extend his sea-level pressure map for July 1816 from Europe to North America, Lamb attributed the coldness of the summer in eastern Canada and northeastern North America, in general, to the prevalence of air drawn directly from the Canadian Arctic and the closeness of a focus of cyclonic activity to Labrador, Newfoundland and off-lying waters (see also Lamb and Johnson 1966). Ludlam (1966) studied the detailed evidence for the 5-10 June cold wave and described the sequence of surface circulation patterns that would have brought such weather. More recently, Wilson has plotted weather information for **east/central Canada and the United States** for summer 1816, and sketched in daily weather maps. These preliminary maps cover 1 June to 13 July (see also Wilson 1985b). Apparently, the exceptionally cold events of early June and early July were associated with blocking highs over Hudson Bay and east of Greenland. In each case, a depression passed across the Great Lakes and northern Ontario, lost speed abruptly over Québec and developed into a large system, gradually drifting eastward toward the Labrador coast. Behind this depression, high pressure extended from the Arctic down over Hudson Bay, and very cold air was pulled unusually far south in the rear of the storms. The storm tracks in early July were unseasonably far south, and evidence suggested a much stronger

---

[1] Deborah Norris Logan's diary, Historical Society of Pennsylvania.

mid-latitude temperature gradient than usual today at this season, and vigorous north-south energy exchanges. Apparently the stalled situations in early June and July, and a third, less persistent one, in mid-June were associated with intensification of a cold surface anticyclone over the Bay ice. Today these Hudson Bay Highs are characteristic of spring (March-May). Johnson's (1948) model shows a ridge at 500 mb (50kPa) in the vicinity of 90°W and a trough down over western Canada and the United States near 112°W; the trough off the east coast is near its normal spring position of 55°W. The map evidence strongly suggests that the atmospheric circulation over eastern Canada and the northeast United States in June and the first half of July 1816 was abnormal in its timing rather than in kind - that spring had been delayed or protracted by as much as six weeks.

A southern extension of the cold in the **eastern United States** is indicated: Deborah Norris Logan reports frost in **Kentucky** (38°N) on 22, 28, 29 August 1816. Frost occurred in **South Carolina** (34°N) in mid-May, and on 29 August. A succession of cold fronts occurred during August, bringing below-normal cold and resulting in poor harvests in parts of South Carolina (Zhang *et al.*). Farther south still, in **Cuba** (20°-23°N) crop failures were attributed to unusually cold weather in 1816, as was the case in the **British West Indies**.

In contrast to the cold in the east, it appears that the **central United States** from the Gulf coast as far north as Iowa and Illinois had a very warm to warm summer in 1816. Tree rings in **Iowa, Illinois and Missouri** (44°-36°N) were well below average index in 1816 -- a year that had the second narrowest average index for these three states since 1640. This implies that it was quite hot and dry[1]. Newspaper accounts from the *Missouri Gazette* (Chenoweth 1986) add detail for the **St. Louis** area. Drought is mentioned on 1 and 15 June. Hardly any rain had fallen since 12 April. There was no further mention of drought or weather until an entry for 14 September described a sudden change in the weather on 1 September from temperatures in the 80s and 90s (°F) through much of August (near or above normal) to the 60s and 70s (°F) in September (below to well-below normal). From an instrumental-temperature record kept at **Natchez** (32°N) on the lower Mississippi River (Chenoweth 1986), summer 1816 was very warm with July close to the warmest on record for the last 175 years. It was also a very dry year. Chenoweth found further evidence of the heat in this region from a newspaper report for **New Orleans** (30°N) of 31 July 1816 quoted in *Niles Weekly Register* describing the unwonted great heat that had prevailed since 10 June. This suggests stronger-than-normal high pressure over the Gulf of Mexico.

Another extraordinary feature of the weather over North America in summer 1816 is the large-scale precipitation pattern - particularly the drought present in the entire eastern half of the continent, with the exception of the windward eastern coast of Hudson Bay. In the Prairies, Allsopp's (1977) wetness series for the **Red River Valley, Manitoba** (50°N) shows a period of sustained drought from 1816 to 1819. In the northeastern United States, the drought was another cause of the extreme hardship in summer 1816 (Baron); 1813-20 saw numerous growing-season droughts here, but they decreased markedly after 1820. The *Missouri Gazette* on 1 June also mentions the unusual drought in the east from Boston, Massachusetts to **Savannah**, Georgia (32°N). In **Texas and southern Oklahoma** (33°-36°N), tree-ring based reconstruction of annual precipitation by Blasing *et al.* (1988) and Stahle and Cleaveland (1988) show near-normal precipitation values. Evidently the western half of the continent had normal to wetter conditions.

---

[1] Unpublished work by D.N. Duvick of Iowa State University, Ames, Iowa; communicated by Malcolm Cleaveland.

Summer 1816 in Europe/Iceland

For Europe there are instrumental records of air temperature (some 46 stations), total precipitation (about 29) and station pressure. Seasonal temperature and precipitation have also been reconstructed from tree-ring width and maximum latewood density indices. In addition there is a wealth of documentary material, official and private, describing the weather and its excesses, as well as its impact on human and animal life, crops, harvests, travel, etc. In the Alps, advances and retreats of the glaciers were recorded not only in geomorphological evidence, but pictorially and in travellers' descriptions.

Briffa and Jones, after checking the homogeneity of the station-temperature series, used gridded data derived from these series to make seasonal temperature anomaly maps over **Europe** for 1816 and the 1810 decade (modern reference 1951-70), and 1816 with reference to 1810-19. Comparable maps were drawn for precipitation (modern reference 1920-60), and for 1816 seasonal anomaly maps of mean sea-level pressure (modern reference 1900-74). Based on tree-ring chronologies, a series of map reconstructions of April-September temperature for each season from 1810 to 1819 put the summer of 1816 in clearer perspective. Results indicated that the 1810 decade was one of the coldest over Europe since 1750. In 1816 all four seasons were generally cooler than today, but it was spring and above all summer (June-August) when the anomalies were most marked both with respect to the modern reference period and the 1810 decade. Exceptions were the European sector of the Soviet Union, which was as warm or warmer than today in spring, summer and autumn, and parts of the Mediterranean in spring and fall. For the most part, the low temperatures were accompanied by greater-than-average rainfall; eastern Europe, the eastern Mediterranean (see Appendix) and southern Italy were dry.

Whereas the negative temperature anomalies were greatest over Scandinavia in spring 1816, it was **west/central Europe** which was most affected in summer. The cold core was the northern alpine regions, **Switzerland and the Austrian Tirol**. The long temperature record for **Basel** (48°N) (Bider *et al.* 1959) shows 1816 as by far the coldest June-August season since the record began in 1755, until present. A fall in summer temperature had begun however in 1812; values did not recover until 1822. The other months of 1816 were also without exception very cold. Pfister describes the 1816 summer in Switzerland and its impact in detail. Precipitation was very heavy in the summer, and the snowline reached down to around 1,000 m. Cattle could not pasture on higher alpine meadows. The weather greatly affected harvests of grain, fruit and vegetables - much of which did not reach maturity before being overtaken by early autumn snowfall. June-July temperatures reconstructed from tree-ring data at the alpine timberline in the **Öztal Alps, Austria** (47°N) (Eckstein and Antõl 1981) also show the 1810 decade to be the coldest of the series (from 1740), with temperatures falling from 1812 to 1816 and not recovering until the 1820s. This period of cold springs and summers was expressed in rapid glacier advances affecting all the Alps (LeRoy Ladurie 1972). Although outside the "core" anomaly area, Bednarz and Trepińska's reconstructed temperatures from trees sampled at the alpine timberline (1,500 m) in the **Tatra Mountains of Poland** (49°N) show a striking fall in temperature after 1814, following a previous period of cool weather (1806-11) (see also Bednarz 1984). The Lwow Newspaper, commenting on the year's harvest in October 1816, noted that the inhabitants had not remembered such bad weather in over 30 years - from autumn 1815 the rains had hardly stopped. In **France**, LeRoy Ladurie (1972) relates the late wine-harvest dates to a series of cold springs and summers from 1812 to 1817, with special mention of 1816; these followed a sequence of hot summers from 1801 to 1811. Manley's (1974) **central England** temperature series marks June-August 1816 as the coldest summer on record since 1659, with temperatures

2.2°C colder than the 1931-60 average. Here again, the decline had begun by the beginning of the decade. Cool summers caused consecutive harvest failures in 1816 and 1817 in **Scotland** (Grove 1988).

In **Prague** (50°N) (Brůžek), monthly mean temperatures in 1816 were lower than the average from May to December, with the largest deviations (June to August) averaging 1.4°C. Set against the absolute extremes registered for summer months, Brůžek does not consider this unusually cold. (The author doesn't mention his reference period, or comment on the homogeneity of the series.) Farther north in the **Baltic Region**, Neumann's (see also Neumann 1990) study of air-temperature data for Copenhagen, Gothenburg, Stockholm, Trondheim, Uppsala, Vöyri and St. Petersburg (together with data on grain harvests and mortality for Denmark, Finland, Norway and Sweden) showed no clear climatic impact following the eruption of Tambora. Although the differences in the April-September mean temperature between 1816 and 1814 were negative at all stations, they were small, whereas 1812 had been a much colder season than 1816. In addition, there was no indication of noteworthy crop failures in 1816 or 1817, but there was a shortfall in 1812 leading to a partial famine in Norway. The mortality data for these countries showed relatively high values in 1812 and 1813, but not in 1816-17. In **Russia**, grain yields were also low in 1812 near Moscow and in Siberia; there was a slight dearth in 1817, but no mention of death or famine in 1816. Grain exports to England from northwestern and northern Russia increased from 1816 to 1817.

For **southwestern Europe and the western Mediterranean**, Guiot's temperatures reconstructed from a combination of tree-ring, oxygen isotope and historical data have an annual rather than a seasonal distribution, but the author suggests that over the southwest any lowering of temperature in summer 1816 would probably have been relatively small - approaching zero toward southern Spain. Summer temperatures in **Morocco** were about normal. In parts of southern France, Corsica, Sardinia and even **northern Tunisia**, perhaps summer temperatures were lower, as a result of a higher frequency of Mistral and Tramontane wind situations, in which cold northerly or northwesterly flow is funnelled into the western Mediterranean down the Rhone Valley and between the Pyrenees and the Central Massif.

Kington's July 1816 rainfall anomaly map and commentary add detail to the summer seasonal map of Briffa and Jones. In **southwestern Britain and Ireland, most of France, parts of Belgium, Holland, western Germany and Switzerland**, it exceeded twice the normal amount, while **northwestern Scotland, Orkney, Shetland, Denmark, Norway and Italy** were drier than usual. Contemporary accounts describe the incessant rain, great storms (England and Ireland), floods and damage to crops and property in these wet areas. Snow remained on some hills in Scotland until mid-July. On the other hand, in **St. Petersburg, Riga and Dantzig**, prayers for rain were being offered up, following a month of drought. In the **Crimea**, the mean thickness of the yearly mud layers in Lake Saki (45°N) (Lamb 1977) suggest that the first two decades of the nineteenth century were generally dry. Autumn-to-spring rainfall in 1816 and 1817 in **Morocco** for humid, sub-humid and semi-arid zones was slightly above the 1924-77 average (Till and Guiot 1990). These values are based on tree-ring width chronologies for 46 sites in the Rif, Middle and High Atlas. From 1809 to 1815, rainfall had been markedly below normal in all three regions; after 1817, totals were above average to the end of the decade. Stockton (1990) using six tree-ring site series suggests that 1816-20 was a period of drought in central Morocco. He notes that problems can arise with spatial resolution in drought history.

The June-to-August mean sea-level pressure anomaly map for 1816 (Briffa and Jones) shows virtually all of Europe affected by anomalously low pressure centred over northern Germany and Denmark. Increased southerly flow brought milder conditions over eastern Europe, while Britain and the rest of western Europe were affected by anomalous northerly and northwesterly flow. Lamb's earlier sea-level pressure map for July 1816 (see also Lamb and Johnson 1966) indicated the prevalence of low pressure over Britain, southern Scandinavia and western continental Europe - unusually far south for summer, and he had drawn similar conclusions. Lamb's work also suggested the probable similarity between this situation and that over northeastern North America then. Kington's analysis and classification (with Lamb) of his series of daily-weather charts over Europe for July 1816 showed the British Isles dominated by cyclonic weather types - a frequency three times the 1868-1967 average, with northwesterly types more than twice as frequent and westerly types down to a third of average. The weather over the British Isles was strongly characterized by blocking and quasi-stationary cyclonic weather systems. Rainfall amount over the British Isles is closely related to the degree of cyclonicity. This marked increase in cyclonicity indicates southeastward displacement of the Icelandic Low over the British Isles. Briffa's map also shows a marked high pressure anomaly southwest of Iceland, again reflecting Lamb's July map.

From documentary accounts for 10 different sites in **Iceland** (Ogilvie), winter 1816 was very severe, and spring severe in most places. In the north, summer was mainly cold, in the east wet, and unfavourable for at least part of the season in most districts, Autumn was variable. Iceland is normally close to the seasonal boundary of the Arctic drift ice, and its presence (most commonly along the northern, northwestern or eastern coasts) lowers both land and sea temperature; rain and mist may also be associated. During 1809-20, heavy ice years occurred in 1811, 1812 and 1817, when ice was present off the northern coast and elsewhere from January until July or August. Very little ice appeared in 1818 and 1819. The year 1816 was a moderate ice year, the sea ice affecting the northern coast from the beginning of March to mid-July, accompanied by persistent northerly winds, frost and cold air. In the twentieth century, sea ice has reached Iceland less frequently. Ogilvie concludes that 1816 was not truly "a year without a summer", considering Iceland as a whole, and in the context of the decade. Nevertheless that summer was unfavourable due to the likely impact of various combinations of excessive cold, wet or drought on the hay harvest, vegetable growing and fishing (poor sea-ice conditions).

Summer 1816 in Asia

For East Asia, we have information derived from historical diaries and other documents for: the eastern plains and basins of China; the Lhasa district of Tibet; Seoul, Korea; and for Japan covering the island of Honshu. A critical insight into climate on the Tibetan Plateau at about the 500-mb (50kPa) level is provided by the ice-core record from Dunde Ice Cap. On the Indian subcontinent, official rainfall records began in 1813 at Madras; 1817 at Bombay; and other written records provide supplementary information. Following the Workshop, I located at the India Office of the British Library a copy of a book written by James Jameson, a surgeon in Calcutta, published there in 1820 (noted and commented upon by Lamb 1982, p. 237). This contains a remarkably detailed account of the climate of Bengal and the Upper Provinces from 1815 to 1819, including a daily diary of barometer and temperature readings, as well as weather notes for Calcutta. Some of this information has been incorporated here.

China

Compared with 1941-70, the ice-core record from Dunde Ice Cap in **northern Tibet** (38°N, elevation 5,325 m) indicates that 1810-20 was colder, much drier, dustier and typical of much of the Little Ice Age (Thompson and Thompson). Cooling, underway just before Tambora, reached the lowest point of the decade in 1819-20, and may have been strengthened by that eruption; but (unlike Peru) many years during the Little Ice Age were colder. With the proximity to the Gobi Desert and dry Qaidam Basin, the dust record is dominated by local sources, and following Tambora shows no significant increase in particle concentration above the high levels of much of the Little Ice Age. However, a significant peak in conductivity (soluble dust) occurred in 1817-18. In **southern Tibet** (30°N, elevation 3,600 m) near Lhasa, precipitation was extremely high in 1816. In June, it snowed for three days and nights so that there was no autumn harvest; a large number of houses collapsed and many others were severely damaged (Zhang *et al.*).

There is a large potential database for studying past climate in China in the written sources from many levels of government, as well as in private diaries (Zhang *et al.*; Wang and Zhang 1988). Annual wetness/dryness indices have already been reconstructed from these sources and mapped for the past 500 years (cf. Huang). At many locations, data exist for precipitation and winds every two hours, offering future possibilities for refining the present results. While the archives are rich in material on precipitation and comments on unusual weather, there are fewer reports of relative temperature as such.

For the lowlands and basins of **eastern China** (Zhang *et al.*) episodes of abnormally cold and stormy weather occurred from autumn 1815 to the end of summer 1817 in provinces ranging from the tropics to 50°N. During winter 1815-16, there was unusual cold from 19° to 30°N. On the tropical plains of **Hainan**, very cold dry weather in the southeast killed half the trees, and cold wet weather in the north destroyed much of the vegetation; rice was killed by frost, where there are no frosts today. In **Taiwan** (23°N), ice an inch thick formed on the ground in December 1815 (and again in December 1816); nowadays, this is unknown. On the **southeast Mainland** (27°N) (Fujian and Jiangxi), snow fell in January, with heavy snowfall and ice in February. Even where it is usual to have very light snow in winter, the reported depths on the ground were much greater than on modern record. Water buffalo were killed by the cold. To the west, in **Sichuan** (30°N), where normally winter is not very cold, there was snow in November 1815 as well as other falls. This was very unusual. Only 10-20% of the Province did not have snow, and in these areas frequent rain was reported. Zhang *et al.* also note that in January 1815, before the Tambora eruption, nearly 1 m of snow fell around **Ningde, Fujian**, where the present-day maximum recorded snow depth is 6 cm.

Summer 1816 was wetter and most likely colder than normal through eastern China (23° - 40°N) (Huang). In addition, this study of the yearly wetness/dryness indices and annals for 20 locations showed it to be by far the wettest year of the decade 1811-20, with severe floods in many parts of China. A change to wetter conditions had taken place in 1815 (Table 3), following a run of dry seasons from 1811 to 1814. These indices refer principally to summer-monsoon rainfall (see also Lough *et al.* 1984), and Huang has found, from recent instrumental records, a negative correlation between heavy precipitation with long runs of raindays and temperature at this season[1]. Much anomalous weather occurred in spring and summer: in the early part of the season, there was heavy rainfall and severe flooding in the **basins of the Yangtze and Hai rivers**; in midsummer, long spells of very wet weather with severe floods in the **lower basin of the Yellow River**. Farther south (23°N), heavy rain in June, followed by frequent rain throughout the season, led to severe and widespread flooding in the **Si River Basin**. The map (Figure 1) also shows those provinces where heavy losses in crop production were reported in autumn 1816 (Zhang *et al.*).

During 1817, summer/autumn drought and early fall frosts were reported in the more temperate regions of China (35° - 46°N) (**Shanxi, Hebei and Heilongjiang**), which together with hail resulted in heavy crop losses. Near **Harbin** frost in July caused widespread damage; at the present time no frost is expected from June through August. Farther south (**Zhejiang**) (29°N), very few sunny days occurred through most of the year, and there was large-scale spoilage of rice. In **Jiangxi Province** (27°N), there were severe thunderstorms in May with high winds, hail and heavy rain; seven or eight people were killed by lightning in 10 days. Many trees were uprooted then and previously in April. In late June and early July the weather was very cold with northerly winds; on 30 June snow lay on a mountain top. Some people wore fur coats, others stayed by the fireside.

The index of wetness or dryness appears to be closely related in summer to the Pacific Subtropical Anticyclone (Huang): wetness is positively correlated with the retreat to the east of the 500-mb high (defined by the most westerly longitude reached by the 588-DM contour); when the strength of the high weakens, there are floods in the basin of the Yellow River. In the decade 1811-20, the average position was farther east than for any other decade from 1771 to 1820, while its intensity reached the lowest value for the period 1471 to 1980. Huang concludes that the circulation over eastern Asia was generally weak at this time. On the other hand, Wang *et al.* (1984, quoted in Lough *et al.* 1987) indicate that an increase in the area and strength of the high at 500 mb is linked with heavy precipitation along the Yangtze River. Xiangding and Lough (1987) suggest that very wet summers in China may be coincident with generally higher pressure over the North Pacific and an intensification of the northern sector of the surface Subtropical Anticyclone. There appears to be a discrepancy here between these views, which a reading of Huang and Wang's 1985 paper might clarify.

Japan/Korea

The analysis of the climate of Japan in summer 1816 is based on weather information in old diaries for 20 locations (including Seoul, Korea), and freezing dates at Lake Suwa. Mikami and Tsukamura have constructed wetness/dryness indices which are comparable to those for China. Weather data are coded (1, rain; 0, no rain), the country was divided into five districts from west to east and daily weather-pattern "calendars" produced. The number of days with rain in each district was indexed by month (very wet to very dry) according to their probability of occurrence. The polar frontal zones around Japan were reconstructed by comparing daily historical weather distribution maps prepared from the diaries with modern daily weather maps to find analogues.

---

[1] Huang was not able to attend the meeting and present his results. During the Workshop, some doubt was expressed concerning the statistical significance of a precipitation/temperature relationship, and China was assessed (in the absence of further information) as having normal temperatures during the summer of 1816. In my revised version of the map (Figure 1), Huang's work has been incorporated with this provision.

Summer 1816 in Japan was rather hot and dry, owing to the northward expansion of the western Pacific Subtropical High, but not unusual compared with other years since the end of the eighteenth century. Over most of the country, the monthly rainfall was normal in July, however August was dry except for northern Japan and Korea, where wetter conditions corresponded with the location of the polar frontal zone from northern Japan to northern China across the Sea of Japan. For both July and August, these zones lay close to their position in "hot summer" months; that is, a few degrees to the north, especially in August. The seasonal northward shift of the fronts in 1816 was earlier than in recent years; and the rainy season in northern China (**Beijing**) was long, suggesting that the frontal zone had stagnated in that position for some time. Apparently midsummer began early and continued long. Few climatic hazards or periods of bad weather were reported this year (Maejima and Tagami 1988), although more than five violent storms - probably typhoons - struck the Japanese islands from June to September.

As a modern analogue for the circulation over East Asia in August 1816, Tsukamura selected August 1964 (the year after the eruption of Agung) which at 500 mb (50kPa) shows an anomalously high ridge over the western Pacific east of Japan, and a trough over northern China. Lamb (1977, p. 633) suggested that there is a strong negative correlation between the annual rainfall at Seoul (37°N) (where an average 77% falls between May and September) and the zonal index for westerly winds in the East Asian sector; 1816 and 1817 were by far the wettest years in the decade, which indicates a low-index situation.

In contrast, winter 1815-16 in Japan was severe, cold and snowy, as in China (Mikami and Tsukamura). The freezing of **Lake Suwa** (36°N) (closely correlated with temperature at Suwa City) occurred on 31 December and the ice bridge had formed on 4 January - far earlier than average at this period. At **Yokohama** (35°N), both the number of snowy days and the ratio of snowfall to total precipitation were higher than average. Again, as in China, snowfall was even greater in 1814-15, before Tambora. Gordon *et al.* (1985) have found that severe winters in Japan are typically associated with a strengthened Aleutian Low.

The Little Ice Age and its several phases have been identified by Maejima and Tagami (1988) in a reconstruction of climate for northern Japan (**Hirosaki**) (40°N) over the past few centuries. Setting 1816 in this context, northern Japan became increasingly rainy and cool in summer from 1800 to 1830, and increasingly snowy and cold in winter. The situation in Hirosaki in 1816 was not unusual compared with adjacent years.

It is useful to recall that no ice was reported in von Kotzebue's (1821) log as he sailed from **Kamchatka** (18 July 1816) to **St. Lawrence Island** and through the Bering Strait to Kotzebue Sound. On his 6 June to early July journey to Kamchatka from California, he reported fog the whole way from 33° to 47°N while sailing along 160°W.

India/Bangladesh/Pakistan

In **Bengal, India/Bangladesh** and the **Upper Provinces of India** (essentially the Ganges Valley from the delta to Saharanpur, 21°-30°N), the normal regularity of the seasons (as experienced for some years before, and described in detail) were suddenly disrupted from the summer of 1815 to end of winter 1818 (Jameson 1820). Further irregularity occurred in 1819. A possible outcome of this unseasonable weather was the first known fully-fledged epidemic of cholera, which severely affected the entire region from August to December 1816. The mortality rate was very high, and the river banks were littered with the dead and dying. A similar mortality, preceded by great scarcity of grain, prevailed at the same time in **Kutch, Sind (Pakistan) and other states bordering western India**. The living were unable to bury the dead, since several cities were depopulated. In **upper Hindustan**, the cattle were sick - their bodies strewn in vast numbers in the pastures. A further epidemic occurred in Bengal and the Upper Provinces in 1817.

The 1815 rainy season in **Bengal** (June-October) was excessively wet with major flooding. The following cool season (November-February 1815-16) was cold, damp and unpleasant with unusually frequent and dense fogs in December and January. In the hot season (March-May 1816), the usual thunderstorms were late and scanty, and great heat and drought ensued. Although several North Westers (storms) occurred in April-May, they were limited in extent and generally unaccompanied by rain. Toward the end of May, it had become so oppressive (the temperatures rising as high as 37°C, unusual in Bengal), that many people, both European and native, fell dead in the streets. Weather was similar in the **Upper Provinces**: a severe cold season, followed by excessive drought in the hot season, led to great sickness among the native people, and by destroying the spring grain crop prepared the way for the general scarcity that followed.

Unbearably sultry weather continued in the **Lower Provinces**, including **Calcutta**, until 14 June 1816, when the rains came. They kept up moderately well through July, but toward the latter part of August and the first week in September, rain became scanty and the days and nights oppressively hot in Calcutta. In **western Bengal** the drought was so severe that rivers dried up and the safety of the rice crop was threatened. The drought was then followed by very heavy rain for the rest of September, which caused greater and more extensive flooding than the oldest inhabitants could recollect. In the **Upper Provinces**, the extraordinary scantiness of the rains was even more remarkable. In July, a few showers fell, but these were local and short-lasting. From **Benares** (Varanasi) upstream, the country was dried up by lengthy unrelieved heat. Instead of the more usual easterlies, parching west winds blew persistently through August and the first half of September. Not a shower fell and the heat was so great that tatties[1] were in use the whole period. No such season had been experienced since 1803. As in Calcutta, the long drought was followed by heavy incessant rain for many days and the whole country lay under water.

In **southeastern India** (Pant et al.[2]), the total June-September rainfall in 1816 at **Madras** (13°N) was, by contrast, about one standard deviation above the 1871-1988 normal (and similarly below normal in 1815). While the rainfall series for **Bombay** (19°N) only begins in 1817, the occurrence of famine (drought?) in Kutch (above) suggests that this **western part of the Peninsula** may also have been drier in 1816. With respect to temperature, Stothers (1984) notes that Madras experienced a remarkable cooling in the last week of April 1815: morning temperature stood at 11°C on Monday dropping to -3°C by Friday. (On the other hand, there was no drop at Canton, China.) In summer 1816, the temperature again dipped below freezing in Madras, which is almost unheard of (Rampino 1989). The annual mean temperatures for Madras (Stothers after Köppen 1873) show that values had fallen below the long-term (1800-40) mean in 1814 to a low in 1818, returning to above normal in 1821.

---

[1] Tatty: matting of cuscus grass hung and kept wet to cool and perfume the air.

[2] The detail given by Jameson (1820) has provided a finer resolution of the 1816 climate within India than was available to Pant et al. The famine in 1816, not signalled in the series of Delhi wheat prices (1803 shows clearly - see above), was probably masked by the danger and high mortality from the cholera epidemic - the high mortality itself being hidden in the 1800-25 mean.

The cold season, 1816-17, was raw, damp and unpleasant through **Bengal and the Upper Provinces** as well as cloudy throughout with frequent falls of rain. Again, fogs were unusually frequent in **Calcutta**. In **western Bengal**, the hot season 1817 was marked by high humidity, much heavy rain and frequent thunderstorms, greatly damaging the spring grain crop and destroying the fruit blossoms. On 25 May, the rainy season set in - two to three weeks earlier than usual - with almost continual heavy rain, close weather and heavy cloud until its end. With the protracted heavy rains (3,048 mm) in the Ganges Delta nearly the whole country was underwater by August. However in **eastern Bengal (Dacca)**, the rains were confined to June and the first 10 days of July; here and elsewhere the Brahmaputra River was some 1.2 m lower than usual. In the **middle reaches of the Ganges River**, the rainy season was near normal, but above Kanpur, rainfall was remarkably scanty once again. At **Bombay** in 1817, the summer monsoon was heavy; at **Madras**, rather below average.

Throughout the **Ganges Valley**, 1818 saw a return to near-normal conditions until the following April. Jameson writes that in 1819 the preceding rains were much heavier and set in earlier in all parts of the **Upper Provinces** than for many years previously; in some parts they were unusually violent and continuous. Other evidence (Pant *et al.*) indicates the late onset of the monsoon in **East Uttar Pradesh** (Upper Provinces) was followed by heavy rains, but preceded in May and June by a heat wave due to lack of rain. The rainy season gave way to an unusually severe cold season in 1819-20. In **northern Bengal** (27°N), temperature fell to freezing on several nights, and did not rise above 2.8°C at 8 a.m. for many days in January 1820 - a degree of cold unknown to that part of the country. Pant *et al.* note that **Rajasthan** and some parts of **Uttar Pradesh** were affected by frost during early spring 1820. From 1818 to 1820, the June-September rainfall for Bombay was near normal, while at Madras, two very wet seasons were followed by a dry one.

The extreme irregularity in timing and progress of the 1816 summer monsoon in the northern plains of India, the extended break in the monsoon over Bengal, and the heavier-than-normal precipitation (including snow) near Lhasa (Tibet) suggest anomalous atmospheric circulation over a very large area, and some affinities with the 1979 season as reported in the literature.

In reporting on the Summer Monsoon Experiment (MONEX) of 1979, Fein and Kuettner (1980) noted the striking role of the variation (low frequency) of the planetary-scale circulation in controlling not only the onset, but also the intensity and variability of the monsoon. For example, in 1979 the onset of the monsoon was held back by travelling waves in the upper tropospheric westerlies over northern India, in the wake of persistent blocking over eastern Europe; this had delayed the establishment of the tropical upper easterlies and their progression northward, and the Tibetan thermal anticyclone in the upper troposphere. Chang (1981) found a negative correlation between daily rainfall anomalies for central Tibet and central India during summer 1979. With an active monsoon over Tibet, there was a break monsoon over India, and vice versa. Chang's study of the 500-mb (50kPa) circulation during the break monsoon over India from 15 August to 20 September 1979 - the longest then on record - indicated a strong stable blocking high west of the Ural Mountains. Wave troughs moved over Tibet within the north/northwest flow ahead of the high-pressure ridge, causing rainfall when meeting warm/moist air from northern India and the Bay of Bengal. Cold air travelling along the leading edge of the blocking high was also advected across Iran and into the western coastal region of the Indian peninsula, pushing the Iranian Subtropical High southward over the northern Arabian Sea, to weaken the northern branch of the Somalia jet. High pressure dominated most of the Indian peninsula. The map of 500-mb (50kPa) height departures showed positive values over central India, and a large area of negative values over central Tibet. On the other hand, when the monsoon was active over India, the thermal anticyclone over Tibet was strongly developed, and the height anomalies reversed.

In winter 1978-79, there was an abnormally prolonged and extensive snowcover over Eurasia, particularly over the mountains and plateau of Tibet between 30° and 40°N (Dewey and Heim 1982). Hahn and Shukla (1976) found evidence of an inverse relationship between winter snowcover south of 52°N over Eurasia and summer monsoon rainfall over India. Unseasonably cold, snowy plateau surfaces interacting with the atmosphere can apparently delay the shift of the upper westerlies to the north of the plateau in early summer and interfere with the establishment and maintenance of the Tibetan High in the upper troposphere. The evidence of a cold, stormy 1815-16 winter throughout China and in Japan, with abnormally heavy snowfall, the extraordinary 72-hour snowfall near Lhasa in June 1816, and the cold recorded in the Dunde Ice Cap suggests that this might well have been the situation in 1816. Further support is offered as follows: Lau and Li (1984) indicate a significant positive correlation between the amount of snowcover on the Tibetan plateau and the magnitude of the spring (Mei Yu) rains over China (mid-May to mid-June), whereas Tao and Ding (1981) report that in the record case of heavy rains and flooding in the Yangtze Basin (June-July 1954), the westerlies were located further south than normal, allowing a sequence of wave troughs to pass over the plateau; a series of depressions, mostly originating over the plateau were then steered eastward to give uninterrupted rain over east/central regions.

Pant *et al.* note a strong relationship between El Niño[1] events and Indian monsoon rainfall, such years showing an average drop of 11% in total fall. They note that the study by Quinn *et al.* (1987) shows a strong El Niño event in 1814 and a moderate one in 1817. However, Lough and Fritts' (1985) Southern Oscillation index reconstructed back to 1600 suggests an extreme low-index (negative) event from June 1815 to May 1816, which may have been comparable in magnitude to that of 1982-83. If this were the case, it may have been a contributing factor - as in the Indonesian drought (see below).

The Tambora Period in the Southern Hemisphere

With a few exceptions, climatic information presently available for the southern hemisphere is based on: (1) tree-ring chronologies from temperate-forest and woodland areas; (2) ice cores from the Quelccaya Ice Cap in Peru, and Siple, Dome C, South Pole and D57 (Adelie Land) in Antarctica. Unfortunately difficulties arise in using conventional tree-ring methodology with tropical and subtropical trees (LaMarche *et al.* 1979), but we include data from a teak chronology for central Java, with the warning that the dating could be out by a year or two for this period. For the Santiago district of Chile, there is a yearly (winter) rainfall series based on an exhaustive survey of historical documents - a reminder that there might well be a wealth of such documents, official and private, associated with European colonization in this hemisphere.

South Africa
The only information seen is the Die Bos[2] (32°S elevation 1,330 m) chronology for **southwestern Africa**. Here the 1815-16 index is small, but the 1816-17 value is extremely large,

---

[1] For an excellent short review article on El Niño and the Southern Oscillation, see Bigg (1990).
[2] Malcolm Cleaveland sent this information together with the publications of tree-ring chronologies for the southern hemisphere.

indicating that this growing season (October-March) was probably quite cool and wet (Dunwiddie and LaMarche 1980). While the mean index for the decade 1811-20 was similar to the 1941-70 average, that for 1816-17 is the sixth largest in the series 1564-1978 and by far the largest from 1789-1899 - standing out among below-average rings from 1815 to 1820 (cf. LaMarche *et al.* 1979, Volume 5).

Southern South America

Although there are some 60 to 70 tree-ring chronologies for southern South America (see also LaMarche *et al.* 1979, Volumes 1, 2), many of these have still to be reviewed for their potential in reconstructing past climate. Work completed with temperature-sensitive chronologies for subtropical, temperate and subantarctic sites (Villalba and Boninsegna) suggest that there were no striking temperature anomalies in the seasons immediately following Tambora. However, in all three areas there was a statistically-significant decrease in tree-ring indices for the three years after the eruption (1815-16 to 1817-18) compared with the three years before. For the subtropical and temperate sites, this indicates cooler late spring and summer weather (November to April); for the southern sites on Tierra del Fuego, a colder late winter/early spring (August to October of the year of growth). Around **27°S** (elevation 1,750 m), the coolest growing season occurred in 1816-17; at **38°S** (elevation 1,454 m), again in 1816-17, but here the temperatures were only marginally low. For **Tierra del Fuego** (54°-55°S), August to October of the 1816-17 season was probably close to normal (in keeping with independent work reported by Jacoby during the Workshop), the coldest season being 1817-18 - also the coldest of the 20 years around the eruption of Tambora (1815). In the subtropics, an overall trend of decreasing temperature had been underway since the beginning of the nineteenth century.

A comparison of these indices with those following the eruption of Agung in 1963 (location and timing almost identical with Tambora, but a very much smaller event) showed a more consistent and well-defined decrease after Agung, statistically significant at all latitudes, but especially notable around 27° and 38°S in the first growing season (1963-64). In seeking an explanation for these differences, differences are invoked in the type and distribution of matter ejected and a possible El Niño/Southern Oscillation event during the Tambora period: (1) the partition of Agung debris between the southern and northern hemispheres was estimated at 2:1 by Legrand and Delmas (1987); (2) a very large optical-depth change in the southern hemisphere, 10 times that in the northern hemisphere, indicated a 0.64°C temperature decrease from 1962 to 1964 between 30° and 60°S (Rampino and Self 1982); (3) the Agung eruption was proportionately richer in sulphur, considered a more critical measure of climatic significance; (4) Aceituno (1987) found that in summers during the mature phase of El Niño (negative Southern Oscillation), anomalously warm water in the eastern Pacific Ocean is consistent with positive air-temperature departures in subtropical and central Argentina (22° to 30°S), and during strong El Niño events (as in 1982-83) as far as 45°S. Any atmospheric cooling resulting from Tambora may have been moderated by El Niño in 1814-17.

In the **Santiago district of central Chile** (33°S) (cf. Lamb 1977, pp. 638-639) 90% of the yearly rainfall comes in winter (May-October), when travelling depressions of the Southern Temperate Zone reach farthest north. Following a dry period at the beginning of the nineteenth century, the rainfall was estimated as normal from 1805 through 1819, with one outstanding exception: the winter of 1818 was very wet, with precipitation well distributed throughout the winter. This suggests a greater expansion or a deformation of the winter vortex in these longitudes.

Tropical South America

Evidence for **tropical South America** (14°S) is provided by two ice cores from the Quelccaya Ice Cap in Peru (Thompson and Thompson). At 5,670 m, the sites are effectively at the 500-mb (50kPa) level of the atmospheric circulation. From oxygen-isotope analysis of the 1,500-year record, in which the Little Ice Age is very apparent, the decade 1810-20 was by far the coldest, although a cooling trend was underway before the eruption. The coldest period of the whole record (both cores) occurred during the wet season of 1819-20, suggesting that the cooling trend may have been strengthened by the eruption of Tambora. Compared with the 1941-70 and 1951-84 normals, the 1810-20 decade was not only much colder, but drier and dustier. There is no obvious sign of Tambora in the dust record, which is not unexpected in view of the predominance of the altiplano as a major local source of dust. High levels of particle concentration characterized much of the Little Ice Age.

New Zealand and Australia

Tree-ring coverage is good for the cool and warm temperate zones, but as yet there are no suitable chronologies from mainland Australia or for the tropical regions (Norton; Palmer and Ogden; LaMarche *et al.* 1979, Volumes 3, 4). For mainly subalpine sites in New Zealand's **South Island** (46°-40°S), Norton's reconstructed summer temperatures (December to March) show falling values from a maximum in 1810-11 (17.5°C) to a minimum in 1817-18 (14.3°C) - the second coldest reconstructed summer from 1750 to 1850. From 1815-16 through 1821-22, the summers were all below the 1800-30 average, but in 1816-17 only marginally so. The period between 1810 and 1825 was one of the cooler episodes between 1730 and the start of instrumental records at Dunedin in 1853. Palmer and Ogden present plots of average chronologies from four other species, covering both **North and South islands** (35°-46°S). Recent summer temperature reconstructions by Palmer (unpublished) from two species on the North Island agree closely with those derived by Norton. The authors found that the majority of chronologies indicate a cooler than average summer in 1817-18. The plots strongly suggest that the effect was most pronounced in the South Island and south/central North Island, and least in the northern part of the North Island, where the coolest season was 1819-20. The summer of 1816-17, near average in the south, became increasingly favourable for growth in the north.

This 1817-18 pattern is not seen in tree rings from **Tasmania** (43° - 40°S). Instead the one feature common to both tree species studied was poor growth in 1814-15 (before Tambora). As in New Zealand, one of the species also showed a decrease in ring width from 1810-11, with a period of low growth from about 1814-15 to 1820-21; the other a second minimum in 1816-17, suggesting a cold winter in 1816 and/or a warm summer in 1815-16. Dendroclimatic temperature reconstructions for Tasmania by LaMarche and Pittock (1982) indicate below-average summer temperatures for the 1816-17 to 1820-21 period. Palmer and Ogden caution that for New Zealand and Tasmania at this time, some uncertainty remains. If a depression of summer temperature did occur in New Zealand as a result of Tambora, it was not evident until two or three years later (1817-18.).

Indonesia

Tree-ring indices from **Java** (7°S) (cf. Lamb 1977, pp. 603-605) are positively correlated with cloudiness and the number of raindays in the year, and negatively with the number of dry months. For the decade 1810-19, the average index is the lowest since the 1720s, and was not as low in the future until the 1910s and 1920s. A low index thus implies less cloud, fewer raindays and more dry months (drought). Within the decade the sharp drop occurred from 1813 to 1814, and values remained very low through 1819; conditions were most severe in 1817, followed by 1818 and 1816. By 1820, the episode had passed. The correlation between the indices and yearly mean temperature is less, but negative, and by implication these years may have been warmer. It is tempting to see the drought as related to El Niño-Southern Oscillation events at this time (see above).

Antarctica

The Tambora deposit has been detected in ice-core records in both eastern and western Antarctica and at the South Pole. (Another eruption has also been identified at about 1809.) The **eastern Antarctic** record (Dome C, 74°S, 125°E) (Legrand and Delmas 1987) and those for **South Pole** and D57, **Adelie Land** (Kirchner *et al.* 1988) have not allowed any climatic change link with Tambora to be detected. In the **western Antarctic** (Thompson and Thompson), two cores from Siple Station (76°S, 84°W, elevation 1,054 m) record what is probably strong physical evidence of the eruption: a major increase in conductivity, sulphate and excess sulphate between 1817 and 1819, peaking in 1817-18. Although two years after the eruption, this is consistent with the time lag in the transport of stratospheric radioactivity to Antarctica (Lambert *et al.* 1977). There was a sharp decrease in net accumulation in 1817-18, while the isotopic temperature actually rose after the eruption, falling in 1817-18 to a minimum for the decade in 1819-20 (as in the Peruvian and Chinese ice cores). Compared with the 1941-70 normals, the 1810-20 decade may have been warmer in this part of Antarctica, as well as drier with less particle dust.

The Silent Regions in 1816

A glance at the map (Figure 1), and the vastness of those areas for which we presently have no climatic information is overwhelming. We may now hope to encourage the search for historical and proxy-weather data in these regions for this period, and where necessary the development of new techniques for extracting climatic information.

The map areas with few or no data are **the Soviet Union, the eastern Mediterranean and the Middle East, Central America, the three southern continents, and the oceans**. In the case of the Soviet Union, perhaps we do not have the information available. Before the Meeting, several attempts were made to contact those who might have been interested in participating, but without success. Many of the other land regions have one disadvantage: with the exception of the sparse areas where temperate forest and woodland are found (for the southern hemisphere, see map), the trees apparently do not lend themselves to current dendroclimatic techniques. Hence information is lacking from this source for the tropical and subtropical lowlands (cf. LaMarche *et al.* 1979). On the other hand, historical records may be available for a number of these "silent" countries. Many had close trading, missionary or colonial relations with European countries in the early nineteenth century, when careful reporting both official and private were often the order of the day; at this time, administrators, civil servants, surgeons and military personnel often kept personal diaries, including weather diaries. A possible source of historical material for some of these areas is the archive of the East India Company (India Office, British Library, London) (Appendix). During the Workshop, we were informed that the Boers had written records in South Africa at this time. We have also been told of weather information in diaries kept by Australian settlers. Some countries may have many years of recorded history (e.g., the 1,000-year records for Ethiopia, from which a history of Ethiopian droughts has been compiled). This was also an age of exploration. Even if weather information is qualitative and spotty, it can often supply key evidence when placed in geographical context.

For the oceans, there is a wealth of unused data. Ships' logs are virtually untapped for the 1816 period. For example, Chenoweth describes some of the routes covered by British Royal Navy logs during 1816 and the type of weather information they contain (Appendix). Besides naval vessels, there are logs of supply and trading vessels, as well as whaling and expedition ships.

In the meantime, it is possible to derive the most probable atmospheric-flow patterns over some of these regions, using methods combining teleconnections and analogues derived from modern climatic records, with the regional data for 1816 collected here. The Canadian Climate Centre has automated a procedure of this kind to specify the 500-mb (50kPa) height-anomaly patterns for the northern hemisphere, using the 500-mb monthly hemispheric data from 1946 to 1984 (personal communication, Peter Scholefield). Based on the known direct relationship between 500-mb height anomalies and surface-temperature anomalies, the temperature-anomaly pattern from our original map for 1816 was used to obtain a hemispheric 500-mb height-anomaly map for July 1816, and the corresponding July 1816 surface-temperature anomaly map for North America. The approach has much to offer, in that it would allow preliminary 500-mb height-anomaly maps to be reconstructed on a hemispheric basis, from relatively few areas of data, for other years in the decade. We have not reproduced the July 1816 500-mb map here, firstly because we now have more information than we had at the Workshop, which might change the configuration; secondly, the corresponding simulated July 1816 temperature map for North America, showed the centre of greatest cold too far west.

Summary and Conclusions

By putting the evidence into geographical context, what conclusions or implications can be drawn concerning the nature of: (1) the global atmospheric circulation in 1816; (2) the possible effects on global climate of the reduction in sunspot number in the early nineteenth century - the Dalton, or Little Maunder Minimum (Eddy; see also Eddy 1977); (3) the possible effects on global climate of the massive injection of ash, dust and sulphur into the atmosphere in April 1815 from the eruption of Mount Tambora, Indonesia (Sigurdsson and Carey; Vupputuri; Skinner; see also Stothers 1984; Lamb 1970)? Was 1816 really "the year without a summer"?

Atmospheric Circulation in 1816

The evidence suggests that in the **northern hemisphere**, the general circulation was marked in summer by a few preferred, persistent, large-scale flow patterns at the 500-mb (50kPa) level, whose stalled surface systems dominated the weather for protracted periods.

Over eastern North America, the North Atlantic and Europe, a characteristic 500-mb summer pattern consisted of a deep, cold trough down over the Québec/Labrador Peninsula and another off western Europe, both extending farther south and east than is normal today, while well-defined ridges of warmer air pushed northward over the east Greenland/Iceland region, eastern Europe, and (at least in the first half of summer), the Hudson Bay region. The ridges frequently developed into blocking highs. At sea level, the tracks of the depressions, steered around the upper troughs, were much farther south than usual, and a very cold north northwesterly flow was drawn down behind them into eastern North America and western Europe. The sea-level centres of maximum cyclonic activity (mean low-pressure centres) were located over the Labrador Sea and western Europe - displaced well to the south and east. Outbreaks of Mistral and Tramontane winds most likely brought some unseasonable storms to the western Mediterranean. The cold waves in eastern North America were felt as far south as South Carolina, if not beyond.

Over the western United States, troughs were more common than usual, bringing cooler, wetter conditions, while ridging over the Mississippi Valley gave hot (even very hot) dry weather. We have evidence of the strengthening and northward extension of the Subtropical High in the western Pacific to the north of Japan into Alaska, and probably northwestern Canada and the Beaufort Sea.

To clarify the situation over central Asia we need more evidence, but from what we know of Dunde and Lhasa, India and East Asia [supported by Scholefield and Shabbar's 500-mb (50kPa) anomaly map for 1816], the mid-troposphere (500-mb level) was generally colder and lower than is normal today over this vast area. Similarity with events in India in 1979 suggest that during much of the summer a deep broad trough was quasi-stationary east of the Ural Mountains, downstream from the persistent ridge of high pressure over European Russia. Typically, strong northwesterly flow penetrated unusually far south, and the most southerly branch of the Jetstream rarely cleared the Tibetan Plateau, to allow the normal summer development of the thermal anticyclone in the upper troposphere over Tibet, and the northward progress of the tropical easterlies over India. The effect was seen in the failure of the monsoon in the Upper Provinces of India, and what was probably the longest break monsoon on record there, through August to mid-September. Depressions originating in the passage of wave troughs over the plateau, or in its lee, then steered eastward into eastern China, plausibly account for much of the heavy rainfall and flooding as in 1954. At least during the second half of the season, there is evidence of a cold trough or closed low over northern China, where it stagnated for some weeks, while northern China and Korea had heavy rainfall and severe flooding. Japan at this time was basking in the northward extension of the Pacific Subtropical High, the polar frontal zone extending from central China across Korea to northern Japan.

There are contrasting situations over the two major oceans. Over the Atlantic, the Subtropical High appears to have been centred several degrees south of its average position today (Lamb). Over the broader expanses of the Pacific, the Subtropical High was strengthened and extended farther north than normal in the west, but possibly weakened in the eastern North American sector, where troughs extended farther to the south in 1816. This does suggest El Niño.

Alt *et al.* have suggested that this period of the Little Ice Age over the eastern Canadian Arctic was dominated by summers similar to that of 1972. The circulation described above for the northern hemisphere has many features in common with the maps illustrated for July 1972 (Alt *et al.*, and Figures 11, 12). The strongly meridional circulation so characteristic of this hemisphere in summer 1816 implies a stronger meridional temperature gradient in mid-latitudes than is usual at this season. Alt *et al.* point out that the July 1972 pattern resembles mean winter conditions, and further suggest that in years of this kind, the winter circulation is never really broken down.

The sparseness of the evidence for the **southern hemisphere** allows for little speculation. Today, the vigorous mid-latitude westerly circulation is predominantly zonal over the unlimited homogenous expanse of ocean, but it does break down into more meridional-flow patterns. Cooler, wetter weather in summer in the southernmost land areas of the three southern continents suggest an increase in meridional flow; this may have been the case in 1816-17 in the longitudes of South America and South Africa, and in 1817-18 for New Zealand. The very long, wet winter in central Chile in 1818 implies frequent blocking between 90° and 150°W, with a deep trough over the west coast of south/central South America.

The Little Maunder Minimum

Lamb (1963) remarked that the **southern hemisphere** seemed to have escaped the cold epoch of the Little Ice Age until about 1800. From our evidence, by far the coldest period of the 1,500-year record from the Quelccaya Ice Cap in Peru was 1810-20; the cooling trend had started before Tambora. In the South Island of New Zealand, the years 1810 to 1825 formed one of the cooler episodes between 1730 and 1853. By contrast in western Antarctica, 1810-20 may have been warmer than the 1941-70 normal.

Our evidence for the **northern hemisphere** indicates that a cooling trend had begun about the first decade of the nineteenth century (before the eruption of Tambora) in northern and east/central Canada, as well as in the Rocky Mountains, in west/central Europe, on the Dunde Ice Cap in China, in northern Japan, and probably in northern Greenland. The cool period frequently lasted into mid-century.

Tambora and World Climate

The eruption of Tambora in April 1815 was not only one of the largest in recorded history, but close enough to us in time to permit detailed study of both the event itself and the climate in this period. The injection of ash, dust and sulphur into the atmosphere was so massive that it must have exerted extremely strong radiative forcing (Vupputuri). But how was that translated into global surface weather?

The acid signal was clear in 1815 in the Greenland Ice Cap, with a maximum acidity in 1816, declining through 1817 and 1818. In China (Dunde Ice Cap), evidence of Tambora was later, in 1817-18. The strongest signal in western Antarctica was also in 1817-18. Lamb (1970) suggested, in keeping with seasonal changes of the stratospheric circulation, that the poleward drift of debris from lower latitudes proceeds with spurts each autumn. Once the material has entered the polar stratosphere, the residence time may be two to three years or more. Given the oblique angle of the sun's rays in the higher latitudes and the concentration and persistence of the debris there, the reduction in solar radiation received at the surface (Vupputuri) must be greatest in those regions. The radiative effects influence the general atmospheric circulation, possibly having far-reaching repercussions. In turn, the three-dimensional circulation associated with the polar vortex and its deformation redistributes the debris in preferred regions of the atmosphere. In the late 1950s and early 1960s (cf. Wilson 1967), a close relationship was found between the tropospheric Jetstream, and associated cyclonic disturbances, and radioactive fallout at the Earth's surface. This suggested the transfer of debris from the stratosphere into the troposphere in the vicinity of the Jetstream (the tropopause break), where subsidence in the rear of the depressions and precipitation processes aid the fall to earth.

To this must be added the differential earth-atmosphere interactions (energy exchanges), which augment or dampen the effects on surface climate. Continental surfaces react faster to cooling and heating than the oceans, where effects are dampened and delayed. To a lesser degree, similar differences exist between large expanses of cleared land versus forest. Where cooling extends the snow and ice cover, climatic effects can be enhanced and prolonged. An abnormal frequency of clear skies (as over northeastern North America in 1816) serves to promote radiation loss at night.

The Meeting and Workshop offered a body of circumstantial evidence for the **northern hemisphere** of rather sudden and often extreme changes in surface weather after the eruption of

Tambora, lasting from one to three years. The nature, intensity and timing of these changes evidently differed from region to region.

In **North America**, the greatest abnormalities occurred in spring and summer 1816 in the northeastern part of the continent - a record lowering of temperature throughout the season, resulting in unseasonable snow and frosts, a devastating shortening of the growing season, generally exacerbated by drought. This was followed by another cold season in 1817. The cold core of the anomaly appeared to be the Québec/Labrador Peninsula. On the eastern coast of Hudson/James Bay, a period of great cold began abruptly in autumn 1815, and ended as abruptly in spring 1818. In western North America, the only evidence of a pronounced irregularity is for the Colorado Plateau/Great Basin region, where it may have been cool and significantly wetter from autumn 1815 through 1816.

Over **western Europe**, including the British Isles, extreme cold through spring and summer in 1816 was accompanied by heavy, incessant precipitation. In Ireland, corn was covered with a reddish powder, destructive to the crop, which may have been volcanic aerosol (Kington). The centre of the cold anomaly was the northern Alpine region of Switzerland and Austria, where unseasonable snow and frost lowered the snowline, truncated the growing season and destroyed the crops. There was some amelioration in many areas in 1817.

China experienced extremes of weather from winter 1815-16 through summer 1817, from the Tropics (where there were winter frosts) to temperate latitudes, and on the plateau of Tibet (where Lhasa had a heavy snowfall in June 1816). Unusually heavy snowfalls in winter, and extremely heavy spring and summer rains affected much of the country, causing severe flooding. An abrupt change to wetter seasons had taken place in 1815, with 1816 by far the wettest year of the decade. The wet summer may also have been associated with colder-than-normal weather. In **Japan**, there was a severe winter in 1815-16 with heavier-than-usual snowfall, but midsummer was warmer and drier than usual in 1816. (Both in China and Japan, snowfall during the 1814-15 winter was even greater in some areas.)

Over the **Indian subcontinent**, the normal regularity of the season was suddenly disrupted in summer 1815, and remained so until the end of the cold season in 1818. The cold season in 1816 was colder, damper and foggier than normal in Bengal and the Upper Provinces. The succeeding hot season was marked by unusual drought and heat. In Bengal, the rains set in on 14 June, but there was a break monsoon during the latter part of August and early September. In the Upper Provinces the extreme heat and drought continued until mid-September, relieved only by a few showers in July. This was followed by many days of incessant rain in September throughout the Ganges Valley, causing severe flooding.

Java probably had a period of severe drought from 1816 to 1818; the sharp decline in yearly rainfall had occurred from 1813 to 1814, and values remained low through 1819. In 1820, the episode was over.

For the **southern hemisphere**, the limited information available suggests that there was a lowering of temperature in southern **South America** in 1816-17, but it was probably very small, and that winter rainfall near Santiago, Chile was exceptionally heavy in 1818, in an otherwise normal period. In southwestern **South Africa**, the 1816-17 growing season seems to have been remarkably cool and wet. For New Zealand, any lowering of temperature occurred mainly in the 1817-18 season - the anomaly being greater in the South Island. The one unequivocal extreme temperature recorded in the hemisphere until now is in subtropical Peru, where at the 500-mb level on the Quelccaya Ice Cap, 1819-20 was the coldest season of a 1,500-year record.

Conclusions

A study of the results of the Workshop has led to the following conclusions:

1. Although the evidence remains circumstantial, there is a strong case for a volcanic influence on the climate. It is highly probable that what we are seeing in 1816 is in part a reaction of the surface climate to the massive injection into the atmosphere of dust and sulphur from the eruption of Tambora the previous year. There is evidence from both hemispheres of a cooling trend underway before the eruption, probably related to the low sunspot number, on which cooling resulting from the eruption of Tambora was superimposed.

2. Owing to the enhanced meridional (north-south) circulation pattern characteristic of stronger meridional temperature gradients in mid-latitudes, the extreme cold of the 1816 summer was **a regional phenomenon**. In the northern hemisphere, the outflow of Arctic air to the south in one region was counterbalanced by poleward flow of tropical air in another (e.g., the normal to above-normal temperatures in parts of western Canada, the hot summer in the Mississippi Valley and the warmth in eastern Europe and Japan). There is also reason to believe that the strengthening of the meridional circulation - in conjunction with El Niño - was the cause of the heat-wave and failure of the monsoon in northern India.

3. Although the occurrence of record cold was confined to a few regions in 1816, the suggestion was made that a more widespread atmospheric cooling may have been masked in warmer areas, as well as magnified in regions of lower temperature. There are few "very hot" symbols on the map (Figure 1). Further, had Tambora erupted during a general warming trend in world climate, would the effects have been so dramatic?

4. In the southern hemisphere, where Tambora debris reached high latitudes a year or more later than in the Arctic, El Niño may have dampened the cooling in northern Argentina in 1816-17, while there is the overwhelming mitigating influence of the ocean with respect to cooling in New Zealand in 1817-18.

5. It was suggested that comparative mapping of other key years might be undertaken (e.g., 1817, considered another critical year in some regions; and 1783, the year of the Laki eruption in Iceland (63°N) - "the year without a summer" in Japan). The mid-1830s also stand out as exceptional in many of the contributions - in Nicaragua, Mount Coseguina erupted in 1835. Of recent eruptions, that of Mount Agung in Bali (8°N) in 1963, although a lesser event, was almost identical in position and timing to that of Tambora. A study of subsequent seasonal circulation patterns and regional weather for both hemispheres might repay closer scrutiny. Another approach is a variant of the superposed epoch method (see Skinner, Villalba and Boninsegna), in which composite seasonal circulation or temperature maps might be produced for several years before and after key eruption years. We were reminded that it is dangerous to assume that the circulation pattern that we have suggested for 1816 was caused by a volcanic eruption, given the possibility that the atmosphere behaves in an intransitive manner, and the patterns could have changed abruptly without an eruption.

6. The year 1816 was perhaps "without a summer" in northeastern North America and western/central Europe, where the extraordinary lowering of temperature truncated or destroyed the growing season. Similarly, in areas near the northern limits for human survival at this time (e.g., Great Whale on Hudson Bay) it forced a migration to the south.

7. Lastly, we hope that research will continue on the summer of 1816 - adding to the data coverage and understanding of the climate in those regions presently silent, and elsewhere refining the resolution of the data. We recognize the pioneering, prescient work of Professor Hubert Lamb (1963) on climate during the Little Ice Age: he has envisaged so much of the map presented here.

Acknowledgements

On behalf of The 1816 Climate Group, I thank: all who took part in the Workshop for their contributions and for their patience as we felt our way through the lengthy mapping procedure; Peter Scholefield and Amir Shabbar (Canadian Climate Centre) for reconstructing the map analogues for July 1816; Malcolm Cleaveland, D.N. Duvick, Joel Guiot, N.M. de Courcy and Michael Chenoweth for contributing information to fill some of the gaps; the many participants who kindly supplied reprints on allied work to help in the compilation of this report; and members of the National Museum of Natural Sciences (now Canadian Museum of Nature) who provided splendid support in fulfilling requirements for materials, equipment, as well as for taping the proceedings so clearly under difficult circumstances, and for supplying copies of all the conference papers. Base maps were kindly provided by the University of Winnipeg. Sharon Helman and Mary Ann Maruska (Canadian Museum of Nature) drafted the map (Figure 1), and Joanne Dinn and Marie-Anne Resiga kindly processed the manuscript.

References

Aceituno, P. 1987. On the interannual variability of South American climate and the Southern Oscillation. Ph.D. thesis. University of Wisconsin, Madison. 128 pp.

Allsopp, T.R. 1977. Agricultural weather in the Red River Basin of southern Manitoba over the period 1800 to 1975. *Fisheries and Environment Canada* CL1-3-77:1-28.

Bednarz, Z. 1984. The comparison of dendroclimatological reconstructions of summer temperatures from the Alps and Tatra mountains from 1741-1965. *Dendrochronologia* 2:63-72.

Bider, M., M. Schüepp and H. von Rudloff. 1959. Die Reduktion der 200 jährigen Basler Temperaturreihe. *Archiv für Meteorologie, Geophysik und Bioklimatologie* B, 9:360-411.

Bigg, G.R. 1990. El Niño and the Southern Oscillation. *Weather* 45:2-8.

Blasing, T.J., D.W. Stahle and D.N. Duvick. 1988. Tree-ring based reconstruction of annual precipitation in the south-central United States from 1750 to 1980. *Water Resources Research* 24:163-171.

Briffa, K.R., T.S. Bartholin, D. Eckstein, P.D. Jones, W. Karlén, F.H. Schweingruber and P. Zetterberg. 1990. A 1,400-year tree ring record of summer temperature in Fennoscandia. *Nature* 346:434-439.

Brooks, C.E.P. 1926. *Climate Through the Ages*. Ernest Benn Limited, London.

Catchpole, A.J.W. and I. Hanuta. 1989. Severe summer ice in Hudson Strait and Hudson Bay following major volcanic eruptions, 1750 to 1889 A.D. *Climatic Change* 14:61-79.

Chang, C.C. 1981. A contrasting study of the rainfall anomalies between central Tibet and central India during the summer monsoon season of 1979. *Bulletin of the American Meteorological Society* 62:20-23.

Chenoweth, M. 1986. The summer of 1816 in North America. *Weather* 41:140-142.

Dewey, K.F. and R. Heim, Jr. 1982. A digital archive of northern hemisphere snow cover, November 1966 through December 1980. *Bulletin of the American Meteorological Society* 63:1132-1141.

Dunbar, M. 1985. Sea ice and climatic change in the Canadian Arctic since 1800. *In: Critical Periods in the Quaternary Climatic History of North America. Climatic Change in Canada 5*. C.R. Harington (ed.). *Syllogeus* 55:107-119.

Dunwiddie, P.W. and V.C. LaMarche, Jr. 1980. A climatically responsive tree-ring record from *Widdringtonia cedarbergensis*, Cape Province, South Africa. *Nature* 286:796-797.

Eckstein, D. and R.W. Antol. 1981. Dendroclimatological reconstruction of the summer temperatures for an alpine region. *Mitteilungen der forstlichen Bundesversuchsanstalt* 142:391-398.

Eddy, J.A. 1977. Climate and the changing sun. *Climatic Change* 1:173-190.

Fein, J.S. and J.P. Kuettner. 1980. Report on the summer MONEX field phase. *Bulletin of the American Meteorological Society* 61:461-474.

Filion, L., S. Payette and Y. Bégin. 1988. Botanical and geomorphic evidence of cold and humid climate in the early nineteenth century in the Hudson Bay area, Quebec. *In: The Year Without a Summer? Climate in 1816*. C.R. Harington (ed.). An International Meeting Sponsored by the National Museum of Natural Sciences, Ottawa, 25-28 June 1988. Program and Abstracts. pp. 26-27.

Filion, L., S. Payette, L. Gauthier and Y. Boutin. 1986. Light rings in subarctic conifers as a dendrochronological tool. *Quaternary Research* 26:272-279.

Gordon, G.A., J.M. Lough, H.C. Fritts and P.M. Kelly. 1985. Comparison of sea-level pressure reconstructions from western North American tree rings with a proxy record of winter severity in Japan. *Journal of Climatology and Applied Meteorology* 24:1219-1224.

Grove, J.M. 1988. The Little Ice Age. Methuen, London, New York. 498 pp.

Guiot, J. 1986. Reconstruction of temperature and pressure for the Hudson Bay region from 1700 to the present. Environment Canada, Atmospheric Environment Service, *Canadian Climate Centre Report* No. 86-11:1-106. (Aussi disponible en édition originale française.)

Hahn, D.G. and J. Shukla. 1976. An apparent relationship between snow cover and Indian monsoon rainfall. *Journal of Atmospheric Sciences* 33:2461-2462.

Hamilton, K. 1986. Early Canadian weather observers and the "Year without a Summer." *Bulletin of the American Meteorological Society* 67:524-532.

Hammer, C.U., H.B. Clausen and W. Dansgaard. 1980. Greenland Ice Sheet evidence of postglacial volcanism and its climatic impact. *Nature* 288:230-235.

Harington, C.R. 1988. Introduction. *In: The Year Without a Summer? Climate in 1816.* C.R. Harington (ed.). An International Meeting Sponsored by the National Museum of Natural Sciences, Ottawa, 25-28 June 1988. Program and Abstracts. pp. 3-5.

Hoyt, J.B. 1958. The cold summer of 1816. *Annals of the Association of American Geographers* 48:118-131.

Huang, J. and S. Wang. 1985. Investigations on variation of the Subtropical High in the Western Pacific during historic times. *Climatic Change* 7:427-440.

Jacoby, G.C., I.S. Ivanciu and L.D. Ulan. 1988. A 263-year record of summer temperature for northern Quebec reconstructed from tree-ring data and evidence of a major climatic shift in the early 1800s. *Palaeogeography, Palaeoclimatology, Palaeoecology* 64:69-78.

Jameson, J. 1820. Report on the epidemick cholera morbus, as it visited the territories subject to the Presidency of Bengal in the years 1817, 1818 and 1819. Calcutta.

Johnson, C.B. 1948. Anticylogenesis in eastern Canada during spring. *Bulletin of the American Meteorological Society* 29:45-55.

Kirchner, S., J. Palais and R. Delmas. 1988. The Tambora years in Antarctic ice. *In: The Year Without a Summer? Climate in 1816.* C.R. Harington (ed.). An International Meeting Sponsored by the National Museum of Natural Sciences, Ottawa, 25-28 June 1988. Program and Abstracts. pp. 38-39.

Köppen, W. 1873. Uber mehrjährige Perioden der Witterung, insbesondere über die 11-jährige Periode der Temperatur. *Zeitschrift für Meteorologie, Vienna* 8:241-248, 257-267.

LaMarche, V.C., Jr., R.L. Holmes, P.W. Dunwiddie and L.G. Drew. 1979. *Tree-ring Chronologies of the Southern Hemisphere.* Volumes 1. Argentina; 2. Chile; 3. New Zealand; 4. Australia; 5. South Africa. Laboratory of Tree-Ring Research, University of Arizona, Tucson, Arizona.

LaMarche, V.C., Jr. and A.B. Pittock. 1982. Preliminary temperature reconstructions for Tasmania. *In: Climate from Tree Rings.* M.K. Hughes, P.M. Kelly, J.R. Pilcher and V.C. LaMarche, Jr. (eds.). Cambridge University Press. pp. 177-185.

Lamb, H.H. 1963. On the nature of certain climatic epochs which differed from the modern (1900-39) normal. *In: Changes of Climate.* Proceedings of the Rome Symposium, UNESCO, Paris. pp. 125-150 (p. 145).

_____. 1970. Volcanic dust in the atmosphere; with a chronology and assessment of its meteorological significance. *Philosophical Transactions of the Royal Society of London*, A266:425-533.

_____. 1977. *Climate: Present, Past and Future*. Volume 2. Methuen, London. 835 pp. (Appendix to Part III, pp. 551-652).

_____. 1982. *Climate, History and the Modern World*. Methuen, London and New York. 387 pp.

Lamb, H.H. and A.I. Johnson. 1966. Secular variations of the atmospheric circulation since 1750. Meteorological Office. *Geophysical Memoir* 110. HMSO, London. 125 pp.

Lambert, G., B. Ardouin, J. Sanak, C. Lorius and M. Pourchet. 1977. Accumulation of snow and radioactive debris in Antarctica: a possible refined radiochronology beyond reference levels. *International Association of Hydrological Sciences (IAHS)* 118:146-158.

Lau, K-m. and M-t. Li. 1984. The monsoon of East Asia and its global associations - a survey. *Bulletin of the American Meteorological Association* 65:114-125.

Legrand, M. and R.J. Delmas. 1987. A 220-year continuous record of volcanic H_2SO_4 in the Antarctic Ice Sheet. *Nature* 327:671-676.

LeRoy Ladurie, E. 1972. *Times of Feast, Times of Famine: A History of Climate Since the Year 1000*. Translated by Barbara Bray. Allen and Unwin, London. 428 pp.

Lough, J.M. and H.C. Fritts. 1985. The Southern Oscillation and tree rings: 1600-1961. *Journal of Climate and Applied Meteorology* 24:952-965.

Lough, J.M., H.C. Fritts and W. Xiangding. 1987. Relationships between the climates of China and North America over the past four centuries: a comparison of proxy records. *In: The Climate of China and Global Climate*. Proceedings of the Beijing International Symposium on Climate, 30 October - 3 November 1984, Beijing. China Ocean Press, Beijing. pp. 89-105.

Ludlam, D. 1966. *Early American Winters, 1604-1820*. American Meteorological Society, Boston. 285 pp.

Maejima, I. and Y. Tagami. 1988. Climatic change during historical times in Japan: some remarks on the weather in 1816. *In: The Year Without a Summer? Climate in 1816*. C.R. Harington (ed.). An International Meeting Sponsored by the National Museum of Natural Sciences, Ottawa, 25-28 June 1988. Program and Abstracts. p. 43.

Manley, G. 1974. Central England temperatures: monthly means 1659 to 1973. *Quarterly Journal of the Royal Meteorological Society* 100:389-405.

Milham, W.I. 1924. The year 1816 - the cause of abnormalities. *Monthly Weather Review* 52:563-570.

Neumann, J. 1990. The 1810s in the Baltic Region, 1816 in particular: air temperatures, grain supply and mortality. *Climatic Change* 17:97-120.

Parker, M.L. and M. Geast. 1988. August temperatures in central Canada from tree rings: implications for the year without a summer. *In: The Year Without a Summer? Climate in 1816*. C.R. Harington (ed.). An International Meeting Sponsored by the National Museum of Natural Sciences, Ottawa, 25-28 June 1988. Program and Abstracts. pp. 54-55.

Parker, M.L., L.A. Jozsa, S.G. Johnson and P.A. Bramhall. 1981. Dendrochronological studies on the coasts of James Bay and Hudson Bay. *Climatic Change in Canada 2*. C.R. Harington (ed.). *Syllogeus* 33:129-188.

Post, J.D. 1977. *The Last Great Subsistence Crisis in the Western World*. Johns Hopkins University Press, Baltimore. 240 pp.

Quinn, W.H., V.T. Neal and S.E. Antunez de Mayolo. 1987. El Niño occurrences over the past four and a half centuries. *Journal of Geophysical Research* 92(C13):14,449-14,461.

Rampino, M.R. 1989. *Horizon: 23. Time of Darkness*. British Broadcasting Corporation. Television script of the programme transmitted 26 June 1989. p. 13.

Rampino, M.R. and S. Self. 1982. Historic eruptions of Tambora (1815), Krakatau (1883), and Agung (1963), their stratospheric aerosols and climatic impact. *Quaternary Research* 18:127-143.

Scoresby, W. 1817. As reported in Minutes of Council, 20 November 1817. Royal Society, London. Minutes of Council, Volume 8:149-153.

Smith, W.P. 1986. Reconstruction of precipitation in northeastern Nevada using tree rings, 1600-1982. *Journal of Climate and Applied Meteorology* 25:1255-1263.

_____. 1988. Reconstruction of precipitation in northeastern Nevada using tree rings, 1600-1982: the case for abnormally wet conditions in 1816. *In: The Year Without a Summer? Climate in 1816*. C.R. Harington (ed.). An International Meeting Sponsored by the National Museum of Natural Sciences, Ottawa, 25-28 June 1988. Program and Abstracts. p. 60.

Stahle, D.W. and M.K. Cleaveland. 1988. Texas drought history reconstructed and analyzed from 1698 to 1980. *Journal of Climate* 1:59-74.

Stockton, C.W. 1990. Climatic variability on the scale of decades to centuries. *Climatic Change* 16:173-183.

Stommel, H. and E. Stommel. 1983. *Volcano Weather, the Story of 1816, the Year Without a Summer*. Seven Seas Press, Newport, Rhode Island. 177 pp.

Stothers, R.B. 1984. The great Tambora eruption in 1815 and its aftermath. *Science* 224(4654):1191-1198.

Tao, S-y. and Y-h. Ding. 1981. Observational evidence of the influence of the Qinghai-Xizang (Tibet) plateau on the occurrence of heavy rain and severe convective storms in China. *Bulletin of the American Meteorological Society* 62:23-30.

Till, C. and J. Guiot. 1990. Reconstruction of precipitation in Morocco since 1100 A.D. based on *Cedrus atlantica* tree-ring widths. *Quaternary Research* 33:337-351.

Vibe, C. 1967. Arctic animals in relation to climatic fluctuations. *Meddelelser om Grønland* 170(5):1-227.

Von Kotzebue, O. 1821. *A Voyage of Discovery into the South Sea and Bering Straits, 1815-1818*. Longman, Hurst, Rees, Orme and Brown, London. (3 volumes).

Wang, P.K. and De'er Zhang. 1988. An introduction to some historical weather records of China. *Bulletin of the American Meteorological Society* 69:753-758.

Wilson, C. 1967. Radioactive fallout in northern regions. United States Army, Cold Regions Research and Engineering Laboratory (CCREL), Hanover, New Hamphire. *Cold Regions Science and Engineering Report* No. 1-A3d:1-35.

_____. 1985a. The Little Ice Age on eastern Hudson/James Bay: the summer weather and climate at Great Whale, Fort George and Eastmain, 1814-1821, as derived from Hudson's Bay Company records. *In: Climatic Change in Canada 5*. C.R. Harington (ed.). *Syllogeus* 55:147-190.

_____. 1985b. Daily weather maps for Canada, summers 1816-1818 - a pilot study. *In: Climatic Change in Canada 5*. C.R. Harington (ed.). *Syllogeus* 55:191-218.

Xiangding, W. and J.M. Lough. 1987. Estimating North Pacific summer sea-level pressure back to 1600 using proxy climate records from China and North America. *Advances in Atmospheric Sciences* 4(1):74-84.

Appendix

Further Information on 1816 in British Archives
Ships' Logs, Public Records Office, Kew Gardens, London, England
This office holds a large quantity of weather data recorded by ships of the British Royal Navy. The majority of the logs are for ships in home port or in waters adjacent to Britain, but dozens cover voyages between duty stations. For the Atlantic there are data for ships travelling to and from Ascension Island (in the equatorial South Atlantic), ships between Britain and the West Indies, between Britain and Canada, and the West Indies and Canada. There are also logs for shipping on the Great Lakes in North America. They are available for the Mediterranean Sea, the Indian Ocean and a few for waters off the South American coast and the Far East. The coverage extends into any areas of the world in which the Royal Navy was present. Very few data exist for vast areas of the Pacific Ocean.

While at sea, weather observations were generally made every two hours and while in port normally four times a day (usually early morning, midday, early evening and midnight); the exact hours are not specified as well in port as when at sea. For 1816, no instrumental data have yet been found, but observations are available on wind direction, wind force, sky conditions, present weather and at times, visibility.

Michael Chenoweth
(For address, see footnote on the second page of this report.)

India Office Records, The British Library, London

This office holds the archives of the East India Company (1600-1858) headquarters, Calcutta. The geographical focus was: India, Pakistan, Bangladesh, Burma, St. Helena and the Cape of Good Hope (to 1830); Zanzibar, Somalia, Ethiopia (mainly nineteenth century); Red Sea, Arabian Peninsula, Persian Gulf, Iraq, Iran (1600-1947); Afghanistan, Russian and Chinese Central Asia, Tibet, Nepal, Bhutan and Sikkim (from the late eighteenth century); China (early seventeenth century to 1947).

There are Factory records from about 1595-1858: letters, diaries and other papers for each of the Factories, Presidencies and agencies established by the Company in India and elsewhere from the seventeenth century onward. Papers continue into the nineteenth century for the Cape of Good Hope, China, Japan, Egypt, Red Sea, Persia and the Persian Gulf, St. Helena, Straits Settlement and Sumatra. With respect to shipping, there are official journals and log books until withdrawal of the Company from shipping in the early nineteenth century.

This summary is based on an India Office brochure. At this point, it is not known how much weather or weather-related information is available for 1816 and the second decade of the nineteenth century. It should be worth investigating.

Historical Records for Greece and the Levant 1816

N.M. de Courcy (15 Market Road, Auckland 5, New Zealand) has documented potential sources of historical phenological information for Greece and the Levant. These include travellers' accounts, diplomatic and consular reports, many of which can be found in the British Library, London and the Cambridge University Library. The Public Records Office, London holds manuscript material on the British Protectorate in the Ionian Islands (1815-) and the papers of the Levant Company.

From the few documents already consulted for 1816, de Courcy notes some comments on hot summer days in Greece, but no evidence of unusual temperatures. At Arta, Epirus (39°N), aqueducts failed in May, preventing the operation of flour mills, which suggests drought. Just to the south in northwestern Peloponnese (38°N), the grain harvest apparently began later in June than usual (famine was reported in 1817), and near Athens (38°N), rye flowering was delayed. Drought in the Levant (35-37°N) from 1816 to 1819 led to high cereal prices and speculation. (Of extreme weather, winter 1812-1813 had been the most severe in Greece for 35 years.)

Isaac H. Jackson's Weather Diary for College Hill, Cincinnati, 1814-1848
Cincinnati Historical Society Archives

Horstmeyer's discovery of this diary, with its daily temperature records, helps to define the southwest limit of the abnormally cold summer in the northeast of the continent in 1816. Data presented by Horstmeyer (1989, *Weatherwise* 42:320-327) suggest that the unusual cold did extend as far midwest as the Ohio Valley (39°N, 85°W). This may have been close to a well-defined western limit. From May through August average sunrise temperature was 2.2°C below the 1951-80 normal minimum, 1.2°C colder than the record in 1917, each month below average. Compared with minimum records on 10 days in 1963, sunrise readings equalled or broke the record on 18 days, 4 each in June and August, 3 in May when frost killed fruit blossom, and 7 in July. These conditions are similar to those in Philadelphia and the northeast. Moreover, the major cold events are synchronous: daily weather data plotted for northeast North America (Wilson 1985b) show cold northern air drawn far south between high pressure just to the west and a semi-stationary trough off the east coast.

Index

A

Aargau, 416
Abies lasiocarpa, 268
acid layers, 71, 311, 326, 547
acidic aerosols, 310
acidity, 393, 403, 414, 497
acidity profiles, 493, 497, 506
Acipenser fulvescens, 208
Aðaldalur, 335, 349
Adelie Land, 541, 544
advection, 175, 177, 325, 360, 384, 425, 540
aerosols, 7, 15, 31-33, 36, 41, 42, 46-48, 49, 53, 54, 70, 97, 115, 119, 175, 305, 393, 403, 493
Afghanistan, 556
Africa, 74, 294, 304, 527, 541
Agassiz Ice Cap, 310, 311, 318, 325, 326, 529
Agathis australis, 511, 516
agriculture, 12, 20, 64, 165, 186, 188, 198, 201, 340, 399, 403, 437, 441, 443
agroclimatic capability classes, 168
Agung, 46-48, 120, 169, 170, 317, 326, 404, 413, 415, 497, 503, 504, 506, 507, 538, 542, 549
air masses, 332
air temperature, 393, 533, 534
air transparency, 420
Akershus, 398, 399
Alaska, 67, 78, 106, 256, 257, 262, 320, 326, 420, 527, 546
Alaska-Beaufort Sea, 529
Albany Factory, 189
Albany, New York, 125, 147, 150, 151, 153, 154, 243, 244, 528
Albany River, 214
albedo, 49, 85, 115
Alberta, 186, 194, 527, 530
Albert, J.M., 124
Alces alces, 206
Alder (Master at Great Whale and Fort George), 171
Aleutian Low, 103, 106, 538
Aleutians, 527
alpine fir, 268
alpine larch, 268
alpine timberline, 418, 516, 533
alpine treeline chronologies, 274
Alpine glaciers, 418
Alps, 304, 384, 418, 419, 533
Altai, central Russia, 61
altiplano, 484, 543

Ambon, 19
American Midwest, 116, 198
American traders, 206
American War of Independence, 199
analogues, 167, 169, 171, 177, 246, 250, 295, 309, 320, 324, 326, 454, 529, 530, 537, 538, 545
Ancaster, 147, 155
Andes, 482
Andhra Pradesh, 432
animals, 14, 167, 187, 194, 197, 200, 206, 341, 345, 349, 529
anomalies, 111, 171, 256, 262, 281, 300, 301, 304, 305, 324, 365, 370, 372, 375, 377, 384, 390, 420, 423, 433, 449, 524, 528, 533, 537, 541, 548
Antarctic ice-core data, 255
Antarctica, 7, 70, 479, 482, 485, 487, 493, 497, 504, 506, 541, 544, 547
Anthropogeography, 196
anticyclones, 175, 177, 180, 443
Antrim, 369
Anyang, 449
Anyuan, 440
apple-tree blossoming, 125, 127, 134
Arabian Peninsula, 556
Arabian Sea, 540
Araucaria, 498-502, 504
Araucaria araucana, 494, 495, 497
Archangel, 402
archives, 7, 163, 165, 355, 536, 556
Archives of Ontario, 145
Arctic, 79, 85, 87, 91, 111, 127, 165, 171, 175, 178, 531, 549
Arctic air, 177, 186, 191, 195, 208, 257, 262, 344, 527, 549
Arctic Canada, 529
Arctic Front, 190, 191, 194, 195
Arctic High, 194
Arctic ice, 178, 249, 345, 535
Arctic Ocean, 320, 324, 326, 529
Argentina, 7, 494, 495, 504, 542, 549
Armagura eruption, 497, 500, 506
Árnes district, 343
Arta, 556
Asama volcano, 71, 73, 504
Asama-Laki eruptions, 497, 498
Ascension Island, 555
ash, 12, 15, 17-22, 25-27, 31, 41, 46, 48, 49, 55, 60, 71, 403, 482, 493, 545, 547
Asia, 7, 63, 68, 74, 99, 103, 106, 451, 452, 527, 546
Askja, 61

---

* To gather all references to plant and animal species, please refer to both their common and scientific names.

Assiniboine River, 187-189, 201
Athabasca Glacier, 165, 270
Athabaska fur trade, 524
Athens, 556
Atlantic, 111, 168, 250, 332, 555
Atlantic air, 332
Atlantic coast, 527
Atlantic Ocean, 6, 305, 332, 355, 436
Atlantic Subtropical High, 529, 546
Atlas Mountains, 534
atmosphere model simulations, 50
atmospheric chemistry, 33
atmospheric circulation, 178, 181, 186, 238, 318, 355, 370, 384, 403, 425, 426, 436, 451, 452, 523, 529, 531, 532, 540, 543, 545, 547
Atmospheric Environment Service of Canada, 155, 163, 291
atmospheric-flow patterns, 545
atmospheric history, 479
atmospheric nuclear explosions, 414
atmospheric phenomena, 41
atmospheric pollution, 422, 425
atmospheric pressure, 168, 523
Auburn, 147, 150
Auður-Þingeyjarsýsla, 335
aurorae, 38, 177
auroral/geomagnetic activity, 177
Australia, 7, 511, 543, 544
Austria, 533, 548
Austrocedrus chilensis, 494
Austur-Húnavatnssýsla, 334
avalanches, 416
Azores High, 360

B

Baffin Bay, 320, 324, 325, 529
Bali, 19, 413, 549
Ball, T.F., 5
Baltic Region, 392, 393, 395-397, 401, 403, 405, 406, 415, 534
Baltic Sea, 415
Baltimore, 66, 147, 151
Banff, 266, 524
Bangka Island, 20
Bangladesh, 556
Banjuwangi, 12, 13, 17, 19, 20
Baoding, 449
Barents Sea, 321
Bariloche, 507
barley, 188, 189, 197
barometer, 168, 169, 524, 527
barometric pressure, 355
Basel, 393, 394, 533
Batavia, 12-14
battles, 188, 189, 200
Bay of Bengal, 540

bears, 200, 206, 211
Beaufort Scale, 169
Beaufort Sea, 321, 325, 326, 530, 546
beavers, 209, 213
begging, 341, 349
Beijing, 437, 461, 538
Belfast, 293
Belgium, 368, 534
Besoeki, 24
Benares, 539
Benares, 17, 19, 20, 27
Bengal, 432, 535, 538-540, 548
Bengkulu, 20
Bennington Glacier, 270
Bering Sea, 524, 527
Bering Strait, 538
Berlin, 62
Berlingske Tidende, 414
Bermuda High, 248
Bern, 67
berries, 194, 211
Besuki, 17, 19
Beverly, Massachusetts, 126
Bezyminanny, 27
Bhutan, 556
Biala, 419
big game, 205, 206, 210, 215, 216
Bima, 17, 18, 20, 21
biologists, 6, 448
biosphere, 17, 33
birds, 208, 211
Bishop, C.A., 206
Bjarnason, Björn, 334
Bjørn, Professor Claus, 399
Björnsson, Sheriff Þorður, 345
black spruce, 67, 266, 289, 528
blight, 369
blocking, 7, 178, 180, 191, 194, 257, 325, 365, 366, 368, 531, 535, 540, 545, 546
Blöndudalur, 334, 338, 343
blossoming, 125, 127, 134, 540
Bobcat Canyon, 116
Boers, 544
Boesand, 71
Boisonneau, Mireille, 5
Bombay, 429, 430, 535, 539, 540
Bonasa umbellus, 210
Bonavista, Newfoundland, 248
bootstrap regression, 296, 297
Border Beacon, Labrador, 292
boreal forest, 199, 527
Borgarfjarðarsýsla, 334, 335, 339
Borneo, 21
Boston, 250, 532
Boyne River, 369
Brahmaputra River, 540
Brandes, Heinrich, 358, 360

Brandon House, 187-189
Brandsstadaannáll, 335, 338, 339, 343
Brandsstadir, 334, 338, 339
Branta canadensis, 209
bread, 398, 402
break monsoons, 7, 540, 546, 548
break-up, 171, 175, 177, 178, 211, 291
Breidafjördur, 345
Breslau, 358
Breughel, P., 202
Briem, Sheriff Gunnlaugur, 343
Briffa, K.R., 365
bristlecone pines, 266, 443
British Archives, 555
British Columbia, 527
British Government, 145
British Isles, 246, 301, 304, 305, 355, 364, 366, 368, 370, 375, 377, 384, 390, 534, 535, 548, 555
British Isles weather types, 365
British Library, 12, 544, 556
British Naval expeditions, 529
British Protectorate in the Ionian Islands, 556
British Royal Navy, 545, 555
British West Indies, 443, 532
British-Russian coalition, 397
Brittany, 67
Brooks Range, 527
Brunei, 21
Brunswick, Maine, 126, 130, 531
Brůžek, Vladimir, 534
Bryson, R., 190
buffalo, 14, 147, 150, 188, 194
Bünden, 416
Burma, 556
Byron, (Lord), ii, iii, 351

C

cabbage, 189, 344
Cairngorms, 198
Calanda massif, 416
Calcutta, 21, 429, 535, 539, 540, 556
caldera, 22, 23
California, 266, 538
Cambridge Bay, 85
Cambridge University Library, 556
Camp Century, 71, 294
Canachites canadensis, 210
Canada, 46, 53, 67, 79, 81, 85, 98, 116, 140, 163, 165, 168, 169, 175, 178, 180, 181, 185-187, 190, 198, 201, 256, 257, 262, 263, 266, 267, 270, 292, 295, 296, 301, 305, 317, 320, 326, 355, 425, 524, 531, 532, 546, 547, 549, 555
Canada Geese, 208
Canadian Arctic, 103, 106, 355, 531, 546
Canadian Arctic Islands, 309, 317, 318, 320, 324-326, 529, 530
Canadian Climate Centre, 5, 163, 545
Canadian Climatological Archive, 145
Canadian Museum of Nature, 6
Canadian Rocky Mountains, 266, 268, 278
Canadian Shield, 162, 178
Canadian weather stations, 81
canoe journeys, 165, 188, 189, 214
canonical regression, 98, 99
Canton, 539
Cape of Good Hope, 556
carbon dioxide, 31
caribou, 206, 216
Carlton House, 187, 189-191, 201
Castine, Maine, 126
Castor canadensis, 209
cataclysmic eruption, 14
catastrophes, 60, 215, 430
Catchpole, A.J.W., 5, 6, 194, 248, 256
catfish, 194
Catostomus, 208
cattle, 14, 20, 36, 39, 60, 126, 341, 416, 533, 539
cedar ring-width series, 295
Cedrela sp., 494
Center for Colorado Plateau Studies, Flagstaff, 141
Central America, 74, 544
central England temperature series, 533
Central Massif (see Massif Central), 534
Central Meteorological Agency, 441
Central Meteorological Institute, 448
Centre d'études nordiques, Université Laval, Québec, 163
Changhua, 441
charity, 216
Charleston, 65, 66
chemistry, 48, 482
Chen caerulescens, 208
Chenoweth, Michael, 524, 532, 545, 555
cherries, 416
Chesapeake Bay, 65, 66
Chile, 541, 542, 546, 548
China, 7, 63, 68, 70, 262, 436, 437, 441, 442, 448, 449, 451, 452, 454, 461, 465, 479, 482, 485, 487, 535-539, 541, 546-548, 556
chlorides, 35, 485
chlorine, 16, 17, 31-33, 41, 42
chlorofluorocarbons, 32
chlorofluoromethanes, 32
Churchill, Manitoba, 85, 199, 234, 238, 243, 244, 250, 282, 283, 289, 291, 292, 304, 340, 528, 530
Churchill River estuary, 237
Cincinnati, Ohio, 116
Cipolleti, 507
circumpolar vortex, 198
Cirebon, 19
cities, 145
Cleaveland, Malcolm, 532, 541
climate, 7, 248, 255, 297, 351, 437

Climate Atlas of China, 441
climatic anomalies, 111, 119, 120, 163, 232, 256, 266, 419, 436, 524
climatic change, 6, 11, 17, 38, 39, 47, 55, 91, 101, 165, 174, 175, 200, 205, 256, 263, 264, 281, 306, 331, 334, 418, 422, 425, 429, 436, 453, 529
climatic data, 237, 340
climatic determinism, 197, 202
climatic disasters, 201
climatic extremes, 201
climatic history, 333, 479
climatic impact, 36, 47, 331, 340, 341, 343, 350, 351, 534
climatic models, 47, 48
climatic proxy data, 530
climatic reconstructions, 106, 111, 218
Climatic Change in Canada Project, 6, 163
Climatic Research Unit, 5, 365
climatologists, 6, 126, 293, 420, 448
climatology, 141, 153, 163, 164, 166, 233
clouds, 47, 55, 116, 124, 168, 169, 177, 324, 325, 339, 358, 368, 510, 540, 543
co-ignimbrite ash falls, 22, 25, 26, 29, 41
coconuts, 21
cod fishery, 246, 349
cold droughts, 186, 191, 194, 195, 339
cold fronts, 132
cold spells, 530, 531
cold surface anticyclone, 532
cold-desert regions, 340
Colen, Joseph, 200
Colorado, 99, 116, 118
Colorado Plateau, 116, 119-121, 524, 548
Concord, New Hampshire, 125, 126
conductivity, 485, 487, 536, 544
conflict, 200
Connecticut, 66, 443
content analysis, 199, 218, 221, 223
contingency tables, 341
Copenhagen, 61, 62, 71, 334, 393, 398, 400, 414, 534
Copenhagen City Archives, 400
Coregonus clupeiformis, 208
corn crops, 125-127, 132, 350, 368, 369, 548
corpses, 20
Corsica, 304, 534
Cortland, 147, 153
Coseguina, 169, 175, 282, 289, 443, 497, 500, 501, 506
Cracow, 419
Craigie, William, 147
Cree, 203
Crete, 266
Crête, Greenland, 393, 414, 529
Cri Lake, 267, 292
Crimea, 357, 534
crises, 246, 341, 434
crop/climatic relationships, 344

crop failure, 140, 198, 395-397, 399-401, 414, 443, 532, 534, 537
crop production, 188, 397, 535, 537
crops, 14, 20, 61, 63, 67, 125-127, 130, 168, 187-190, 194, 208-211, 213-215, 248, 257, 339, 343, 344, 357, 368, 369, 397-400, 416, 419, 432, 434, 442, 533, 535, 548
cross-checking, 270, 482, 516, 530, 539, 540
Cuba, 443, 532
cultures, 198, 216
curling, 198
cycle (nine-year), 209
cyclic crash, 215
cyclones, 431
cyclonic activity, 355
cyclonic storms, 191, 198, 432
cyclonicity, 370, 535

D

D57 (Adelie Land), 541, 544
Dacca, 540
Dade, (Reverend), 146, 151
daily-weather records and maps, 163, 168, 181, 233, 234, 339, 360, 365, 453, 531, 535
Dalton Minimum, 11, 38, 545
damage to crops and property, 438, 441, 442, 534
dams, 190
Danish authorities, 71, 344
Danish grain, 399
Danish merchants, 341
Danish ships, 246
Danish Sound, 400
Danske Statstidende, 400, 414
Danzig, 370, 397, 402, 534
Darkness (poem by Lord Byron), ii, iii, 351
Datong, 441
Davis Strait, 248, 324, 529
dearth, 333, 344, 396, 398, 402, 534
death, 21, 36, 199, 204, 211, 432
death rates, 305, 397, 398
debt, 349
deciduous forest, 198
de Courcy, N.M., 556
deep-sea core, 27
Defence Research Board of Canada, 529
degassing, 17, 31, 35
Delaware River, 66
Delhi, India, 63, 432, 434
Delhi wheat prices, 429, 433, 434, 539
delta ^{18}O values (δ^{18}O), 311, 313, 317, 326, 482, 487, 488
delta-Eddington method, 47
dendrochronologists, 420
dendrochronology, 270, 516
dendroclimatology, 98, 101, 390, 418, 510, 511, 516, 543, 544

dendroclimatologists, 418
dendroecological analysis, 497
Denmark, 246, 341, 350, 368, 384, 397-400, 534, 535
Dennysville, 125
depressions, 332
Deqing, 440
desertion of farms, 349
desiccation, 283
Devon Island, 325
Devon Island Ice Cap, 310, 315, 317, 318, 324-326, 529
Dey, B., 194
diaries, 60, 124, 125, 127, 165, 186, 437, 443, 453, 462, 465, 466, 524, 536, 537, 544, 556, 564
diarrhoea, 21
diet, 395, 399
Ding'an, 441
Dinn, Joanne, 5
Direct Solar Radiation (DSR), 403, 405, 406
diseases, 14, 21, 186, 198, 203, 334, 349, 395
Dispatch, 20, 21
Ditch Canyon, 116
documentary evidence, 98, 332, 333, 436, 533
Doilungdeqen, 437, 438
Dome C (Antarctica), 497, 506, 541, 544
Dome glaciers, 270
Domesday Book, 197
Dompu, 21
Douglas, Thomas, 5th Earl of Selkirk, 199, 200
Douglas-fir chronologies, 524
Douglas-fir trees, 67, 116, 118, 199
Drake, M., 399
drift ice, 332
driftwood, 345
droughts, 63, 67, 121, 125-127, 134, 186, 188-191, 194, 197, 201, 215, 400, 402, 414, 429, 432-434, 441, 450, 465, 532, 534, 535, 537, 539, 544, 548, 556
dry fogs, 61, 68, 70, 71, 73, 74
dryness, 177, 198, 335, 338, 339, 343
dryness/wetness grades, 448-450, 452
Duchang, 440
ducks, 208, 211
Dunbar, Moira, 529
Dunde Ice Cap, 479, 482, 484, 485, 487, 535, 536, 541, 546, 547
Dunde ice-core records, 488
Dunedin, 516, 543
dust, 13, 14, 32, 47, 55, 61, 70, 177, 181, 317, 487, 544, 545, 547, 549
dust records, 482, 484, 485, 543
dust veils (see volcanic dust veils), 81, 186, 266, 267
Dust Veil Index (DVI), 46, 48, 55, 79, 91, 115, 395, 487, 497
Dutch Harbour, 527
Duvick, D.N., 532
Dye 2, Greenland, 315, 326, 529
Dye 3, Greenland, 311, 315, 415, 529

E

earthquakes, 58
East Asia, 461, 462, 465, 475, 535, 538, 546
East Bridgewater, 125
East Greenland Sea, 246
East Greenland current, 332, 529
East India Company, 544, 556
Eastmain, 165-167, 170, 171, 177, 234, 240, 243, 244, 527, 528
ecology, 206, 214
economic factors, 140, 165, 201, 202, 340, 341, 350, 351, 397, 419, 437
Edinburgh, 62
Edo era, 462
Egypt, 556
eigenvalues, 450, 451
eigenvector matrix, 299
eighteen hundred and froze to death, 126
El Chichón, 32, 35, 46-49, 54, 55, 78, 169, 177, 415
El Niño, 7, 47, 53, 55, 120, 121, 263, 429, 432, 434, 504, 506, 508, 541, 542, 544, 546, 549
Eldgja eruption, 33
electrical conductivity, 311, 482, 485, 488
eleven-year solar cycle, 38
Ellesmere Island, 310, 324
Emei, 441
Emerald, 240
Engelmann spruce, 268, 270
England, 67, 146, 165, 171, 197, 219, 246, 250, 262, 293, 294, 368, 369, 384, 399, 403, 414, 415, 533, 534, 555
Environment Canada, 250, 253
environmental data, 17, 42, 189, 218, 219, 487, 506
epidemics, 14, 16, 20, 203, 343, 344, 538, 539
Epirus, 556
Equatorial Current, 21
Escabachewan, 213
Esox lucius, 208
Espólín, Sheriff Jón, 344, 350
Esquel, 507
Ethiopia, 544, 556
Ethiopian droughts, 544
Eurasia, 454, 541
Eureka, 313, 318
Europe, 6, 7, 46, 53, 58, 61, 62, 64, 68-71, 73, 74, 97, 116, 124, 145, 164, 168, 180, 185, 186, 190, 197, 198, 200, 206, 250, 255, 262, 274, 293, 295-297, 299, 301, 304-306, 355, 357, 358, 360, 365, 368, 370, 372, 375, 377, 384, 390, 392, 393, 395, 399, 402, 403, 405, 406, 419, 420, 423, 425, 426, 453, 462, 475, 482, 506, 524, 531, 533-535, 539, 540, 544, 545, 547-550
European annual-temperature series, 296
European parts of the Soviet Union, 375, 403, 533, 546
European proxy series, 296
evapotranspiration, 524

extreme events, 166
Eyjafjarðarsýsla, 334, 335, 343
Eyjafjord, 345

F

Fairbanks, 78
Fairfield, 147, 153
Fairlie, 21
Falbe-Hansen, V., 399
famines, 16, 63, 126, 204-206, 209, 213-216, 392, 397, 398, 400-402, 414, 429, 432, 434, 462, 465, 475, 534, 539, 556
Far East, 555
farming, 125-127, 130, 140, 197, 343, 349, 350, 449, 530
fascism, 202
Faurer, M. 248, 256
Featherstonhaugh, George W., 125
feuds, 187
Fidler, Peter, 187-189, 190, 194, 201
Finland, 63, 384, 393, 395, 400, 401, 534
Finnish mortality data, 401
fiords, 60
fir, 274
fir ring-width series, 293
fire, 212
firestorm, 18
firn, 311
first-blooming, 416, 437
first-breaking, 233, 234, 237, 238, 240, 243, 244, 528
first-freezing, 233, 234
fish, 208-210, 213, 215, 246, 349, 350
fisher, 209
fishing, 60, 188, 208, 209, 211, 213-215, 340, 341, 345, 350, 529, 535
Fitzroya, 498-502, 504
Fitzroya cupressoides, 494, 495, 497
flatfish, 349
flies, 61
floods, 63, 67, 125, 141, 338-340, 368-370, 429, 431, 440, 448-452, 465, 534, 536, 537, 539, 541, 546, 548
flora, 345
Flores, 20
Florida, 99
flour, 396, 556
fluorides, 35
fluorine, 16, 17, 31-33, 36, 41, 42
fluorosis, 36, 42
fodder, 206, 341, 349, 399
fogs, 61, 69, 325, 339, 538-540, 548
folklore, 531
food, 141, 185, 186, 194, 197, 200-202, 208, 211, 213, 341, 345, 349, 350, 369, 400, 419, 436, 437, 440

food-grain, 429, 432-434
forcing functions, 263, 301
forests, 168, 211, 270, 281, 289, 547
forest fires, 61, 206, 215
forest-tundra, 256, 282, 283, 288, 292, 530
Forintek, Vancouver, 270
Fort Albany, 234, 237
Fort Chimo, 292
Fort Chipewyan, 191
Fort Churchill, 530
Fort George, 165-167, 171, 174, 177, 527, 528
Fort Prince of Wales, 234, 237
Fort York, 145, 146, 150
fossil fuels, 31
Foxe Basin, 250
France, 61, 71, 206, 293, 294, 304, 358, 368, 384, 390, 403, 419, 533, 534
Franklin, Benjamin, 58, 61, 69, 71
Fredonia, 147, 153
freeze-thaw cycles, 211
freezing, 66, 171, 208, 233, 238, 291, 437, 466, 537
French Revolution, 360
French wine-harvest data, 293
Fritts, H.C., 98, 101, 238
frontal systems, 169
frontal zones, 168, 527
frost damage, 127, 443
frost rings, 266
frosts, 60, 66, 125, 126, 132, 178, 187-189, 198, 202, 209, 214, 215, 248, 257, 335, 338, 339, 345, 349, 350, 398, 431, 432, 441, 443, 530, 531, 535-537, 540, 548
frozen ground, 345
fruit, 125, 194, 211, 533
Fuan, 440
Fuego eruption, 36, 41
Fujian, 440, 536
fungal rust, 369
fur trade, 165, 186, 194, 197, 199, 201, 202, 206, 211, 215, 530
furbearers, 206, 209
furs, 187, 197, 199, 200, 211-213
Fusarium nivale, 67

G

gales, 214, 443
Ganges River, 538, 540, 548
gardens, 187-189, 214, 248, 344, 530
Garður, 334, 335, 339, 345, 349
Gazeta Lwowska, 419
geese, 208, 211-214, 530
Geneva, 62
geography, 6, 197, 333, 437, 448
geology, 58, 98
geomorphology, 524, 528, 533
Georgia, 532

geosphere, 17
Germany, 268, 293, 358, 369, 370, 384, 402, 419, 534, 535
Gexia, 437, 438
glaciology, 6, 174, 270, 339, 340, 448, 482, 533
Gladman (Master at Eastmain), 170, 171
glass elutriation, 25
glass inclusions, 30, 41
glassy tephra, 36
global climate, 6
global warming, 195, 198
goats, 14
Gobi Desert, 484, 536
Godthaab (Nuuk), 165
goose harvest, 208, 209
goose migration, 215
Gorham, Maine, 126
Gothenburg, 393, 394, 534
grain, 14, 67, 126, 369, 393, 395, 397-402, 406, 415, 534, 539, 540, 556
Grande Armée, 414
Grande Rivière de la Baleine (Great Whale), 528
Grant, Cuthbert, 201
grapes, 294, 416
grass, 60, 188, 198, 334, 343, 344, 349, 350, 368
grasshoppers, 190
Great Basin, 443, 524, 548
Great Lakes, 103, 178, 531, 555
Great Plains, 186, 194, 198
Great Whale, 163, 165-167, 170, 171, 174, 175, 177, 527, 528, 550
Greece, 266, 556
Greenland, 33, 70, 71, 116, 165, 178, 198, 246, 248, 294, 315, 320, 324, 325, 333, 393, 403, 414, 482, 524, 529, 531, 545, 547
Greenland Ice Cap, 529, 547
Greenland ice cores, 32, 33, 74
Greenwich Meridian, 529
Gresik, 20
Griffier, Jan, 202
Grimsvotn caldera, 60, 73
Gröf, 334, 339, 343, 345, 349
Grosswetterlagen classification, 422, 423, 425
groundwater, 60
grouse, 210
growing seasons, 67, 118-121, 124, 127, 130, 132, 134, 140, 141, 198, 257, 283, 288, 396, 397, 418, 530-532, 542, 548, 550
grubs, 209
Grund, 334, 338, 343, 344
Guan Xian, 441
Guangze, 440
Gudlaugsson, Sheriff Sigurdur, 343, 345, 349
Gudmundsson, Sheriff Lydur, 339
Gujarat, 432
Gulf of Bothnia, 384
Gulf of Genova, 304
Gulf of Mexico, 65, 67, 532
Gullbringusýsla, 334

Gunung Gunter, 14
Gunung Klut (see Klut), 14

H

Hai River Basin, 449, 537
hail, 67, 349, 368, 369, 431, 432, 440, 441, 537
Hainan Island, 441, 442, 536
Halley, Edmund, 198
Hallowell, Maine, 125, 443
halogen gases, 33, 36, 41
Hamilton, 147, 153
Handbook of Geophysics and Space Environments, 405
Harang bollong, 13
Harbin, 441, 537
hardships, 199, 532
hares, 209-211, 215
Harington, C.R., 7, 360, 523
Hartford, 66
harvests, 63, 67, 127, 198, 215, 338, 343, 344, 350, 360, 369, 395, 397, 399-402, 415, 419, 438, 441, 442, 532-534, 536
Haryana, 432
Havrebjerg, 399
Hawaiian volcanoes, 70
hay crops, 126, 338, 339, 341, 344, 349, 350, 369, 400, 535
Hayes River estuary, 189, 190, 234
haze, 61, 62, 70
heat-unit concept, 168
heat waves, 432, 540, 549
Hebei, 441, 537
Hecla, 69
Heilongjiang, 441, 442, 537
Heimaey, 71
Hekla eruption, 36
Helman, Sharon, 5
Helsinki, 403
hemispheric temperatures, 317
Hesse-Brezowski classification, 426
Heze, 449
High Arctic Canada, 309, 321, 524
high-latitude eruptions, 370
high-pass filter, 270, 297
Highlands, 198
Hill, Leonard, 125, 126
Hindustan, 539
Hirosaki, 538
historians, 6, 124, 126, 197, 199, 202, 293, 399, 448
historic records, 68
Historical Climate Records Office, 141
historical climates, 181, 202, 465
historical climatology, 197
historical records, 166, 178, 197, 218, 234, 237, 238, 293, 294, 332, 341, 360, 429, 431, 436, 437, 439, 441, 442, 448, 449, 453, 454, 462, 465, 475, 485, 528, 534, 535, 541, 544, 556
Historical Society of Pennsylvania, 178, 531

Historisk Statistik för Sverige, 395-397
history, 163, 166, 197, 202, 246, 482, 547
Hitler, Adolf, 196
Hodgins, (Dr), 145
Hodgins Papers, 145
Holland, 61, 368, 534
Holocene pollen, 198
Honshu, 535
Hookamarshish, 212
horses, 14, 20, 21, 60, 61, 341, 349
hot droughts, 194
hot weather, 70
Hoyt, J.B., 124, 185
Hsinchu and Miaoli, 441
Huairen, 441
Hudson Bay, 6, 103, 163, 175, 177, 178, 181, 186-188, 199, 208, 219, 223, 234, 237, 238, 240, 243-245, 250, 257, 266, 268, 278, 289, 304, 305, 324, 527, 528, 530-532, 550
Hudson Bay Highs, 180, 532
Hudson Bay Lowlands, 208
Hudson Bay region, 165, 169, 545
Hudson Bay sea-ice processes, 223
Hudson/James Bay, 163, 164, 169, 171, 175, 178, 181, 291, 527, 548
Hudson Strait, 67, 219, 233, 238, 240, 243, 246, 248, 528
Hudson's Bay Company (HBC), 6, 67, 162-166, 169, 171, 174, 175, 177, 178, 181, 186-189, 194, 195, 199, 202, 203, 205, 206, 208, 215, 218, 219, 221, 223, 233, 238, 246, 248, 256, 524, 527, 528, 530
Hukao, 440
Hulan, 441
human activities, 7, 167, 196, 209, 340, 341
Húnavatn, 349
Húnavatnssýsla, 343
hunger, 20, 21, 186, 349
hunting, 188, 194, 209, 211, 213, 216
Huang Ho River (see Yellow River), 63
hydrocarbons, 48
hydrology, 437

I

ice, 60, 67, 70, 165, 167, 171, 174, 175, 177, 178, 180, 181, 187-189, 191, 194, 198, 208, 211, 213, 233, 248, 253, 270, 311, 339, 340, 345, 349, 350, 393, 440-442, 527-530, 532, 538, 547
ice caps, 309, 311, 320, 325, 479, 485, 530
ice cores, 39, 41, 97, 115, 294, 309, 311, 315, 318, 326, 414, 479, 485, 524, 529, 535, 541, 543, 544
ice sheets, 479, 485
ice skating, 65, 66
ice-out, 124, 125
ice-severity indices, 237

Iceland, 7, 27, 36, 58, 60, 64, 68-71, 197, 198, 246, 250, 293, 311, 331-335, 338-341, 344, 345, 350, 351, 368, 370, 414, 423, 475, 535, 545, 549
Icelandic eruptions, 61
Icelandic Low, 535
ignimbrite, 22, 29, 31, 36, 41
Île-à-la-Crosse, 191
Illinois, 532
impact on human and animal life, 7, 215, 341, 533
India, 15, 430, 432, 433, 535, 538-541, 546, 548, 549, 556
India/Bangladesh, 538
India Office of the British Library, 535, 544
India Office Records, 556
Indian corn, 126, 127
Indian monsoon rainfall, 434, 541
Indian Ocean, 21
Indian subcontinent, 548
Indians, 186, 188, 194, 199-201, 213
indices of Alexandre, 300
Indonesia, 12, 15-17, 42, 46, 205, 309, 317, 543, 545
Indonesian drought, 541
infrared (IR) radiation, 46, 49, 50, 53, 177
injuries, 211, 368
insects, 189, 190
insolation, 147, 267
Institute of Geography, Academia Sinica, 436
instrumental records, 62, 98, 106, 115, 116, 120, 124, 126, 127, 166, 237, 331, 332, 372, 390, 448, 449, 465, 482, 516, 524, 527, 531-533, 536, 543, 555
interest rates, 415
international date line, 434
inundations (see floods), 368-370
Iowa, 198, 532
Iowa State University, 532
Iran, 540, 556
Iranian Subtropical High, 540
Iraq, 556
Ireland, 293, 368, 384, 534, 548
Irish bog oaks, 266
Irminger current, 332
isotopic series, 294
Israel, 17
Italy, 61, 293, 368, 375, 377, 533, 534

J

jackknife replication, 296
Jacoby, G.C., 5, 542
Jakarta, 24
James Bay, 174, 175, 189, 238, 240, 324, 528
Jameson, James, 535, 539, 540
Japan, 7, 16, 63, 68, 70, 71, 73, 262, 370, 453, 454, 460-462, 465, 466, 475, 535, 537, 538, 541, 546-549, 556

Jasper National Park, 266, 524
Java, 12, 15, 17-20, 24, 41, 541, 543, 548
Jemima, 245
jetstream, 191, 257, 546, 547
Jiangxi, 440, 536, 537
Jianyang, 440
Jiaxing, 440
Jintang, 441
Jiujiang, 440, 449
Johnson, A.I., 6
Jones, John Paul, 199
Jones, P.D., 365
Jónsson, Trausti, 333
journals, 127, 186, 203, 213, 556
Juglans, 504
Juglans australis, 494, 495, 498
Jujuy, 507

K

Kadilangu, 13
Kali, 7
Kaluga, 402
Kamchatka, 527, 538
Kanpur, 540
Karnataka, 432
Katmai, 120, 317, 326, 420
Keewatin, 324
Kellogg, Elijah, 125
Kentucky, 532
Ketilsstaðir, 334, 335, 338, 343, 349
Kiev, 414
Kilkenny, 369
killing frosts, 125, 127, 132, 141
Kimball, Benjamin, 126
King Frederick VI, 398
Kingston, 531
Kington, J.A., 168, 535
Kjós district, 343
Klementinum temperature series, 426
Klimageschichte der Schweiz 1525-1860, 416
Kloet, 13
Klut (Kelut), 14
Koerner, R.M., 5
Köppen, W, 171
Korea, 466, 535, 537, 538, 546
Korean Peninsula, 466
Kotzebue Sound, 527, 538
Krakatau, 7, 16, 33, 46, 61, 69, 70, 79, 119, 169, 267, 317, 326, 393, 394, 413, 414, 497, 501, 506, 508
Krakatau Committee of the Royal Society of London, 415
Krippendorff's agreement coefficient, 228
Kristianstad, 397
Kutch, 432, 539
Kuujjuaq, 85, 292
Kuujjuarapik, 292, 304, 528

L

La Marche, Valmore C., Jr., 351
Laboratorio de Dendrocronología, 494
Laboratory of Tree-Ring Research, Tucson, Arizona, 98, 270
Labrador, 145, 245, 246, 248, 250, 253, 262, 355, 528, 529, 531
Labrador Sea, 245, 246, 248, 250, 253, 528, 529, 545
Labrador-Ungava, 324-326, 529
Lac Seul, 206, 209, 211
lag effects, 283, 485, 510, 544
Lago Argentino, 507
Lake Biel (Bielersee), 416
Lake Erie, 180
lake ice, 198, 212
Lake Louise, 270, 274
Lake Murten (Murtensee), 416
Lake Neuenberg (Neuenbergersee), 416
Lake Ontario, 145
Lake Saki, 534
Lake Superior, 191
Lake Suwa, 67, 466, 537, 538
Lake Winnipegosis, 188
lakes, 168, 188, 190, 194, 197, 200, 208, 210, 211, 213, 214, 233
Laki, 27, 46, 58, 60-62, 70, 71, 73, 74, 119, 266, 274, 278, 311, 313, 315, 318, 326, 350, 414, 423, 504, 549
Laki/Grimsvotn system, 74
Lamb, Hubert (Professor), 6, 168, 197, 198, 360, 365, 370, 531, 535, 550
land-fast ice, 345
land-ocean surface cooling, 55
Landsberg, H.E., 124
landslide, 528
larch, 270, 274
larch ring-width, 292, 293
Larch Valley, 270, 274
Larix laricina, 282
Larix lyallii, 268
last glaciation, 6
last ice age, 46
lateral moraine, 270
Laurentian Ice Sheet, 165
lava, 14, 58-60, 70, 71
Leirá, 334, 336, 339, 343, 344, 349
lemmings, 200
Leningrad, 414
Lepus americanus, 209
Leshan, 441
letters, 66, 334, 335, 343, 349, 350, 556
lettuce, 188
Levant Company, 556
Lewiston, 147, 153
Lhasa, 437, 535, 536, 540, 541, 546, 548
Libau, 402

Libocedrus, 516
lightning, 19, 125, 127, 139, 440, 537
Lincoln, Theodore, 125
Linde, Helle (Dr), 400
Literary and Philosophical Association of Manchester, 69
lithology, 24, 25
Little Climatic Optimum, 197-199, 300
Little Ice Age, 7, 11, 39, 198, 199, 201, 253, 268, 301, 305, 320, 326, 365, 416, 418, 419, 482, 484, 485, 487, 529, 536, 538, 543, 546, 547, 550
Little Maunder Minimum, 11, 38, 545, 547
livestock, 20, 21, 60, 341, 343, 345, 349
Livingstone, Bonnie, 5
local gazette, 437
local histories, 124, 127
log books, 165, 218, 219, 221, 233, 248, 545, 556
Logan, Deborah Norris, 178, 531, 532
Lombok, 12, 17, 21, 23
London, 198, 200, 294, 390, 555, 556
Long Island, 66
long-wave radiation, 177
Longfellow, Stephen, 125
longwave pattern, 257
Lough, J.M., 238
low sunspot activity, 124, 175
low-pass filter, 297
Lower Provinces (India), 539
Lowville, 147, 153
Lund, 397, 400
Luoyang, 449
Lutra canadensis, 209
Luzhou, 441
Lwow Newspaper, 533
lynx, 209
Lynx canadensis, 209

M

Macassar, 17, 19, 20, 27
Macdonell, Miles, 200
Mackenzie Delta region, 67
MacKenzie Collection, 12
macroclimate, 510
Madhya Pradesh, 432
Madison, James, 66
Madras, 15, 429, 430, 535, 539, 540
Madura Island, 19, 29
magma, 16, 22-27, 29-31, 36, 41, 70
magmatic water vapour, 31, 35, 41, 42, 506
Maharashtra, 432
Maine, 65, 125, 127, 132, 443
Malmöhus, 397
malnutrition, 194, 199
Manhattan, 66
Manitoba, 85, 194, 325, 530
Manley series (see central England temperature series), 384

Mannheim, 360
Mansell Island, 240
Maozhou, 441
Maquinchao, 507
marine plants, 345
maritime air, 332
marten, 209
Martes americana, 209
Martes pennanti, 209
Martin, Richard, 5
Martins Falls, 189
Massachusetts, 125, 127, 132, 139, 443, 532
Massif Central, 304, 305, 534
Maunder Minimum, 11, 30, 175, 198
maximum-density chronologies, 278
mean sea-level pressure, 372, 375
mean sensitivity statistic, 118
meat, 188, 206, 210, 341
Mecklenburg, 415
medieval annals, 332-334
medieval warm epoch, 197
Mediterranean climate, 306
Mediterranean region, 300, 301, 304, 355, 375, 390, 533, 534, 544, 545
Mediterranean Sea, 305, 555
Mein Kampf, 196
Melsted, Sheriff Páll Þorðarson, 338, 343
meltwater, 175, 188, 311, 318, 325, 326, 530
meltwater electrolytic conductivity records, 313
Melville Island, 529
memoranda, 437, 438
Mercantour, 293
meridional atmospheric circulation, 238, 546, 549
Mesa Verde National Park, 116
meteor, 69
meteorites, 422
meteorological regions of India, 431
Meteorological Registers, 163, 165, 166, 168, 171, 174
meteorological stations, 270, 291, 414
meteorology, 145, 197, 218, 248, 292, 295, 358, 360, 403, 430, 434, 453, 479, 506, 507
Métis, 186, 201
Mexico, 78, 415
microclimate, 270, 274, 530
microprobe analysis, 32
Mid-Atlantic Ridge, 71
mid-latitude westerlies, 257, 325, 365
Middle Ages, 300, 301, 332
Middle East, 544
migrations, 21, 187, 188, 198, 200, 208, 550
Milham, W.I., 124
milk products, 341
Miller, Alexander, 125
Mindelberg, Captain, 71
Minnesota, 198
Minoan civilization, 266
misery, 350

Mission reports, 165
missions, 246
Mississippi River, 65-67, 198, 524, 532, 546, 549
Missouri, 524, 532
Missouri Gazette, 532
mist, 345
Mistral wind, 304, 305, 534, 545
Mittelland, 416
models, 53, 69, 97-99, 166, 255, 263, 292, 311, 517
Möðruvellir, 334, 335, 338, 343-345
moisture stress, 118, 119
monsoon, 430-432, 434, 536, 540, 541, 546, 549
monthly wetness indices, 166
Montreal, 81, 206, 531
Mont Ventoux, 293
moose, 206, 209, 213, 215, 216, 234, 240, 244, 528
Moose Factory, 171, 237, 240
Moravian records, 245, 246, 248, 528, 529
Morocco, 293, 295, 301, 304, 305, 534
mortality, 14, 60, 198, 341, 349, 393, 397-401, 406, 534, 538, 539
Moscow, 402, 414, 534
Mossop, S.C., 404, 415
Mould Bay, 85
Mount Agung (see Agung), 46, 317, 413, 549
Mount Coseguina (see Coseguina), 549
Mount Laki (see Laki), 311
Mount St. Helens, 16, 26, 32, 36, 70, 169
Mount Tambora (see Tambora), 11, 46, 48, 115, 165, 169, 186, 309, 315, 326, 331, 515, 545
Mount Vernon, 66
Moyo, 21
multiple regression, 295, 296
Munaut, A., 295
muskrats, 209, 211
Mýrasýsla, 334, 335, 339, 343

N

Nagasaki, 462
Nain, Labrador, 292
Nanchuan, 441
Napoleon, 397, 414
Napoleonic Wars, 399
Natchez, 532
National Archives in Reykjavík, 334
National Center for Atmospheric Research, 5
National Museum of Natural Sciences, Ottawa, 6, 163, 196
National Research Council of Canada, 163
native people, 205, 208, 209, 213-215
Near East, 70
Nelson Encampment, 201
Nelson River, 189, 190
Nepal, 556
Nevada, 67, 99, 524
New Bedford, Massachusetts, 126, 127, 139, 443
New Brunswick, New Jersey, 127

New England, 46, 53, 64, 116, 125-127, 132, 134, 139-141, 178, 180, 246, 250, 442, 443, 530, 531
New Hampshire, 125, 132, 443, 530
New Haven, Connecticut, 67, 116, 127, 130, 151, 153, 154, 443, 531
Newfoundland, 180, 246, 248, 250, 355, 370, 529, 531
Newfoundland Archives, 246
New Jersey, 64, 127, 530
New Mexico, 116
New Orleans, 65-67, 524, 532
New Quebec/Labrador, 165, 174
Newell, J.P., 248
newspapers, 66, 124, 127, 146, 147, 165, 400, 419, 531, 532
New York City, 66, 147, 153, 154
New York State, 147, 530
New York State grammar school records, 147
New Zealand, 7, 27, 510, 511, 515, 516, 518, 520, 543, 546-549
New Zealand palaeotemperatures, 520
Nicaragua, 443, 549
Niemen River, 414
Nilahue, 497, 503, 506, 507
Niles Weekly Register, 532
Ningde Meteorological Station, 440, 536
Nitchequon, 528
nitrates, 482
Nordic countries, 414
Norfolk, Connecticut, 125, 126
North Africa, 301, 304
North America, 6, 7, 46, 55, 58, 61-64, 69, 70, 97-99, 101, 108, 111, 116, 124, 185, 186, 189, 190, 194, 195, 197, 198, 202, 246, 255-257, 262, 263, 266, 278, 355, 370, 392, 393, 405, 419, 436, 442, 443, 453, 462, 475, 482, 506, 531, 532, 535, 540, 545, 547, 548, 550, 555
North American boreal treeline, 262
North American temperatures, 98
North American tree-rings, 98, 238, 262, 274
North Atlantic, 97, 98, 180, 246, 250, 253, 332, 393, 442, 545
North Atlantic-European region, 365
Northeast Environmental Research Group, 141
Northern Baffin Bay Low, 324
northern boreal forests, 190, 256, 263
northern hemisphere, 7, 37-39, 42, 61, 68, 79, 87, 253, 256, 262, 310, 365, 370, 405, 482, 506, 523, 530, 542, 545-547, 549
northern limit of trees, 165, 198
North Greenland Ridge, 325
North Pacific, 98, 101, 103, 106, 111
North Pacific Anticyclones, 466
North Pole, 324, 529
North Saskatchewan River, 187, 189
North Sea, 365, 384
North Water, 529
North West Company, 165, 186-189, 194, 199, 201, 524

North Westers (storms), 539
North-West Passage, 529
Northwest Territories, 85, 262, 527
Norway, 198, 332, 333, 368, 397, 400, 402, 414, 534
Norwich, 6
Nothofagus, 494, 495, 497, 500, 504, 510, 511, 516-518
Nothofagus menziesii, 516, 518
Nothofagus solandri, 516, 518
Nuuk (Godthaab), 165
Nyey, 71

O

oak leaves, 188
oak ring-width series, 293
oats, 188, 197, 198, 416
oceanography, 32, 52, 233, 238, 332, 434
oceans, 30, 35, 47, 197, 544-547, 549
official histories, 437
Ogilvie, A.E.J., 60, 293
Ohio, 443
Ohio State University, 482
Ojibwa, 203, 205, 206, 208, 209, 211, 213
Okkak, 248, 528
Oklahoma, 532
Old Norse colony, 198
omega blocking, 191, 194
Ondatra zibethicus, 209
Oneida, 147, 153
onions, 188
Ontario, 178, 200, 203, 205, 206, 208, 209, 530, 531
open forest, 282, 283, 288, 292, 530
optical depth, 48, 49, 506, 542
optical phenomena, 493
Orgère, 293
Orkney, 368, 534
Oslo, 398, 403, 414
Osnaburgh House, 203, 205, 206, 208-211, 213, 214, 216, 530
otter, 209
Otteson, Sheriff Pétur, 349
outbreaks, 257
oxygen isotopes, 318, 482, 484, 485, 487, 488, 534, 543
ozone, 32, 33, 35, 42, 48, 50, 54, 55
Öztal Alps, Austria, 533

P

Pacific High (subtropical), 191, 194, 537, 538, 546
Pacific Northwest, 103, 116
Pacific Ocean, 7, 47, 103, 111, 434, 504, 506, 538, 542, 546, 555
Pacific Polar Frontal Zone, 453
paintings, 202

Pakistan (Sind), 539, 556
palaeoclimatic studies, 6, 7, 516, 517
Paris, 360, 370, 414
Parker, M.L., 6, 267
paroxysmal event, 25, 41
Parry, M.L., 198
particle cloud, 392
pasture, 343, 533
Patagonian Andes, 494
Pearson International Airport, 151, 153, 155
peasant revolts, 202
Pedersen, Søren, 399, 400, 415
Pedioecetes phasianellus, 210
Peloponnese, 556
Pekat, 22
Pengze, 440
people, 189, 194, 195, 197, 200, 202, 210, 211, 213-216, 344, 345, 349
permafrost, 198
Persia, 556
Persian Gulf, 556
Peru, 434, 479, 482, 484, 536, 541, 543, 547, 549
petrology, 17, 30, 41
petty crime, 349
Pfister, C., 167, 168, 293, 533
phenology, 124, 167, 187, 437
Philadelphia, 66, 67, 127, 178, 530, 531
Philipps, Lieutenant Owen, 21
Phillipstown, Massachusetts, 125, 126
Philosophical Transactions of the Royal Society, 198
photochemical system, 48
photochemical transport model, 47
photodissociation, 48, 54
photosynthates, 283, 289
photosynthetic efficiency, 283
phreatomagmatic activity, 22, 23, 25, 60
Phyllocladus, 516
Phyllocladus aspleniifolius, 511
Picea engelmannii, 268
Picea glauca, 256, 282
pickerel, 208, 211
pike, 208
pine tree-ring series, 293, 301
Pingchan, 440
Pinus albicaulis, 268
Pinus cembra, 418
Pittsburgh, 443, 530
planetary albedo, 53, 60
plants, 127, 167, 171, 188, 197, 200, 341, 393, 531
plinian events, 22, 24, 25, 27, 29, 31, 41
plumes, 25, 36, 41, 70, 393
poisoning, 36
Poland, 384, 419
polar air masses, 256, 332, 360
polar bears (white bears), 200
polar frontal zones, 190, 199, 256, 453, 454, 460, 461, 466, 537, 538, 545, 546
polar regions, 332, 479

polar vortex, 547
Polish Carpathian Mountains, 419
political conditions, 340, 341, 350, 351, 360, 433, 437
populations, 185, 203, 397
Portland, Maine, 125, 126
Post, John D., 6, 185
Post Journals (Hudson's Bay Company), 162, 163, 166, 174, 177, 187, 527, 528, 530
potatoes, 188, 189, 208, 209, 211, 213-215, 344, 369, 416
pottery, 14
Prague, 375, 422, 534
Prague Klementinum station, 422, 423, 425
prairies, 162, 186, 189-191, 199, 201, 326, 532
precipitation, 99, 106, 108, 119, 134, 139, 166, 168, 169, 171, 187, 191, 200, 248, 301, 324, 372, 377, 397, 405, 406, 416, 418, 438, 465, 482, 495, 524, 528, 530, 532, 533, 536-538, 540, 542, 548
predation pressure, 216
pressure, 106, 108, 111, 166, 169, 186, 360, 429
prices, 141, 185, 246, 341, 368, 399, 400, 402, 415, 419, 429, 432-434, 440, 539, 556
Prince Edward Island, 200
Prince Frederick of Hessen, 398
Prince of Wales, 240
principal component analysis, 99, 296, 365, 375, 450, 517
Provincial Archives of Manitoba, 188
proxy climatic data, 63, 64, 98, 115, 163, 166-168, 187, 246, 266, 291, 293, 295, 297, 306, 309, 332, 416, 482, 485, 516, 523, 528, 544
Pruden, John, 189, 190
PSCM indices, 365, 368
psychological impact, 215
ptarmigan, 200
Public Records Office, 555, 556
pumice, 12, 20, 21, 25, 27, 30, 71
Punta Arenas, 507
Puttie (a disease), 14
Pyrenees, 304, 305, 358, 534
pyroclastic deposits, 17, 18, 21, 22, 25, 27, 34, 35, 41

Q

Qaidam Basin, 484, 536
QBO, 55
Qinghai-Tibetan Plateau, 482
Qiqihar, 441
quasi-stationary cyclonic weather systems (see blocking), 535
Quaternary, 6
Quebec, 67, 85, 178, 199, 266, 274, 289, 292, 524, 530, 531
Quebec City, 81
Québec/Labrador Peninsula, 528, 545, 548

Quelccaya Ice Cap, 479, 482, 484, 485, 487, 488, 541, 543, 547, 549
quiet sun, 177

R

rabbits, 209
radiation (see solar radiation), 32, 46, 47, 53-55, 132, 147, 177, 186, 255, 547
radiative-convective-photochemical diffusion model (RCPD model), 47, 55
radioactive fallout, 175, 414, 547
radiocarbon dating, 313
Raffles, Thomas Stamford, 12
rain, 20, 60, 63, 67, 174, 189, 190, 201, 213, 214, 248, 332, 338, 339, 343-345, 349, 368-370, 397, 400, 414, 419, 429-431, 434, 440, 441, 448, 449, 452, 462, 464, 528-530, 532-537, 539-543, 546, 548
rainshadow effect, 198
Rajah of Sanggar, 21
Rajasthan, 432, 540
Rangifer tarandus, 206
Ratzel, Friedrich, 196
rawinsonde data, 41
Rayleigh scattering, 48
Red River, 165, 189, 201, 291, 524, 532
Red River Carts, 201
Red River settlement, 186, 201
Red Sea, 556
refugees, 441
reliability tests, 218
Rembang, 19
remote-sensing techniques, 414
reports, 163, 360, 529
Resiga, Marie-Anne, 5
resources, 200
response function analyses, 511, 517
Reval (Tallinn), 414
Reykanes Peninsula, 71
Reykjavík, 35, 333, 334, 339, 343
Reynolds number, 404
Rhineland, 294
Rhine Valley, 293, 304, 305, 416, 534
rice, 14, 21, 213, 440, 441, 475, 536, 537, 539
Rice, Gail, 5
Richmond Gulf, 171
ridge, 180, 186, 320, 321, 325, 326, 529, 532, 538, 540, 545, 546
Rif, 534
Riga, 370, 402, 414, 534
Rigi (mountains), 416
Rinjani volcano, 17
Rio Gallegos, 507
Río Alerce, 495
Río Horqueta, 495
Rivadavia, 507

river-ice, 124, 171, 237, 240, 244, 528
rivers, 63, 194, 197, 208, 213, 233, 234, 238, 243, 340, 443
roads, 201
Robbins, B.F., 125
Robbins, Thomas, 126
Robson Glacier, 268
Rochester, 147, 150, 153
Rochester College, 147, 150
Rocky Mountains (see Canadian Rocky Mountains), 191, 198, 267, 443, 527, 547
Royal Society, 166, 529
Rudloff, Hans Von, 360
Russia (see Soviet Union), 201, 397, 401, 402, 405, 406, 414, 419, 534, 556
Russian Economic History, a Guide to Information Sources, 401
rye, 214, 402, 416, 556

S

sagas, 333
Saharanpur, 538
Salem, Massachusetts, 126
saline lakes, 188
Salta, 507
salts, 35
Samarang, 20
Sanbornton, New Hampshire, 125, 126
Sanderson, Daphne, 5
Sanggar, 18, 21, 25, 35
Santa Maria, 119
Santa María-Soufriere-Pelée eruptions, 497, 506, 507
Santai, 441
Santiago, 541, 542, 548
Santorini, 266
Sardinia, 534
Saskatchewan, 194, 530
satellite imagery, 248
Sault Ste. Marie, 206
Saurashtra, 432
Savannah, 532
Scandinavia, 355, 375, 377, 384, 390, 533
Scania, 397
Schefferville, 528
Scheving, Sheriff Jónas, 343, 344
Schneider, S., 257
Scholefield, Peter, 545
Schweingruber, F.H., 420
Scoresby, W., 529
Scotland, 61, 186, 197, 198, 200, 201, 206, 293, 357, 368, 369, 375, 384, 534
Scott, P., 292
sea ice, 6, 60, 64, 85, 219, 221, 225, 229, 232, 233, 237, 238, 240, 243, 246, 248, 253, 257, 332, 333, 335, 343, 345, 350, 485, 528, 535, 538
sea mammal, 345
sea temperature, 527, 535

sea-level rise, 18
Sea of Japan, 454, 461, 538
sea-surface temperatures, 120, 121, 506
sealing, 165, 168, 246, 248, 345, 350, 555
seas, 233, 384
seasonal conditions, 124, 372, 375, 533
seaweed, 345
Second World War, 197, 341
seeds, 189, 198, 208, 344
Selkirk, Lord, 199, 200
Selkirk Settlers, 186, 194, 200, 201
Seoul, 462, 465, 466, 535, 537, 538
settlement, 201, 333, 340
Seven Oaks massacre, 186, 194
Severn, 234, 240, 243
Shanxi, 441, 537
shark fishing, 345
sheep, 14, 60, 126, 198, 341, 349
Shehong, 441
shellfish, 345
Shepherd, Kieran, 5
Shetland, 357, 368, 534
Shijiazhuang, 441
Shikotsu eruption, 16
ships, 66, 67, 165, 171, 219, 233, 240, 245, 246, 345, 545, 555
ships' log-books, 6, 223, 237, 238, 360, 524, 545, 555
short-wave radiation balance, 177
shortages, 401, 402, 436, 441, 443, 534
Si River Basin, 537
Siberia, 402, 534
Sichuan, 441, 442, 536
sickness, 204, 211, 213, 539
Síðumúli, 334, 335, 339, 343, 349
Siggental, 416
Sikkim, 556
Silesia, 358
silica, 35
silicate ash, 36, 97
silicic magmas, 36
simulation procedure, 355
Sind (Pakistan), 539
Siple Station, Antarctica, 479, 482, 485, 487, 488, 541, 544
Sjælland, 399, 400
Skaftafellsjökull, 338, 339
Skagafjarðarsýsla, 334, 335, 338
Skagafjord, 349
Skeiðará (river), 338, 339
Skeiðarárjökull, 339
Skeiðarásandur, 340
skin cancer, 32
Slade and Kelson, 248
Slatter, James, 213
slaves, 21
sleet, 60, 335, 339, 349
smallpox, 203

Smithsonian Institution, 147
smoke, 21, 60, 70, 71
smoky fog, 61
Smolensk, 402
Snæfellsnes, 350
Snæfellsnessýsla, 334, 339, 349
snow, 64-67, 69, 85, 124-127, 134, 139, 165, 167, 168, 171, 174, 177, 178, 181, 189-191, 194, 197, 198, 200, 201, 206, 208, 213-216, 248, 310, 311, 313, 325, 335, 338, 339, 344, 349, 369, 398, 416, 437, 440-442, 449, 466, 475, 528, 530, 531, 533, 534, 536, 538, 540, 541, 547, 548
Snow Geese, 208
snow mold, 67
snowshoe hares, 209
snowshoes, 206
social problems, 198, 340, 341, 349, 350, 360, 419, 437
Societas Meteorologica Palatina, 358, 360
Société Royale de Médecine, 360
socio-economic conditions, 186
soils, 22, 177, 344, 402
solar activity, 7, 11, 38, 39, 46-48, 50, 53, 70, 78, 79, 115, 177, 256, 263, 266, 325, 393, 403, 420, 547
Solo, 17, 19, 20
Somalia, 556
Somalia jet, 540
Sommarin, 397
Souracarta, 12, 13
South Africa, 544, 546, 548
South America, 7, 74, 493, 494, 497, 504, 506-508, 542, 543, 546, 548, 555
South Atlantic, 555
South Carolina, 66, 443, 532, 545
southeast trade winds, 21
southern hemisphere, 7, 39, 256, 310, 485, 487, 506, 510, 523, 541, 542, 544, 546-549
Southern Oscillation index, 541
Southern Temperate Zone, 542
South Korea, 462
South Pacific, 429
South Pole, 32, 414, 541, 544
Soviet Union (see Russia), 384, 414, 544
Spain, 301, 304, 534
Spanish Armada, 168
Spannmålshandel och Spannmålspolitiki Sverige 1719-1830, 395
specific volume increment (SVI), 282
spectral decomposition, 297, 299
Spessart forest, 293
Spruce Canyon, 116
spruce chronologies, 270, 274
St. Augustine eruption, 30
St. Helena, 556
St. John's, 246, 248
St. Lawrence Valley, 178, 530, 531
St. Lawrence Island, 527, 538

St. Louis, 524, 532
St. Petersburg [Leningrad], 370, 393, 402, 403, 414, 534
stagnation, 461, 466, 538, 546
stalled surface systems (see blocking), 532, 545
standard mean ocean water (SMOW), 311
stars, 360
starvation, 21, 60, 194, 199, 200, 203, 204, 208-211, 213-215, 341, 402
Staten Island, 66
statistical analysis, 126, 155, 296, 341, 360, 365, 504, 508, 510
statistics, 99, 248, 270, 295-297, 517, 542
Steel River, 189
Steen, S., 398
Stewart, Arch, 5
Stizostedion vitreum, 208
Stockholm, 62, 393, 394, 396, 399, 403, 534
Stokes' equation, 404
Stommel, H. and E., 6, 124, 443
stone pines, 418, 420
storms, 66, 126, 139, 140, 171, 178, 189, 198, 213, 338, 339, 343, 344, 349, 350, 369, 436, 527, 529, 530, 531, 534, 536, 538, 541, 545
Stothers, R.B., 124
Straits Settlement, 556
Strathglass, 198
stress, 199, 200, 215
Strutton Sound, 171
Student t-test, 87, 91, 504
sturgeon, 188, 208, 211, 213
Stykkishólmur, 333
subalpine zone, 270, 517, 543
subantarctic zone, 498-500, 507, 542
subarctic zone, 166, 177, 199, 203, 233, 530
submarine seismic activity, 178
subtropical zone, 487, 499, 507, 541, 542, 544
subtropical high pressure system, 178, 191, 360, 405, 451, 452, 466, 546
sucker, 208
Suður-Múlasýsla, 334, 335
Suður-Þingeyjarsýsla, 334, 345
Sulawesi, 27
sulphates, 31, 35, 97, 115, 255, 482, 485, 544
sulphur, 16, 17, 30-32, 38, 41, 47, 55, 69-71, 177, 436, 493, 506, 508, 542, 545, 547, 549
sulphuric-acid aerosols and solutions, 31-33, 36, 41, 42, 49, 69, 70, 97, 115, 119, 404-406, 413
sulphurous rain, 60
Sumatra, 20, 41, 556
Sumbawa, 12, 16, 17, 20-22, 46, 422
Sumenep, 19, 20
Summer Monsoon Experiment (MONEX), 540
sun, 11, 13, 60, 117, 357
sunburn, 32
Sunda Islands, 17
sunspots, 38, 39, 55, 177, 186, 198, 256, 545, 549
Sunwapta Valley, 270

superposed epoch analysis, 79, 81, 85, 116, 238, 255, 549
Surabaja, 41
Surakarta, 12
Surtsey, 71
Sutherland, James, 189
Suwa, City, 466
Swan River, 191
swans, 200
Sweden, 384, 395-397, 400-402, 534
Swedish Pomerania, 397
Switzerland, 63, 64, 67, 167, 293, 294, 369, 533, 534, 548
synoptic weather studies, 163, 165, 168, 186, 191, 194, 248, 304, 309, 320, 324-326, 358, 360, 422, 423, 426, 454, 524, 529
Syria, 61

T

Tafí del Valle, 507
Tainjing, 449
Taiwan, 441, 442, 536
tamarack, 282
Tambora, 6, 7, 11, 12, 14, 16-22, 25-27, 30-35, 41, 46, 48, 50, 54, 70, 74, 97, 115, 119, 121, 165, 169, 175, 177, 181, 186, 200, 205, 243, 256, 262, 266, 267, 274, 278, 281, 282, 289, 301, 305, 309, 311, 313, 315, 326, 331, 393, 395-397, 403, 405, 414, 420, 422, 423, 440, 442, 475, 479, 482, 485, 487, 497, 504, 511, 515, 524, 527, 534, 536, 538, 542-545, 547-549
Tambora acid layer, 529
Tambora Coffee Estate, 24
Tambora debris, 549
Tambora difference, 394
Tambora eruption, 37, 42, 47-50, 52-55, 111, 115, 116, 121, 124, 214, 255, 310, 370, 392, 393, 403, 405, 406, 415, 425, 436, 462, 466, 484, 488, 499, 500, 504, 506, 508, 510, 518, 520, 528
Tambora fluorine, 36
Tambora period, 7, 313
Tambora village, 21, 22, 24
Tambora volcanic aerosol, 7, 39
Tambora volcanic signal, 311, 318, 320, 326
Tamilnadu, 432
Tarawera, 497, 501, 506
Tasmania, 510, 511, 515, 543
Tatra Mountains, 418-420, 533
tatty, 539
Taupo eruption, 27, 29
teak chronology, 541
technology, 197, 433
Teignmouth, 18
teleconnections, 545

temperature, 50, 52, 99, 103, 106, 108, 116, 119, 126, 127, 145, 163, 166, 168, 169, 171, 178, 246, 293, 300, 301, 306, 317, 325, 333, 360, 372, 377, 399, 423, 429, 443, 454, 495, 507, 517, 518, 528, 531, 533
Tempo, 22
tephra, 22-25, 29, 32, 33, 35, 36, 41, 42, 60
terminal velocity, 404, 405
Ternate, 18, 24
Terumon, 20
Texas, 415, 532
Thames River, 198, 245
thaws, 125, 335, 349
the Enlightenment, 360
the year without a summer (various combinations), 6, 7, 39, 46, 55, 97, 111, 116, 124-126, 130, 141, 160, 200, 201, 214, 331, 339, 350, 351, 360, 372, 393, 399, 403, 429, 448, 453, 475, 482, 535, 545, 549, 550
The Year Without a Summer? Climate in 1816 (conference), 5, 214
the year of the great frost, 198
The British Library, 556
The Economic Development of the Scanian Agriculture, 397
The Ice Observer's Training Manual, 225
The King's Mirror, 333
The Last Great Subsistence Crisis in the Western World, 406
The Modern Encyclopedia of Russian and Soviet History, 401
The Round Tower (Copenhagen), 400
thermal forcing, 97
thermal indices, 238, 293
thermal radiation, 53, 54
thermometers, 62, 146, 147, 393, 416
Thomas, David, 443
Thomas, Isaiah, 125, 126
Thorarinsson, S., 60
Þórarinsson, Stefán, 338, 343, 344, 345
Þorsteinsson, Jón, 333
thunderstorms, 124, 125, 127, 139, 189, 201, 358, 368, 369, 431, 432, 443, 536, 537, 539, 540
Tibet, 437, 535, 540, 546, 548, 556
Tibetan High, 540, 541
Tibetan Plateau, 535, 541, 546
tidal waves, 20, 432
Tierra del Fuego, 495, 497, 500, 542
Till, C., 295
timber, 20, 190
Timor, 20
Tipperary, 369
Tobias, 21
Toronto, 81, 145-147, 150, 151, 153-155, 159, 531
trace gases, 48

trade, 194, 197, 219, 334, 341, 345, 349, 350, 393, 395-397, 399-402, 415, 529, 534
Tramontane wind, 304, 305, 534, 545
transportation, 197, 341, 433
trapping, 197, 206, 208, 209
travel accounts, 187, 194, 333, 533, 556
tree growth, 262, 282, 418, 494, 495, 507, 510, 524, 527, 530
treeline, 67, 256, 266-268, 270, 278, 281, 510, 524
tree-ring studies, 6, 7, 67, 98, 99, 103, 106, 108, 111, 115, 116, 118, 121, 237, 256, 262, 266-268, 270, 274, 278, 281-283, 292, 293, 301, 372, 390, 418, 443, 448, 493, 494, 497-504, 506-508, 510, 511, 515, 516, 524, 527, 528, 530, 532-534, 541-543
trees, 20, 22, 25, 108, 116, 119, 125, 237, 256, 267, 268, 270, 281-283, 288, 390, 418, 420, 440, 441, 510, 533, 536, 537, 544
Trenton, New Jersey, 125, 126, 530
Trier, Germany, 293
Trinity, Newfoundland, 248, 529
Trondheim, 393, 394, 398, 399, 414, 534
tropical air, 549
tropical easterlies, 546
tropical regions, 434, 485, 487, 511, 536, 541, 543, 544, 548
troughs, 180, 248, 250, 324, 325, 423, 524, 529, 532, 538, 545, 546
Tucumán, 507
Tunisia, 301, 304, 305, 534
turnips, 189, 344
typhoon, 538

U

Udshan abu (dust), 13
Ulster, 198
ultraviolet (UV) radiation, 32, 53, 54
Umfreville, Edward, 199
United Kingdom (see British Isles), 390
United States, 37, 64, 67, 68, 70, 74, 98, 99, 103, 106, 111, 116, 118, 119, 124, 125, 140, 145, 165, 178, 180, 181, 185, 198, 245, 256, 257, 414, 443, 524, 527, 530-532, 546
United States Department of Energy, 436
United States National Oceanic and Atmopsheric Administration, 127
United States weather stations, 150, 151, 154, 159
United States Weather Bureau, 147
University of Copenhagen, 399, 482
University of East Anglia, 5
University of King's College, 145
University of Maine, 141
University of Toronto, 145
University of Washington, 482
Upper Canada College, 146
Upper Canada Gazette, 145
Upper Provinces (India), 535, 538-540, 546, 548

Uppsala, 393, 394, 534
Ural Mountains, 358, 540, 546
urban heat-island effect, 145, 159, 161
Ursus americanus, 206, 211
Ushuaia, 507
Utah, 67
Utica, 147, 153
Uttar Pradesh, 432, 540

V

Vaasa, 401
Valdivia, 507
Vallée des Merveilles, 293
Vancouver Island, 267
vegetable crops, 125, 190, 208, 209, 248, 339, 343, 344, 533, 535
vegetation, 14, 22, 126, 188, 340, 341, 441, 536
Vermont, 443
vertical lapse rate, 47
Vestmann Islands, 71
Vestur-Skaftafellssýsla, 334, 335, 339, 344
Vesuvius, 27, 425
Viðvík, 334, 335, 338, 339, 343, 344, 350
Vienna, 62
Viipuri, 401
Vík, 334, 335, 338, 339, 349
Villa Nogués, 507
Vilnius, 414
vineyards, 197, 368, 416
Virginia, 66
Vistula River, 397
visual extinction curve, 50
viticulture (see vineyards and grapes), 64
volcanic-acidity layers, 33
volcanic activity, 174, 420
volcanic aerosols, 32, 36-39, 41, 46, 55, 69, 74, 115, 177, 369, 504, 548
volcanic ash, 21
volcanic cooling, 120, 253
volcanic cloud, 41, 48
volcanic debris, 506
volcanic degassing, 30
volcanic dust, 11, 14, 61, 124, 415, 422, 425, 426
volcanic dust veils (see dust veils), 78-80, 85, 87, 91, 443
volcanic eruption - climate relationships, 58
volcanic eruption, columns and plumes, 27, 29, 31, 34-36, 41
volcanic eruptions, 12, 14, 16, 25, 30, 34, 35, 38, 39, 46, 53, 61, 69, 70, 79, 81, 85, 97, 98, 111, 118, 120, 121, 169, 200, 232, 238, 255, 266, 282, 289, 301, 305, 313, 318, 331, 338-340, 370, 395, 403, 405, 413, 414, 420, 422, 423, 425, 475, 482, 493-495, 497, 502, 504, 506, 508, 515, 524, 543, 544, 549
volcanic events, 262, 309

575

Volcanic Explosivity Index, 36, 115
volcanic forcing, 52, 121
volcanic gases, 32, 34-36, 61
volcanic glass, 32
volcanic halogens, 41
volcanic haze, 61
volcanic hypothesis, 70
volcanic indices, 263
volcanic particles, 393, 395, 403-406, 415
volcanic pollution, 17, 31
volcanic volatiles, 30-32, 41
volcanically-induced cooling, 47
volcanism, 7, 58, 63, 255, 263
Volcano Weather, 185
volcanoes, 14, 17, 70, 413, 423
volcanologists, 6, 448
von Kotzebue, O., 524
vortex, 321, 324, 325, 529, 542
Vöyri, 393, 414, 534

W

Wales, 368
Wallingford, Vermont, 125, 126
Wanning, 441
War of 1812, 246
warm drought, 191
wars, 395, 399
Warsaw, 375, 414, 419
Washington, George, 66
washout, 405
water, 208, 213, 214, 324, 349
water buffalo, 440, 536
water levels, 215
water vapour, 35, 36, 42, 47
waterfowl, 200
wave troughs, 546
weather and proxy data, 163, 399
weather diaries, 248, 333, 524, 527, 529, 531, 544
weather fluctuations, 437
weather journals, 124, 163
weather maps, 358, 453
weather records, 165, 416
weather-pattern calendars, 465
Weggis, 416
Wenjiang, 441
Wenkchemna Glacier, 270
Wenzai, 440
West Indies, 555
wet/dry-pattern maps, 461, 465, 466
wetness/dryness indices, 163, 168, 171, 536, 537
whaling, 246, 345, 529, 545
wheat, 63, 125, 188, 369, 416, 434
whirlwind, 18
White, Gilbert, 61, 70
whitebark pine, 268, 270
whitefish, 208, 211
White Mountains, 266

white spruce, 256, 266, 282, 283, 289, 292, 527
Wicklow, 369
Wigley, T.M.L., 365
wildfowl, 208, 213, 215
wildlife, 186, 189, 194, 195, 199
wildrice, 208, 210, 211, 213-215
Wilkerson, Stephen, 199
Williamstown, 127, 443
Wilson, C., 5, 6, 194, 238, 324, 531
winds, 61, 127, 139, 166, 168, 169, 175, 177, 178, 181, 187, 189, 199, 201, 208, 211, 218, 238, 246, 248, 250, 294, 304, 305, 335, 339, 344, 345, 360, 405, 440, 454, 527-531, 535, 537-539, 555
wine-harvest dates, 533
Winnipeg, 81, 203, 291
Wisconsin, 198
wolves, 200
wood, 345
wool, 341
Worcester, Massachusetts, 125, 126
Workshop on World Climate in 1816, 5, 523, 535, 542, 544, 545, 547, 549
World Meteorological Organization, 5
Wuxing, 440
Wyoming, 67

X

X-ray densitometry, 116, 274
Xingzi, 440
Xuyong, 441

Y

Yangtze River, 449, 450, 452, 537
Yangtze region, 63, 449, 541
years without summers, 350
Yellow River (see Huang Ho River), 63, 449, 450, 452, 537
Yichun, 440
Yogyakarta, 17
Yokohama, 466, 538
Yongning, 441
York, Ontario, 145
York Factory, Manitoba, 199, 201, 234, 238, 240, 243, 244, 291, 530
Yukon Territory, 262

Z

Zanzibar, 556
Zhang River, 449
Zhejiang, 440, 537
Zollinger, H. 17, 21, 22
Zongdui Weiga, 438